THERMOPLASTIC ELASTOMERS

Edited by **Adel Zaki El-Sonbati**

Thermoplastic Elastomers
Edited by Adel Zaki El-Sonbati

Published by InTech
Janeza Trdine 9, 51000 Rijeka, Croatia

Publishing Process Manager Oliver Kurelic

Technical Editor Teodora Smiljanic

Cover Designer InTech Design Team

Additional hard copies can be obtained from orders@intechopen.com

Thermoplastic Elastomers, Edited by Adel Zaki El-Sonbati
p. cm.
ISBN 978-953-51-0346-2

For EU product safety concerns:
IN TECH d.o.o., Prolaz Marije Krucifikse Kozulić 3, 51000 Rijeka, Croatia, info@intechopen.com or visit our website at intechopen.com.

Meet the editor

Professor Adel Z. El-Sonbati received his M.Sc. from the University of Mansoura, Egypt in 1978 and a Ph.D. in coordination chemistry from the University of Tanta, Egypt in 1983. He worked as a lecturer at the Faculty of Science, Mansoura University, Demiatta, Egypt. In 1987 he was promoted to an assistant professor. Since 1991, he is a full professor of inorganic chemistry. He is one of the editorial board members in the Journal of Spectroscopy and Dynamics. He has supervised more than 35 M.Sc. and 20 Ph.D. students. Prof. El-Sonbati has published over 150 scientific papers. His current research interests are synthesis and characterization of supramolecular polymers, supramolecular polymer complexes, conducting polymers and polymer complexes, preparation and characterization of heterocyclic/polymers of uranyl complexes by using the El-Sonbati equation, thermal stability and mechanism of degradation of homopolymers and copolymers, and application of potentiometer and spectrometer for the determination of the dissociation and stability constant of some polymers and their complexes with transition elements.

Contents

Preface XI

Part 1 Modifications of Thermoplastic Starch 1

Chapter 1 **Melt Blending with Thermoplastic Starch 3**
Francisco J. Rodríguez-González

Chapter 2 **Thermoplastic Cassava Flour 23**
Diana Paola Navia and Héctor Samuel Villada

Chapter 3 **Physical and/or Chemical Modifications of Starch by Thermoplastic Extrusion 39**
Maria Teresa Pedrosa Silva Clerici

Chapter 4 **Properties and Biodegradation Nature of Thermoplastic Starch 57**
Redouan Saiah, Richard Gattin and P.A. Sreekumar

Chapter 5 **Starch Protective Loose-Fill Foams 79**
Marcin Mitrus

Chapter 6 **Thermoplastic Starch 95**
Robert Shanks and Ing Kong

Chapter 7 **Retrogradation and Antiplasticization of Thermoplastic Starch 117**
Yachuan Zhang and Curtis Rempel

Part 2 Modifications of Thermoplastic Elastomers 135

Chapter 8 **Thermoplastic Elastomers 137**
Robert Shanks and Ing Kong

Chapter 9 **Modification of Thermoplastics with Reactive Silanes and Siloxanes 155**
Jerzy J. Chruściel and Elżbieta Leśniak

Chapter 10 **Advantages of Low Energy Adhesion PP for Ballistics 193**
J. Gonzalo Carrillo-Baeza,
W.J. Cantwell and Ricardo A. Gamboa-Castellanos

Chapter 11 **Microinjection Molding of Enhanced Thermoplastics 213**
Mónica Oliveira, Victor Neto,
Maria Fonseca, Tatiana Zhiltsova and José Grácio

Chapter 12 **Investigation of the Physical Characteristics of Polypropylene Meltblown Nonwovens Under Varying Production Parameters 243**
Deniz Duran

Chapter 13 **Thermoplastic Extrusion in Food Processing 265**
Caroline Joy Steel, Maria Gabriela Vernaza Leoro,
Marcio Schmiele, Reinaldo Eduardo Ferreira and Yoon Kil Chang

Chapter 14 **Lightweight Plastic Materials 291**
Marek Kozlowski

Chapter 15 **The Performance Envelope of Spinal Implants Utilizing Thermoplastic Materials 319**
Daniel J. Cook, Matthew S. Yeager,
Shamik Chakraborty, Donald M. Whiting and Boyle C. Cheng

Chapter 16 **Application of Thermoplastics in Protection of Natural Fibres 329**
Fauziah Ahmad, Farshid Bateni,
Mohammadreza Samadi Tavana and Ahmad Shukri Yahaya

Chapter 17 **Characterization of Thermoplastic Elastomers by Means of Temperature Scanning Stress Relaxation Measurements 347**
Vennemann Norbert

Chapter 18 **New Thermoplastic Ionic Elastomers Based on MA-g-EPDM with Advanced Characteristics 371**
Anton Airinei, Mihaela Homocianu,
Daniel Timpu and Daniela Maria Stelescu

Chapter 19 **Electroactive Thermoplastic Dielectric Elastomers as a New Generation Polymer Actuators 399**
Chong Min Koo

Preface

Nowadays it is difficult to imagine a life without thermoplastic elastomers. The uses of thermoplastic elastomers have increased because of its specific properties, such as low cost, light weight, high strength, non-biodegradability, durability, non-corrosive nature, process ability and high energy effectiveness. Hence, these thermoplastics elastomers can be used for various applications, which range from household articles to the aeronautic sector. This book is comprised of nineteen chapters, written by specialized scientists dealing with physical and/or chemical modifications of thermoplastic elastomers and thermoplastic starch. Such studies will provide a great benefit to the specialists in food, electric, telecommunication devices, and plastic industries.

The book "Thermoplastic Elastomers" is comprised of two sections.

Section I: Modifications of Thermoplastic Starch

The first section has seven Chapters on physical and/or chemical modifications of thermoplastic starch. Chapter 1 studies the relationship between morphology and biodegradation of thermoplastic starch blends.

Chapter 2 presents some excerpts related to thermoplastic cassava flour as a raw material useful for packaging applications. Cassava flour is viable material for use as part of a processable thermoplastic matrix by molding technique, which allows obtaining of materials with acceptable mechanical and thermal properties for agro-industrial applications.

The main starch modification obtained with the use of the thermoplastic extrusion process is addressed in Chapter 3. The thermoplastic extrusion process is capable of causing changes in starch, making it present a large variety of applications, both in the food industry and in other industries.

The modification of starch using various plasticizers, cross linking agent, as well as its biodegradation nature is discussed in Chapter 4.

Chapter 5 provides information on physical properties of starch foams such as cell structure, foam unit density and bulk density, compressive stress, resiliency and friability.

Chapter 6 provides a framework for transition from native starch, its primary molecular structure, secondary structure and tertiary granules to thermoplastic starch, with its properties that parallel and contrast with synthesis thermoplastics.

Chapter 7 summarizes the current knowledge of thermoplastic starch to its plasticization, retrogradation and antiplasticization. Starch retrogradation mechanisms are discussed at a molecular level. Methods to measure the retrogradation degree, such as differential scanning calorimetry, differential thermal analysis, X-ray, etc. are also reviewed. Changes in thermoplastic starch (TPS) property, such as tensile strength, elongation, gas permeability are due to the retrogradation of starch polymers and these are described.

Section II: Modifications of Thermoplastic Elastomers

The second section has twelve Chapters on physical and/or chemical modifications of thermoplastic elastomers. Chapter 8 compiles the updated knowledge on thermoplastic elastomers in general, practically their structures, syntheses, processing methods, mechanical properties and applications.

The subject of Chapter 9 observes a continuously growing interest in applications of reactive silanes and polysiloxanes in many different fields of material science and the chemical technology.

Chapter 10 shows the advantage of low interfacial adhesion thermoplastic matrix, such as polypropylene (PP), and how it can result in improvements to an aramid fabric for ballistics.

The challenges that involve the micro-injection of enhanced thermoplastics are discussed in Chapter 11. Special attention is given to the microinjection technology, tooling and injection materials.

The influence of some crucial production parameters, namely die air pressure, extruder pressure, collector drum speed, and collector vacuum on the thickness, basis weight, air permeability, fiber diameter and tensile properties of polypropylene meltblown nonwoven webs are investigated in Chapter 12.

Thermoplastic extrusion is considered a high-temperature, short-time (MTST) process in the food industry. The use of thermoplastic extrusion in food processing is studied in Chapter 13.

Lightweight plastic materials increasingly used in automotive, aerospace and construction sectors are studied in Chapter 14.

Thermoplastics demonstrating a track record of success in medical device applications are discussed in Chapter 15.

Application of thermoplastics in protection of nature fibers are described in Chapter 16.

Chapter 17 describes the characterization of thermoplastic elastomers by means of temperature scanning stress relaxation measurements.

Chapter 18 gives an overview of the research on ionic thermoplastic elastomers based on maleated ethylene propylene diene terpolymer (EPDM-g-MA), to obtain some new generations of ionic thermoplastic elastomers with high technical and processing characteristics, intended to be processed on the injection moulding machines, resulting in high quality products complying with the international market requirements.

Lastly, chapter 19 is about thermoplastic dielectric elastomers as a new generation polymer actuator with a high auction performance. Electroactive thermoplastic dielectric high actuation performance is owed to their ultra high electrostriction coefficients. Such big electrostriction coefficient was attributed to their ultra high density of the dielectric mismatched interfaces.

The editor of this book would like to express his gratitude to **Prof. M.A. Diab** and **Prof. A.A. El-Bindary,** Chemistry Department, Faculty of Science (Demiatta), Mansoura University, Demiatta, Egypt, for their useful advice in the process of preparing the book.

Adel Zaki El-Sonbati
Chemistry Department, Faculty of Science (Demiatta),
Mansoura University,
Egypt

Part 1

Modifications of Thermoplastic Starch

1

Melt Blending with Thermoplastic Starch

Francisco J. Rodríguez-González
Centro de Investigación en Química Aplicada
México

1. Introduction

Starch is a natural polymer synthesized by green plants as energy source. In comparison with low-cost synthetic polymers, starch is inexpensive, abundant and renewable raw material for the development of polymeric sustainable materials. It has be used in its native granular form as rigid filler or transformed in a thermoplastic material for melt blending with synthetic or natural polymers. Polymers filled with dry starch granules behave as typical composite materials where modulus increases and ductility decreases due to the stiffening effect of the starch granules (Willett, 1994, Kim et al, 1995, Chandra & Rustgi, 1997). An important disadvantage showed by polymeric composites filled with granular starch is the low starch content that can be added, especially for application where high ductility is required (Griffith, 1977). In contrast to the ordered structure of starch molecules in granular starch, thermoplastic starch (TPS) is an amorphous material that can flow and be deformed as any synthetic polymer (St.-Pierre et al., 1997). Crystallinity of starch granules is destroyed by the application of heat and shear in the presence of moisture during the gelatinization process. The addition of a good plasticizer, such as glycerol, allows TPS to be extruded at the processing temperatures of most commodity polymers (St.-Pierre et al., 1997). Mechanical performance of TPS material blended with synthetic polymers depends on a series of parameters including blend morphology (particle size and shape, and particle dispersion and distribution), interfacial adhesion and the intrinsic characteristics of TPS (Rodriguez-Gonzalez et al., 2003b). It has been reported that melt blending of TPS with synthetic polymer is an excellent alternative for the development of sustainable and more environmentally friendly product (Rodriguez-Gonzalez et al., 2003b).

2. Thermoplastic starch preparation

The gelatinization of starch is a process that permits the release of starch macromolecules from granules. It can be carried out by exposing starch granules to heat and shear in the presence of moisture. In the gelatinization of starch during extrusion, it is important to have strict control of the energy applied and the moisture content. The gelatinization process is depicted in Figure 1. The application of excessive heat and shear, such as that observed during extrusion processing of starch at low moisture content, leads to its thermo-mechanical degradation (Gomez & Aguilera, 1983, 1984, Lai & Kokini, 1991). Products of starch degradation are mainly dextrin, and in more extreme cases oligomer and sugar (Gomez & Aguilera, 1983, 1984). Once starch granules are disrupted, the resulting gelatinized starch (GS) can be mixed with a suitable plasticizer to reduce its melting temperature and improve its processabillity. This material is known as thermoplastic starch (TPS).

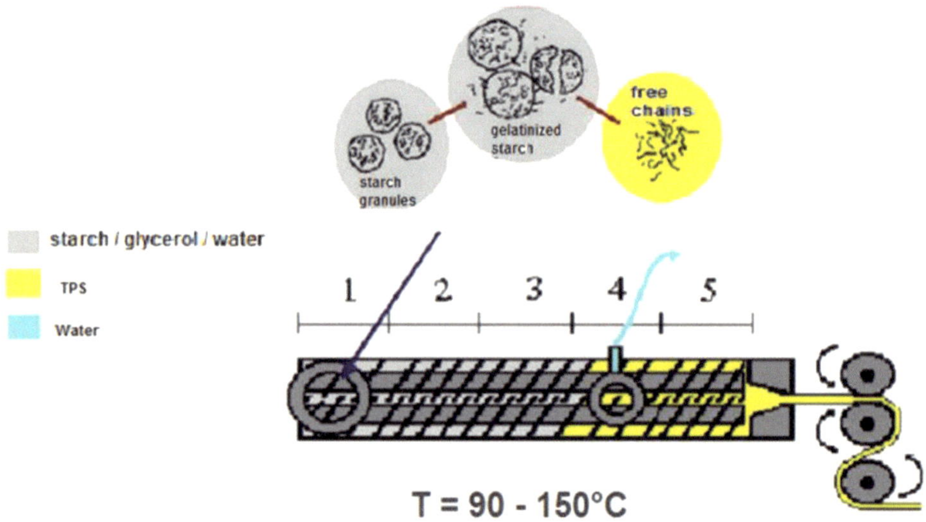

Fig. 1. Schematic representation of starch gelatinization and plasticization processes during extrusion.

Water is a good plasticizer for TPS but its use leads to a high dependence of final properties to environmental conditions of humidity. Utilization of plasticizers other than water helps to stabilize the properties of TPS. The main plasticizer used in TPS composition is glycerol (Forssell et al. 1997; Mathew & Dufresne, 2002; Souza & Andrade, 2002; Ma & Yu, 2004a; Ma & Yu, 2004b; Parra et al., 2004; Rodriguez-Gonzalez et al., 2003a; Rodriguez-Gonzalez et al., 2003b; Rodriguez-Gonzalez et al., 2004; Mali et al., 2005; Chand et al., 2006; Ma et al, 2006; Teixeira et al., 2007; Talja et al., 2007; Talja et al., 2008; Tena-Salcido et al., 2008; Chaudhary, 2010; Mendez-Hernandez et al., 2011) but other alcohols (Da Roz et al., 2006), polyols (Mathew & Dufresne, 2002; Parra et al., 2004; Mali et al., 2005; Da Roz et al., 2006; Talja et al., 2007; Chaudhary, 2010), sugars (Da Roz et al., 2006; Teixeira et al., 2007; Talja, 2008) or nitrogen compounds such as ethanolamine (Ma et al, 2006), formamide (Ma & Yu, 2004a; Ma & Yu, 2004b), acetamide (Ma & Yu, 2004a) or urea (You et al., 2003; Ma et al, 2006) have also been successfully employed. TPS materials have been prepared using casting process (Mathew & Dufresne, 2002; Parra, et al., 2004; Mali et al., 2005; Chand et al., 2006; Talja et al., 2007; Talja, 2008) or by melt mixing in batch, internal mixer (Forssell et al. 1997; Da Roz et al., 2006; Teixeira et al., 2007), or continuous equipment such as single (Souza & Andrade, 2002; Ma & Yu, 2004a; Ma & Yu, 2004b; Ma et al, 2006) or twin-screw extruders (Rodriguez-Gonzalez et al., 2003a; Rodriguez-Gonzalez et al., 2003b; You et al., 2003; Rodriguez-Gonzalez et al., 2004; Tena-Salcido et al., 2008; Chaudhary, 2010; Mendez-Hernandez et al., 2011). In the case of melt mixing processes, starch, plasticizer and water have been fed as dry blends (Ma & Yu, 2004a; Ma & Yu, 2004b; Da Roz et al., 2006; Ma et al, 2006; Chaudhary, 2010) or slurries (Rodriguez-Gonzalez et al., 2003a; Rodriguez-Gonzalez et al., 2003b; Rodriguez-Gonzalez et al., 2004; Tena-Salcido et al., 2008; Mendez-Hernandez et al., 2011). In some cases, TPS materials prepared by melt mixing have a significant water content which limits the processing temperature far below the processing conditions of most synthetic polymers, i.e. >150˚C, in order to avoid water vapor bubbles into TPS extrudates (Souza & Andrade, 2002; Farhat et al., 2003; Ma et al, 2006; Chaudhary, 2010). The development of an extruder configuration having a venting zone after both starch gelatinization and plasticization processes were accomplished and before exiting from the die allowed the preparation of water-free TPS (Favis et al., 2001; Favis et al., 2003, Favis et al., 2005).

3. Rheological and thermal properties of water-free TPS

The rheological and thermal properties of water-free TPS materials having high glycerol contents (29, 36 and 40%) were evaluated by DSC analysis and rheological measurements in shear and oscillatory modes (Rodriguez-Gonzalez et al., 2004). TPS materials were labeled according to their glycerol content. Hence, TPS29,33, TPS36 and TPS40 have 29, 33, 36 and 40% of glycerol.

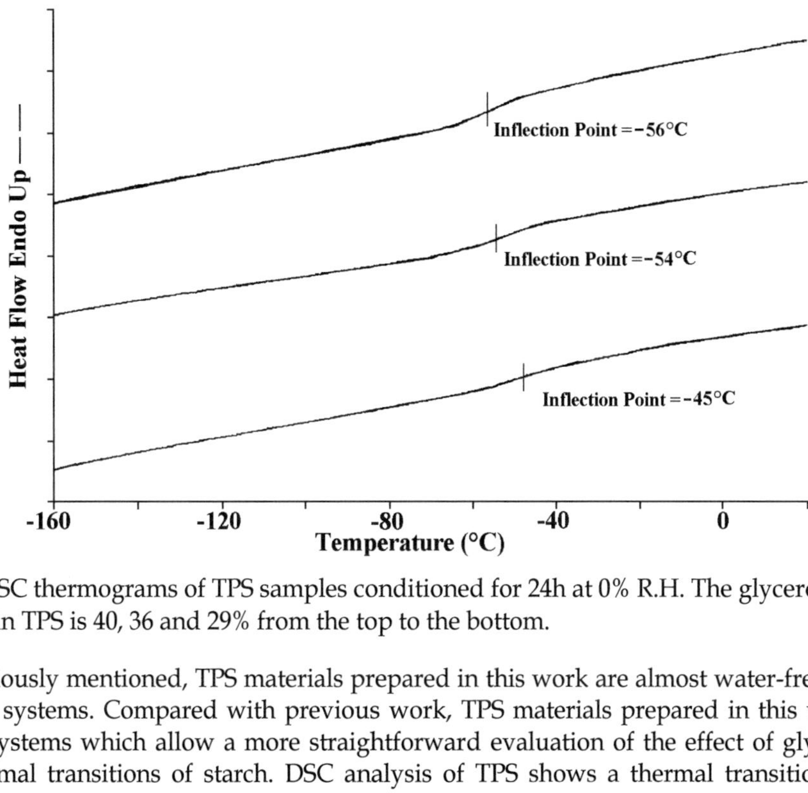

Fig. 2. DSC thermograms of TPS samples conditioned for 24h at 0% R.H. The glycerol content in TPS is 40, 36 and 29% from the top to the bottom.

As previously mentioned, TPS materials prepared in this work are almost water-free starch-glycerol systems. Compared with previous work, TPS materials prepared in this work are binary systems which allow a more straightforward evaluation of the effect of glycerol on the thermal transitions of starch. DSC analysis of TPS shows a thermal transition below ambient temperature that decreases as glycerol content increases (Figure 2). On the other hand, no thermal transitions are observed between 25 and 200ºC (not shown). The Tg of TPS decreases from -45 to -56°C as glycerol content increases from 29% to 40%. Van Soest et al. have reported the Tg of extruded TPS materials containing a starch/water/glycerol ratio of 100:27:5 of ≈+59°C (Van Soest et al., 1996). Forssell et al. (1997) studied the thermal transition of TPS materials prepared in a melt mixer as a function of glycerol and water content. Depending upon the composition, TPS materials presented one or two thermal transitions. In that work, at the lowest water content (ca. 1%) the upper transition of TPS decreases from

145 to 70°C as the glycerol content is increased from 14 to 29% while only TPS compounded with 29 and 39% glycerol showed lower transitions both at ≈ -50°C. The upper transition was attributed to starch-rich phase while the lower transition was related to a starch-poor phase. Lourdin and coworkers prepared TPS cast films by mixing starch with different amounts of water and glycerol (Lourdin et al., 1997a; Lourdin et al., 1997b). Films having around 13% water content showed a reduction of Tg from 90 to 0°C when glycerol content increased from 0 to 24% (Lourdin et al., 1997a). In that case they observed a glassy to rubbery transition of TPS at around 15% glycerol. In a further paper, they compared the Tg of TPS films having around 11% water with respect to glycerol content and they found that Tg decreased from 126 to 28°C when glycerol content was increased from 0 to 40% (Lourdin et al., 1997b). Discrepancies in Tg values as a function of glycerol content can be related, as mentioned by Kalichevsky to the mixing history during TPS preparation (Kalichevsky et al., 1993).

During on-line measurements, TPS extrudates did not present bubbles due to the almost absence of water. The pressure readings of TPS36 and TPS40 at 150ºC were quite regular while those of TPS29 were mostly irregular. For this reason only TPS36 and TPS40 were evaluated. As observed by other authors (Aichholzer and Fritz, 1998; Della Valle et al., 1992; Lai and Kokini, 1990; Senouci and Smith, 1988; Willett et al., 1995; Willett et al., 1998), the viscosity (η) of both TPS and PE1 melts display a power-law (shear thinning) behavior at the shear rate ($\dot{\gamma}$) interval developed over die extrusion conditions (Figure 3). The η of TPS materials depends on the plasticizer content. An increment of glycerol content from 36% to 40% results in a reduction of 20% of η of TPS36 (at $\dot{\gamma}$ ~ 130 s^{-1}).

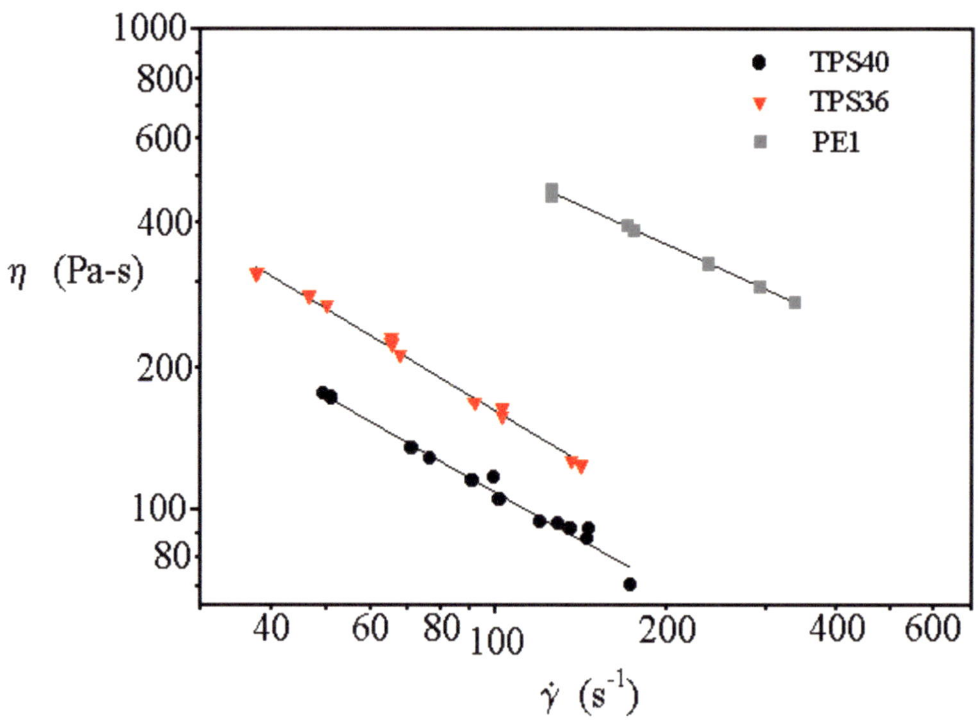

Fig. 3. Comparison of the viscosity of TPS40, TPS36 and PE1 measured on-line in the TSE at 150ºC.

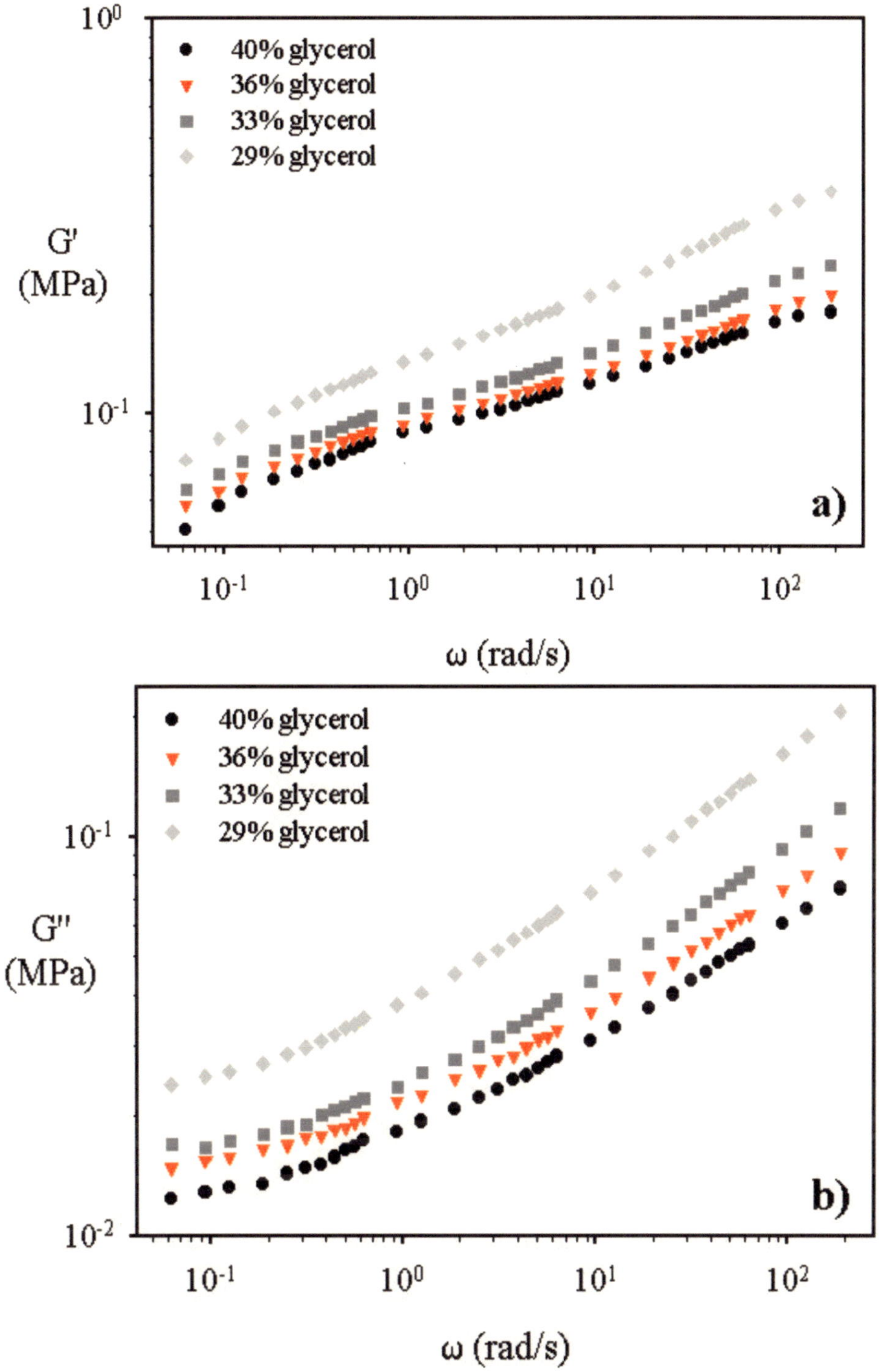

Fig. 4. Effect of glycerol content on (a) elastic modulus (G′) and (b) loss modulus (G″) of TPS materials evaluated at 150°C

TPS exhibits the rheological behavior of a typical gel as characterized by a storage modulus (G′, Figure 4a) larger than the loss modulus (G″, Figure 4b) and with both moduli largely independent of frequency over the amplitude of the experimental window (Ross-Murphy, 1995). This behavior is produced by the presence of an elastic network embedded in a softer matrix. The rigidity in those regions can be produced by chemical or physical crosslinking. The structure of the elastic network has been related to the crystallinity derived from the complexation reaction between amylose and lipids (Conde-Petit & Escher, 1995; Della Valle et al., 1998) and the physical entanglement of the high molecular weight polysaccharides (Della Valle et al., 1998; Ruch and Fritz, 2000).

As expected, the augmentation of the glycerol content in TPS results in a reduction of both G′ and G″. However, the trend in the modulus curves was nearly the same, regardless of the glycerol content. From the study of low-concentration starch dispersions, Conde-Petit and Escher (1995) showed that the formation of amylose-emulsifier complexes modifies the viscoelastic response of potato starch dispersions. Crystalline regions produced during the amylose-emulsifier complexation form an elastic network, which is responsible for the liquid-like to solid-like viscoelastic modification. From the similarity of the trend of the G′ curves shown in Figure 4a, it can be inferred in this work that glycerol variation does not affect the nature of the hypothetical crystalline elastic network, it just plasticizes the amorphous fraction of starch.

The study of the viscoelasticity of starch-based materials has mainly focused on concentrated gels and dispersions (≤ 5% starch). In this work, the viscoelastic behavior of water-free TPS at high glycerol contents has been evaluated at 150°C. G′ decreases as glycerol content increases and the changes are similar at both low and high frequencies. Della Valle and co-workers also studied the behavior of a water-free TPS at 150°C and found that the decrease of G′ with glycerol content was dependent on frequency (Della Valle et al., 1998). However, that material was obtained by subjecting the TPS to a separate drying step, a process which can induce structural changes in the starch. The proportional reduction of G′ as a function of glycerol content observed in this work is similar to that observed in starch gel systems (Kulicke et al., 1996). Figure 6a shows that the reduction of the glycerol content from 40% to 33% results in a quasi-linear increment of G′, while the reduction from 33% to 29% glycerol produces a larger variation in G′. In the case of the elastic modulus of polymer composites, percolation theory explains the non-linearity produced by the phase inversion effect at high filler content (Willett, 1994). The limit of glycerol plasticization that produces the non-linearity observed in the G′ of TPS at a concentration around 30% glycerol can be explained in a similar way. TPS can be considered as a homogeneous system composed of a hard elastic network and soft amorphous regions. Amylose complex crystallites, highly entangled starch molecules, poorly plasticized starch-rich sites, or a combination of them could compose the hard elastic network. Soft amorphous regions could be composed of well-plasticized glycerol-rich starch. Even though the elastic network is present at 33% glycerol, the soft amorphous regions dominate the viscoelastic response. Increasing glycerol content, beyond this concentration, produces a relatively small reduction in the rheological parameters. On the other hand, below 30% glycerol the phase inversion of a soft to a hard matrix occurs resulting in the domination of the viscoelastic response by the hard elastic network, which is in good agreement with percolation theory. That suggests a glycerol plasticization threshold at a concentration around 30%.

4. Blending with polyethylene

Blending TPS with synthetic polymers have shown the typical characteristics of immiscible polymer blends (St-Pierre et al, 1997). The melt blending of TPS with synthetic polymers has given place to a series of scientific and technologic developments. Such works differed in the mixing protocol and the type of additives used. Some authors proposed the use of two steps for the preparation of TPS-based blends (Aburto et al., 1997, Bikiaris et al., 1997a, 1997b, 1998, Prinos et al., 1998, Averous et al., 2000a, 2000b, 2001a, 2001b, Martin & Averous, 2001) while other preferred just one-step processes (Dehennau & Depireux, 1993, St-Pierre et al., 1997). Starch-based blends prepared in two steps are generally characterized for the preparation of TPS in a separated extrusion step. St-Pierre and coworkers presented a one-step blending process for TPS-based polymer blends (St-Pierre et al., 1997). They developed an extrusion system combining a TSE with a single-screw extruder (SSE). TPS was prepared in the SSE, and then it was blended with LDPE in the last sections of the TSE. Using such an extrusion system, they demonstrated experimentally that a certain morphological control of PE/TPS blends could be achieved by varying the TPS concentration from 0 to 22 wt%. Those blends showed an unusual high level of ductility.

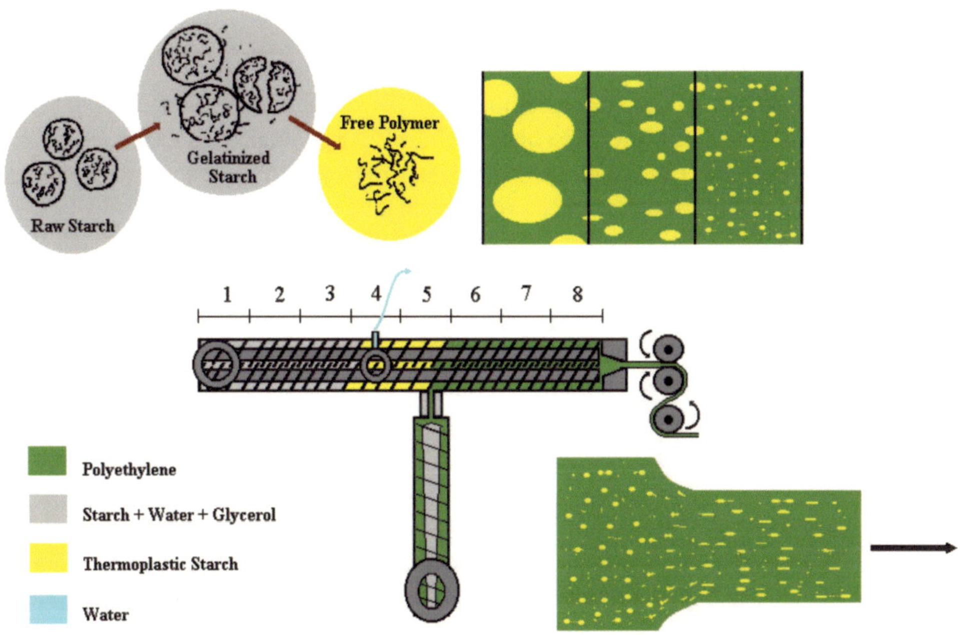

Fig. 5. Schematic representation of the one-step extrusion system designed for the melt blending of LDPE with water-free TPS.

An improved approach for LDPE/TPS blends in a one-step process was developed by Rodriguez-Gonzalez and coworkers (Rodriguez-Gonzalez et al., 2003b). It consisted of an extrusion system equipped with a single-screw extruder, from which molten LDPE is fed to the middle section of a twin-screw extruder. Suspensions of starch, glycerol and water were fed to the hopper of the twin-screw extruder and, as described in section 3, water-free TPS having 29, 36 and 40% glycerol (TPS29, TPS36 and TPS40, respectively) were prepared and melt blended with the LDPE as depicted in Figure 5. In order to evaluate the effect of PE and TPS viscosities on the morphology of LDPE/TPS blends two commercial LDPE resins, LDPE2040 (PE1, MFI = 12g/10min) and LDPE2049 (PE2, MFI = 20g/10min), and the three TPS were used.

4.1 Effect of glycerol content on morphology

PE/TPS blends display a discrete morphology where LDPE is the matrix, especially at low TPS content. The combined effect of glycerol content and the elongational flow exerted on PE/TPS blends (TPS concentration ≈ 30 wt%) during quenching can be observed in Figure 6. PE1 blends prepared with TPS40 and TPS36 (Figures 6a and 6b) show a high level of deformation in the machine direction. Conversely, blends compounded with TPS29 show very little deformation (Figure 6c) and even less when prepared with PE2 (Figure 6d). The singular morphologies displayed by PE/TPS blends are closely related to the differences in viscosity of both TPS and PE. As mentioned in section 3, it was found that 30% glycerol is required to effectively plasticize starch (Rodriguez-Gonzalez et al., 2004). From Figure 6, it can be seen that below that limit, the viscosity and elasticity of TPS are too high to allow the LDPE matrix to greatly deform the TPS dispersed phase. When the Low-viscosity PE2 is used, it can be seen (Figure 6d) that the dispersed particles of TPS are of a spherical nature and that the particle size has increased compared to those of PE2/TPS29 blends (Figure 6c). These results clearly demonstrate that a high degree of morphological control is possible for this system and that the full range from spherical dispersed phase to that of a highly deformed fibrillar phase can be obtained at a given TPS concentration level. In fact, it is apparent that the control of the glycerol concentration allows one to modify the state of the starch from that of a solid particle to that of a quasi crosslinked dispersed phase to that of a highly deformable material.

4.2 Effect of TPS concentration on morphology

The axial direction morphology of PE1/TPS36 blends was a combination of large fiber-like structures with small spherical-like particles (Figure 7). Increasing the TPS concentration reduces the number of small spherical particles due to particle-particle coalescence. The larger particle size of the TPS domains plus particle coalescence leads to the lengthening of TPS fibers in the machine direction. At high TPS loadings (above 45 wt%), it was difficult to distinguish whether LDPE or TPS constituted the matrix. Both components appear to be fully continuous in the axial draw direction. The orientation imposed by the elongational flow field at the die exit plays an important role in the continuity development of starch in these PE/TPS blends.

The starch domain size increases in PE1/TPS29 as the TPS29 content increases (Figure 8). In contrast to the high continuity observed for the low-viscosity low-elasticity TPS36, TPS29 particles remain dispersed in a PE1 matrix, even at high loadings (conc. of TPS ≈ 49 wt%). It can be observed from Figure 8 that increasing the concentration of the TPS at low glycerol contents has little effect on the particle shape.

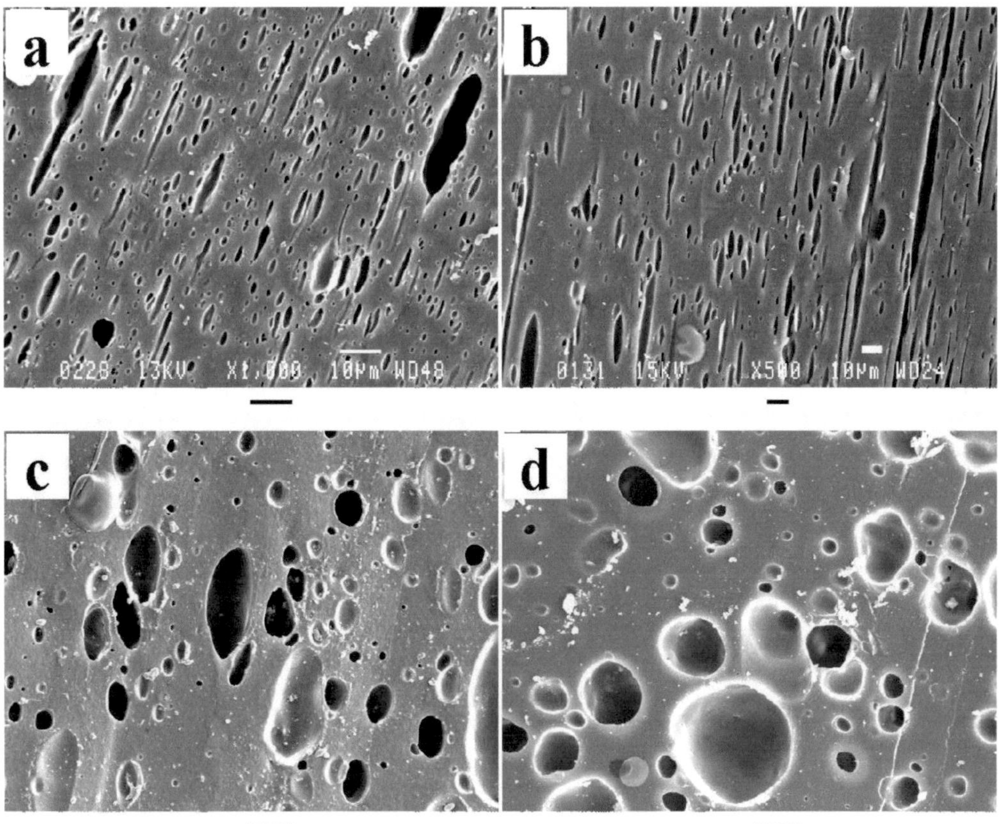

Fig. 6. Effect of glycerol content and LDPE viscosity on the morphology of microtomed PE/TPS (70/30) blends. PE1/TPS blends: a) 40% glycerol, b) 36% glycerol, and c) 29% glycerol. d) PE2/TPS at 29% glycerol content. The black bar below the micrographs represents 10μm.

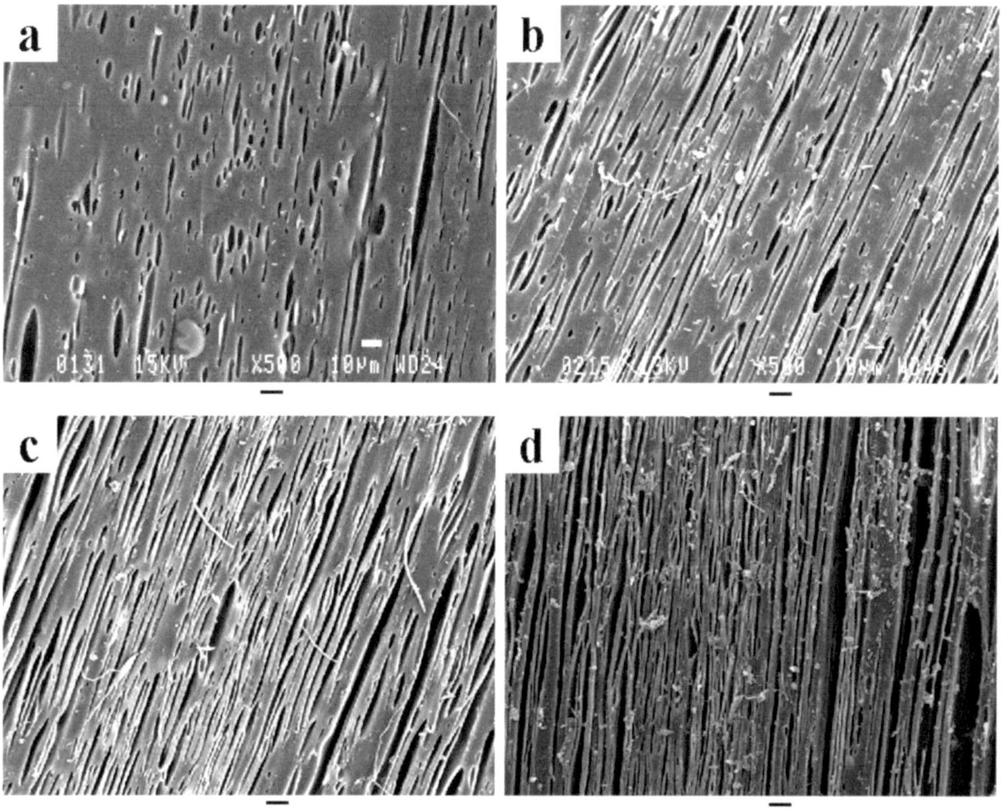

Fig. 7. Influence of TPS concentration on the morphology of PE1/TPS36 blends. a) 29 wt% TPS, b) 36 wt% TPS, c) 45 wt% TPS, and d) 53 wt% TPS. The black bar below the micrographs represents 10μm.

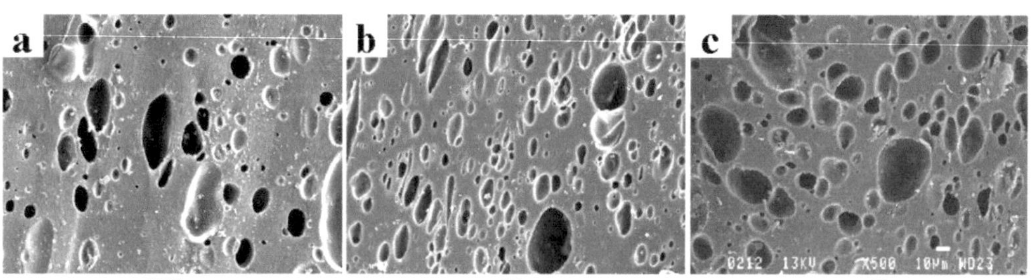

Fig. 8. Influence of TPS concentration on the morphology of PE1/TPS29 blends. a) 30 wt% TPS, b) 41 wt% TPS, and c) 49 wt% TPS. The black bar below the micrographs represents 10μm.

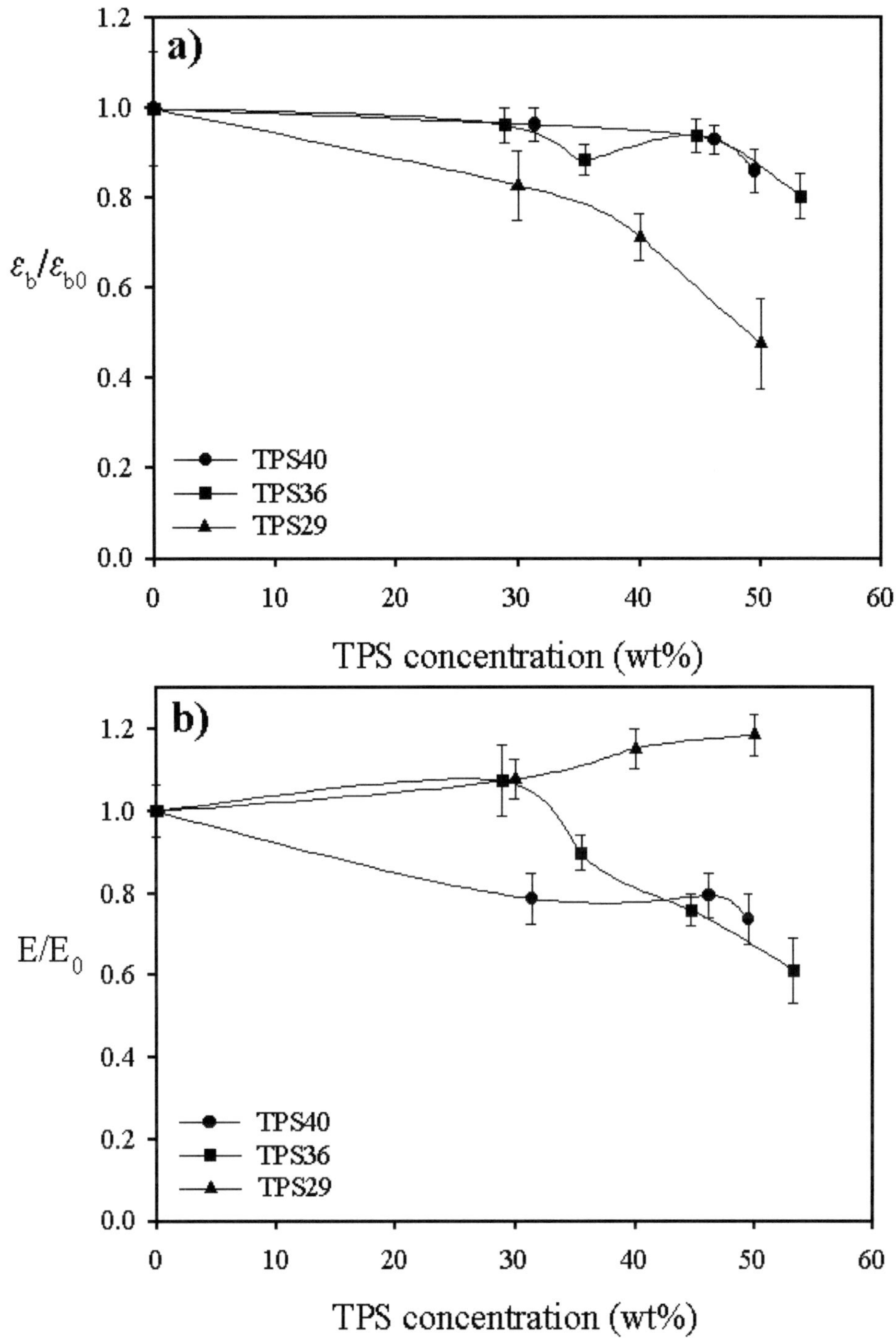

Fig. 9. (a) Relative elongation at break ($\varepsilon_b/\varepsilon_{b0}$) and (b) relative Young's Modulus (E/E_0) of PE1/TPS blends as a function of TPS concentration (wt%). Terms with subscript 0 refer to the pure LDPE.

4.3 Mechanical properties

4.3.1 Elongation at break (ε_b)

The relative elongation at break ($\varepsilon_b/\varepsilon_{b0}$) in the machine direction of PE1/TPS blends is shown in Figure 9a. The results are excellent and demonstrate that at high glycerol contents (36% and 40%), the blends have an ε_b comparable to the virgin polyethylene (ε_{b0}) even at 53 wt% TPS. The ε_b values of PE1 blends drop with the addition of TPS29. If these data are compared with the morphology results from the previous section, it is clear that the high ε_b for blends with TPS36 and TPS40 is closely related to the ability to deform the TPS phase.

In St-Pierre's work (St. Pierre et al., 1997), PE/TPS blends presented a maximum in the ε_b at around 10 wt% TPS followed by a dramatic drop at 22 wt%. In this work, the improved extrusion process and the controlled deformation of the TPS phase yields an important improvement in the ε_b of PE/TPS blends as a function of composition, as observed in Figure 9a. Such an improvement in ε_b is also, in part, due to a highly effective removal of water by venting before blending with polyethylene. In St-Pierre's process, TPS was blended with LDPE and then passed through the venting section. At low concentration, TPS was probably encapsulated into a LDPE matrix, which impeded proper water removal. The presence of water at the blending temperature (150°C) can lead to the formation of bubbles in the extrudate, which weakens the final product (Verhoogt et al., 1995). In the present system, water was completely devolatilized from TPS before mixing with polyethylene (Favis et al., 2003).

4.3.2 Young's modulus

The relative Young's modulus (E/E_0) is demonstrated in Figure 9b. Once again the results are excellent. The E can be maintained at high levels even at high loadings of TPS36 and TPS40. At lower levels of glycerol (TPS29) the E of the blend can be seen to even exceed that of the neat polyethylene. These are unusual results considering the high levels of immiscibility between PE and TPS. The results also indicate the potential of tailoring the mechanical properties of the blend through an appropriate glycerol content. This unexpected result can be explained by good interfacial contact. Leclair and Favis found that the compression exerted by a crystalline matrix (HDPE), during crystallization, on an amorphous dispersed phase (PC) can result in good interfacial contact and a higher elastic modulus (Leclair and Favis, 1996). They also observed that this effect had a positive influence on the modulus only when the contraction took place on a smooth, non-deformable surface.

4.4 Connectivity of TPS particles

4.4.1 Hydrolytic degradation of LDPE/TPS blends

It is well known that acid hydrolysis of starch involves the random cleavage of glycoside bonds producing from oligosaccharides fractions to glucose units (Leach, 1984). In order to quantitatively determine the extent of continuity of TPS blends, samples were exposed to hydrolytic extraction. Figure 10 shows the percent continuity of starch as a function of TPS content for PE1/TPS40 and PE2/TPS40 blends. In both cases there is a monotonic increase in continuity as the concentration of TPS increases. At concentration of 43% or lower, blend morphology plays an important role on percent continuity of LDPE/TPS40 blends. Blends depicting elongated particles show higher percent continuity at comparative concentrations than those displaying spherical morphology. For instance, PE1/TPS40 blends containing 32% TPS40 have 66% continuity while PE2/TPS40 blends composing of 31% TPS40 have only 38% continuity. Above 50% TPS40, at almost 95% continuity, blend morphology does not make any significant difference. At 62 wt% TPS40 the percent continuity of starch domains reaches 100% and the starch phase could be completely extracted. This is indicative of the full connectivity of starch particles through the entirely sample (Figure 10). The use of hydrolytic degradation as previous technique to biodegradation studies could be an important tool to predict enzymatic and bacterial biodegradation.

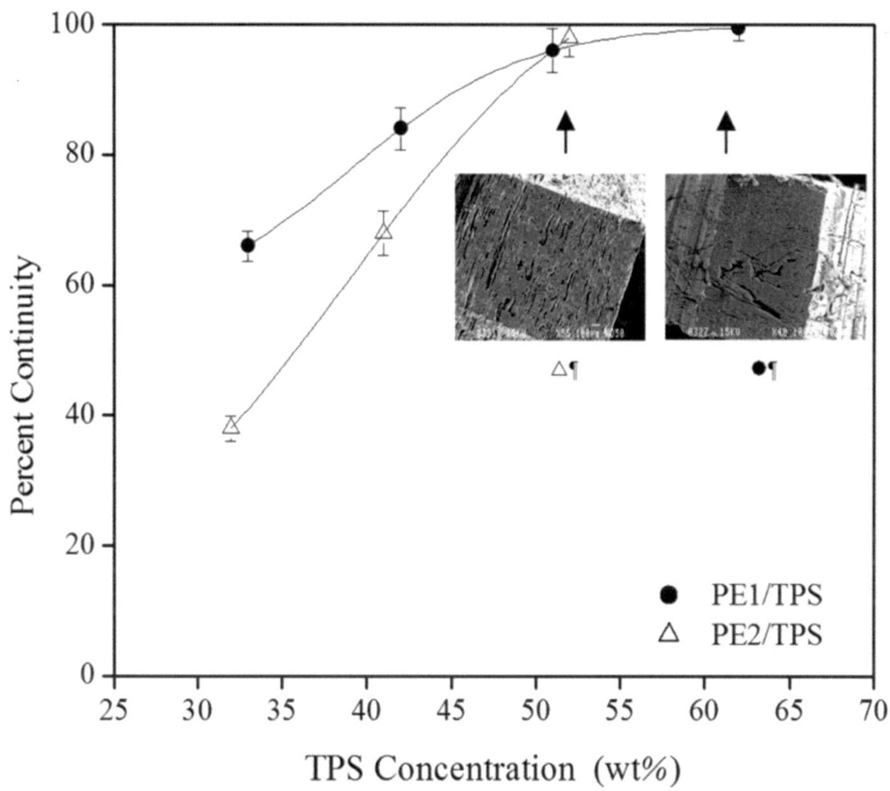

Fig. 10. Accessibility of starch domains LDPE/TPS40 blends exposed in solution of HCl 6N for 72 hours.

4.4.2 Enzymatic degradation of LDPE/TPS40 blends

Numerous studies have been done to investigate the enzymatic hydrolysis of starch-based materials. These works involve blends system with synthetic polymers like LDPE (Danjaji, 2002), ethylene vinyl acetate (EVA) (Simons & Thomas, 1995; Araujo et al., 2004) and polycaprolactone (PCL) (Seretoudi et al., 2002). The kinetic of enzymatic degradation of TPS40 and LDPE/TPS40 blends is shown in Figure 11. Amylase from the enzymatic cocktail triggers the cleavage of 1-4 acetal link while glucoamylase attacks the 1-6 links of amylopectin (Chaplin & Kenedy, 1986), which results in starch solubilization and, consequently, weight loss. The extent of enzymatic degradation of starch is depended on TPS40 concentration. As expected, raw TPS40 is completely degraded during the first 36 hours. Blends of PE1/TPS40 having 62% and 32% and PE2/TPS40 (69:31) result in weight losses of TPS40 of 97%, 65% and 32%, respectively at 72 hours. Therefore, weight loss percent is related to the total amount of TPS40 in the blends. Percolation theory is concerned with the connectivity of one component (in our case, TPS40) randomly dispersed in another (Peanaski et al., 1991). Peanansky showed that below an apparent percolation threshold of 30% by volume (40 wt%) of granular starch, only small amounts were accessible for removal. Granular starches are compact particles, such as those observed in the PE2/TPS40 blends. Fiber-like particles observed in PE1/TPS40 blends could be responsible for a lower apparent percolation threshold in this system and, consequently, higher enzymatic degradation values (Li et al., 2005). Extent of enzymatic degradation of LDPE/TPS40 blends is very similar to that obtained by acid hydrolysis.

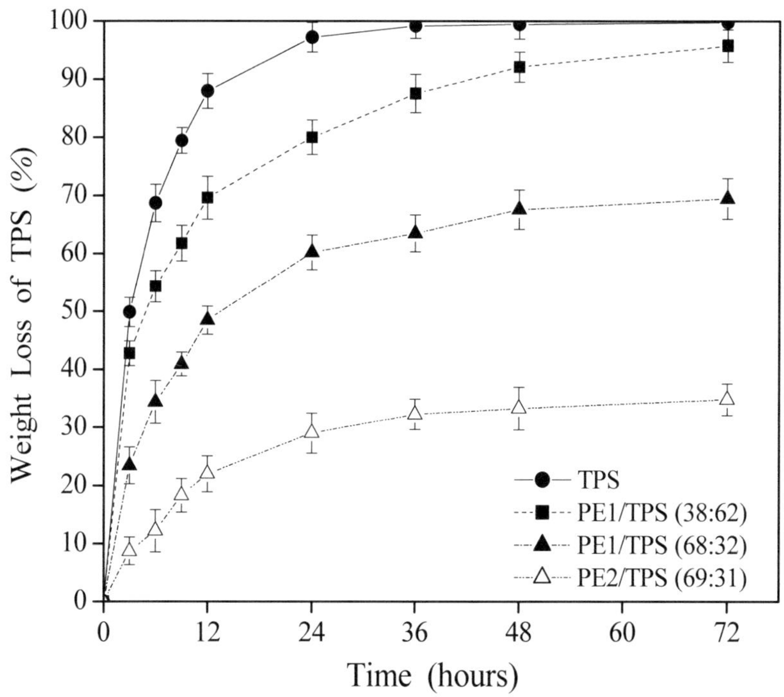

Fig. 11. Enzymatic degradation kinetic expressed as weight loss for raw TPS40 (●), PE1/TPS40 blends: (■) 62 wt % TPS40, (▲) 32 wt % TPS40 and PE2/TPS40 blends with (△) 31 wt % TPS40 as a function of incubation time.

On the other hand, TPS40 enzymatic degradation rate is depended on starch concentration and the accessibility of starch domains as is in the case of LDPE/TPS40 blends. TPS40 is almost insoluble in cold water. When TPS40 is exposed to cold water, it swells and glycerol and low molecular fractions become soluble, but the specimen shape remains intact. Enzymatic hydrolysis of insoluble polymers is known to be affected by the mode of interaction between the enzymes and the polymeric chains and typically involves four steps: (i) enzyme diffusion from the bulk solution to the solid surface, (ii) enzyme adsorption on the substrate, resulting in the formation of enzyme-substrate complex, (iii) catalysis of the hydrolysis reaction, and (iv) diffusion of the hydrolyzed fraction from the solid substrate to the solution (Azevedo et al., 2003). Blends with high loadings of TPS40 show an enzymatic degradation rate as fast as that of the raw TPS40 during the first 3 hours of exposure. This is probably due to the large amount of TPS40 observed on the surface of LDPE/TPS40 blends. Similarly, blends containing about 30% of TPS40 have less starch available on the surface and, consequently, the initial enzymatic degradation rate is slower than the others. As the soluble degradation products of TPS40 diffuse out of the sample, the number of active enzyme units available for starch degradation decreases resulting in a reduction of degradation rate. TPS40 is completely degraded in 36 hours, whereas PE1/TPS40 having 62% and 32% TPS40 and PE2/TPS40 compounded with 31% TPS40 reach their maximum degradation in 72 hours. Conversely, the 69:31 PE2/TPS40 stabilizes at a short period of about 20 hours, whereas the 68:32 PE1/TPS40 blends reaches its plateau at 48 hours. This is likely due to the connectivity of PE2/TPS40 (69:31) blend of starch from the surface; therefore the path of the enzyme is less obstructive.

4.4.3 Microbial biodegradation

Weight loss as a function of time is the most useful method employed to monitor biodegradation (Swanson et al., 2003; Bikiaris et al., 1997b). Figure 12 shows the weight loss of LDPE/TPS40 blends exposed to activated sludge as a function of degradation time. As expected, raw PE1 remains unchanged after 45 days. On the contrary, raw TPS40 is completely consumed within 21 days of exposure. For the LDPE/TPS40 blends, the maximum biodegradation extent is observed at times longer than the raw TPS40. If TPS40 particles are present only on the surface, and not interconnected with particles inside the LDPE/TPS40 blends, then it could be expected that starch domains would be completely biodegraded like the raw TPS40. Percent continuity observed in Figure 10 shows that TPS40 particles are interconnected one to another. At TPS40 concentration of about 30%, interconnection increases when the morphology of starch domains changes from spherical (PE2/TPS40 blend) to fiber-like particles (PE1/TPS40 blend). The extent of biodegradation of TPS40 at 45 days of extraction for PE1/TPS40 blends at 62%, 32% of TPS40 and PE2/TPS40 (69:31) were 92%, 39% and 22%, respectively. However, when the maximum biological extraction is compared with the maximum enzymatic degradation, important difference is noticeable, especially in blends with ca. 30 wt% TPS40.

Kinetic of biodegradation of TPS40 and LDPE/TPS40 blends shows two stages (Table 1). In all cases, there is a fast weight loss during the first 1.5 days, followed by another stage where biodegradation rate decreases progressively. The fast stage could be related to the combined effect of biodegradation and diffusion of glycerol and low molecular starch fractions out of the sample. Diffusion of water soluble components can be accelerated by starch swelling, as observed in raw TPS40, during the first 6 hr. Weight loss during this period is almost 4 times faster than the following 30 hr. In the case of LDPE/TPS40 blends, starch swelling is limited by polyethylene matrix, which results in longer diffusion time. Decrease of biodegradation rate observed after 3 days could be explained by the lower degradability of TPS40 domains that remain in the material.

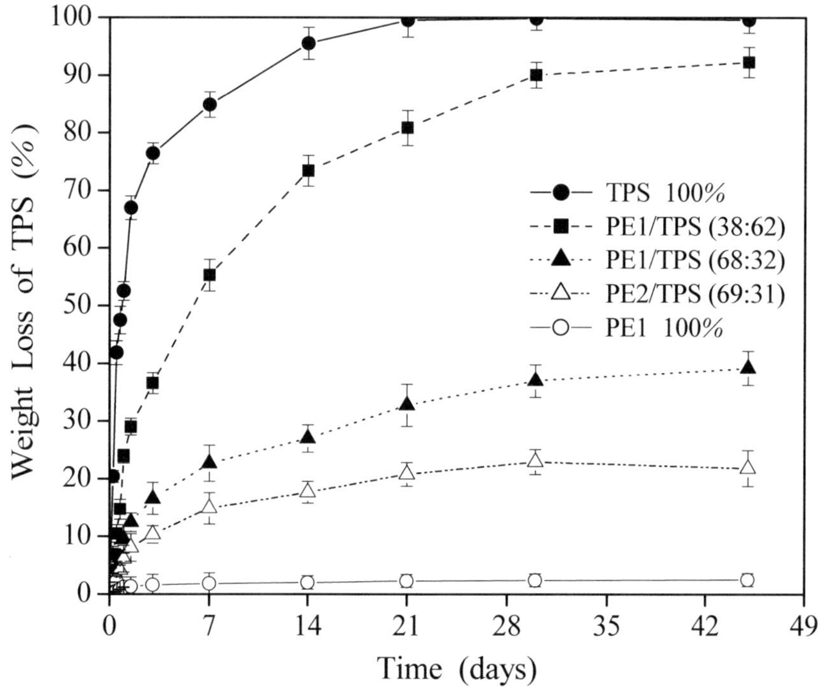

Fig. 12. Bacterial biodegradation kinetic expressed as weight loss for TPS40 (●), PE1/TPS40 blends with: (■) 62 wt% TPS40, (▲) 32 wt% TPS40 and PE2/TPS40 blends (△) 31 wt% TPS40 during exposure in activated sludge.

From comparison of the three degradation techniques, it can be inferred that some phenomenon is taking place during the bacterial degradation of LDPE/TPS40 blends. Weight losses for acid hydrolysis and biodegradation were 100% and 92%, 66 and 39%, and 38% and 22%, respectively for PE1/TPS40 (38:62), PE1/TPS40 (68:32), and PE2/TPS40 (69:31). In the case of PE1/TPS40 (38:62), the difference can be neglected due to the possibility of bacterial waste accumulation inside polyethylene cavities. At around 30% TPS40, however, differences are more prominent. This could be related to other phenomena. Micrographs of the surface of PE1/TPS40 and PE2/TPS40 blends (reported elsewhere) show that pores on PE1 matrix left after TPS40 extraction are below 1 μm, while those observed on PE2 ranged between 3 to 10 μm (Tena-Salcido et al., 2008). On the other hand, different microorganisms have a length between 0.4 and 14 μm and width of 0.2 to 12 μm (Gibbon, 1997). In the case of blends having about 30% TPS40, it is possible that microorganisms or their colonies can restrict starch diffusion by obstructing the polyethylene pores to result in a significant reduction of the final extent of biodegradation.

		dC/dt	(g/l.days)	
Time (days)	TPS40	PE1/TPS40 (38:62)	PE1/TPS40 (68:32)	PE2/TPS40 (69:31)
0.25	81.6	27.6	5.9	6.4
0.5	85.8	14.4	11.6	5.2
0.75	22.6	16.9	7.7	5.2
1.5	26.0	19.1	8.2	5.2
3	6.3	5.0	2.7	1.5
7	2.1	4.7	1.5	1.1
14	1.5	2.6	0.8	0.5
21	0.6	1.1	0.6	0.4
30	0.3	1.0	0.5	0.2

Table 1. Biodegradation rate for TPS40 and LDPE/TPS40 blends as a function of exposure time in activated sludge.

5. Conclusions

The analysis of thermal properties of water-free TPS materials prepared in a TSE showed that granular starch was completely disrupted and that TPS shows a thermal transition below room temperature corresponding to the glass transition temperature and this Tg is dependent on glycerol content. As was observed for the thermal properties, the rheological properties were also highly dependent on glycerol content. η of TPS36 at shear rate ~ 130 s^{-1} decreases by 20% when the glycerol content is increased from 36 to 40%. In the same way, G′ and G″ also decrease as glycerol content increases. However, a particularly dramatic variation is observed when the glycerol content is varied from 29 to 33%. These latter results suggest a phase inversion from a hard elastic network matrix to a soft amorphous one. The glycerol plasticization threshold thus occurs at a content of approximately 30%. This result concerning a critical plasticization threshold is very important for morphology control strategies.

The PE/TPS blends prepared using the one-step process demonstrated levels of ductility and modulus similar to the virgin polyethylene even at very high loadings of TPS without the addition of any interfacial modifier. The excellent properties are a combination of both the melt blending process and a sophisticated morphology control. Through a control of the glycerol content and thermoplastic starch volume fraction, the above process can result in morphological structures, which run the full range of those observed in classical blends of synthetic thermoplastics. Spherical, fiber-like and co-continuous morphologies are observed. Control of the glycerol content of the starch allows one to control the properties of starch from that of a solid filler through to that of a highly deformable thermoplastic material. A wide range of potential properties can be exploited for this type of material.

This material has the added benefit of containing large quantities of a renewable resource and hence represents a more sustainable alternative to pure synthetic polymers. Since the starch can be fully interconnected through morphology control, it is also completely accessible for biodegradation as opposed to the case of starch particles dispersed in a synthetic polymer matrix.

In this work, a relationship between morphology and biodegradation of LDPE/TPS blends was discussed. Percent continuity of the blends is monitored by means of hydrolytic degradation, from which the results show that at TPS concentration below 50%, it is depended on LDPE viscosity and above that value it is independent. Enzymatic degradation is a technique that is closer to the actual biodegradation than acid hydrolysis but we have demonstrated both to have an excellent correlation. However, a correlation of these two techniques with bacterial biodegradation is difficult because of the accumulative deposit of bacteria through empty pores left by the loss of TPS. This difference is more pronounced for the two blends we investigated which contain ca. 30% TPS. In these two blends, the extent of bacterial biodegradation was 39% and 22%, respectively which are less than 60% of the available TPS, as demonstrated by hydrolytic degradation.

6. References

Aburto J, Thiebaud S, Alric I, Borredon E, Bikiaris D, Prinos J, and Panayiotou C. (1997) *Carbohyd. Polym.* 34: 101-112.

Aichholzer W, Fritz H-G. (1998). *Starch/Stärke* 50: 77-83.

Araujo MA, Cunha AM, Mota M (2004) *Biomaterials* 25:2687.

Averous L, Moro L, Dole P, and Fringant C. (2000a). *Polymer* 41: 4157-4167.

Averous L, Fauconnier N, Moro L, and Fringant C. (2000b). *J. Appl. Polym. Sci.* 76: 1117-1128.

Averous L and Fringant C. (2001a). *Polym. Engng. Sci.* 45 (5): 727-734.

Averous L, Dole P, Martin O, Schwach E, and Couturier Y. (2001b) *Symposium de la société Polymérique du Québec*, Nancy (France).

Azevedo HS, Gamma FM, Reis RL. (2003). *Biomacromolecules* 4: 1703.

Bikiaris D, Prinos J, and Panayiotou C. (1997a). *Polym. Degrad. Stab.* 56: 1-9.

Bikiaris D, Prinos J, and Panayiotou C. (1997b). *Polym. Degrad. Stab.* 58: 215-228.

Bikiaris D, Prinos J, Koutsopoulos K, Vouroutzis N, Pavlidou E, Frangis N, Panayiotou C. (1998). *Polym. Degrad. Stab.* 59: 287-291.

Chandra R and Rustgi R. (1997) *Polym. Degrad. Stabil.* 56: 185.

Chang YP, Karim AA, Seow CC. (2006). *Food Hydrocolloids* 20: 1-8.

Chaplin MF, Kennedy JF (1986) *Carbohydrate analysis a practical approach.* IRL Press.

Chaudhary DS. (2010). *J. Appl. Polym. Sci.* 118: 486-495.

Conde-Petit B, Escher F. (1995). *J. Rheol.* 39: 1497-1518.

Da Roz AL, Carvalho AJF, Gandini A, Curvelo AAS (2006). *Carbohydrate Polymers* 63: 417-424.

Danjaji ID, Nawang US, Ishiaku H, Mohd ZA. (2002). *Polymer Testing* 21:75.

Dehennau C and Depireux T (1993). European Patent EP 0 554 939 A2.

Della Valle G, Vergnes B, Tayeb J. (1992). *Entropie* 169 : 59-63.

Della Valle G, Buleon A, Carreau PJ, Lavoie P-A, Vergnes B. (1998). *J. Rheol.* 42, 507-525.

Farhat IA, Mousia Z, Mitchell JR. (2003) *Carbohydrate Polymers* 52: 29-37.

Favis BD, Rodriguez F, Ramsay BA. (2001). International Application WO01/48078 A1.

Favis BD, Rodriguez F, Ramsay BA. (2003). US Patent 6,605, 657.

Favis BD, Rodriguez F, Ramsay BA. (2005). US Patent 6,844, 380.

Forssell PM, Mikkilä JM, Moates GK and Parker R. (1997). *Carbohydrate Polymers* 34: 275-282.

Gibbon D (1997) *Aeration of activated sludge in sewage treatment.* Pergamon Press NY.

Gomez, M. H., & Aguilera, J. M. (1983). *J. Food Sci.* 48: 378.

Gomez, M. H., & Aguilera, J. M. (1984). *J. Food Sci.* 49: 40.

Griffith GJL. (1977). US Patent 4 021 388.
Kalichevsky MT, Jaroszkiewicz EM, Blanshard JMV. (1993). *Polymer* 34: 346-358.
Kim YJ, Lee HM, and Park OO. (1995). *Polym. Engng. Sci.* 35 (20): 1652.
Kulicke WM, Eidam D, Kath F, Kix M, Kull AH. (1996). *Starch/Stärke* 48: 105-114.
Lai LS, Kokini JL. (1990). *J. Rheol.* 34: 1245-1266.
Lai LS, Kokini JL. (1991). *Biotechnol. Prog.* 7: 251-266.
Leach HW (1984) *Starch chemistry and technology*. Academic Science, New York
Leclair A and Favis BD. (1996). *Polymer* 37 (21): 4723.
Li J, Ma PL, Favis BD (2002) *Macromolecules* 35:2005.
Lourdin D, Bizot H, Colonna P (1997a). *J. Appl. Polym. Sci.* 63 : 1047-1053.
Lourdin D, Coignard L, Bizot H, Colonna P. (1997). *Polymer* 38: 5401-5406.
Ma X, Yu J. (2004a). *Carbohydrate Polymers* 57: 197-203
Ma X, Yu J. (2004b). *J. Appl. Polym. Sci.* 93: 1769-1773.
Ma XF, Yu JG, Wan JJ. (2006). *Carbohydrate Polymers* 64 (2): 267-273.
Mali S, Sakanaka LS, Yamashita F, Grossmann MVE. (2005). *Carbohydrate Polymers* 60: 283-289.
Martin O and Averous L. (2001). *Polymer* 42: 6209-6219.
Mathew AP, Dufresne A. (2002). *Biomacromolecules* 3: 1101-1108.
Mendez-Hernandez ML, Tena-Salcido CS, Sandoval-Arellano Z. Gonzalez-Cantu MC, Mondragon M, Rodriguez-Gonzalez FJ. (2011). *Polym. Bull.* 67: 903-914.
Parra DF, Tadini CC, Ponce P, Lugão AB. (2004). *Carbohydrate Polymers* 58: 475-481.
Peanasky JS, Long JM, Wool RP (1991) *J. Polym. Sci., Part Polym. Phys.* 29:565.
Prinos J, Bikiaris D, Theologidis S, and Panayiotou C. (1998). *Polym. Engng. Sci.* 38 (6): 954-964.
Rodriguez-Gonzalez FJ, Virgilio N, Ramsay BA, Favis BD. (2003a). *Adv. Polym. Technol.* 22 (4): 296-305.
Rodriguez-Gonzalez FJ, Ramsay BA, Favis BD. (2003b). *Polymer* 44: 1517-1526.
Rodriguez-Gonzalez FJ, Ramsay BA, Favis BD. (2004). *Carbohydrate Polymers* 58: 139-147.
Ross-Murphy SB. (1995). *J. Rheol.* 39: 1451-1463.
Ruch J, Fritz HG. (2000). *Proceedings of Sixteenth Annual Meeting of Polymer Processing Society* (pp. 350-351), Shanghai (China).
Senouci A, Smith AC. (1988). *Rheol. Acta* 27: 546-554.
Seretoudi G, Bikiaris D, Panayiotou C (2002) *Polymer* 43:5405.
Simons S, Thomas EL (1995) *J. Appl. Polym. Sci.* 58:2259
Souza RCR, Andrade CT. (2002). *Adv. Polym. Technol.* 21 (1): 17-24.
St-Pierre N, Favis BD, Ramsay BA, Ramsay JA, and Verhoogt H. (1997). *Polymer* 38 (3): 647-655.
Swanson CL, Shogren RL, Fanta GF, Imam SH. (1993). *J. Environ. Polym. Degrad.* 1: 155.
Talja RA, Helén H, Roos YH, Jouppila K. (2008). *Carbohydrate Polymers* 71: 269-276
Talja RA, Helén H, Roos YH, Jouppila K. (2007). *Carbohydrate Polymers* 67: 288-295.
Teixeira EM, Da Roz AL, Carvalho AJF, Curvelo AAS. (2007). *Carbohydrate Polymers* 69: 619-624.
Tena-Salcido CS, Rodriguez-Gonzalez FJ, Mendez-Hernandez ML, Contreras-Esquivel JC. (2008). *Polym. Bull.* 60: 677-688.
Van Soest JJG, Benes K, de Wit D. (1996). *Polymer* 37: 3543-3552.

Verhoogt H, Truchon FS, Favis BD, St-Pierre N, and Ramsay BA. (1995). *Proceedings of the Annual Technical Conference ANTEC'95* 2028-2032.
Willett JL. (1994). *J. Appl. Polym. Sci.* 54: 1685-1695.
Willett JL, Jasberg BK, Swanson CL. (1995). *Polym. Engng. Sci.* 35: 202-210.
Willett JL, Millard MM, Jasberg BK. (1998). *Eur. Polym. J.* 34: 1477-1487.
You X, Li L, Gao J, Yu J, Zhao Z. (2003). *J. Appl. Polym. Sci.* 88: 627-635.

2

Thermoplastic Cassava Flour

Diana Paola Navia[1] and Héctor Samuel Villada[2]
[1]*Universidad de San Buenaventura Seccional Cali*
[2]*Universidad del Cauca*
Colombia

1. Introduction

Stocks of oil in the world are limited and most synthetic plastics can not be degraded by the environment, whereby people are investigating other sources of raw materials aimed at the production of materials less aggressive to the environment, to serve to decrease the amount of plastic waste.

The food industry plays an important role in the use of plastic for protection before, during and after food harvest to ensure the integrity of these (Weber, 2001; De Graff et al, 2003; Tharantahan, 2003, Halley, 2005); for example the common consumer products such as plates, cups, spoons, knives, disposable, film or coating films and bags (Wang et al, 2003; De Graff et al, 2003; Averous & Boquillon, 2004; Bastioli, 2001, Garcia et al, 2000, Shogren et al, 2003; Thompson, 2003, Wang et al, 2003).

The plastics produced from fossil sources have contributed significantly to increasing environmental pollution caused by the accumulation of solid waste that can not degrade in landfills, so that has prompted the search for new biodegradable materials not only in the food, but also in the medical, automotive, among others. Renewable natural raw materials become an important alternative, including flour, starch, natural fibers, proteins, and others prominent in developing options for bioplastics (Tharantahan, 2003). Currently there is growing interest to use raw materials and agricultural byproducts in obtaining biodegradable plastics, such as from corn and cassava and potato tubers. However, plastics developed from these sources have certain drawbacks of structural stability compared to conventional plastics, caused by its stiffness or weakness due to its high hygroscopicity and rapid aging (Villada, 2005). Therefore, the research efforts must be maintained and increased in this field, taking into account the use of local products in the region such as cassava, which are being studied in research projects, through which it is intended that the methodology production of biodegradable plastics is reproducible on an industrial scale, taking into account the specific functional requirements for various applications.

This document presents some excerpts related to thermoplastic cassava flour as a raw material useful for packaging applications.

2. Cassava

Cassava (*Manihot esculenta* Crantz), also known as mandioca, is a starchy root belonging to the Euphorbiaceae family and is one of the most important energy sources for tropical areas of the world. Although cassava thrives in fertile soil, its comparative advantage to other more profitable crops is their ability to grow in acidic soils of low fertility, with sporadic rainfall or long periods of drought. The crop is widely adapted as it is planted from sea level up to 1800 meters, at temperatures between 20 and 30°C with an optimum of 24°C, relative humidity between 50 and 90% with an optimum of 72% and an annual rainfall between 600 and 3000mm with an optimum of 1500mm (Casaca, 2005). It is widely cultivated in tropical Africa, Asia and Latin America, and it is the fourth most important global crop in developing countries, with an estimated production in 2006 of 226 million tonnes. It is characterized by great diversity of uses both roots and leaves can be eaten by humans and animals. Cassava products can be used in industry mainly from the starch (Ceballos, 2002).

The bulk of cassava produced in farms of small farmers and marginal areas, so that a significant proportion of production is not recorded in the statistics accurately, in addition, these farmers are generally isolated from the distribution channels and the product processing industries, mainly in areas that have little or no access to improved varieties, fertilizers and other production inputs. Governments have not yet made the necessary investments to boost its value added, that would make cassava starch products uncompetitive internationally (FAO, 2007).

This root is composed of three tissues: the periderm (husk), the cortical parenchyma (cortex) and inner parenchyma, where approximately 80% of fresh weight of the root, corresponds to the parenchyma or pulp, which is the tissue that plant stores starch. The dry matter content of cassava root fluctuates between 30% and 40%, dry matter is composed of parenchyma, mostly (90% to 95%), the nitrogen fraction, ie, carbohydrates (starch and sugars), the rest of this dry matter corresponds to fiber (1% to 2%), fats (0.5% to 1.0%), ash and minerals (1.5% to 2.5%) and protein (2.0%); starch also represents most of the carbohydrates (96%) and it is, therefore, the main component of root dry matter (FAO, 2007).

Moreover, cassava varieties can be classified as bitter, which contain a cyanogenic glycoside called linamarin, which are hidrolized in the presence of enzymes and acids linamaraza, resulting in the formation of hydrocyanic acid. This acid is under the skin of the roots, inside a layer of viscous-looking latex, white and with characteristic odor, usually this variety is used for industrial processes.

Sweet varieties have low or no presence of hydrocyanic acid, therefore its use is safe after cooking (Aristizabal & Sanchez, 2007). If cassava use is for human consumption, it will be called culinary quality, but it used for the manufacture of products such as flour, starch or dry pieces it be called industrial cassava, and finally it can be called dual purpose if it is intended to human consumption or industrial use (Aristizabal & Sanchez, 2007). In Colombia, CIAT, Corpoica and CLAYUCA have developed improved varieties of cassava for growth in certain areas of the country, taking into account biological and climatic factors. The Cauca region is located within zone which it can grow varieties such as HMC-1, MPER 183, MBRA 383, among others (Cadavid, 2005).

Cassava roots are rich in calories but deficient in protein, fats, minerals and vitamins. It also presents several secondary compounds in the root tissues as polyphenols, tannins, carbohydrates and cyanogenic compounds (Sanchez & Alonso, 2002).

2.1 Cassava flour

After harvest, fresh cassava roots can deteriorate very quickly, since they have a 65% water content (Ceballos, 2002). In order to preserve the fresh cassava, an alternative is by drying to obtain dried pieces, which can be obtained flour, whether for food, feed or industrial use. The cassava flour for human consumption can be classified as fermented (gari) or not fermented. Unfermented flours are made by grinding the roots peeled or cut into small pieces, then the resulting material is dried and milled (Ceballos, 2002). According to the Codex standard for edible cassava flour (*Manihot esculenta* Crantz), it is the product obtained from cassava chips or pasta with a grinding process, followed by screening to separate the fiber from flour. In the case of edible meal prepared with the bitter cassava (*Manihot utilissima* Pohl) it shall be made by soaking the tubers detoxification in water for several days prior to drying in the form of milled whole tuber (pasta) or small pieces.

Cassava flour is obtained by grinding dried cassava chips, as explained in detail below (Alvarado & Cornejo, 2009):

- Receiving. A visual inspection of the raw material is performed. The performance process is determined by the cassava weight.
- Washing. It removes dirt and other debris present in cassava roots with water.
- Peeling. When the process is done on a small scale, cassava peel is removed by abrasive equipment or knives. When the shell is not removed you get cassava meal.
- Cutting. Cassava is cut into small pieces.
- Crushed. The roots will dry quickly, so, is necessary to increase the surface exposed to hot air grinding cassava chips until slurry.
- Drying. It can be accomplished by hot air dryers or in yards by sunlight.
- Milling. Once dried cassava chips, reducing the particle size using a hammer mill generally.
- Sieving. The ground flour is passed through a series of mesh to determine particle size. According to the CODEX STAND 176-1989, cassava flour is fine when at least 90% passes through a 0.6mm mesh sieve and it is thick when 90% passing through a 1.20mm mesh sieve.

Cassava flour consists mainly of starch (80 - 90%) and fiber (1.5-3%) depending on the variety of cassava from which one obtains (Charles, Sriroth, & Huang, 2005). It is considered a potential raw material in the field of developing new materials, including biocomposites, because of its high concentration of starch, (Martínez et al, 2007).

Cassava flour should have a moisture content no greater than 13% for easy storage and transport conditions (Codex Stand 176, 1989). Cassava flour can also be classified as integral or bakery. The integral flours are the result of grinding the dried cassava chips with bark, which is used as a substitute for carbohydrates in cereals (maize, wheat, sorghum) and it is usually used in food formulations for animals. The bakery is obtained by grinding the dried pieces of peeling cassava, passing the product through a fine sieve. If the product of grinding through a sieve less dense, it is obtained granulated or cassava semolina (Montaldo, 1985).

2.1.1 Aplications

In Latin America and other continents, cassava flour is sold primarily as a potential substitute for cereal flours (wheat and sorghum) in the field of baking (Shittu et al, 2008, Benitez et al, 2008). However, cassava flour has cyanogenic content (specific processing technology) that limit their applications in this niche market, because the standards set in these components in relation to human consumption (Ceballos, 2002).

It is possible to develop higher value-added products based on cassava flour in order to expand production and processing and open new markets, promoting the establishment of rural industries and providing the opportunity to expand the income of small farmers (Garcia et al, 2005).

The probability that cassava flour can increase its demand in industrial applications such as adhesives plant, plywood, corrugated cardboard, thread cones, packaging materials, among others, is very high, due to environmental concerns that have emerged in recent decades, not only in Colombia, but also globally, so it requires implementation in the various products of renewable materials, cassava flour being one of the alternatives with great prospects in this area (Ceballos, 2002). However, the materials sector based on cassava flour is unexplored and little known in the market for this power plant.

2.1.2 Termoplastic cassava flour

Cassava flour can be a thermoplastic material through the disruption of the molecular chains under specific conditions of temperature and presence of plasticizer (Rahman and Brazel, 2004, Martinez et al., 2007), so the TCF (Thermoplastic Cassava Flour) could become one of the important raw materials in the development of biodegradable plastics.

Several studies report the use of glycerol as a plasticizer in TCF (Ma, Yu & Wang, 2008; Ma et al, 2009; Chang et al, 2010), however the use of such TCF in the development of biodegradable materials has been limited due to problems associated with poor viability in the market, poor processability, low strength, low stability to humidity and retrogradation. For that, there have been implementing strategies to minimize these characteristics. Among them is the chemical modification of the flour (crosslinking and esterification) and TCF mixture with other polymers and biodegradable polyesters. However the industry has found a less expensive and more environmentally friendly through the use of natural fibers (Curvelo, Carvalho & Agnelli, 2001; Oksman, Skrifvars & Selin, 2003; Guan & Hanna, 2004; Avella et al, 2005; Ma, Yu & Kennedy, 2005; Teixeira et al, 2009). They have desirable characteristics such as low density, low cost, biodegradability, non-abrasive, good ductility and thermal properties (Wambua, Ivens & Verpoest, 2003; Martins et al, 2009), besides acting as a reinforcing material due to the strong bond that develops between the fiber-matrix interface due to the chemical similarity of starch in the flour and cellulose fibers (Luna, Villa & Velasco, 2009).

Within this area several studies indicate that when natural fibers are mixed with the TCF, improve mechanical properties (Lee & Wang, 2006; Martinez et al, 2007; Nair, Wang & Hurley, 2010). Other studies such as Carvalho and others (2005) investigated the influence of plasticizer (glycerol) and fiber (cellulose) in TCF, finding that the use of plasticizer, significantly reduced the degradation of starch, while that the increase in fiber content, increased it.

There are few investigations made in TCF biodegradable packaging. At the international level have been evaluated physical and chemical properties of varieties of cassava flour for different applications, however, in the field of biocomposites studies are scarce. Researchers in Ghana, concluded that 31 varieties of cassava studied should not be wasted by their low quality cooking or high cyanogenic, however recommend its use in industrial applications such as extraction of starch and/or flour, sugar production, adhesives, among others (Aryee et al, 2006). In Nigeria, a study on the effect of type of material (pellets and cassava flour) and drying method (solar-oven) of material on the yield and physicochemical properties of starch in these materials and found that drying oven yielded the most appropriate results (Olomo & Ajibola, 2003). Venezuelan authors studied the effects of heat treatments on cassava meal, finding that pre-gelatinized flour decreased the tendency to retrogradation, consistency and rate of absorption and suggested the use of these conventional products and new product development (Pérez et al, 2007).

3. Related research

As part of the implementation of the program "Use By-Products of Cassava", cofinanced by the Agriculture and Rural Development Ministry, the Cauca University and the Cauca Productivity and Innovation Regional Center (Popayán, Cauca, Colombia),we have carried out some important technical developments for the use of cassava flour thermoplastic in the biodegradable materials manufacture. Next, we present some results of these studies of both raw materials and product obtained.

3.1 Characterization of cassava flour

In this regard, we have used the following techniques of analysis:

3.1.1 High Resolution Optical Microscopy- HROM

Samples were spread on glass sheets to be observed with optical microscope (Nikon Eclipse 80i, Japan) coupled to digital camera (Nikon DS-2MV 2Mp, Japan), through the bright field technique.

Photomicrographs of cassava flour in Figure 1 were captured with 10x (down) and 40x (up) objectives; in them may be observed longitudinal and amorphous fibers, and round /truncated starch granules.

Fibers, stained blue-violet because dye used (toluidine blue), it were longitudinally elongated and not elongated amorphous. As shown in Figure 1, there is a greater number of starch granules by number of fiber (s) found in each catch, this is because the fiber content in the flours studied is very low, between 1.7 and 2.7% in wet basis, compared with the starch content, between 75 and 85% wet basis, according to proximate analysis results (not shown).

Moorthy, 2002, reported that the fiber content of cassava flour is usually between 2-3% and starch content accounts for 84% (Rodriguez et al., 2008). A recent study on cassava roots (Teerawanichpan et al., 2008) discusses the anatomy by light microscopy using toluidine blue and shows four types of tissues, sclerenchyma, parenchyma, secondary and primary xylem. Starch granules have spherical and semi-spherical forms, as reported by other studies of cassava starch morphology (Alvis et al., 2008) - some of them truncated features shown as flat surfaces on one or more sides of the granule.

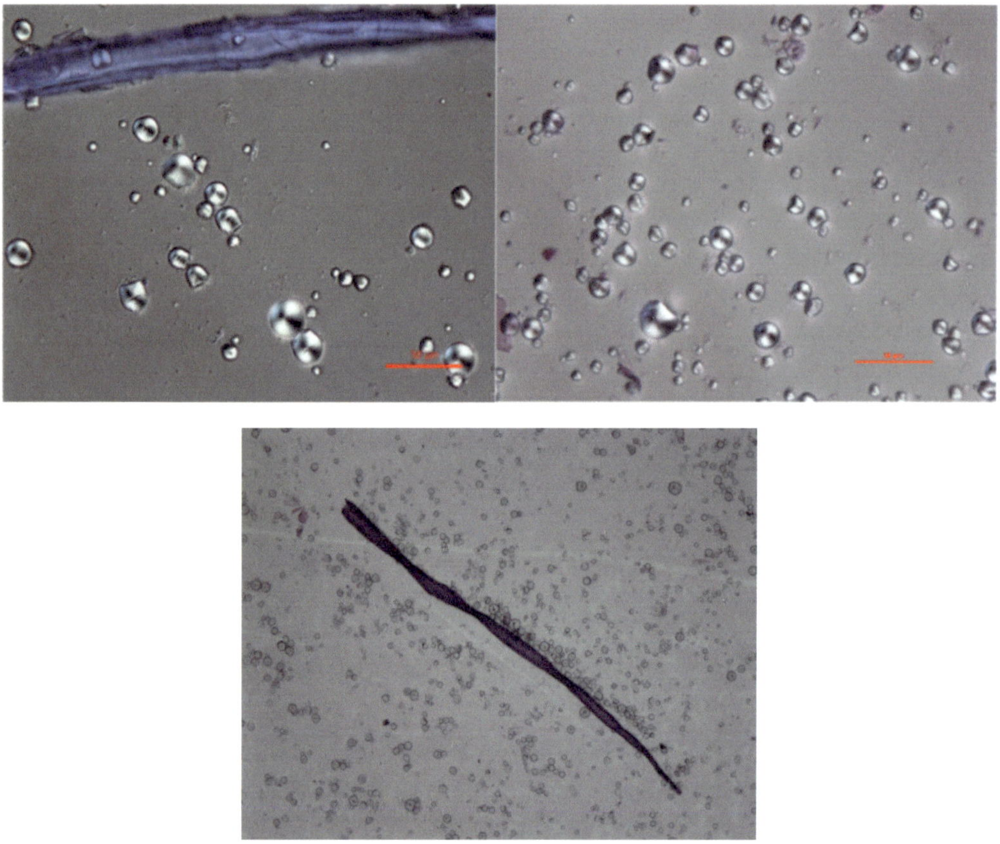

Fig. 1. HROM microphotographs cassava flour

3.1.2 Scanning Electron Microscopy- SEM

The samples were spread out on cylindrical specimens with carbon tape and subjected to gold bath (JEOL JSM-6490) for 200 seconds with a gap of 40-60 mTorr.

Figure 2 shows starch granules of variable diameter that does not exceed 25μm, and truncated spherical shapes typical of cassava starch. Fractures were also observed in its structure, possibly by the effects of grinding, especially when you consider that the particles are broken because the process to reduce particle size at the beginning of the characterization process of raw material.

Fig. 2. SEM microphotographs cassava flour

3.1.3 X-Ray diffaction

Samples were analyzed between $2\theta = 2°$ and $2\theta = 35$ °, in an X-ray diffractometer (Rigaku 2002, Japan), (wavelength = 0.15405 nm) at 40 kV and 30 mA. The scanning speed was 5°C/min.

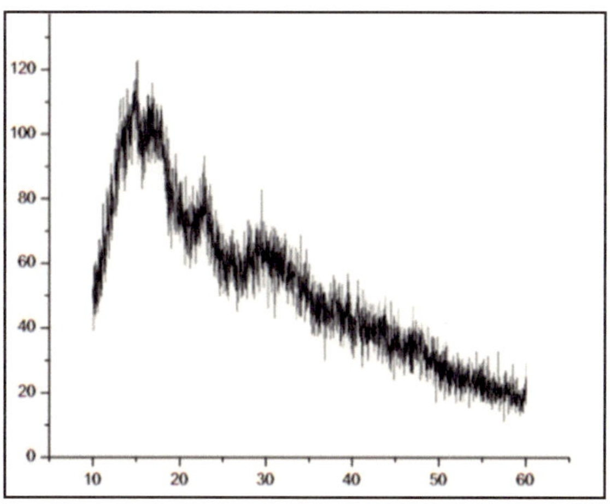

Fig. 3. X-Ray Diffractogram cassava flour

The samples showed a peak at 2θ: 20°, which according to Zobel (1988, quoted in Singh, 2006), is attributed to the presence of amylose-lipid complexes in starches, whose intensity could be related to the proportion of them. There were strong peaks at 2θ: 15°, 2θ: 17°, 2θ: 18° and 2θ: 23°, characteristic of type A pattern (Van Soest et al., 1996; Cheetham & Tao , 1998; Rodriguez et al., 2007; Leblanc et al. 2008; Perdomo et al, 2009), which indicates that the crystalline arrangement is monocyclic.

3.2 Characterization of TPCF material

The material was obtained by manual molding technique. To determine the operating conditions, we developed an experimental factorial design as shown in Table 1. The results of this design, indicated that there were significant differences among the factors evaluated ($p < 0.05$), so an response surface analysis was developed using the software matlab (R2008a) to determine the process conditions: the content of plasticizer, drying time and particle size of cassava flour. Figures 4 and 5 show the results of the analysis.

Factor	Level	Response
Plasticizer content (%)	High:25 Low: 10 Center: 17.5	Tensile strenght (MPa)
Drying time at 45°C (hours)	High: 26 Low: 8 Center: 17	
Particle size of cassava flour (μm)	High: 600 Low: 250 Center: 425	

Table 1. Experimental design moulded material process

Response surface analysis established that the particle size of 600 microns was the one who reported the highest value on the strength of tensile strength of the material, the same way, drying times above 20 hours at 45°C, interacting with a content of 15% plasticizer favor mechanical properties in tension, bearing in mind that this property is important because it will identify the functional applications of the material.

Figures 4 and 5 show the optimization of the variables evaluated (drying time, particle size and concentration of plasticizer). This optimization is valuable because it indicates the values of variables in which the response evaluated had the highest value. Small particle sizes cassava flour could be contributing to a greater absorption of water by the increased free volume between particles, which probably caused a lower response to stress. Water acts as a plasticizer, resulting in intermolecular mobility, therefore, a drier material (longer drying time), the higher rigidity and resistance to a tensile stress.

Some of the techniques to characterize the material valued based TPCF were:

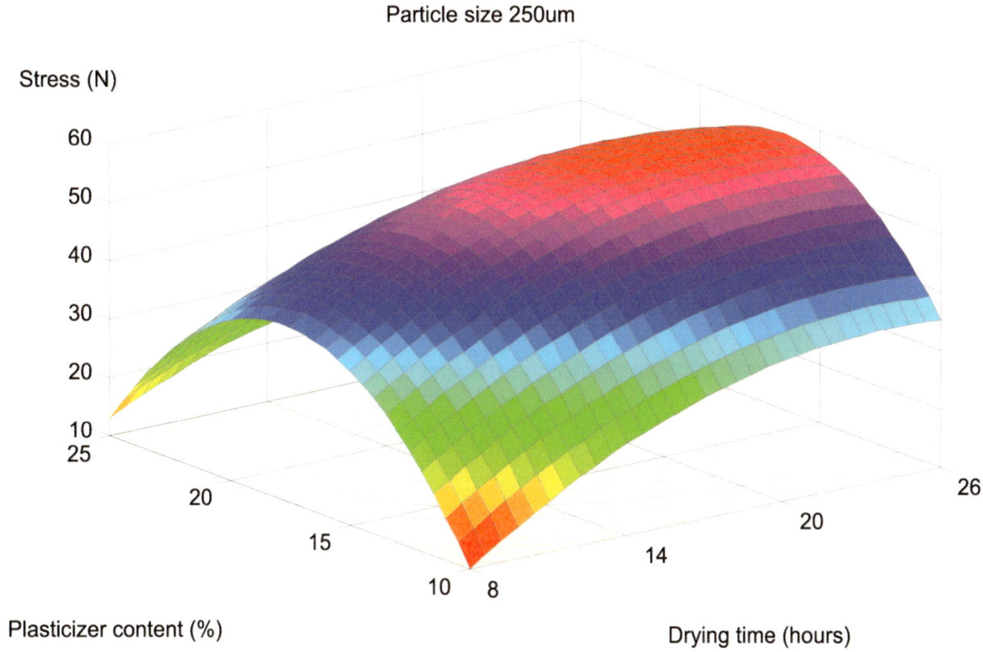

Fig. 4. Response surface of TPCF material with particle size 250 μm

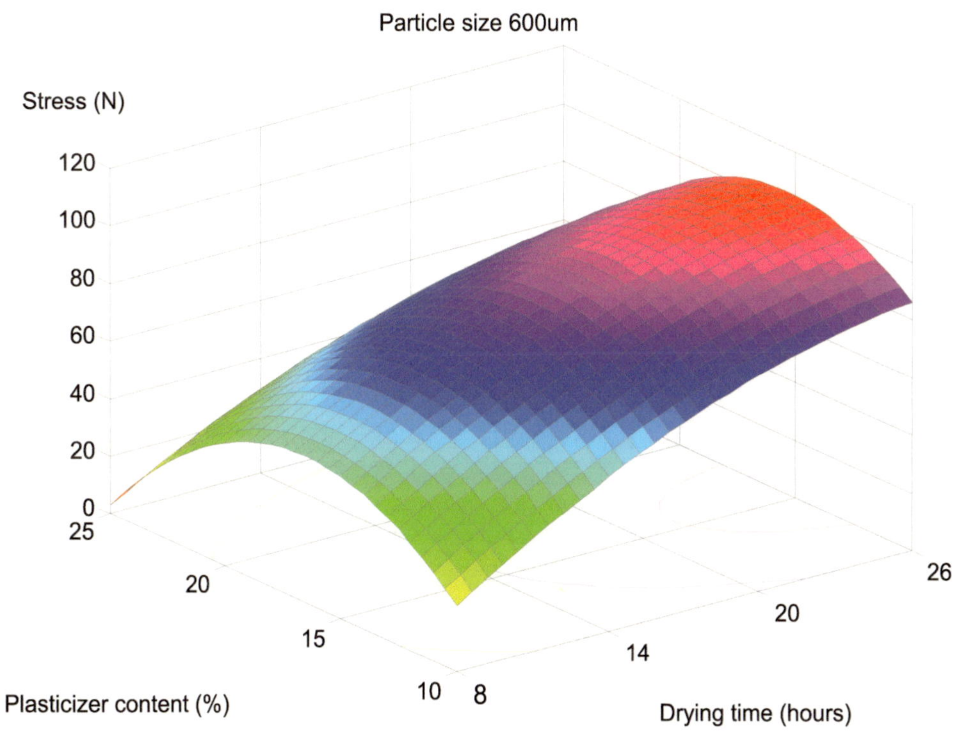

Fig. 5. Response surface of TPCF material with particle size 600 μm

3.2.1 Differential scanning calorimetry

Samples were evaluated according to ASTM D3418-08 applied to the analysis of polymeric materials, with some modifications. Equipment used was a DSC (TA Instruments, Q20, USA). The samples were stored in hermetically sealed aluminum pans and subjected to heating from -50°C to 225°C at a heating rate of 20°C/min, then cooling to -50°C and a final heating similar to the first.

Figure 6 shows the three cycles that were submitted material samples in DSC. In the first heating scan showed a first endothermic peak before 0°C, then a glass transition and a second endothermic peak of melting of the material close to 150°C. Sample was then cooled with the drop from the flow of heat and finally heated in the third cycle, evidencing only an endothermic peak below 0°C, as presented in the first cycle.

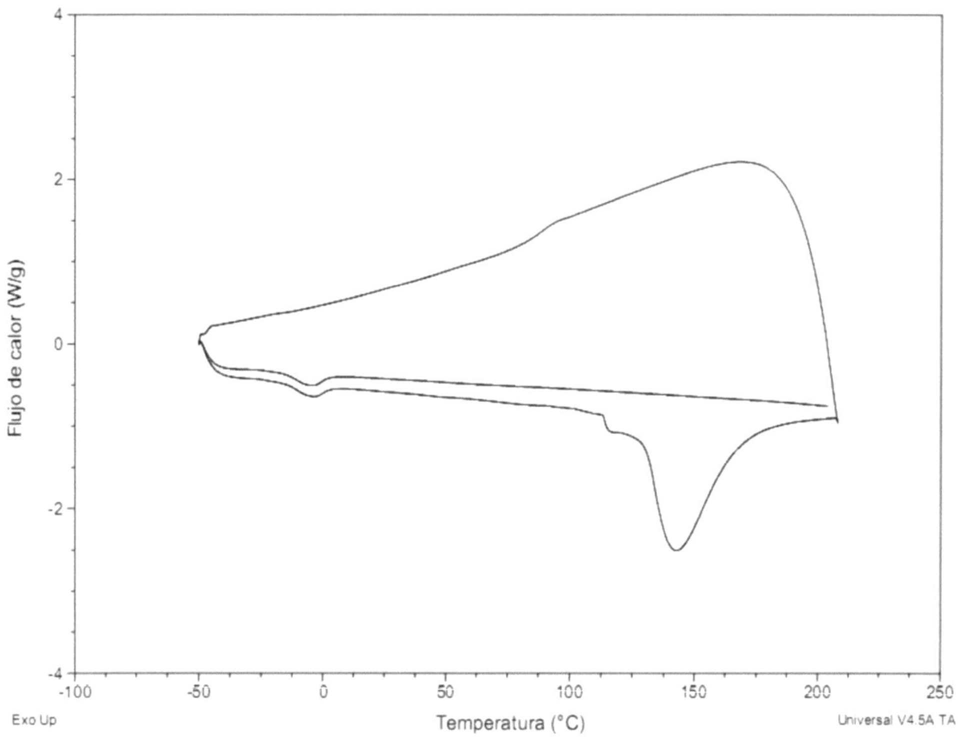

Fig. 6. Termogram DSC material of TPCF

The first scan was performed to remove the thermal history of material, that is, to prevent abnormal results because thermal processes to which the material was subjetc, which can alter the phase transitions characteristic of the sample, so it is expected that with the initial heating the material molecules were casted, and then, with the cooling the material molecules get to organize freely (second cycle) to obtain the real phase transitions of the material in the third cycle. No clutch, the first cycle performed (Figure 6) shows the typical transitions of the material, which was not presented in the third or second heating cycle, as usually (Mohamed et al, 2010). Some authors report that the Tg (glass transition temperature) was obtained only in the first heating, and melting temperature in the first and second heating in polylactide films samples mixed with thermoplastic cassava starch extruded (Lee, Chen & Hanna, 2008). Possibly, the material molecules can not organize freely during cooling, maybe because they require a much slower process for ordering, which resulted in no evidence of the glass transition and fusion of material on third cycle, further, cooling after first heating is probably causing irreversible structural changes and transformations at the molecular level, which prevents phase transitions evident in the first heating cycle, but it was evident by the second one.

In Figure 6, can be seen that the glass transition temperature is above 110°C, this is an indication of the high stability of the material, since in future applications, it may be subjected to temperatures near 100°C maintaining its stability because it is in their glassy solid state.

3.2.2 Thermogravimetric Analysis - TGA

TGA equipment was used (Q50, TA Instruments). Samples between 10 and 20 mg were assessed of molded material, according the ASTM E1131-08. The sample was placed in a platinum tray open and subjected to heating from 25°C to 500°C at a rate of 10°C/min. The tests were carried out in a controlled environment using nitrogen level UAP.

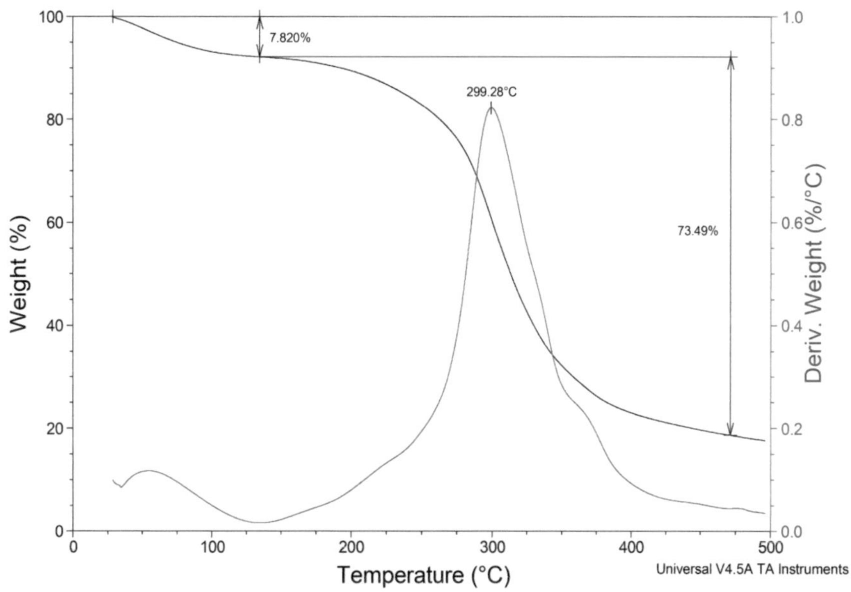

Fig. 7. Termogram TGA material of TPCF

The TGA curve has four main areas, namely: the area of highly volatile material (200°C or less) represented by moisture, plasticizers, residual solvents and other components of low boiling point, the material area average volatility (between 200 and 750°C) represented by compounds of polymer degradation, oil; area where combustible material degrades oxidized material such as coal nonvolatile (temperature depends on the material), and the area of ash corresponds to non-volatile residue in an oxidizing atmosphere including metal components, inert fillers or reinforcements (ASTM E1131). Figure 7 shows the first two zones in the TGA curve, represented by 7.820% in highly volatile material, and the area average volatility material that starts at 134.89°C and ends at 470.6°C and a residue of 18.69%. The degradation temperature (Td) for the material presented was 299.28°C, meaning adequate thermal stability, comparable with thermoplastic cassava starch nanoreforced. Td which occurred between 309 and 327°C (Schlemmer, Angelica & Sales, 2010), and compounds near extruded and injection molded thermoplastic starch reinforced with lignocellulosic fibers whose degradation temperatures were between 335 and 339°C (Averous & Boquillon, 2004).

4. Conclusion

Cassava flour is a viable material for use as part of a processable thermoplastic matrix by molding technique, which allows to obtain materials with acceptable mechanical and thermal properties for agro-industrial applications.

5. Acknowledgment

The authors acknowledge the support of the Ministry of Agriculture and Rural Development, University of Cauca, and University of San Buenaventura Sectional Cali, in this publication.

6. References

Alvarado, G. & Cornejo, F. (2009). Obtención de harina de yuca para la obtención de productos dulces destinados para la alimentación de celíacos. Escuela Superior Politécnica del Litoral. [on line]. Available from: http://www.dspace.espol.edu.ec/bitstream/123456789/6391/1/Obtenci%C3%B3n%20de%20harina%20de%20yuca%20para%20el%20desarrollo%20de%20productos%20dulces.pdf

Alvis, A.; Vélez, C.; Villada, H. & Rada, M. (2008). Análisis físico-químico y morfológico de almidones de ñame, yuca y papa y determinación de la viscosidad de las pastas. Información Tecnologica, Vol.19, No. 1, pp. 19-28. ISSN 0718-0764

American Society for Testing and Materials. ASTM E1131. (2008). Standard Test Method for Compositional Analysis by Thermogravimetry. West Conshohocken: ASTM International.

Aristizábal, J. & Sánchez T. (2007). Guía técnica para producción y análisis de almidón de yuca. Boletín de servicios agrícolas de la FAO 163. Organización de las naciones unidas para la agricultura y la alimentación, ISSN, 1020-4334

Aryee, F.; Oduro, I.; Ellis, W. & Afuakwa, J. (2006). The physicochemical properties of flour samples from the roots of 31 varieties of cassava. Food Control, Vol. 17, No.11, pp. 916 - 922, ISSN 0956-7135

Avella, M.; De Vlieger, J.; Errico, M.; Fischer, I.; Vacca, P. & Grazia, M. (2005). Biodegradable starch/clay nanocomposite films for food packaging applications. Food Chemistry, Vol. 93, No.3, pp. 467-474, ISSN 0308-8146

Avérous, L. & Boquillon, N. (2004). Biocomposites based on plasticized starch: thermal and mechanical behaviours. Carbohydrate Polymers, Vol. 56, No. 2, (August 2011), pp. 111-122, ISSN 0144-8617

Bastioli C. (2005). Handbook of Biodegradable Polymers. 517 p, ISBN 978-1-85957-389-1

Benitez, B.; Archile, A.; Rangel, L.; Ferrer, K.; Barboza, Y. & Márquez, E. (2008). Composición proximal, evaluación microbiológica y sensorial de una galleta formulada a base de harina de yuca y plasma de bovino. Interciencia, Vol. 33, No.1, pp. 61-65, ISSN 0378-1844

Cadavid, L. (2005). Producción de Yuca. Consorcio Latinoamericano y del Caribe de Apoyo a la Investigación y al Desarrollo de la Yuca; Centro Internacional de Agricultura Tropical; USAID.

Carvalho, A.; Zambon, M.; Da Silva, A. & Gandini, A. (2005). Thermoplastic starch modification during melt processing: Hydrolysis catalyzed by carboxylic acids. Carbohydrate Polymers, Vol. 62, No.4, pp. 387-390, ISSN 0144-8617

Casaca, Á. (2005). Documento técnico. El Cultivo de la Yuca. 2005. Available from: http://www.sag.gob.hn/files/DICTA/DICTA%20WEB/Yuca.pdf

Ceballos H. (2002). La yuca en Colombia y el Mundo: Nuevas perspectivas para un cultivo milenario. In: Centro Internacional de Agricultura Tropical CIAT., editor. La yuca en el tercer milenio. Sistemas modernos de producción, procesamiento, utilización y comercialización. Chapter 1, p. 1-13. (Palmira, Colombia), ISBN 958-694-043-8

Chang et al. (2010). Preparation and properties of glycerol plasticized-starch (GPS)/cellulose nanoparticle (CN) composites. Carbohydrate Polymers, Vol. 79,No.2, pp. 301-305, ISSN 0144-8617

Charles, A.; Sriroth, K. & Huang, T. (2005). Proximate composition, mineral contents, hydrogen cyanide and phytic acid of 5 cassava genotypes. Food Chemestry, Vol. 92, No.4, (September 2011), pp. 615-620, ISSSN 0308-8146

Cheetham, N. & Tao, L. (1998). Variation in crystalline type with amylose content in maize starch granules: an X-ray powder diffraction study. Carbohydrate Polymers, Vol.36, No.4, pp. 277- 284, ISSN ISSN 0144-8617

Codex Stand 176-1989. Norma del codex para la harina de yuca comestible. Available from: http://www.codexalimentarius.org

Curvelo, A.; Carvalho, A. & Agnelli, J. (2001). Thermoplastic starch-cellulosic fibers composites: preliminary results. Carbohydrate Polymers, Vol. 45, No. 2, pp. 183-188, ISSN 0144-8617

De Graaf, R.; Karman A. & Janssen, L. (2003). Material properties and glass a thermoplastic starches after extrusion processing. Starch/ Stärke, Vol. 55, No.2, (August 2011), pp. 80-86, ISSN 1521-379X

FAO -Food and Agriculture Organization-. (2007). Guía técnica para producción y análisis de almidón de Yuca. Roma: FAO. ISSN, 1020-4334. Available from: http://www.fao.org/docrep/010/a1028s/a1028s00.htm

Garcia, M.; Martino, M. & Zaritzky, N. (2000). Microstructural Characterization of Plasticized Starch-Based Film. Starch/Stärke, Vol. 52, No (August 2011), pp. 118-124, , ISSN 1521-379X.

García, J.; Alonso, L.; Barona, S.; Fernández, A.; Isaza, L. & Londoño, J. (2005). Tecnología para la extracción de harina de yuca. Clayuca. Available from: http://www.clayuca.org/clayucanet/edicion08/refinacion.pdf

Guan, J. & Hanna, M. (2004). Functional properties of extruded foam composites of starch acetate and corn cob fiber. Industrial Crops and Products, Vol. 19, No. 3,pp, 255-269, ISSN 0926-6690

Halley P. (2005). Thermoplastic starch biodegradable polymers, Part 1. In: Smith R., editor. *Biodegradable polymers for industrial applications*. Cambridge: Woodhead,pp. 140-159, ISBN 1 85573 934 8

Leblanc, N.; Saiah, R.; Beucher, E.; Gattin, R.; Castandet, M.; & Saiter, J. (2008) Structural investigation and thermal stability of new extruded wheat flour based polymeric materials. Carbohydrate Polymers, Vol. 73, No.4, pp. 548–557, ISSN 0144-8617

Lee, S.; Chen, H. & Hanna, M. (2008). Preparation and characterization of tapioca starch-poly(lactic acid) nanocomposite foams by melt intercalation based on clay type. Industrial Crops and Products, Vol. 28, No.1, pp. 95-106, ISSN 0926-6690

Lee, S.; & Wang, S. (2006). Biodegradable polymers/bamboo fiber biocomposite with bio-based coupling agent. Composites: Part A, Vol. 37. No.1, pp. 80-91, ISSN 1359835X

Luna, G.; Villada, H. & Velasco, R. (2009). Fiques fiber reinforced thermoplastic starch of cassava: preliminary. Dyna. Vol. 76, No. 159, pp. 145-151, ISSN 0012-7353.

Martínez, A.; Velasco-Santos, C., de-Icazaa, M., & Castaño, V. (2007). Dynamical-mechanical and thermal analysis of polymeric composites reinforced with keratin biofibers from chicken feathers. Composites: Part B, Vol. 38. No. 3. (September 2011) pp. 405-410, ISSN 1359-8368

Martins, I.; Magina, S.; Oliveira, L.; Freire, C.; Silvestre, A.; Neto, C. & Gandini, A. (2009). New biocomposites based on thermoplastic starch and bacterial cellulose. Composites Science and Technology, Vol. 69, No. 13,pp. 2163-2168, ISSN 0266-3538

Ma, X.; Chang, P. & Yang, J. (2009). Preparation and properties of glycerol plasticized-pea starch/zinc oxide-starch bionanocomposites. Carbohydrate Polymers., Vol. 75, No 3, pp. 472-478, ISSN 0144-8617

Ma, X.; Y, J. & Kennedy, J. (2005). Studies on the properties of natural fibers-reinforced thermoplastic starch composites. Carbohydrate Polymers, Vol. 62, No.1, pp. 19-24, ISSN 0144-8617

Ma, X.; Yu, J. & Wang, N. (2008). Glycerol plasticized-starch/multiwall carbon nanotube composites for electroactive polymers. Composites Science and Technology., Vol. 68, No.1, pp. 268-273, ISSN 0266-3538

Mohamed, A.; Finkenstadt, V.; Gordon, S. & Palmquist, D. (2010). Thermal and Mechanical Properties of Compression-Molded pMDI-Reinforced PCL/Gluten Composites. Journal of Applied Polymer Science, Vol.118, No.5, pp. 2778-2790, ISSN 1097-4628

Montaldo, A. (1985). La yuca o mandioca: cultivo, industrialización, aspectos económicos, empleo en la alimentación animal, mejoramiento. San José, Costa Rica: Instituto Interamericano de Cooperación para la Agricultura. ISSN 92-9039-053-0

Nair, S.; Wang, S. & Hurley, D. (2010). Nanoscale characterization of natural fibers and their composites using contact-resonance forcé microscopy. Composites. Vol.41, No.5, pp. 624-631, ISSN 1359-835X

Oksman, K.; Skrifvars, M. & Selin, J. (2003). Natural fibres as reinforcement in polylactic acid (PLA) composites. Composites Science and Technology, Vol. 63, No.9, pp. 1317–1324, ISSN 0266-3538

Olomo, V. & Ajibola, O. (2003). Processing factors affecting the yield and Physicochemical properties of starch from cassava chips and flour. Starch/Stärke, Vol.55, No. 10, pp. 476-481, ISSN 1521-379X

Perdomo, J.; Cova, A.; Sandoval, A.; García, L.; Laredo, E. & Müller, A. (2009). Glass transition temperatures and water sorption isotherms of cassava starch. Carbohydrate Polymers, Vol.76, No.2, pp. 305–313, ISSN 0144-8617

Pérez, E.; Amaiz, M.; González, Z. & Tovar, J. (2007). Production and characterization of Cassava (Manihot esculenta crantz) flours using different termal treatments. Interciencia, Vol. 32, No.9, pp. 615-619, ISSN 0378-1844

Rahman, M. & Brazel, C. (2004). The plasticizer market: an assessment of traditional plasticizers and research trends to meet new challenges. Progress in polymer science, Vol. 29, No.12, pp. 1223-1248, ISSN 0079-6700

Rodríguez, E.; Sandoval, A. & Fernández, A. (2007). Evaluación de la retrogradación del almidón de harina de yuca precocida. Revista Colombiana de Química, Vol. 36, No.1, pp 13-30, ISSN 01202804

Rodríguez-Sandoval, E.; Fernández-Quintero, A.; Cuvelier, G.; Relkin, P. & Bello-Pérez, L. (2008). Starch Retrogradation in Cassava Flour From Cooked Parenchyma. Starch - Stärke, Vol.60, No.3-4, pp. 174–180, ISSN 1521-379X

Sánchez, T. & Alonso, L. (2002). Conservación y Acondicionamiento de las Raíces Frescas. Cali, Colombia. En: La yuca en el tercer milenio. Centro Interamericano de Agricultura Tropical; Consorcio Latinoamericano y del Caribe de Apoyo a la Investigación y Desarrollo de la Yuca; 2002, ISBN 958-694-043-8

Schlemmer, D.; Angélica, R. & Sales, M. (2010). Morphological and thermomechanical characterization of thermoplastic starch/montmorillonite nanocomposites. Composite Structures, Vol. 92, No.9, pp. 2066-2070, ISSN 0263-8223

Shittu, T.; Dixon, A.; Awonorin, S.; Sanni, L. & Maziya-Dixon B. (2008). Bread from composite cassava-wheat flour. II: Efect of cassava genotype and nitrogen fertilizer on bead quality. Food Research International, Vol. 41, No. 6, pp. 569-578, ISSN 0963-9969

Shogren, R.; Doane, W.; Garlotta, M., Lawton, J. & Willett, J. 2003. Biodegration of starch/polylactic acid/poly (hydroxyester-ether) composite bars in soil. Polymer Degration and Stability, Vol. 79, pp. 405-411, ISSN 0141-3910

Singh, V.; Ali, S.; Somashekar, R. & Mukherjee, P. (2006). Nature of crystallinity in native and acid modified starches. International Journal of Food Properties, Vol. 9, pp. 845–854, ISSN 1094-2912

Teerawanichpan, P.; Lertpanyasampatha, M.; Netrphan, S.; Varavinit, S.; Boonseng, O. & Narangajavan, J. (2008). Influence of Cassava Storage Root Development and Environmental Conditions on Starch Granule Size Distribution. Starch - Stärke, Vol.60, No.12, pp. 696–705, ISSN 1521-379X

Teixeira, E.; Pasquini, D.; Curvelo, A.; Corradini, E., Belgacem, M. & Dufresne, A. (2009). Cassava bagasse cellulose nanofibrils reinforced thermoplastic cassava starch. Carbohydrate Polymers, Vol. 78, No. 3, pp. 422-43, ISSN 0144-8617

Tharanathan R. (2003). Biodegradable films and composite coatings: past, present and future. Trends in Food Science and Technology, Vol. 14, No.3, (August 2011), pp. 71–78, ISSN 0924-2244

Thompson, D. (2003). Strategies for the manufacture of resistant starch. Trends in Food science and Technology, Vol. 11, No. 7, (August 2011); pp. 245-253, ISSN 0924-2244

Van Soest, J.; Hulleman, S.; De Wit, D. & Vliegenhart, J. (1996). Crystalinity in starch bioplastics. Industrial Corps and Products, Vol.5, No. 1, pp. 11-12, ISSN 0926-6690

Villada H. (2005). Influencia de mezclas de almidón agrio, perfil de temperatura y velocidad de tornillo de un extrusor sencillo en la producción de almidón termoplástico, su caracterización físico-química, mecánica y microestructural y comportamiento durante el almacenamiento [doctoral thesis]. [Cali (CO)]: Universidad del Valle, 170 p. Spanish.

Wambua, P.; Ivens, J. & Verpoest, I. (2003). Natural fibres: can they replace glass in fibre reinforced plastics?. Composites Science and Technology, Vol. 63, No.9, pp. 1259-1264, ISSN 0266-3538

Wang, Y.; Rakotonirainy, M. & Padua, W. (2003) Thermal behavior of Zeína-base biodegradable films. In: Starch/Stärke, Vol. 55, No.1, (August 2011); pp. 25-29, ISSN 1521-379X

Weber, C. (2001). *Biobased Packaging Materials for the Food Industry,* ISBN , 87-90504-07-0, Rolighedsvej, Denmark

3

Physical and/or Chemical Modifications of Starch by Thermoplastic Extrusion

Maria Teresa Pedrosa Silva Clerici
Federal University of Alfenas-UNIFAL-MG
Brazil

1. Introduction

Starch makes up the nutritive reserves of many plants. Starch biosynthesis is a complex process, which may be summarized as during the growing season, the green leaves collect energy from the sun, this energy is transported as a sugar solution to the starch storage cells, and the sugar is converted into starch in the form of tiny granules occupying most of the cell interior. The conversion of sugar into starch takes place through enzymes (Corn Production Source, 2011; Tester et al., 2006).

Starch granules are composed of two types of alpha-glucans, amylose and amylopectin, which represent approximately 98–99% of the dry weight. The ratio of the two polysaccharides varies according to the botanical origin of the starch and classifies starch as the 'waxy' starches contain less than 15% amylose, 'normal' 20–35% and 'high' amylose starches greater than about 40% (Tester et al., 2006; Wurzburg, 1989).

Amylopectin is a much larger molecule than amylose with a molecular weight and a heavily branched structure built from about 95% (α 1- 4) and 5% (α 1- 6) linkages. Amylopectin unit chains are relatively short compared to amylose molecules with a broad distribution profile. They are typically, 18–25 units long on average (Tester et al., 2006; Wurzburg, 1989).

Nutritionally, starch is consumed as an energy source, and it is the most abundant energy source in the human diet as it is present at high amounts in cereals, roots and tubers, which products range from breads, cookies, pastes to consumption as snacks, porridges, or as processed cooked grains (white rice, corn grain) or whole grains (whole grain of rice, wheat, popcorn, etc.).

It may be technologically considered as a nutrient of great versatility, with numerous applications in the food industries and others such as in cosmetics, pharmaceutical, paints, children toys, glues and adhesives, and nowadays as a biopolymer for the production of packagings.

This versatility is related to some characteristics of starch such as being of low cost and easy to undergo physical and chemical modifications which allow changes to the physical-chemical and rheological properties of starches.

The main starch modifications obtained with the use of the thermoplastic extrusion process will be addressed in this chapter.

2. Starch

The different kinds of starch with commercial importance are corn, wheat, potato and cassava starch:

- Corn, *Zea Mays*, is grown in most countries throughout the world. It requires, however, warmer climates than those found in the temperate zones to grow to maturity. It is classified into several kinds, with different uses (Table 1), and it has the largest production.
- Wheat starch granules are divided in two groups by size, B-starch (15 - 20 % and 2 - 15μ) and the larger A-starch (80 - 85 %, 20 - 35μ in diameter) granules. B-starch is contaminated with pentosans, fibers, lipids and protein to an extent requiring special treatment in the factory.
- Cassava is the term usually applied to the roots, and tapioca is the name given to starch and other processed products. There are many varieties of cassava, but they fall into two main categories, namely bitter and sweet cassava (*Manihot palmata* and *Manihot aipi*), depending on their content of cyanohydrins. For industrial purposes bitter varieties are most frequently used because of their higher starch content. Sweet cassava is preferred for food because of its taste and dough forming ability.
- Potato starch: 75% of the potato crop is grown for industrial processing and Danes produce per capita more starch than any other nation (Corn Production Source, 2011).

Corn	Scientific name	Use	Characteristics
Dent	*Zea mays Indentata*	Food, animal feed, and industrial products	Hard and soft starch Become indented at maturity
Flint	*Zea mays indurata*	Food, animal feed, and cornstarch manufacturing	Hard, horny, rounded, or short and flat kernels soft and starchy endosperm completely enclosed by a hard outer layer
Waxy	*Zea mays*	Special starches for cornstarch manufacturing	A waxy appearance (only branched-chain starch)
Sweet	*Zea saccharata* or *Zea rugosa*	Green corn is eaten fresh, canned, or frozen	High percentage of sugar
Popcorn	*Zea mays everta*	Popcorn	They are popped on exposure to dry heat
Indian	*Zea mays*	Food and animal feed	White, red, purple, brown, or multicoloured kernels
Flour	*Zea mays amylacea*	Tortillas, chips, and baked goods	Soft corn

* Corn Production Source (2011).

Table 1. Kinds of corn*

These four types of starches are industrially produced and present great applicability in the food industry. Other starch sources, such as other cereals, tubers and fruits are produced at smaller amounts, limiting their wide utilization, and some are produced with research purposes for viability of new starch sources (Govindasamy, 1996, 1997; Alves et al., 1999).

Starch must be gelatinized in the human diet in order to be digested by the amylolytic enzymes of the human digestive system. The classic model of obtaining gelatinized starches, where starch granules are slowly heated in a medium with little agitation and much water, which promotes imbibition, swelling and polymer release (Leach, 1965) for a prolonged time, such as in the obtaining of cooked rice, corn flour porridges, was replaced by other methods, such as extrusion, spray-drier and drum dryer, which promote fast starch gelatinization and followed by drying may obtain flours and/or pregelatinized starches of long-term stability and quick preparation.

Pregelatinized starches or flours are paste-forming products in the presence of cold water or (partially or totally) soluble products in cold water (Colonna et al.,1984) and present the following characteristics: they disperse more easily and absorb more water than their untreated matches, they form gel at room temperature and are less prone to deposit (Powell, 1965).

The use of gelatinized starch in food products affects their characteristics and qualities, such as bread volume and crumb (Williams & Lesselleur, 1970); elasticity and softness of pastas (pasta), digestibility and palatability, tolerance in the properties of beating and cake mixtures, ice creams, doughnuts, growth of sugar crystals in food products (Powell, 1965); texture, volume, shelf-live and stability during thawing of cakes and breads (Michael & Brown, 1968).

In the food industry, pregelatinized starches are used to achieve thickening or water retention without employing heat, for example, puddings, instant lactic mixtures and breakfast foods; to prepare ready-to-use bread mixtures, where the increased absorption and retention of water improves the quality of the product, to work as an agglutinant in the meat industry; and, as a filling for fruit pies, as they make the use of heat dispensable and increase flavor retention. They also have a non-food use, such as in the industries of textiles and in drugs, paper, metallurgy, etc (Powell, 1965).

Pregelatinized flours may be obtained on an industrial basis through extrusion or drum-drying and through the use of atomizers (spray-drier).

The use of atomizers is economically limited as starch pastes are highly viscous and require drying at a low content of solids (Chiang & Johnson, 1977).

Drum-dryers are simple and commonly used, but they present the disadvantage of high-cost products due to low efficiency, low production, difficult operation, constant need for drum maintenance and adjustment (Greenwood, 1976).

The extrusion process presents the advantages of versatility, high productivity and low cost (Smith, quoted by Harper, 1979) and more strict control of the desired gelatinization degree, where small modifications in the equipment and/or in the raw material may lead to different final results (El-Dash, 1982). According to Lorenz & Jansen (1980), the low cost of gelatinization through extrusion is due to the fact that it efficiently converts electrical energy into thermal energy and also the manpower and space per kilogram of cooked product required are lower than any other cooking method.

The market of fast preparation products has grown and many varieties of pregelatinized flours are available in the market nowadays. The main trends in the use of thermoplastic extrusion process applied to starchy ingredients will be approached in this chapter.

3. Thermoplastic extrusion process

El-Dash (1982) defined the process of thermoplastic extrusion as being a continuous process in which mechanical friction is combined with thermal heating in order to continuously mix, plasticize and gelatinize the starch, denature protein materials, restructuring them for the obtaining of products with new textures and shapes.

Single-screw cooking extruders were developed in the 1940's to make puffed snacks from cereal flours or grits. An expanding demand for precooked cereals and starches required machines with larger capacity, so extruders with a nominal capacity of 5 ton per hour were developed in the 1960's, with numerous news applications: snacks, infant feeding, pet foods, etc. In the 1970's products containing more than one component were developed, such as egg rolls and ravioli for coextrusion. Then, the use of two extruders in series, the first for cooking and the second one for forming and structuring, resulted in several products. At the end of the 1970's, the use of twin-screw extruders for food processing that expanded the range of application began (Mercier & Feillet, 1975; Linko et al., 1981; Harper, 1979). Finally, the extruders are meant for specific markets, as it will be seen for extruders intended for the productions of biodegradable packagings (Flores et al., 2010; Mandrogón et al., 2009).

Food extruders are generally available with segmented screws and barrel section, facilitating total control over the configuration of the machine to get a combination of various process parameters. Extrusion cooking is a high-pressure operation that provides sudden expansion of the processed products. The physical characteristics of the extrudate reflect the effectiveness of the process and suitability of ingredients (Patil et al., 2005).

Beneficial effects include destruction of antinutritional factors, gelatinization of starch, protein denaturation/texturization, increased soluble dietary fibre and reduction of lipid oxidation. But Maillard reactions between protein and sugars can reduce the nutritional value of the protein, depending on the raw material types, their composition and process conditions. Heat-labile vitamins may be lost to varying extents (Singh et al., 2007, Patil et al., 2005).

Some of the use applications of the extrusion process to physically and/or chemically modify starch both in the areas of food and biopolymers will be presented below.

3.1 Ready-to-eat cereals and snacks

They are produced with cereal flour and starch ingredients, the extruded products are highly expanded and have several shapes and textures.

Ready-to-eat cereals are manufactured from mixtures of cereal flour and starch combined with small amounts of malt, fat, sugars, emulsifiers and salt. The extrusion process requires moisture of 20% and temperature >150°C, after the extrusion, they are dried and toasted.

Much research has been made to obtain ready-to-eat cereals, as the control of the extrusion process is rather complex, due to the large number of variables affecting it (Ostergard & Bjorck, 1989, Stojceska et al., 2009).

Some early works, such as that of Lawton et al. (1972) considered extrusion temperature and initial moisture in the raw material as the variables with greater effect on starch gelatinization, and the maximum degree of gelatinization and shear occur when these variables act with opposite end values and when both values are high or low, low degrees of gelatinization occur. Other works showed that the extrusion process destroys the organized crystalline structure of the starch granule, with different degrees of intensity which depend on the ratio of amylose to amylopectin and on the independent variables used such as humidity and shear (Charbonnieri et al., 1973). Thus, starch may be gelatinized, which is what occurs at the extruder with humidity lower than 20% (Linko et al.,1981), dextrinized, which happens in more severe conditions and with low humidity content (Gomez & Aguillera, 1983, 1984). The process may also lead to starch liquefaction and partial hydrolysis of starch molecules (Faubion et al., 1982).

The production of snacks through extrusion represents a great achievement for the Food Technology area as it efficiently converts crude cereal flours into products with different shapes, flavors and long shelf-life.

At first, snacks were obtained from whole grains combined with moisture content, cooking temperature and drying, considered the first generation (Huber & Rokey, 1990).

The second generation snacks presented more expansion ability, and were obtained with flours refined from cereals and cereal and tuber starches. These snacks have a large volume and require appropriate packagings to avoid humidity, light and heat, and they must protect the product against mechanical shocks, in order to avoid breaking during transport and storage. These facts boosted the development of third generation snacks (Huber & Rokey, 1990).

Third generation snacks are not expanded through the extrusion process, for this reason they are known as pellets or half-products, for they will be expanded through a process of deep-frying or hot air, or with the use of microwaves, during consumption. Although they have an additional process for expansion, these products present great advantages in transport and storage (Huber & Rokey, 1990).

With the worldwide tendency of weight gain by the population, studies showed that gelatinized starch, if excessively consumed, contributes to an increased number of diseases, such as obesity, diabetes and increased blood triglycerides, which may lead to serious heart diseases (Jenkins et al., 1980).

A ranking for starch has been made:

- Rapidly digestible starch (RDS): amount of glucose release after 20 min.
- Slowly digestible starch (SDS): amount of glucose released between 20 and 120 min of *in vitro* digestion.
- Resistant starch (RS): total starch minus amount of glucose released within 120 min of *in vitro* digestion (Singh et al., 2010).

Studies evaluating the digestibility of starches in snacks showed that they presented RDS type starch (Singh et al., 2007; Goni et al., 1997). Thus, the tendency is to promote the return of the use of the whole grain, addition of fibers and ingredients leading to increased SDS and RS in snacks, in addition to ingredients beneficial to health, such as antioxidants, and omega oils. That is, the new generation of snacks must include benefits to the consumer's health, in addition to being nutritious.

3.1.1 Oils

Abu-Hardan et al. (2011) investigated the addition of three commercial vegetable oils (1 to 8%) to extruded wheat starch, namely: palm oil, soybean oil and sunflower oil. The effects of the addition of the oils on the sectional expansion of extrudates was complex in which a significant increase up to a 5% oil concentration was reached and further increase of oil quantities resulted in a drastic reduction and no significant differences between oils were noticed. However, the crystallinity indexes indicated no interference from the three oils in the complex formation. These authors suggested that the endogenous lipids naturally present in wheat starch were sufficient to complex the starch.

3.1.2 Protein

Limón-Valenzuela et al. (2010) obtained a functional third-generation snack food with good expansion characteristics using a microwave oven, and this snack has health benefits due to the addition of milk protein concentrate (0-10%) and quality protein maize (20%) in a blend of corn starch (80%). These authors used a laboratory single-extruder with a 3:1 compression ratio, feed moisture (20-30%), a rectangular die, and a central composite non-routable model with two variables

According to the increased demand for new healthy snacks as an alternative for fried starch-based snacks with low nutrient density, Cho & Rizvi (2010) showed the potential of supercritical fluid extrusion (SCFX) technology for healthy snack food production containing high whey protein concentration. SCFX chips had uniform cellular microstructure that cannot be obtained using conventional steam-based extrusion.

Lobato et al. (2011) developed a functional puffed product for extrusion containing 250 g/kg corn starch, 375 g/kg soy flour, and 375 g/kg oat bran extruded under the process conditions of 250 g/kg moisture; 45 g/kg inulin and 130ºC). The puffed product had 212.6 g/kg fiber, 281.0 g/kg protein, and a caloric value of 319.1 kcal/100 g and it was well accepted by the panelists in the sensory evaluation, mainly in terms of texture.

3.1.3 Fibers and resistant starch

Céspedes et al. (2010) obtained extruded orange pulp using a Brabender laboratory single-screw extruder (20:1 Céspedes 2010) and observed that it could also be added to high-fiber foods as a low-calorie bulk ingredient to reduce the calorie level, since it showed potential hypoglycemic effects.

Souza & Lionel (2010) verified that the mixture of cassava starch and dried orange pulp extruded under different conditions (0 to 20% of fibers in the mixture), (14.6 to 21.4% moisture), and (60.8 to 129°C temperature of extrusion), could aim at the use in high-fiber instant products.

Bello-Pérez et al. (2006) extruded starches isolated from unripe banana (*Manguifera indica* L.) and mango (*Musa paradisiaca* L.) fruits to obtain a product with high content of resistant starch (RS) and verified that RS formation in the extruder for banana starch was affected positively by temperature and inversely by moisture. Moisture did not significantly affect RS formation in mango starch.

Two types of products were produced: pure whole meal products and breakfast cereals made from whole meal/maize blends were processed by pilot-plant extrusion and the enzyme-resistant starch (RS) content and hydrolysis index (HI) were not correlated to the extrusion temperature, but with whole meal products (Chaunier et al., 2007).

Yanniotis et al. (2007) verified the effect of pectin alone or in combination with wheat fiber on the physical and structural properties of extruded cornstarch, under specific moisture content, barrel temperature and screw speed conditions were studied using a laboratory single-screw extruder. These authors observed that fibers reduced the size of the cells and increased their number and pectin increased porosity and reduced expansion ratio and hardness of the snacks.

Stojceska et al. (2009) investigated the use of brewer's spent grain and red cabbage barley and red cabbage in wheat flour and corn starch extruded in co-rotating twin-screw extruder, under the conditions studied, the results were promising towards the increase of total dietary fibre and the level of total antioxidant capacity and total phenolic compounds of the snacks.

3.1.4 Antioxidant activity

The blends of various formulations of durum wheat flour (8-20%), partially defatted hazelnut flour (PDHF) (5-15%), fruit waste blend (3-7%) and rice grits were extruded using single-screw extruder and when higher, PDHF and fruit waste content caused an increase in the total phenolic content and antioxidant activity of the extruded samples, whereas percentage starch gelatinization and digestibility values decreased (Yaĝci & Göĝûş, 2009).

Limsangouan et al. (2010) demonstrated the effect of extrusion processing on the functional properties of extruded snack foods developed from cereal and legumes, and the by-products from herbs and vegetables, and the extrusion process slightly decreased the antioxidant capacity and phenolic content.

Table 2 shows new starches that are been used in thermoplastic extrusion researches.

3.2 Chemical modification of starches

Starch may be modified through physical or chemical methods and its use relates to improved quality and decreased cost of the products.

According to Light (1990), modified starch is used in foods, for 3 main reasons:

- to provide functional features in food applications which native starches may not normally provide. For example: pudding mixture,
- it is readily available, and
- provides economical advantages in many applications where more expensive additives are used. For example: gums.

Some researchers used simultaneously a chemical reagent and the extrusion process to obtain modified starch for various purposes, such as production of expanded extruded products (Lai et al., 1989); starch phosphate production (Chang & Lii, 1992); alcohol production (Chang, 1989), extruded rice flour (Clerici & El-Dash, 2006) and acidic extruded rice flour (Clerici et al., 2009) for production gluten-free bread, and lactic beverage production (Lee et al., 1992) . The simultaneous use of these modifications poses advantages such as saving reagent, and absence of effluent formation, low reaction time, processing at lower moisture content, and the elimination of drying the starch dispersion.

Starch or flour	Process	Application	Reference
Sago +Alfa-amylase	Single screw extruder	Hydrolylisis or dextrinization	Govindasamy (1997)
Sago	Co-rotating twin-screw extruder	Limited degradation	Govindasamy (1996)
Yam (*Dioscorea alata*)	Single screw extruder	Decrease of retrogradation tendency, high cold thickening capacity, high gel strength	Alves et al. (1999)
Foxtail millet grains	Flaked, extrusion cooked and roller-dried products	Ready-to-eat , popped millet	Ushakumari et al. (2004)
Unripe banana (*Manguifera indica L.)*	Single-screw equipment	No effect in the resistant starch formation	Bello-Pérez et al. (2006)
Mango (*Musa paradisiaca L.*)	Single-screw equipment	No effect in the resistant starch formation	Bello-Pérez et al. (2006)
Amaranth flour *Amaranthus cruentus* L	Single-screw equipment	Potential for use in instant meal products	Menegassi et al. (2011)

Table 2. News source extruded starches

3.2.1 Starch phosphates

Native starches usually contain small amounts of phosphorus (0.1%). In tubers and roots, phosphorus is covalently bound to starch (Hodge et al., 1948), while in cereal starches, it occurs mainly as a phospholipid contaminant (Lim et al., 1994).

Starch phosphates are esters derived from phosphoric acid. When only a hydroxyl is involved in the starch phosphate binding, the product is a monoester. The other starch phosphate class is the *cross-linked* type which contains mono-, di- and triester starch phosphate (Hamilton & Paschall, 1967). Approximately 60%-70% of total phosphorus of starch monophosphate is located at C-6 while the rest is located at C-3 of anhydroglucose units (Tobata & Hizukuri, *apud* Wurzburg, 1986). Most phosphate groups (88%) are on chain β of amylopectin (Wurzburg, 1986)

Cross-linked starch is obtained by introducing an agent capable of reacting with the hydroxyl groups of two different molecules within the granule. These synthetic bridges reinforce the natural hydrogen bonds, delaying the speed of granule swelling and reducing the rupture of the swollen granule (Wurzburg, 1986). Its main use is as filling in fruit pies and canned goods.

Chang & Lii (1992) compared the conventional process to that of extrusion for phosphatation of cassava and corn starches and they verified that, in order to prepare starch phosphate with a similar degree of substitution, the extrusion process requires less reagent than the conventional method, and the latter requires an excessive amount of reagents and causes water pollution, increasing the cost of production.

Salay & Ciacco (1990) also found that it is possible to obtain starch phosphate with a low degree of substitution (DS) value through the extrusion process and observed that the extrusion temperature of 200°C, concentration ≥ 1.4 g/100 mL of sodium tripolyphasphate and pH 8.5 were the conditions that resulted in a higher value of DS.

Nabeshima & Grossmann (2001) obtained cassava cross-linked starch with different DS and degree of gelatinization for use in food by using cassava starch with sodium trimetaphosphate (STMP) processed on a Cerealtec single-screw extruder at different extrusion temperatures and concentrations of STMP and NaOH.

Seker et al. (2003, 2004, 2005a, 2005b) mixed starch with sodium hydroxide and sodium trimetaphosphate, the mixture was then extruded in both single- and twin-screw extruders and it was verified that the mixing elements did not change the amount of phosphorus incorporated into the starch in both processes. They developed works showing the phosphatation process through starch extrusion and its changes in rheological properties.

3.2.2 Acid-modified starch

Acid-modified starch suffers hydrolysis of some glucosidic bonds, which occurs first in the amorphous regions of the starch containing branch points and α-D bonds (1→6), reducing the molecular size and diminishing the viscosity of the paste. Depending on the treatment intensity, there is formation of dextrins (Wurzburg,1986). Kerr, quoted by Wurzburg (1986) showed that during acid modification, the amount of starch amylose increases, indicating that acid preferably hydrolyses amylopectin.

Acid-modified starches are normally made out of a starch paste (about 36% to 40% of solids) heated at a temperature below the starch gelatinization temperature (about 40°-60°C) and the addition of mineral acid, agitation for a varied period (about one to several hours). When viscosity or degree of conversion desired is reached, the acid is neutralized and the starch is retrieved through filtration or centrifugation, washing and drying. The type of mineral acid, its concentration, temperature, starch concentration and reaction time influence starch properties (Wurzburg, 1986).

Acid-modified starches differ from granular starch in lower viscosity of the paste (under cold and hot conditions) and other properties. However, they have the same physical form, insolubility in cold water and similar birefringence (Shildneck & Smith, 1967).

The literature indicates, most of the times, starch modification through mineral acid. However, Mehltretter (1967) used organic acids to modify starch and found that some carboxylic acids such as formylic acid react with starch at room temperature and in the presence of water; whereas other acids, such as acetic acid and citric acid do not react in an aqueous medium and require heating to force the reaction.

3.3 Biodegradable polymers

Due to increasing environmental awareness and the environmental legislations, scientists around the world have made strong efforts to develop methods using natural polymers as an alternative to the petroleum synthetic polymers for industrial and consumer applications. (Liu et al., 2009).

Biodegradable polymers represent a promising solution to the environmental problem of plastic waste disposal. Among the candidate polymers, starch, a low-cost natural polymer, can be processed as a thermoplastic (Rosa et al., 2008).

Starch is being researched both for the production of biodegradable packagings with the use of extrusion blow molding machine, where bags are made even for the formation of edible films to coat industrialized foods or fruits.

The shelf-life of foods is governed by their numerous interactions with their surroundings and can be extended by using protective films. The deterioration of packaged foodstuffs largely depends on the transfers that may occur between the internal environment of the packaged food and the external environment. Edible films can be used to reduce water vapor, oxygen, lipid, and flavor migration between components of multicomponent food products, and between food and the surroundings. Many proteins and polysaccharides have good film-forming properties and can be used in the preparation of edible films (Torres, 1994).

For packagings, which will be used for food transporting, such as bags and boxes, plasticized starches that are commonly called thermoplastic starches (TPS) are used (Stepto, 2003).

TPS can be processed through traditional processing conditions (extrusion, blow molding and injection molding increase the properties of the blends and the content of TPS in TPS/PE blends (Rodriguez-Gonzalez et al., 2003).

Carvalho et al. (2005) used carboxylic acids to decrease the viscosity of TPS by controlling the macromolecules of starch. After melt processing in the presence of glycerol, water and carboxylic acids, both the molar masses and viscosity of TPS are decreased. In their previous work (Yu, Wang & Ma, 2005) citric acid (CA) is also used as an additive to modify TPS during melt processing. CA can effectively restrain starch re-crystallization, except for increasing the fluidity of TPS. But the mechanical properties of TPS are decreased, especially the tensile strength (below 4 MPa).

Chaudhary et al. (2008) obtained thermoplastic resin with a mixture of different concentrations of unmodified starches with 0%, 28%, 50% and 80% amylose; 80% amylose hydroxypropylated starch using extrusion process with several variations of screw speed, die pressure, motor torque, mean residence time and specific mechanical energy.

Pea starch-based composites reinforced with citric acid-modified pea starch (CAPS) and citric acid-modified rice starch (CARS), respectively, were prepared through screw extrusion to obtain biodegradable polysaccharide (CA-modified granular starch/TPS) composites to be used with a potential replacement for edible films, food packaging, biodegradable packaging, etc. The CARS/TPS composites exhibited better storage modulus, tensile strength, elongation at break and water vapor barrier than CAPS/TPS composites because of the smaller size of granular CARS (Ma et al., 2009).

Flores et al. (2010) used a mixture of experimental design to study the physical and microbiological properties of tapioca starch-based glycerol edible films with the addition of xanthan gum (XG) and potassium sorbate (PS) and obtained through extrusion technology. The results showed that PS presence decreased the ultimate tensile strength and elastic modulus and increased strain at break. XG produced a reinforcing effect on the films and also enhanced solubility in water and decreased moisture content.

Guan & Hanna (2006) have extruded biodegradable composite foams based on starch acetate and poly (tetraethylene adipate-co-terephthalate) (EBC). It was reported that low EBC contents in the blends favored the miscibility of the two polymers, as characterized by an increase of the glass transition temperature of starch acetate, a decrease in the melting point temperature of starch and EBC in a differential scanning calorimetry (SEM) analysis and the formation of a homogeneous morphology observed with SEM. Large amounts of EBC decreased the miscibility of these two polymers.

Multifunctional epoxy-based copolymers can be used as chain-extender (CE) to increase the molecular weight and create branching in polylactides (PLA). Li & Huneault (2011) studied the effect of a multifunctional epoxy-acrylic-styrene copolymer on the properties of PLA/Thermoplastic Starch (PLA/TPS) blends that were prepared by twin-screw extrusion. The plasticizers were mixed together in the first half of the extruder to complete starch gelatinization. Water was removed by devolatilization at midex-truder and the PLA matrix was mixed with the water-free TPS in the latter portion of the compounding process. The standard blends comprised 27% TPS in the PLA matrix. The TPS phase itself comprised 36% plasticizer in the form of glycerol or sorbitol. The blends were injection molded into standard test bars and their tensile properties were measured. It was found that the combination of interfacial modification and chain-extension strategies led to greatly improved ductility. The viscosity of the PLA/TPS blends was also dramatically increased by adding a small amount of epoxy-based chain extender. This is of great interest for polymer processing techniques (such as foaming or film blowing) that require high melt strength.

Nabar et al. (2006) produced the cylindrical starch foam shapes on a small scale (~11-12 kg/hr) Werner Pfleiderer ZSK-30 twin-screw extrusion (TSE) process using water, which functions as a plasticizer as well as a blowing agent. The properties of the starch foams depend on the type of starch used (hydroxypropylated high amylose corn starch, 70% amylose), the amount of water and additives (poly(hydroxyamino ether)) (PHAE) used, and extrusion conditions such as temperature and the screw configuration.

Table 3 summarizes some other works with biopolymers using starch as base.

Starch+ Polymers or another composite	Process	Authors
Potato starch + polypropylene	Corotating twin-screw extruder	Roy et al. (2011)
Tapioca starch+ glycerol	Twin screw extrusion at 150°C	Yunos & Rahman (2011)
Starch + polylactides (PLA)/ Thermoplastic starch (TPS)	Twin-screw extrusion	Li & Huneault (2011)

Starch+ Polymers or another composite	Process	Authors
TPS + PLA	Co-rotating twin-screw extruder	Świerz-Motysia et al. (2011)
Thermoplastic acorn starch(TPAS)/ Polycaprolactone (PCL) + different plasticizers:ethylene glycol, glycerol, monoethanolamine, iminobisetnanol and triethanolamine	Hot-melt extrusion	Li et al. (2011)
TPS + poly (butylene adipate-co-terephthalate - PBAT) + organically modified nano-clays	Melt intercalation technique	Mitrus & Mościcki (2011)
Starch+ chitosan + oregano essential oil	Single screw extruder	Pelissari et al. (2011)
Cassava starch+ glycerol + vegetable fibres	Single screw extruder	Debiagi et al. (2010)
(hydroxypropylated high amylose corn starch + amylase+ water and additives (poly-hydroxyamino ether- PHAE)	Twin-screw extrusion and cylindrical starch foam shapes	Nabar et al. (2006)
High amylose potato starch and normal potato starch+ glycerol	Buss co-kneading extruder	Thuwall et al. (2006)
TPS /natural rubber/montmorillonite type clay nanocomposites	Twin-screw extrusion	Mondragón et al. (2009)
Corn starch+ low density polyethylene (LDPE)	Singlestep twinscrew extrusion or by a twostep process involving compounding (pelleting) of the ingredients before film formation	Pushpadass et al.(2010)
Cassava TPS + PBAT+ surfactant Tween 80	Extrusion	Brandelero et al. (2010)
TPS–silica (sio2) PVOH composite films + tetraethyl orthosilicate (TEOS)	Extrusion	Frosta et al. (2011)
High amylose hydroxypropylated starch + films nanosilicate additives+ montmorillonite	Extrusion	Dean et al. (2011)
Wheat flour+ glycerol	Extrusion followed by compression molding	Sreekumar (2010)
Tapioca starch–glycerol+ xanthan gum (XG) + potassium sorbate (PS)	Extrusion technology	Flores et al. (2010)
TPS+ glycerol	One-step combined twin-screw/single screw extrusion	Rodriguez-Gonzalez et al. (2003)
TPS + glycerol	Extrusion	Rosa et al. (2008)

Table 3. Biodegradable polimers obtained with thermoplastic extrusion

4. Conclusion

The thermoplastic extrusion process is capable of causing changes in starch, making it present a large variety of applications both in the food industry and in other industries; and, as a matter of fact, the use of starches in packagings has increased due to the easy process in modifying and promoting interactions between starch and other polymers.

5. References

Abu-Hardan, M.; Hill, S.E. & Farhat, I. (2011). Starch conversion and expansion behaviour of wheat starch cooked with either; palm, soybean or sunflower oils in a co-rotating intermeshing twin-screw extruder. *International Journal of Food Science and Technology*, Vol. 46, No.2, 268-274.

Alves, R.M.L.; Grossmann, M.V.E. & Silva, R.S.S.F. (1999). Gelling properties of extruded yam (*Dioscorea alata*) starch. *Food Chemistry*, Vol. 67, 123-127.

Bello-Pérez, L.A.; González-Soto, R.A. ; Sánchez-Rivero, M.M. ; Gutiérrez-Meraz, F. & Vargas-Torres, A. (2006). Extrusion of starches from non-conventional sources for resistant starch production *Agrociencia*, Vol. 40, No. 4, 441-448.

Brandelero, R. P. H.; Yamashita, F.; Grossmann, M.V.E. (2010). The effect of surfactant Tween 80 on the hydrophilicity, water vapor permeation, and the mechanical properties of cassava starch and poly (butylenes adipatecoterephthalate) (PBAT) blend films. *Carbohydrate Polymers*, Vol. 82, 1102–1109.

Carvalho, A. J. F.; Zambon, M. D.; da Silva Curvelo, A. A. & Gandini, A. (2005). Thermoplastic starch modification during melt processing: Hydrolysis catalyzed by carboxylic acids. *Carbohydrate Polymers*, Vol. *62*, 387–390.

Céspedes, M.A.L. ; Martínez-Bustos, F. & Chang, Y.K. (2010). The effect of extruded orange pulp on enzymatic hydrolysis of starch and glucose retardation index . *Food and Bioprocess Technology*, Vol. 3, No.5, 684-692.

Chang, Y. & Lii, C. (1992).Preparation of starch phosphates by extrusion. *Journal of Food Science*, Vol. 57, No.1, 203-5.

Chang, Y. K. (1989). *Efeito da concentração de ácido, umidade e temperatura na hidrólise de amido de mandioca por extrusão termoplástica, visando a produção de álcool*. Tese de Doutorado, FEA, UNICAMP.

Charboniere, R.; Duprat, F. & Guilbot, A. (1973).Change in various starch by cooking-extrusion processing, part II, physical structure of extruded products. *Cereal Science Today*, Vol.18, No.9, 286.

Chaudhary, A.L.; Miler, M.; Torley, P.J.; Sopade, P.A. & Halley, P.J. (2008). Amylose content and chemical modification effects on the extrusion of thermoplastic starch from maize. *Carbohydrate Polymers*, Vol. 74, 907–913.

Chaunier, L.; Valle, G. D.; & Lourdin, D. (2007). Relationships between texture, mechanical properties and structure of cornflakes. *Food Research International*, Vol. 40, 493–503.

Chiang, B-Y. & Johnson, J.A. (1977). Gelatinization of starch in extruded products. *Cereal Chemistry*, Vol.54, No.3, 436-443.

Cho, K.Y. & Rizvi, S.S.H. (2010). New generation of healthy snack food by supercritical fluid extrusion. *Journal of Food Processing and Preservation*, Vol. 34, No.2, 192-218.

Clerici, M. T. P. S.; Airoldi, C. & El-Dash, A. A (2009). Production of acidic extruded rice flour and its influence on the qualities of gluten-free bread. *LWT - Food Science and Technology*, Vol. 42, 618-623.

Clerici, M. T. P. S.; El-Dash, A. A (2006). Farinha extrusada de arroz como substituto de glúten na produção de pão de arroz. *Archivos Latinoamericanos de Nutrición*, Vol. 56, 288-294.

Colonna, P.; Doublier, J.L.; Melcion,J.P.; Monredon, F. & Mercier, C. (1984). Extrusion cooking and drum drying of wheat starch, part I, physical and macromolecular modifications. *Cereal Chemistry*, Vol.61, No.6, 538-43.

Corn production Source: U.S. Department of Agriculture, Foreign Agricultural Service. At, http://www.starch.dk/isi/stat/worldstat.asp. Access October 10th, 2011.

Dean, K. M.; Petinakis, E.; Goodall, L.;Miller, T.; Yu, L. & Wright, N. (2011). Nanostabilization of thermally processed high amylose hydroxylpropylated starch films, 2011. *Carbohydrate Polymers*, doi:10.1016/j.carbpol.2011.05.003.

Debiagi, F.; Mali, S.; Grossmann, M.V.E. & Yamashita, F. (2010). Effects of vegetal fibers on properties of cassava starch biodegradable composites produced by extrusion. *Ciencia e Agrotecnologia*, Vol. 34, No.6, 1522-1529.

El-Dash, A.A. (1982).Aplication and control of thermoplastic extrusion of cereals for food and industrial uses. In. Pomeranz, Y. & Munch, L. *Cereal a Renewable Resource: Theory and Practice.* American Association of Cereal Chemists.

Faubion, J.M.; Hoseney, R.C. & Seib, P.A. (1982). Functionaly of grain components in extrusion. *Cereal Foods World*, Vol.27, No.5, 212-16.

Flores, S. K.; Costa, D.; Yamashita, F.; Gerschenson, L. N. & Grossmann, M. E.V. (2010). Mixture design for evaluation of potassium sorbate and xanthan gum effect on properties of tapioca starch films obtained by extrusion. *Materials Science and Engineering*, Vol. 30, 196–202.

Frosta, K.; Barthes, J.; Kaminski, D.; Lascaris, E., Niere, J. & Shanks, R. (2011). Thermoplastic starch–silica–polyvinyl alcohol composites by reactive extrusion. *Carbohydrate Polymers*, Vol. 84, 343–350.

Gomez, M.H. & Aguilera, J.M. (1983). Changes in starch fraction during extrusion-cooking of corn. *Journal of Food Science*, Vol.48, 378-381.

Gomez, M.H. & Aguilera, J.M. (1984).A physicochemical model for extrusion of corn starch. *Journal of Food Science*, Vol. 49, No.1, 40-3.

Goni, I.; Garcia-Alonso, A. & Saura-Calixto, F. (1997). A Starch hydrolysis procedure to estimate glycemic index. *Nutrition Research*, Vol.*17*, 427-37.

Govindasamy, S. (1997). The single screw extruder as a bioreactor for sago starch hydrolysis. *Food Chemistry*, Vol. 60, No. 1, 1-11.

Govindasamy,S.; Campanella & Oates, C.G. (1996). High moisture twin-screw extrusion of sago starch,1, Influence on granule morphology and structure. *Carbohydrate Polymers*, Vol. 30, 275-286.

Greenwood, C.T. Starch. In: *Advances in Cereal Science and Technology*, ed. Pomeranz, AACC, 1976.

Hamilton, R.M. & Paschall, E.F. (1967).Production and uses of starch phosphates. In.: *Starch Chemistry and Technology*, vol.II., ed. Whistler, R.L. & Paschall, E.F. Academic Press, New York.

Harper, J.M. (1979). Food extrusion. *CRC - Critical Reviews in Food Science and Nutrition*, Vol.11, No.2, 155-215.

Hodge, J.E.; Montgomery, E.M. & Hilbert, G.E. (1948). Hydrolysis of the amylopectins from various starches with beta-amylase. *Cereal Chemistry*, Vol..25, 19-30.

Huber, G. F., & Rokey, G. J. (1990). Extruded Snacks. In: *Snack Food*. AVI: New York.

Jenkins, D. J. A.; Wolever, T. M. S.; Taylor, R. H.; Ghafari, H.; Jenkins, A. L.; Barker, H.& Jenkins, M.J. (1980). Rate of digestion and postprandial glycaemia of foods in normal and diabetic subjects. *British Medical Journal*, Vol.281, No. 6232, 14-17.

Lai, C.S.; Guetzlaff, J. & Hoseney, R.C. (1989). Role of sodium bicarbonate and trapped air in extrusion. *Cereal Chemistry*, Vol. 66, No.2, 69-73.

Lawton, B.T.; Hendersonn, G.A. & Delatka, E.J. (1972).The effect of extruder variables on the gelatinization of corn starch. *The Canadian Journal of Chemical Engineering*, Vol.50, No.2, 168-72.

Leach, H.W. (1965). Gelatinization of starch. In: *Starch Chemistry and Technology*, Vol. I, ed. Whistler, R.L. & Paschall, E.F. Academic Press, New York.

Lee-Ch; Min-Kc; Souane-M; Chung-Mj; Mathiasen-Te & Adler-Nissen-J. (1992). Fermentation of prefermented and extruded rice flour by the lactic acid bacteria from skhae. *Food Biotechnoloy*, Vol.6, No.3, 239-255.

Li, H. & Huneault, M.A. (2011). Effect of chain extension on the properties of PLA/TPS blends. *Journal of Applied Polymer Science*. Article in Press

Li, S.-H.; Zhuang, X.-W.; Wang, C.-P. & Chu, F.-X. (2011). Renewable resource-based composites of thermoplastic acorn starch and polycaprolactone: Preparation and FTIR spectrum analysis. *Guang Pu Xue Yu Guang Pu Fen Xi/Spectroscopy and Spectral Analysis*, Vol. 31, No.4, 992-996.

Light, J.M. (1990). Modified food starches: why, what, where, and how. *Cereal Foods World*, Vol.35, No.11, 1081-92.

Lim, S.J.; Kasemsuwan, T. & Jane, J.L. (1994).Characterization of phosphorus in starch by ^{31}P-Nuclear Magnetic Resonance Spectroscopy. *Cereal Chemistry*, Vol.71, No.5,.488-493.

Limón-Valenzuela, V.; Martínez-Bustos, F.; Aguilar-Palazuelos, E.; Caro-Corrales, J.J. & Zazueta-Morales, J.J. (2010). Physicochemical evaluation and optimization of enriched expanded pellets with milk protein concentrate. *Cereal Chemistry*, Vol. 87, No.6, 612-618.

Limsangouan, N.; Takenaka, M.; Sotome, I.; Nanayama, K.; Charunuch, C. & Isobe, S. (2010). Functional properties of cereal and legume based extruded snack foods fortified with by-products from herbs and vegetables. *Kasetsart Journal - Natural Science*, Vol.44, No. 2, 271-279.

Linko, P.; Colonna, P. & Mercier, C. (1981). HTST-extrusion of cereal based materials. In: *Advances in Cereal Science and Technology*, AACC, Vol.4, pp.145-235.

Liu, H.; Xie, F.; Yu, L.; Chen, L. & Li, L. (2009). Thermal processing of starch-based polymers. *Progress in Polymer Science*, Vol. 34, 1348–1368.

Lobato, L.P.; Anibal, D.; Lazaretti, M.M. & Grossmann, M.V.E. (2011). Extruded puffed functional ingredient with oat bran and soy flour. *LWT - Food Science and Technology*, Vol. 44, No. 933-939.

Lorenz, K. & Jansen, G.R. (1980).Nutrient stability of full-fat soy flour and corn-soy blends produced by low-cost extrusion. *Cereal Foods World*, Vol.25, No.4, 161-2, 171.

Ma, X.; Chang, P. R.; Yu, J. & Stumborg, M. (2009). Properties of biodegradable citric acid-modified granular starch/thermoplastic pea starch composites. *Carbohydrate Polymers*, Vol. 75, 1–8.

Mehltretter, C.L. (1967). Production and use of dialdehyde starch. In: *Starch, Chemistry and Technology*, ed. by Whistler, R.L. & Paschall, E.F. Vol. II, cap.18, Academic Press, New York.

Menegassi, B.; Pilosof, A.M.R. & Arêas, J.A.G. (2011). Comparison of properties of native and extruded amaranth (*Amaranthus cruentus* L. e BRS Alegria) flour. *LWT - Food Science and Technology*, Vol. 44, 1915-1921.

Mercier, C. & Feillet, P. (1975). Modification of carbohydrates components by extrusion-cookinng of cereal products. *Cereal Chemistry*, Vol.52, No.3, 283-297,

Michael, E. & Brown, J.A. (1968). Improve texture, consistency, shelf life and freeze-thaw stability at reduced cost. *Food Processing*, Vol.28, No.6, 436-443.

Mitrus, M. & Mościcki, L. (2011). Physical properties of thermoplastic starches. *International Agrophysics*, Vol. 23, No. 3, 305-308.

Mondragón, M.; Hernández, E.M.; Rivera-Armenta, J.L. & F.J. Rodríguez-González. (2009). Injection molded thermoplastic starch/natural rubber/clay nanocomposites: Morphology and mechanical properties. *Carbohydrate Polymers*, Vol. 77, 80–86.

Nabar, Y.; Narayan, R. & Schindler, M. (2006). Twin-screw extrusion production and characterization of starch foam products for use in cushioning and insulation applications. *Polymer Engineering and Science*, Vol. 46, No. 4, 438-451.

Nabeshima, E.H. & Grossmann, M.V.E. (2001). Functional properties of pregelatinized and cross-linked cassava starch obtained by extrusion with sodium trimetaphosphate. *Carbohydrate Polymers*, Vol. 45, 347-353.

Ostergard, K. & Bjork, I. (1989).Effects of extrusion cooking on starch and dietary fibre in barley. *Food Chemistry*, Vol.34, No.3, 215-27.

Patil, R.T.; De Berrios, J.J.; Tang, J.; Pan, J.& Swanson, B. (2005). Physical characteristics of food extrudates - a review. *2005 ASAE Annual International Meeting*, 17p.

Pelissari, F.M.; Yamashita, F.; Grossmann, M.V.E. (2011). Extrusion parameters related to starch/chitosan active films properties. *International Journal of Food Science and Technology*, Vol. 46, No.4, 702-710.

Powell, E.L. (1965). Production and uses of pre-gelatinized starches. In: *Starch Chemistry and Technology*, vol.II., ed. Whistler, R.L. & Paschall, E.F. Academic Press, New York.

Pushpadass, H. A.; Bhandari, P. & Hanna, M. A. (2010). Effects of LDPE and glycerol contents and compounding on the microstructure and properties of starch composite films. *Carbohydrate Polymers*, Vol. 82, 1082–1089.

Rodriguez-Gonzalez, F.J.; Ramsay, B.A. & Favis, B.D. (2003). High performance LDPE/thermoplastic starch blends: a sustainable alternative to pure polyethylene. *Polymer*, Vol. 44, 1517–1526.

Rosa, D.S.; Carvalho, C.L.; Gaboardi, F.; Rezende, M.L.; Tavares, M.I.B.; Petro, M.S.M. & Calil, M.R. (2008). Evaluation of enzymatic degradation based on the quantification of glucose in thermoplastic starch and its characterization by mechanical and morphological properties and NMR measurements. *Polymer Testing*, Vol. 27, 827–834.

Roy, S.B.; Ramaraj, B.; Shit, S.C. & Nayak, S.K. (2011). Polypropylene and potato starch biocomposites: Physicomechanical and thermal properties. *Journal of Applied Polymer Science*, Vol. 120, No.5, 3078-3086.

Salay, E. & Ciacco, C.F. (1990). Production and properties of starch phosphates produced by the extrusion process. *Starch*, Vol.42, No.1, 15-17.

Seker, M. & Hanna, M. A. (2005a). Sodium hydroxide and trimetaphosphate levels affect properties of starch extrudates. *Industrial Crops and Products*, Vol. 23, 249–255.

Seker, M. & Hanna, M. A. (2005b). Cross-linking starch at various moisture contents by phosphate substitution in an extruder. *Carbohydrate Polymers*, Vol. 59, 541–544.

Seker, M.; Sadikoglu, H. & Hanna, M. A. (2004). Properties of crosslinked starch produced in a single-screw extruder with and without a mixing element. *Journal of Food Process Engineering*, Vol.27, 47–63.

Seker, M.; Sadikoglu, H.; Ozdemir, M. & Hanna, M.A. (2003). Phosphorus binding to starch during extrusion in both singleand twin-screw extruders with and without a mixing element. *Journal of Food Engineering*, Vol. 59, 355–360.

Shildneck, P. & Smith, C.E. (1967). Production and uses of acid-modified starch. In: *Starch, Chemistry and Technology*, ed. by Whistler, R.L. & Paschall, E.F. Vol. II, cap.IX, Academic Press, New York,.

Singh, J.; Dartois, A. & Kaur, L. (2010). Starch digestibility in food matrix: a review. *Trends in Food Science & Technology*, Vol. 21, 168-180.

Singh, S.; Gamlath, S. & Wakeling, L. (2007). Nutritional aspects of food extrusion: A review. *International Journal of Food Science and Technology*, Vol. 42, No.8, 916-929.

Souza, L.B. & Leonel, M. (2010). Effect of the fiber content and operational extrusion parameters on the pasting characteristics of cassava starch and orange pulp mixture . *Ciência e Tecnologia de Alimentos*, Vol. 30, No.3, 686-692.

Sreekumar, P.A.; Gopalakrishnan, P.; Leblanc, N. & Saiter, J.M. (2010). Effect of glycerol and short sisal fibers on the viscoelastic behavior of wheat flour based thermoplastic. *Composites: Part A*, Vol. 41, 991–996.

Stepto, R.F.T.(2003). Predicting the modulus of end-linked networks from formation conditions. *Macromolecular Symposium 2003*; 200:255–64.

Stojceska, V.; Ainsworth, P.; Plunkett, A. & Ibanoglu, S. (2009). The effect of extrusion cooking using different water feed rates on the quality of ready-to-eat snacks made from food by-products. *Food Chemistry*, Vol. 114, 226–232.

ŚWierz-Motysia, B.; Jezióṙska, R.; Szadkowska, A. & Piotrowska, M.(2011). Synthesis and properties of biodegradable polylactide and thermoplastic starch blends . *Polimery/Polymers*, Vol. 56, No. 4, 271-280.

Tester, R. F.; Karkalas, J. & Qi, X. (2004). Starch—composition, fine structure and architecture. *Journal of Cereal Science*, Vol. 39, 151–165.

Thuwall, M.; Boldizar, A.& Rigdahl, M. (2006). Extrusion processing of high amylose potato starch materials. *Carbohydrate Polymers*, Vol. 65, 441–446.

Torres, J. A. (1994). Edible films and coatings from proteins, in protein functionality. In Hettiarachchy, N. S., & Ziegler, G. R. (Eds.), *Foodsystems* New York: Marcel Dekker. (pp. 467–507).

Ushakumari, S.R.; Latha, S.& Malleshi, N.G. (2004). The functional properties of popped, flaked, extruded and roller-dried foxtail millet (*Setaria italica*). *International Journal of Food Science and Technology*, Vol. 39,No.9, 907-915.

Williams, P.C. & Lesselleur, G.C. (1970). Determination of damaged starch in flour, a comparative study of present day procedures. *Cereal Science Today*, Vol.15, 4.
Wurzburg, O.B. (1989). *Modified starches: properties and uses*. CRC Press, Boca Raton, 277.
Guan, J.& Hanna, M.A. (2006). Selected morphological and functional properties of extruded acetylated starch–cellulose foams. *Bioresource Technology*, Vol. 97, 1716–1726.
Yaĝci, S. & Göĝûş, F. (2009) Effect of incorporation of various food by-products on some nutritional properties of rice-based extruded foods. *Food Science and Technology International*, Vol. 15, No.6, 571-581.
Yanniotis, S.; Petraki, A.& Soumpasi, E. (2007). Effect of pectin and wheat fibers on quality attributes of extruded cornstarch. *Journal of Food Engineering*, Vol. 80, 594–599.
Yunos, M.Z.B. & Rahman, W.A.W.A. (2011). Effect of glycerol on performance rice straw/starch based polymer. *Journal of Applied Sciences*, Vol. 11, No.13, 2456-2459.

4

Properties and Biodegradation Nature of Thermoplastic Starch

Redouan Saiah[1], Richard Gattin[1] and P.A. Sreekumar[2]
[1]*Laboratoire de Génie des Matériaux (LGMA),*
Ecole d'Ingénieurs en Agriculture Esitpa, 3 rue du Tronquet,
Mont Saint Aignan Cedex,
[2]*Department of Chemical Engineering,*
King Fahd University of Petroleum & Minerals, Dhahran,
[1]*France*
[2]*Saudi Arabia*

1. Introduction

For the last few decades, the usage of plastic increased because of its specific properties such as low cost, light weight, high strength, non-biodegradability, durability, non corrosive nature, process ability and high energy effectiveness. Hence these plastics can be used for various application which includes household articles to aeronautic sector. Now a day it's difficult to imagine a life without plastic which are mostly derived from crude oils and natural gas. Among the various polymers, polyethylene, polypropylene and polystyrene are used greatly for food packaging, biomedical field and in agriculture. According to statistics, from 1950 onwards, 9% of growth can be seen globally, in the production and consumption of plastics. In 1950 the overall production of plastic was 1.5 million tones while it reached 245 million tones in 2008.

In these polyethylene is one of the most dominant packaging material, creating the real problems in the disposal of one-trip packaging. These polymers will take millions of years to degrade under natural weathering conditions. Hence careless dumping of these plastics after its usage creates severe problems to the environment. Also during combustion it produces toxic materials which eventually pollute the atmosphere. The land filling results in the contamination of water, thereby adversely affecting the soil's biological balance. 'Recycling' is another solution for reducing the amount of waste polyolefin materials. But recycling has its own limitation in regard to compatibility of different polyolefins which adversely affects the processability and final properties. Subsequently the problems created by plastic wastages to the environment triggered the interest in the development of biodegradable disposable plastics. So that the onetime use items can be disposed off with the hope that they will not remain for centuries in a landfill, or as litter, which is one of the tenets driving the recent interest in "green" technologies. The current biodegradable plastics, such as PLA, PHBV, Mater-Bi etc are very costly and the processing and mechanical properties of these materials are not good enough for the production of consumer products. Hence several studies were conducted to modify the current commodity plastics such as polyethylene, polypropylene into biodegradable type. One method to achieve this goal was blending of plastics with biodegradable agricultural feed stocks to meet the requirements of responsible and ecologically sound utilization of resources. This will reduce our dependence on depleting petrochemical resources.

2. Fabrication of thermoplastic starch

The fusion of mixture of native starch granules with sufficient amount of plasticizer, at lower temperature than their degradation, leads to a starchy material consisting of entangled polysaccharide chains. This material is called thermoplastic starch (TPS). Most common methods used for the preparation of TPS are extrusion and solvent casting method. For that various types of plasticizers such as glycerol, water, urea formamide etc are used during the formulation. The characteristics of mechanical work and flow (temperature, pressure, residence time, energy) during the preparation of TPS are known and modeled [Agassant et al., 1996, Vergnes et al., 1998]. A specific mechanical energy (SME) higher than 300kJ/g is necessary to achieve a complete destructuration of starch granules. The presence of low molecular mass (plasticizers) raises the threshold energy similar to the action of sugars in the cereal products [Fan et al., 1996]. The SME transmitted as shear, leads to the breakdown of starch grains by fragmentation, and once the melt phase obtained, it is accompanied by a moderate depolymerization, particularly in the amylopectin [Barron et al., 2002].

The native starch can be transformed to TPS by several treatments which destroys its granular structure. Physical treatment requires simultaneous action of temperature, shear and lower water content during extrusion. Initially amorphous regions of starch granules are more accessible to water followed by its crystalline zones. The amount of water should be sufficient to hydrate these starch molecules which ultimately results in its gelatinization. Also for obtaining the melt phase of plasticized TPS, additional energy is required to destroy the residual crystalline structures. This mechanical treatment reduces the crystallinity of starch granules. Figure 1 shows that the viscosity of plasticized starch is of the same order and magnitude as those of synthetic thermoplastics at lower temperatures [Martin et al., 2003].

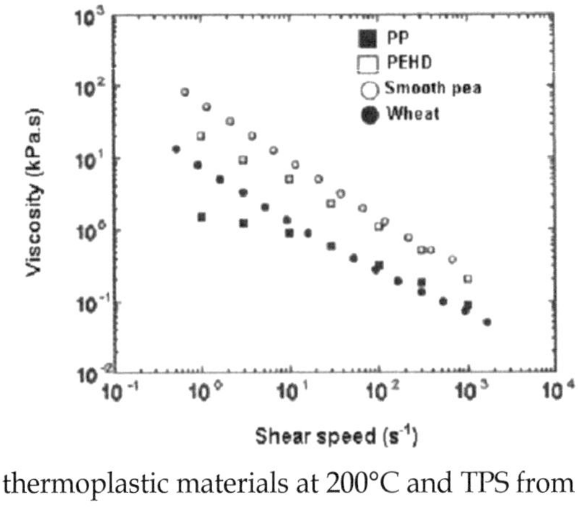

Fig. 1. Flow curves of thermoplastic materials at 200°C and TPS from smooth pea and wheat at 125°C

3. Properties of thermoplastic starch

3.1 Morphology

Studies show that the structures of native and thermoplastic wheat starch are totally different. Scanning Electron Micrograph (SEM) reveals that the native wheat starch has a granular structure (Figure 2) [Leblanc et al., 2008].

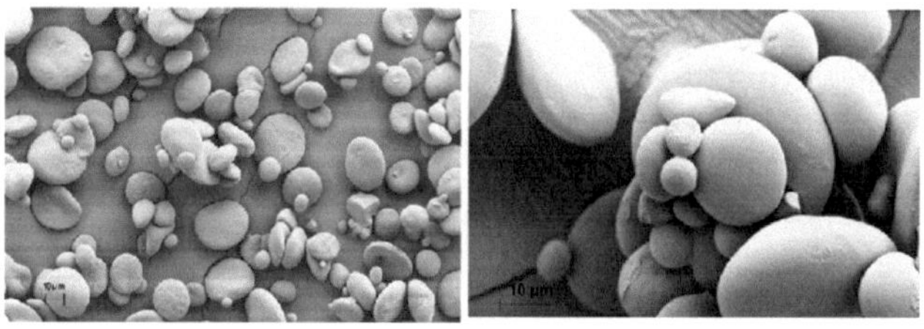

Fig. 2. SEM micrographs of native wheat starch

These granules are spherical or oval and have different domain sizes. They are smooth, free from pores, cracks, or fissure around to lenticular and polyhedral shapes and are relatively thick. For both starch granules there exists a wide distribution of granule size. This wide distribution is a common feature of cereal starch [Buleon et al., 1998; Charles et al., 2003]. During plasticization (the transformation of granular morphology into a homogeneous polymeric film), the destruction of hydrogen bonds between the starch molecules occurs synchronously with the formation of the hydrogen bonds between the plasticizer and starch molecules [Yang et al., 2006].

The extrusion method is a combination of thermal and mechanical input. During this process, starch was plasticized and a homogeneous molten phase characteristic of thermoplastic polymeric material was obtained (Figure 3) [Saiah et al., 2009].

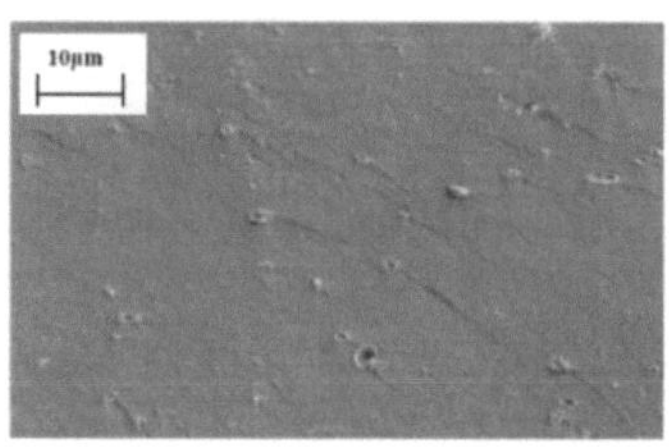

Fig. 3. SEM of extruded wheat starch based thermoplastic film

3.2 Structure

The X-ray diffraction (XRD) patterns for native and plasticized starch based on wheat flour (with 9% water and 12.8% glycerol) are displayed in Figures 4 A and B, respectively [Saiah et al., 2007]. The signal obtained from XRD, for native starch shows the semi-crystalline nature of this material. The diffraction peaks were obtained at 2θ values equal to 11.3, 15.2, 17.3, 18.1, 20.1, 23.3, and 26.7°, leading to the conclusion that these raw materials present A-type crystalline structure. This general characteristic of cereal starches was already observed in many other studies [Katz and Van Italie, 1930; Le Bail et al., 1993; Krogars et al., 2003]. For the extruded thermoplastic films, the peak appears at 2θ = 7.2°, 12.9°, 19.8°, and 22.6° (Figure

4 B); which are characteristics of a V_h-type structure [Le Bail, 1995; Fanta et al., 2002] indicating a change in the initial crystalline structure of native starch wheat flour. This structure was obtained by complexation of amylose with lipids. However, another peak at $2\theta = 17.3°$ was observed which corresponds to an A-type structure. This A-type residual crystallinity is due to incomplete destructuration and fusion during the transformation [Van Soest et al., 1996(c); Willett and Doane, 2002].

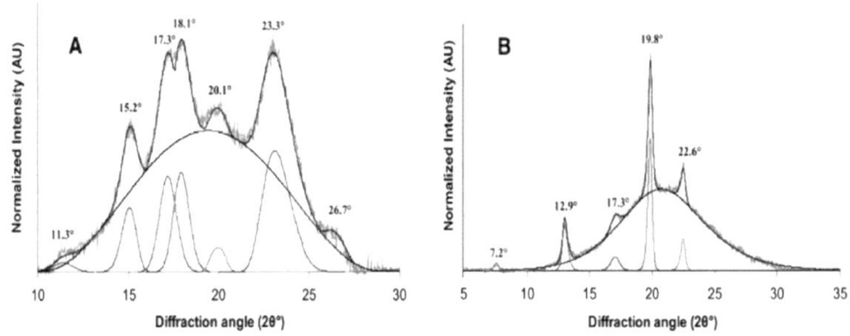

Fig. 4. XRD pattern of native (A) and extruded (B) wheat flour showing different contributions to rebuild experimental signals

The allomorph V_h (Figure 5) obtained with lipid unicycles and linear alcohols is the most common and most studied due to its involvement in many transformations of the starch. It is characterized by an orthorhombic lattice (a = 1.37 nm, b = 2.37 nm and c = 0.805 nm) and P212121 space group type.

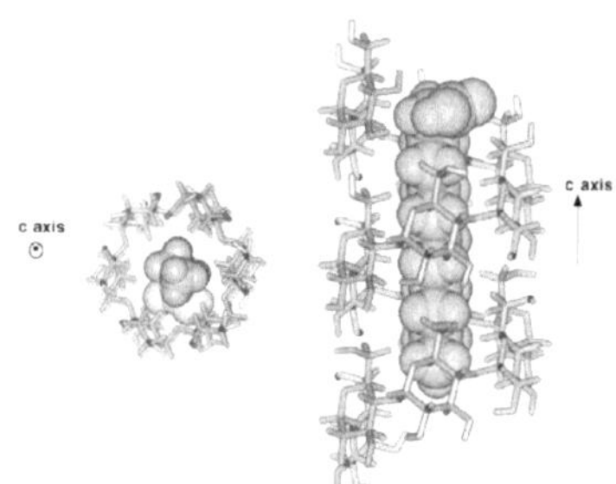

Fig. 5. Conformation of inclusion model of a fatty-acid in an amylose helix (Structure of V_h-type)

The V_a structure is another form of crystalline amylose-lipid complex. The characteristic angles of XRD peaks are respectively at $2\theta = 7.8$, 13.5 and 20.9° [Paris et al., 2001]. It is characterized by an orthorhombic lattice (a = 1.30 nm, b = 2.25 nm, c = 0.79 nm) with a P212121 space group [Winter and Sarko (a), 1974; Zaslow et al., 1974; Winter and Sarko (b), 1974]. In the V_a crystal structure, the helices of amylose are more contracted and there is less water compared to the V_h crystal structure. The transition V_h form to V_a is observed when the form V_h is dried to a water activity less than 0.6 [Rappenecker and Zugenmaier, 1981; Hinkle and Zobel, 1968; Murphy et al., 1975]. This transition is reversible because it is possible to transform the V_a type to V_h type by hydration in the vapor phase. It is usually interpreted as a change of form during the hydration with increasing distance between helices of 1.30 to 1.37 nm [Van Soest et al., 1996; Zaslow and Miller, 1961] due to the introduction of water molecules between amylose helix. The V_a form is never obtained by direct crystallization of solutions of amylose, which is not the case of structures of type A, B and V_h [Buléon et al., 1984].

To have more information regarding the effect of plasticizers on the structure [Saiah et al., 2007], the amount of one plasticizer was kept constant (i.e., water initially fixed at 9%), and the percentage of glycerol varied from 12.8% to 16.5%, keeping the total amount as 20%. For all these materials, the same 2θ values peaks were observed and also the characteristics peaks for V_h and A-type crystalline structure. Also an increase in the intensity of the amorphous halo can be seen as a function of glycerol content (Figure 6).

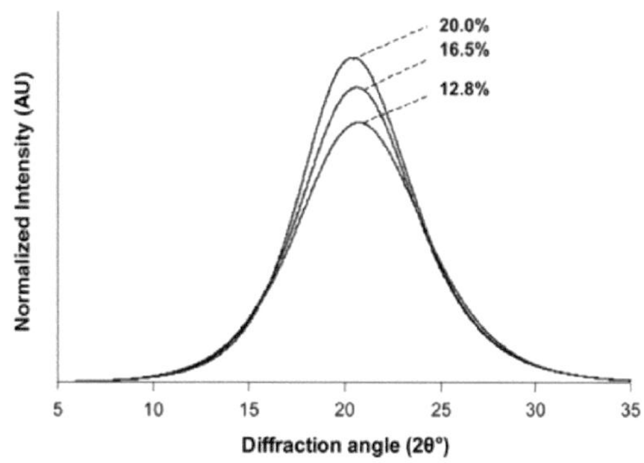

Fig. 6. XRD pattern showing variations in the amorphous halo as a function of glycerol content

In other words, the increase in percentage of glycerol decreases the percentage of crystallinity of the material from 14 to 11%. Table 1 gives the characteristics of the amorphous halo (the maximum diffraction angle, θ_m), which allows calculation of an average intermolecular distance (d_m) according to the Bragg formula (1):

$$2d_{hkl}\sin\theta = n\lambda \tag{1}$$

where λ is the wavelength of X-rays ($\lambda_{K\alpha Cu}$ = 1.54A°) and n is a positive whole number called order of diffraction.

Glycerol (%, w/w)	θ_m	Sin θ_m	d_m (A°)
12.8	10.36 ± 0.02	0.1798	4.28 ± 0.008
16.5	10.30 ± 0.02	0.1788	4.31 ± 0.008
20.0	10.20 ± 0.02	0.1770	4.35 ± 0.008

Table 1. Maximum diffraction angle and average intermolecular distance observed from XRD

The amorphous halo shifts toward the smaller angles when the percentage of glycerol increases, which increases the average distance between two molecular chains. The position of the amorphous halo reflects the average density (or specific volume) of the material [Arrighi et al., 1998]. When the average intermolecular distance increases from 4.28 to 4.35A°, the average density (or the specific volume) decreases (Figure 7).

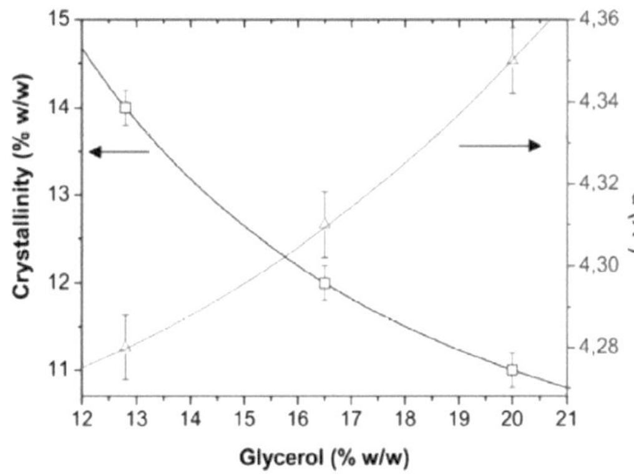

Fig. 7. Variations in crystalline fraction of samples and modification of average intermolecular distance in amorphous domains, when glycerol content changes

As a consequence, it is expected that the amount of free volume linked to existence of an amorphous or a vitreous phase increases in the extruded materials.

3.3 Mechanical behavior

The mechanical properties of TPS are influenced by the botanical origin of starch, more specifically the proportion of amylose and amylopectin. The materials obtained from wheat, corn and potatoes starch, have failure stress above that of a material based on waxy maize starch (rich on amylopectin) (Figure 8 a) [Hulleman et al., 1998]. Tensile tests on films of pure amylose and amylopectin highlights the difference in behavior between these two materials (Figure 8 b) [Lourdin et al., 1995]. The amylopectin based material has a ductile behavior (high failure strain), while that based on amylose has a typical behavior of a brittle material.

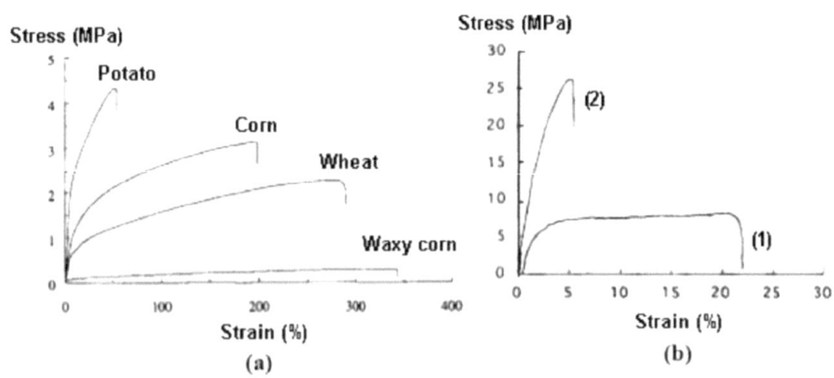

Fig. 8. Stress-strain curves for: (a) based materials starch potato, corn, wheat and waxy maize (b) materials based pure amylopectin (1) and pure amylose (2)

The effect of plasticizers on film resulting from agro-resources, generally leads to a decrease in the modulus and failure stress, and an increase in failure strain (Figure 9) [Poutanen and Forssell, 1996 ; Gontard and Guilbert, 1993; Lourdin et al., 1997].

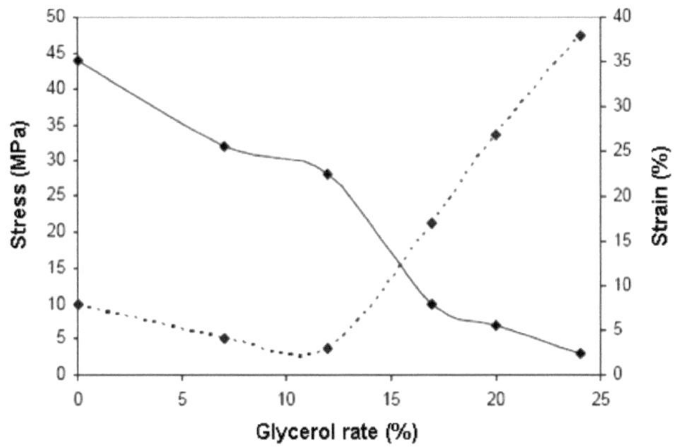

Fig. 9. Influence of glycerol rate on the strain at break (—) and stress at break (---) for potato starch films [Lourdin et al., 1997]

For glycerol level below 12%, a phenomenon "antiplastification" is observed, resulting in decreased stress and strain at break. This phenomenon is due to strong interactions between the polymer and the plasticizer, which forms a hydrogen network reinforcing material. From 12%, the strain at break increases quickly, against the stress at break decreases, and that up to 25% glycerol. This change in mechanical behavior is due to displacement of the glass transition temperature (Tg) of the system below the ambient temperature. The values of stress at failure, strain at failure, and tensile modulus are summarized in Table 2 [Saiah et al., 2007].

Glycerol (%)	σ_{max} (MPa)	ε_{max} (%)	E (MPa)
12.8	3.2 ± 0.10	17.0 ± 1.0	125 ± 6
16.5	2.7 ± 0.20	19.4 ± 1.2	102 ± 6
20.0	2.1 ± 0.12	22.9 ± 0.5	57 ± 4

Table 2. Tensile properties of wheat flour based TPS having different glycerol content

Adding glycerol during the formulation of TPS, decreases the stress at failure from 3.2 to 2.1 MPa and the tensile modulus from 125 to 57 MPa, while an increase from 17 to 22.9% in strain at failure. These results indicate that the ductility of material increases and, as expected, a plasticization effect is obtained by introducing glycerol in the sample composition. These plasticization effects due to glycerol have been observed in TPS made of starch from other sources [Mali et al., 2006]. In these introducing plasticizer reduces direct interaction between starch chains, thus facilitating movement of starch chains under tensile forces [Garcia et al., 1999; Mali et al., 2002)]. The same scenario occurs for wheat-flour-based TPS.

Similar to glycerol, the addition of sorbitol will lead to a significant change in the mechanical behavior of TPS and antiplasticization effect is also observed at sorbitol content less than 27% [Gaudin et al., 1999]. Beyond this amount sorbitol acts as a plasticizer in TPS. During the storage of films, changes in mechanical properties can occur [Van Soest and Knooren, 1997 ; Van Soest et al., 1996 (d); Forssell et al., 1999]. The strain at break decreases, while the stress at break increases. The changes observed over time are due to several concurrent factors: reduction in water content [Van Soest et al., 1996 (b); Van Soest et al.,

1996 (a)], increased Tg and crystallinity. Ageing causes reorientation and/or crystallization of molecules of amylose and amylopectin [Vergnes et al., 1998]. The crystallites acting as physical nodes, generate as stress concentrations and thus weaken the material [Van Soest et al., 1996 (b)].

3.4 Thermal stability

The activation energy calculated for the degradation using TGA, for native and thermoplastic wheat flour is 223 kJ.mol^{-1} and 90 kJ.mol^{-1} respectively [Saiah et al., 2009]. The value of activation energy of native wheat flour is greater than the TPS. This difference is related to the transformation linked to the modifications of crystalline and amorphous phases obtained by extrusion process. The earlier studies on the percentage of crystallinity, using XRD proved that the extrusion causes a reduction in the degree of crystallinity of native wheat flour from 30 to 14% [Saiah et al., 2007]. It indicates that the extruded films have greater amorphous or vitreous phase content. Hence lower amount of energy is enough for its degradation.The effect of lipids (monoglyceride) on thermal stability of TPS is clearly displayed in Figure 10 [Saiah et al., 2009].

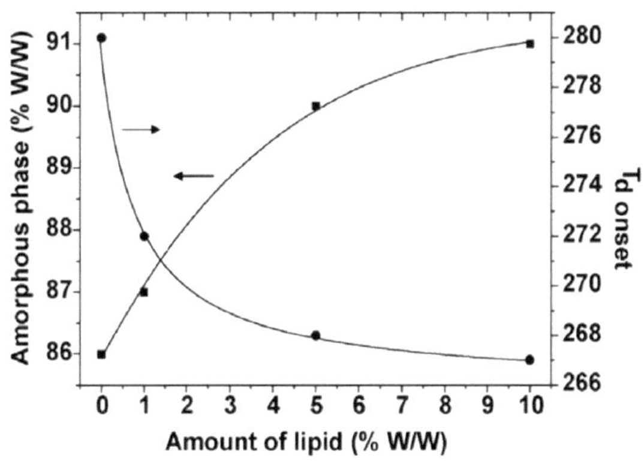

Fig. 10. Variations of amorphous phase content (%, w/w) (■) and the $T_{d\ onset}$ (•) of the polymeric material having varying lipid content (0, 1, 5 and 10%, w/w)

It reveals that as the lipid content increases, the $T_{d\ onset}$ value decreases drastically and is more prominent for the materials having lipid content more than 5% (w/w). Here, the monoglyceride has low thermal stability and they start to degrade before the degradation of the matrix. This degradation is more prominent in the matrix having lipid content greater than 5% (w/w). Moreover, the residue at 800°C for all the materials are almost similar (≈12%, w/w) since variation in the formulation of the matrix is the amount of the wheat flour and lipid content.

The kinetic aspect of the thermal degradation of polymeric material can be obtained by using Broido method [Broido, 1969]. Here, the assumption is that degradation is a first order or a superposition of first order process and the polymeric material will follow the first order type reactions, i.e. (n = 1). The assumption of Broido leads to equation 2:

$$\ln\left[-\ln(1-\alpha)\right] = \ln K - \frac{\Delta E}{RT} \tag{2}$$

where α is the amount of material degraded at time *t*, ΔE is the change in the activation energy which can also be calculated using the Arrhenius equation, *R* is the universal gas constant and *T* is the temperature in Kelvin scale. In this, α can be calculated using the following equation 3:

$$\alpha = \frac{(W_0 - W)}{(W_0 - W_\infty)} \quad (3)$$

where *W* is the mass at time *t*, W_0 is the initial mass and W_∞ is the mass after infinite time. The result obtained does not depend upon the value of heating rate and also gives us the activation energy independently of the value of T_m at which the reaction is maximum. Figure 11 represents the Broido plot for the polymeric materials having various lipids content [Saiah et al., 2009].

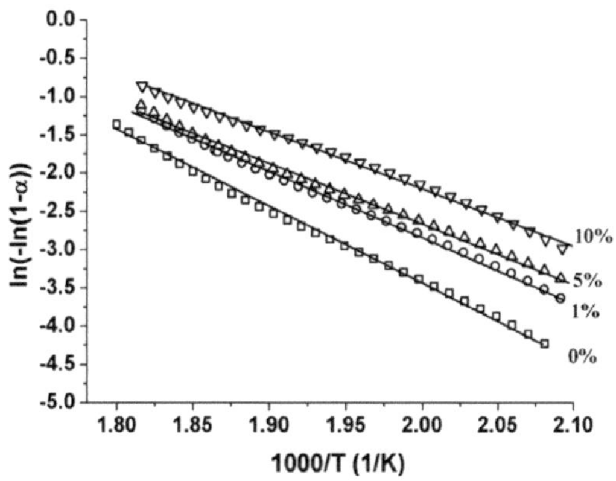

Fig. 11. Broido plot for the polymeric materials having various lipid content (0, 1, 5, and 10%, w/w)

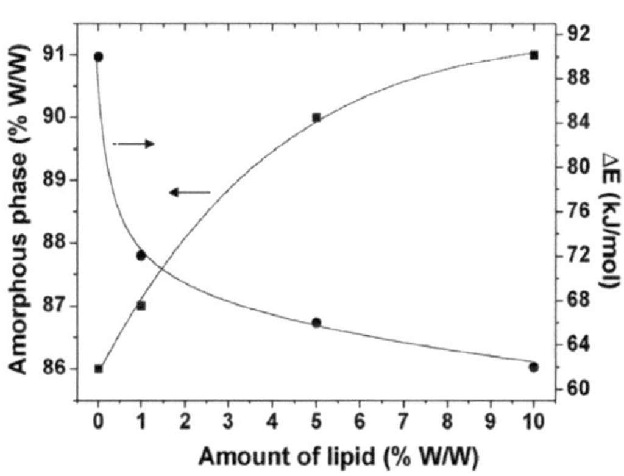

Fig. 12. Variation of amorphous phase content (%, w/w) (■) and change in activation energy (•) of the TPS having various lipid content (0, 1, 5 and 10%, w/w).

From this, variation in activation energy were calculated and displayed in Figure 12.When the lipid content increases, the change in activation energy decreases from 90 to 62 kJ.mol^{-1}.This decrease in the thermal stability and change in activation energy is due to the variation in the degree of crystallinity (Figures 10 and 12) of the polymeric material when the lipids are incorporated.

The influence of the degree of crystallinity on the thermal stability is much clearer by studying the thermal stability of the native and thermoplastic wheat flour. The T_{donset} degradation temperature for native wheat flour is 291°C while that for the thermoplastic wheat flour is 280°C [Leblanc et al., 2008] showing that the thermal stability decreases after the extrusion.

3.5 Glass transition temperature

Figure 13 display the DSC curves obtained for the wheat flour based thermoplastic materials [Saiter et al., 2010].

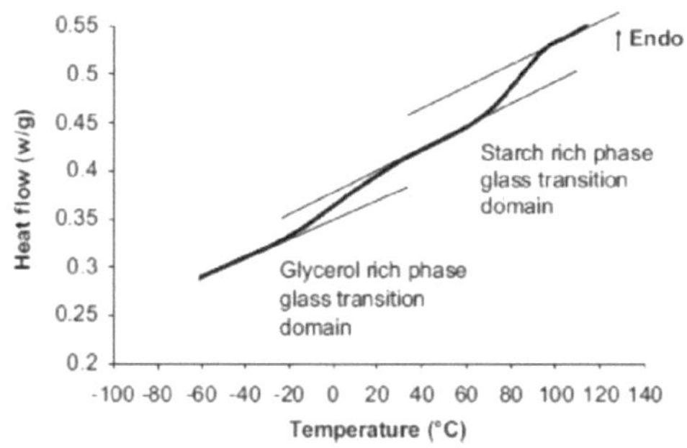

Fig. 13. DSC curves obtained for thermoplastic wheat starch

Only two endothermic steps showing glass transitions are observed at -56 and 10°C, while no peaks for exothermic transitions of crystallization and for degradation. i.e. vitreous and/or the amorphous fraction of the samples are not modified by heat. When the amount of plasticizer changes, the T_g of the starch rich phase is shifted toward the lower temperature. The first transition observed at -56°C is due to the T_g in the glycerol rich phase and the second one at 10°C, due to T_g of the starch rich phase [Saiah et al., 2011].

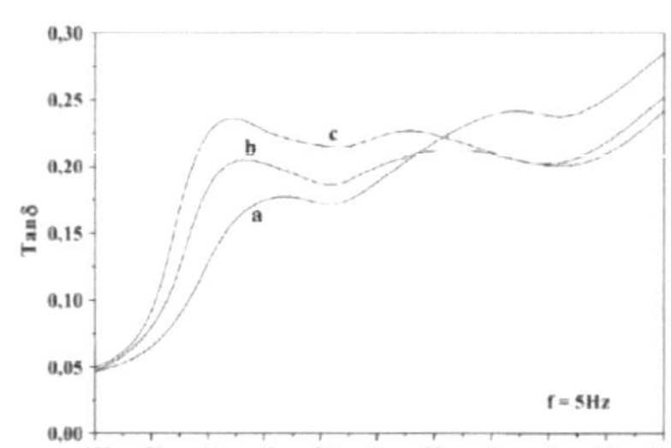

Fig. 14. Dynamic mechanical analysis of wheat-flour based TPS with a: 12,8%; b: 16,5%; c: 20% of glycerol (w/w)

The study of molecular relaxations of TPS having different glycerol content (Figure 14) [Terrie et al., 2010] shows that there are two phases in the TPS: one at lower temperature region corresponding to a glycerol rich phase; another at higher temperature region corresponding to a starch rich phase. The percentage of glycerol in the matrix modifies the characteristic temperature of these two phases.

Silicon dioxide also affects the molecular relaxations of the matrix (Figure 15), but only the second phase transition of the material i.e. the starch rich phase [Saiah, 2007].

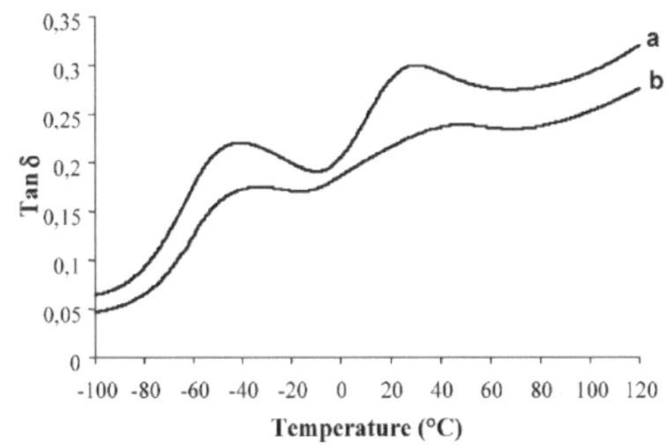

Fig. 15. Dynamic mechanical analysis of wheat-flour based TPS with a: 0%; b: 1%; of silicon dioxide (w/w)

This difference in relaxation is due to the obstacles imparted by silica particles for the starch chain in the starch rich phase.

4. Biodegradation of thermoplastic starch-based materials

Conventional polymeric materials (plastics commonly called) are generally resistant to degradation in the environment due to their high molecular weight and their hydrophobic character. The treatment of plastic waste has therefore become a major environmental concern and programs to recycle, incinerate, or convert waste plastics have been developed. Moreover, the ban on landfill accompanied by concerns related to gaseous emissions from incineration and the difficulty of collection and recycling of certain waste plastics (food packaging, diapers, hospital waste, etc...) stimulated the development of new materials made from biodegradable polymers. These new biodegradable polymers have the advantage of being recycled by composting. However, under industrial composting, the degradation of a product can lead to undesirable compounds and cause a drift of the process. Here the composting process should be carefully validated to avoid disrupting production and protect the environment. Compostability of such materials must first be checked as indicated for instance by the EU directive on packaging (EN 13432), based on the pattern of acceptance below (Figure 16).

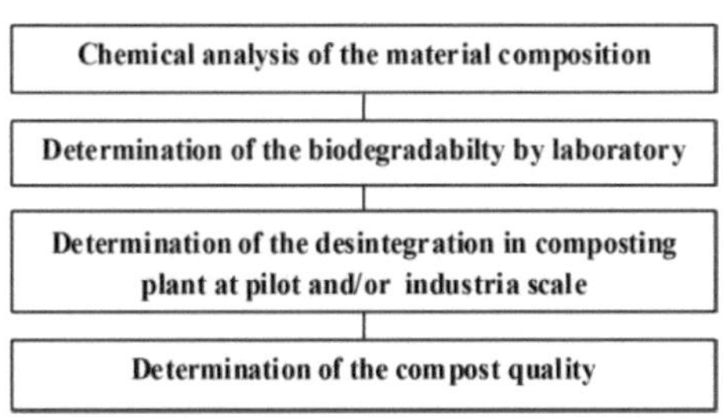

Fig. 16. Scheme of DIN/CEN test used to determine compostability of materials (according to [Pagga et al., 1996]).

In terms of laboratory tests, the standards available about compostability as well ASTM 5 338 than ISO/CEN 14 855 measure CO_2 released. But to ensure the compostable property of materials, it is necessary to do carbon balance in solid medium, where:

$$Ct = \underbrace{\boxed{Cg + Cb + Cs}}_{Cd} + Cnd$$

Where Ct: amount of carbon material introduced into the degradation medium

Cg: amount of carbon material converted into CO_2 by micro-organisms (mineralization)

Cb: amount of carbon material assimilated by biomass (bio-assimilation)

Cs: amount of carbon material converted into soluble by-product of degradation

Cg + Cb + Cs represent all the carbon degraded (Cd) by micro-organisms.

Cnd: fraction of the material which is not degraded or residual material

This approach also allows access to by-products of degradation of the material, which, depending on the material used, can be toxic and persist in the environment.

The materials studied were obtained by extrusion. Once extruded, the films were stored at room temperature (23°C) and controlled humidity (50%). The results presented are those obtained with a co-extruded material, combining PLA and starch. The way of co-extrusion of starch/PLA provides a blend, where the heart of starch (80.6%) is protected on both sides by the PLA (19.4%) which is its outer layer. Thus, the hydrophilic nature of starch is masked by the PLA, which in turn could undergo degradation more important caused of the presence of starch.

4.1 Mineralization in liquid and solid media

The first step was to choose an inert solid medium that simulates the compost and allow the growth of microorganisms, while being devoid of organic matter. The solid inert support used is vermiculite ($(Mg,Fe,Al)3(Al,Si)_4$ $O10(OH)_2.4$ $H2O$): a silicate of aluminum-magnesium-iron that holds water very well. The water activity (aw) is equal to 1 when the water content is adjusted to 70%, so the moisture of the medium was adjusted to this value [Spitzer and Menner, 1996]. Therefore, vermiculite, simulates the structure of aerated compost, allows the proliferation of microorganisms and is completely free of carbon products. The airy structure of the vermiculite also avoids the anaerobic zones and allows simulating, at best, the composting process. The figure 17 illustrates the possibility to use vermiculite to realize composting studies.

Indeed, the difference between the kinetics of mineralization of different media (liquid, vermiculite and compost) is solely due to the moisture content of these environments, humidity higher in the liquid and vermiculite (70%) favoring the degradation of hydrolytic substrate and consequently its availability to the microflora. For cons, whatever the medium, we obtain a final percentage of mineralization is comparable to the order of 74%. Vermiculite can be used to make carbon balance in solid medium [Gattin et al., 2000].

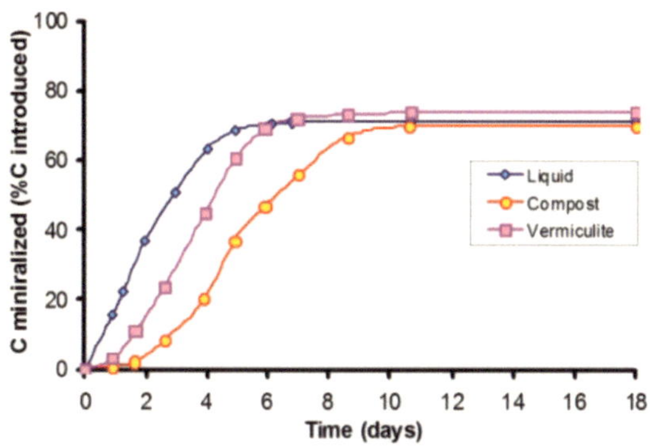

Fig. 17. Mineralization of a ground starch film in liquid and solid media [ASTM D-5209-92 and D-5338-92]

However, it is important to note that the nature of the microorganisms involved in the degradation may be different from one medium to another. Thus, in a liquid medium, the microbial population is predominantly bacterial, while in solid medium, the fungal population is far from negligible. This partly explains the observed mineralization rates since each organism has an efficiency of assimilation of its own. The amount of CO_2 released for the same amount of organic carbon degraded will therefore vary from one organism to another. A faster rate of mineralization induces earlier onset of the plateau phase on the curves of mineralization. Indeed, in this study, the plateau phase of mineralization was obtained from 6 and 8 days in liquid medium, 8 and 9 days in the vermiculite medium and 9 and 10 days in compost. These results are in the same direction as those of Starnecker and Menner [Starnecker and Menner 1996 (a)] who observed the plateau phase after 15 and 40 days respectively in liquid and compost for a starch-based film and those of Van der Zee et al. [Van der Zee et al., 1998] who obtained a plateau phase after 40 and 45 days in liquid and compost in the study of cellulose degradation.

In fact, throughout the study of compostable material, only the final result of mineralization is very important. Therefore, liquid or vermiculite media are relevant approaches to assess the compostability of a material which are readily biodegradable like starch, as the mineralization rate at the end of experiment are the same as those obtained with the compost. These results are consistent with those of Starnecker and Menner [Starnecker and Menner 1996 (b)] and Bellia et al. [Bellia et al., 1999] which showed similar results of mineralization between a compost environment and inert solid medium (vermiculite) inoculated with an extract of the same compost, respectively, for a plastic bag of starch and a thermoplastic material based on polyurethane, polycaprolactone and starch. However, Van der Zee et al. [Van der Zee et al., 1998] found in their study and questioned the relevance of the tests in liquid medium (modified Sturm test) to predict the biodegradation of polymers for biological treatment of waste. These authors have shown that cellulose acetates with a degree of substitution less than 2.5 are easily mineralized into CO_2 in the test compost while no degradation was observed in the test in liquid medium.

Figures 18 and 19 presents the results from mineralization obtained for co-extruded material in each degradation medium according to ASTM and ISO/CEN standards.

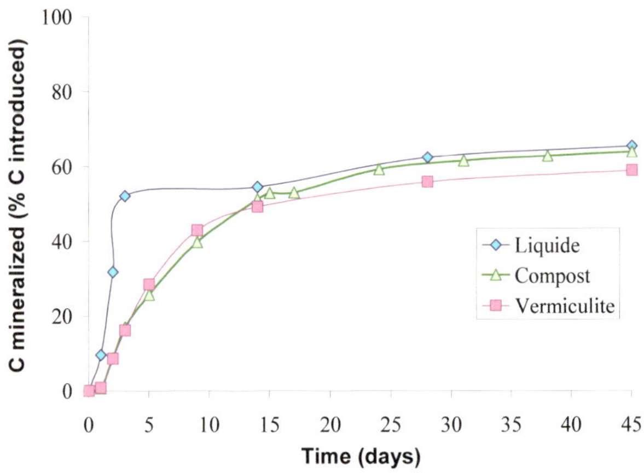

Fig. 18. Mineralization of co-extruded starch/PLA films in liquid (norm ASTM D-5209-92), vermiculite and compost (norm ASTM D-5338-92) media

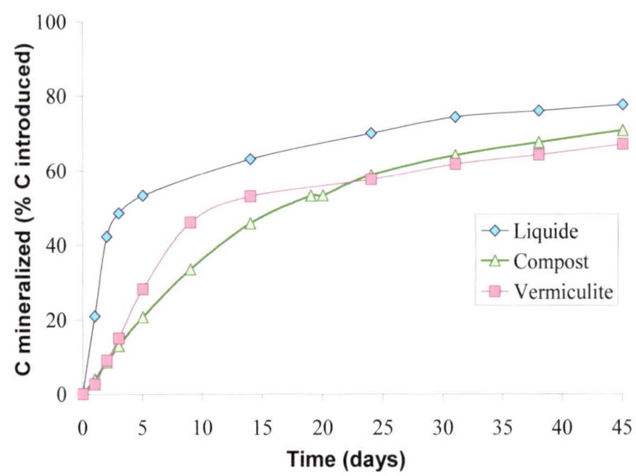

Fig. 19. Mineralization of co-extruded starch/PLA films in liquid (norm ISO/CEN 14852), vermiculite and compost (norm ISO/CEN 14855) media

Whatever the standard used the comparison of degradation in different media shows that the mineralization begins earlier in liquid medium, in agreement with the results for starch alone. In fact, in liquid medium, the mineralization of the co-extruded material is faster than in solid medium and the rate of mineralization obtained after 45 days of incubation is higher than in solid medium for both standards. With ASTM, the difference in mineralization rate between the liquid medium and solid media is small. Indeed, after 45 days of incubation, mineralization rates are 65%, 59% and 63 % for liquid medium, vermiculite and compost. The degradation rate of mineralization of the co-extruded in different environments is as follows: Liquid> compost> vermiculite. With ISO/CEN, this classification is the same, but the differences are much more pronounced since the mineralization rate at 45 days were 78%, 67% and 71% respectively for liquid media, vermiculite and compost.

The degradation of co-extruded starch/PLA was influenced by the standard used to perform the test. This influence comes from the temperature applied during the degradation. Nature of the environment of degradation also affects the biodegradation of the co-extruded. Indeed, biodegradation of the material was greater in liquid than in other environments, because of hydrolytic degradation. High temperatures coupled with high humidity, are the optimal conditions of abiotic degradation, a prerequisite for biodegradation. However, in solid medium, compost produced results mineralization greater than the vermiculite medium, which was not the case with starch alone. This is probably due to the presence of micro-organisms in the compost which was absent in the vermiculite medium. Indeed, it was shown that the degradation of PLA was predominantly of fungal origin. Fungi, many in the compost, are often more closely attached to the substrate than bacteria. This may explain why in the preparation of the inoculum, the total fungal population was not extracted. Therefore, differences in microbial and fungal compost and vermiculite medium are likely to be different which could explain a higher mineralization in urban compost.

The mineralization rate observed for the co-extruded starch/PLA that were 65, 59 and 63% for liquid media, vermiculite and compost with ASTM and 78, 67 and 71% for the same media with ISO/CEN show that this material is compostable. Therefore, the incorporation of a material which is not readily biodegradable in a biodegradable load promotes biodegradation. This result is consistent with that obtained by Bastioli et al. [Bastioli et al., 1995 (a) and (b)] for a mixture starch/PCL (Mater-Bi class Z) degraded in compost. These authors explain that the degradation of starch increases the surface area of the PCL available to microbial attack. Similarly, Park et al. [Park et al., 1994] showed that the rate of degradation of a mixture injected/molded starch and poly(vinyl alcohol) in the presence of activated sludge (25°C) was similar to that of easily degradable materials like starch alone or cellulose. Chen et al. [Chen et al., 1997] also confirm the positive effect of the incorporation of starch on the biodegradation of poly (vinyl alcohol) in soil at 24°C, the biodegradation of the mixture being more important than that of poly (vinyl alcohol) alone. However, when the polymer coupled with starch is not degradable, Wool et al. [Wool et al., 1990] and Gilmore et al. [Gilmore et al., 1992] showed that the material was only bio-fragmented.

Finally, and regardless of the standard and the degradation medium used, the co-extruded starch/PLA is compostable. All these results thus show that a liquid or solid inert simulates satisfactorily the final result in urban compost. Therefore, if the objective is to conclude positively or not on the compostable property of a material, studies in liquid or vermiculite media, easy to implement, may be sufficient. For cons, the kinetics of mineralization obtained for the co-extruded starch/PLA is different depending on the environment degradation. Therefore, if one is interested in the dynamics of carbon during the degradation of materials, then the use of liquid and vermiculite is not sufficient to simulate what happens in urban compost.

4.2 Application of the carbon balance methodology in solid medium

Carbon balance methodology has been applied to various materials. The carbon balance of co-extruded material has been studied. Figure 20 shows the carbon balance of co-extruded starch/PLLA obtained in the vermiculite medium according to ISO/CEN 14855. Regarding the results of the carbon balances obtained for the co-extruded starch/PLLA, it appears that the choice of the standard on the biodegradation of the material is critical. While 95% of the carbon ends up in degraded after 45 days with ISO/CEN, the value is less with the ASTM (85%). In addition, moisture of degradation medium favors the rate of degradation of the material. It's interesting to note that the rate of mineralization of this material according to ASTM standard, in vermiculite medium is 59% while the carbon balance indicates that 85% of the material undergoes degradation.

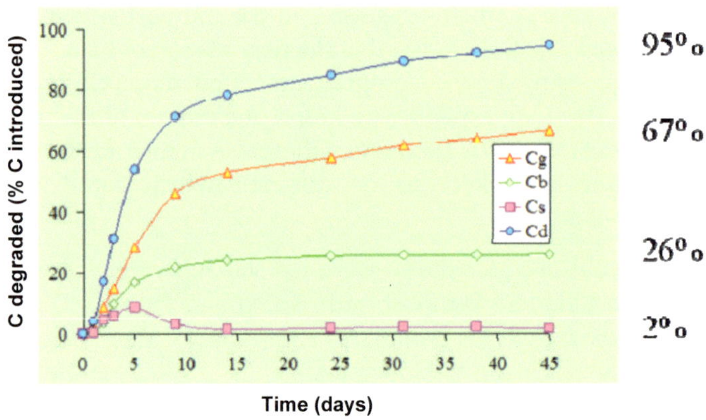

Fig. 20. Carbon balance in vermiculite medium with co-extruded starch/PLLA (ISO/CEN 14855)

In conclusion, the vermiculite medium can be used to make carbon balances and to approach more precisely the degradation process. The recovery rate of carbon was satisfactory regardless of the degradation medium and the substrate studied. At the end of experiment, from 92 to 97% of the carbon has been identified for starch (data not shown), and 100% (liquid medium) for the co-extruded starch/PLA. The study shows that the combination of starch/PLA is an easily compostable material. Moreover, the choice of technology implementation (co-extrusion) allows preparing starch easily accessible due to the grinding before composting process. Considering the results obtained for the biodegradation of co-extruded starch/PLA, this type of complex material is an interesting way to make a compostable material. It remains to optimize the conditions for formatting this type of material so that it can fulfill its function such as packaging and then to verify that the optimized material remains compostable and its degradation by-products as they exist, do not create environmental problems.

4.3 By-products of degradation

The study of by-products of degradation has been proposed for the complex material: co-extruded starch/PLA. In liquid medium, only the presence of glucose, maltose and lactic acid were detected as shown in Figure 21. Initially, only by-products of starch degradation (glucose and maltose) are present. Then, from the 28th day is lactic acid, a byproduct of degradation of PLLA, while we no longer detected by-products associated with the starch. This result suggests that degradation occurs in two stages, first the starch and then the PLA. By cons, for the vermiculite medium, no by-products of degradation have been detected. This would suggest that degradation of by-products is done as they arise.

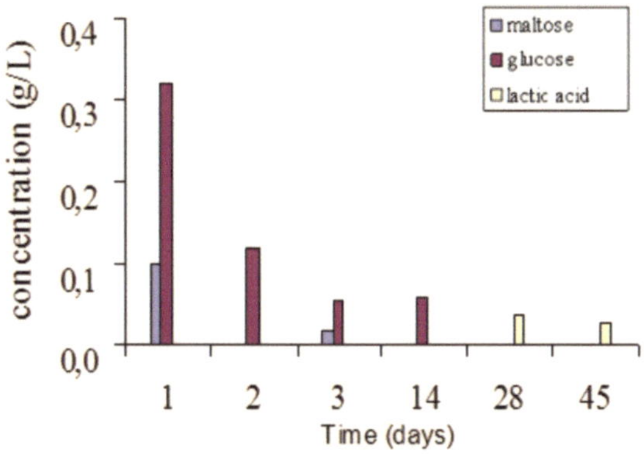

Fig. 21. Evolution of the by-products for co-extruded starch/PLA during degradation in liquid medium using the ASTM standard

4.4 Residual material

Visual observation for starch plates was not possible due to the complete disappearance of the material in the process of biodegradation. Consequently only results obtained with the co-extruded material are presented here (Figure 22).

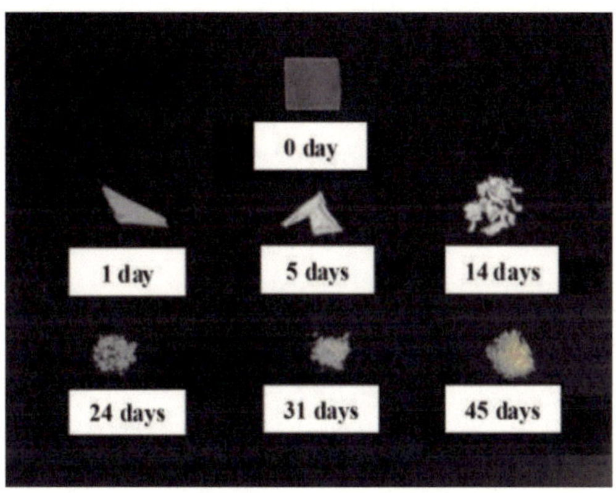

Fig. 22. Physical state of starch /PLA plates during the degradation in liquid medium (ISO/CEN 14852)

The results of our study suggest that the readily biodegradable fraction is degraded first. Indeed, visual observations show that quickly co-extruded material contains only the PLA. This is consistent with the results of Scandola et al. [Scandola et al., 1998], who studied the degradation of Mater-Bi in liquid medium. The compost has showed that after 45 days, the remaining fraction of the material contains only the PCL. In addition, the mineralization rate of the material obtained in this study is 75%, which is similar to our results.

In conclusion, the fraction of non-degraded material can be considered only when its recovery is possible. This shows the important of combining techniques to characterize the material to carbon balances, when the material is with difficulty degraded, particularly to clarify the degradation mechanisms brought into play.

4.5 Mechanisms of biodegradation

The results show that when conditions, especially temperature (ISO/CEN standard) permit it, the two polymers can be degraded together. However, they can also be, as in liquid medium according to ASTM standard, degraded by a fractionated way (Figure 23).

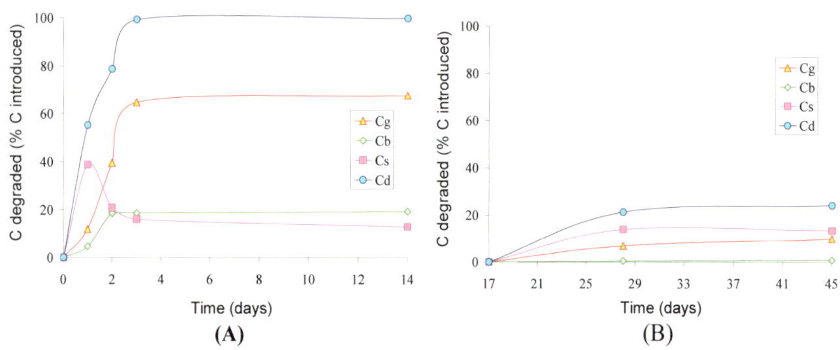

Fig. 23. Carbon balance from starch degraded (A) and PLA degraded (B) during the study of degradation in liquid medium of co-extruded starch/PLA (ASTM D 5209-92 standard)

This behavior is similar to that observed by Mayer et al. [Mayer et al., 1995] who studied the biodegradation in soil and compost of a material combining cellulose acetate from degree of substitution 2.5 and propylene glycol. Indeed, the authors had to extend the incubation to detect weight loss on cellulose acetate in the blend. This fractionated degradation is confirmed by the results obtained for the by-products of degradation (Figure 21). Indeed it appears that the first by-products are related to the degradation of starch, followed by degradation of lactic acid.

This work has also highlighted the influence of starch on the degradation of PLA. Indeed, the percentages of biodegradation were between 82 and 93% for the co-extruded introduced in the form of pieces in liquid and vermiculite media, while the starch is only 80% of the material. Thus, the approach to make compostable materials based on starch by co-extrusion worth pursuing because it would (i) to develop a way to value non-food starch which is a renewable resource abundant and cheap (ii) to limit the fraction of less degradable material by a polymer also derived from renewable resources (PLA), but more expensive than a conventional polymer material (PS, PE, PVC, PET, etc ...) and (iii) to obtain a compostable material for an attractive price.

5. Conclusions

During the last few decades, the studies regarding thermoplastic based on starch have continued due to the severe problems created by the plastic wastages to the surroundings. Up to day, the biodegradable polymers produced are costly and the properties are not competitive with the polymers derived from petroleum. Hence more attention should be given to produce polymers from the agricultural products and from nature. Even though starch can be used for the preparation of thermoplastic, its hydrophilic nature and poor mechanical properties restrict its usage in various applications. Therefore more effective methods, to increase the properties of the starch thermoplastic by physical or chemical method should be explored. This can result in an increase in the usage of these materials in various applications where life span time and mechanical properties are not too high. Hence the current researches are focused on the modification of starch and also to blend the starch with other thermoplastic polymers to have superior properties. Biodegradation studies indicate that the successful formulation of polymers based on starch can alleviate the problems created by synthetic polymers. Moreover the results from the laboratory scale should be industrialized to have more benefits for the human being as well as to maintain a green environment.

6. References

Agassant J. F., Avenas P., Sergent J. P., Vergnes B., Vincent M., "La mise en forme des matières plastiques", Tec et Doc, Lavoisier, Paris., 613, 1996.

American Standard ASTM D-5209-92, Standard test method for determining the aerobic biodegradation of plastic materials in the presence of municipal sewage sludge. In: Annual book of ASTM standards, Vol. 8.03., Philadelphia, Pa, pp. 372-375, 1994.

American Standard ASTM D-5338-92, Standard test method for determining the aerobic biodegradation of plastic materials under controlled composting conditions. In Annual book of ASTM standards, American Society for Testing and Materials, Philadelphia, Pa, Vol 8.03, pp. 439-443, 1994.

Arrighi V., Higgins J. S., Burgess A. N., Floudas G., Local dynamics of poly(dimethyl siloxane) in the presence of reinforcing filler particles., Polymer, 39, 6369, 1998.

Barron C., Della Valle G., Colonna P., Vergnes B., Energy balance of low hydrated starches transition under shear, J. Food. Sci, 67, 1426, 2002.

Bastioli C., Cerutti A., Guanella I., Romano G. C. and Tosin M., Physical state and biodegradation behavior of starch-polycaprolactone systems. J. Environ. Polym. Degrad. 3(2), 81-95, 1995 (a).

Bastioli C., Degli Innocenti F., Guanella I. and Romano G. C., Compostable films of Mater-Bi Z grades. J. Macromol. Sci.-Pure Appl. Chem. A32(4), 839-842, 1995 (b).

Bellia G., Tosin M., Floridi G. and Degli-Innocenti F., Activated vermiculite, a solid bed for testing biodegradability under composting conditions. Polym. Degrad. Stab. 66, 65-79, 1999.

Broido A., A simple sensitive graphical method of treating thermogravimetric analysis data. J. Polym. Sci. Part A2: Polym. Phys., 7, 1761-1773, 1969.

Buléon A., Colonna P., Planchot V., Ball S., Starch granules: structure and biosynthesis, Biol. Macromol., 23, 85, 1998.

Buléon A., Duprat F., Booy F. P., Chanzy H., Single crystals of amylose with a low degree of polymerization, Carbohydr. Polym, 4, 161, 1984.

Charles, A. L., Kao, H.-M., & Huang, T.-C., Physical investigations of surface membrane-water relationship of intact and gelatinized wheat-starch systems. Carbohydrate Research, 338, 2403-2408, 2003.

Chen L., Imam, S. H., Gordon S. H. and Greene R. V., Starch-polyvinyl alcohol crosslinked film-performance and biodegradation. J. Environ. Polym. Degrad. 5(2), 111-117, 1997.

European Standard 13 432, Requirements for packaging recoverable in the form of composting and biodegradation. Test scheme and evaluation criteria for the final acceptance of packaging, 2000.

Fan J., Mitchell J. R., Blanshard J. M. V., The effect of sugars on the extrusion of maize grits: II. Starch conversion, Journal of food Science and Technology, 31, 67, 1996.

Fanta, G. F., Felker, F. C., and Shogren, R. L., Formation of crystalline aggregates in slowly-cooled starch solutions prepared by steam jet cooking. Carbohydr. Polym., 48, 161-170, 2002.

Forssell P. M.,Hulleman S. H. D., Myllarinen P. J., Moates G. K., Parker R., Ageing of rubbery thermoplastic barley and oat starches, Carbohydr. Polym., 39, 43, 1999.

Garcia M. A., Martino M. N., and Zaritzky N. E., Edible starch films and coatings characterization: Scanning electron microscopy, water vapor transmission and gas permeabilities. Scanning., 21, 348, 1999.

Gattin R., Poulet C., Copinet A. and Couturier Y., Comparison of mineralization of starch in liquid, inert solid and compost media according to ASTM and CEN norms for the composting of packaging materials, Biotechnol. Lett., 22, 1471, 2000.

Gaudin S., Lourdin D., Le Botlan D., Ilari J. L., Colonna P., Plasticisation and mobility in starch-sorbitol films, Journal of Cereal Science, 29, 273, 1999.

Gilmore D. F., Antoun S., Lenz R. W., Goodwin S., Austin R. and Fuller C., The fate of biodegradable plastics in municipal leaf compost. J. Ind. Microbiol., 10, 199, 1992.

Gontard N., Guilbert S., Cuq J. L., Water and glycerol as plasticizers affects mechanical and water vapor barrier properties of an edible wheat gluten film, J. Food. Sci., 58, 206, 1993.

Hinkle M. E., Zobel H. F., X-ray diffraction of oriented amylose fibers. III. The structure of amylose-n-butanol complexes, Biopolymers, 6, 1119, 1968.

Hulleman S. H. D., Janssen F.H.P., Feil H., The role of water during plasticization of native starches, Polymer, 39, 2043. 1998.

International Standard ISO/CEN 14852, Plastics. Evaluation of the ultimate aerobic biodegradability of plastic materials in an aqueous medium. Method by analysis of released carbon dioxide, 1998.

International Standard ISO/CEN 14855, Plastics. Evaluation of the ultimate aerobic biodegradability and disintegration of plastic materials under controlled composting conditions. Method by analysis of released carbon dioxide, 1998.

Katz, J. R., and Van Itallie, T. B. Z., The physical chemistry of starch and bread making. All varieties of starch have similar retrogradation spectra., Physik. Chem., A150, 90, 1930.

Krogars, K., Heinämäki, J., Karjalainen, M., Rantanen, J., Luukkonen, P., and Yliruusi, J., Development and characterization of aqueous amylose-rich maize starch dispersion for film formation., Eur. J. Pharm. Biopharm., 56, 215, 2003.

Le Bail P., Bizot H., and Buléon A., 'B' to 'A' type phase transition in short amylose chains., Carbohydr. Polym., 21, 99, 1993.

Le Bail P., Bizo H., Pontoir B., Buléon, A., Polymorphic transitions of amylose-ethanol crystalline complexes induced by moisture exchanges, Starch/Starke, 47, 229, 1995.

Leblanc N., Saiah R., Beucher E., Gattin R., Castandet M., Saiter J-M., Structural investigation and thermal stability of new extruded wheat flour based polymeric materials, Carbohydrate. Polymer., 73, 548, 2008.

Lourdin D., Bizot H., Colonna P., Antiplasticization in starch-glycerol films, J. Appl. Polym. Sci., 63, 1047, 1997.

Lourdin D., Della Valle G., Colonna P., Influence of amylose content on starch films and foams, Carbohydr. Polym., 27, 261, 1995.

Mali S., Grossmann M. V. E., Garcia M. A., Martino M. M., Zaritzky N. E., Microstructural characterization of yam starch films., Carbohydr. Polym., 50, 379, 2002.

Mali S., Grossmann M. V. E., Garcia M. A., Martino M. M., Zaritzky N. E., Effects of controlled storage on thermal, mechanical and barrier properties of plasticized films from different starch sources., J. Food Eng., 75, 453, 2006.

Martin O., Avérous L., Della Valle G., In-line determination of plasticized wheat starch viscoelastic behavior: impact of processing, Carbohydr. Polym., 53, 169, 2003.

Mayer J. M., Elion G. R., Buchanan C. M., Sullivan B. K., Pratt S. D. and Kaplan D. L., Biodegradable blends of cellulose acetate and starch: production and properties. Plast. Eng. 29, 183, 1995.

Murphy V. G., Zaslow B., French A. D., The structure of V amylose dehydrate: A combined X-ray and stereochemical approach, Biopolymers, 14, 1487, 1975.

Pagga U., Beimborn D. B., Yamamoto M., Biodegradability and compostability of polymers-test methods and criteria for evaluation, J. Environ. Polym. Degrad., 4, 173, 1996.

Park E. H., George E. R., Muldoon M. A., Flammino A., Thermoplastic starch blends with poly(vinyl alcohol):processability, physical properties, and biodegradability. Polym. News, 19, 230, 1994.

Paris M., Bizot H., Emery J., Buzaré J. Y., Buléon A., NMR local range investigations in amorphous starchy substrates I. Structural heterogeneity probed by 13C CP-MAS NMR, Biological Macromolecules, 29, 127, 2001.

Poutanen K., Forssell P., Modification of starch properties with plasticizers, TRIP, 4, 128, 1996.

Rappenecker G., Zugenmaier P., Detailed refinement of the crystal structure of V_h-amylose, Carbohydr. Res., 89, 11, 1981.

Saiah R., Propriétés physiques et comportement dans le temps des agro-matériaux vitreux issus de la farine de blé, Ph.D. Thesis., Rouen, France, 2007.

Saiah R., Sreekumar P.A., Leblanc N., Castandet M., Saiter J-M., Study of wheat flour based agropolymers: Influence of plasticizers on the structure and ageing behaviour, Cereal Chemistry, 84, 276, 2007.

Saiah R., Sreekumar P.A., Gopalakrishnan P., Leblanc N., Gattin R., Saiter J-M., Fabrication and characterization of 100% green composite: Thermoplastic based on wheat flour reinforced by flax fibers, Polymer Composites, 30, 1595, 2009.

Saiah R., Sreekumar P.A., Leblanc N., Castandet M., Saiter J-M., Investigation of ageing and influence of lipids on the structure and thermal stability of thermoplastic based on wheat flour", "Industrial Crops and Products, 29, 241, 2009.

Saiter J. M., Dobircau L., Saiah R., Sreekumar P. A., Galandon A., Gattin R., Leblanc N., Adhikari R., Relaxation map of a 100% green thermoplastic film. Glass transition and fragility, Physica B, 405, 900, 2010.

Saiah R., Sreekumar P. A., Selvin Thomas P., Thermoplastic and Thermosetting Polymers and Composites, Chapter 2: Biofiber Reinforced Starch Composites, Nova Science Publishers Inc, ISBN: 978-1-61209-264-5, 37-84, 2011.

Scandola M., Finelli L., Sarti B., Mergaert J., Swings J. and Ruffieux K., Biodegradation of a starch containing thermoplastic in standardized test systems., J. Macromol. Sci.-Pure Appl. Chem., 35, 589, 1998.

Spitzer B., Menner M., Simulation of a compost environment in a mono-substrate fixed-bed system. DECHEMA Monogr., 133, 681, 1996.

Starnecker A., Menner M., Assessment of biodegradability of plastics under simulated composting conditions in a laboratory test system., Int. Biodeter. Biodegrad., 7, 85, 1996 (a).

Starnecker A., Menner M., Kinetics of aerobic microbial degradation of aliphatic polyesters. DECHEMA Monogr., 133, 221, 1996 (b).

Terrie C., Dobircau L., Gopalakrishnan P., Galandon A., Saiah R., Gattin R., Leblanc N., Saiter J-M., Biodegradable Materials from Agro-Based By-Products, Macromol. Symp., 290, 132-136, 2010.

Van der Zee M., Stoutjesdijk J.H., Feil H. and Feijen J., Relevance of aquatic tests for predicting degradation of polymeric materials during biological solid waste treatment. Chemosphere, 36, 461-473, 1998.

Van Soest J. J. G., De Wit D., Vliegenthart J. F. G., Mechanical properties of thermoplastic waxy maize starch, J. Appl. Polym. Sci., 61, 1927, 1996 (b).

Van Soest J. J. G., Benes K., De Wit D., Vliegenthart J. F. G., The influence of starch molecular mass on the properties of extruded thermoplastic starch, Polymer, 37, 3543, 1996 (c).

Van Soest J. J. G., Hulleman S. H. D., De Wit D., Vliegenthart J. F. G., Crystallinity in starch bioplastics. Indus. Crops Prod., 5, 11, 1996 (a).

Van Soest J. J. G., Hulleman S. H. D., De Wit D., Vliegenthart J. F. G., Changes in the mechanical properties of thermoplastic potato starch in relation with changes in B-type crystallinity, Carbohydr. Polym., 29, 225, 1996 (a).

Van Soest J. J. G., Knooren N., The influence of glycerol and water content on the structure and properties of extruded starch plastic, J. Appl. Polym. Sci., 64, 1411, 1997

Vergnes B., Della Valle G., Delamare L., A global computer software for polymer flows in corotating twin screw extruders, Polym Eng Sci, 38, 1781, 1998

Willett J. L., Doane W. M., Effect of moisture content on tensile properties of starch/poly(hydroxyester ether) composite materials. Polymer, 43, 4413-4420, 2002.

Winter W. T., Sarko A., Crystal Molecular Structure of the Amylose-DMSO Complex, Biopolymers, 13, 1461, 1974 (a).

Winter W. T., Sarko A., Crystal and Molecular Structure of V-Anhydrous Amylose, Biopolymers., 13, 1447, 1974 (b).

Wool R. P., Penasky J. S., Long J. M. and Goheen S. M., Degradation mechanisms in polyethylene-starch blends. In Degradable materials :Perspectives, Issues and Opportunities, Barenberg S. A., Brash J. L., Narayan R. and Redpath A. E. (eds), CRC Press, Boca Raton, FL pp 515-537, 1990.

Yang J. H., Yu J. G., Ma X. F., Preparation and properties of ethylenebisformamide plasticized potato starch (EPTPS), Carbohydr. Polym., 63, 18, 2006.

Zaslow B., Miller R. L., Hydration of "V" amylose helix, J. Am. Chem. Soc., 83, 4378, 1961.

Zaslow B., Murphy K. V. G., French A. D., The V amylose-H2O system: Structural changes resulting from hydration, Biopolymers, 13, 779, 1974.

5

Starch Protective Loose-Fill Foams

Marcin Mitrus
University of Life Sciences in Lublin
Poland

1. Introduction

Shredded newsprint and cardboard, popcorn, flour, starch and expanded polystyrene (EPS) are the most common materials used to protective packing products. Their function is to provide cushioning, protection, and stabilization of articles packaged for transport. EPS-based loose-fill foam products have enjoyed a steady growth in this application over the last decades, but became targeted recently in the solid waste disposal debate. Huge quantities of plastic is used for refuse and retail bags, egg cartons, packaging film, and loose fillers.

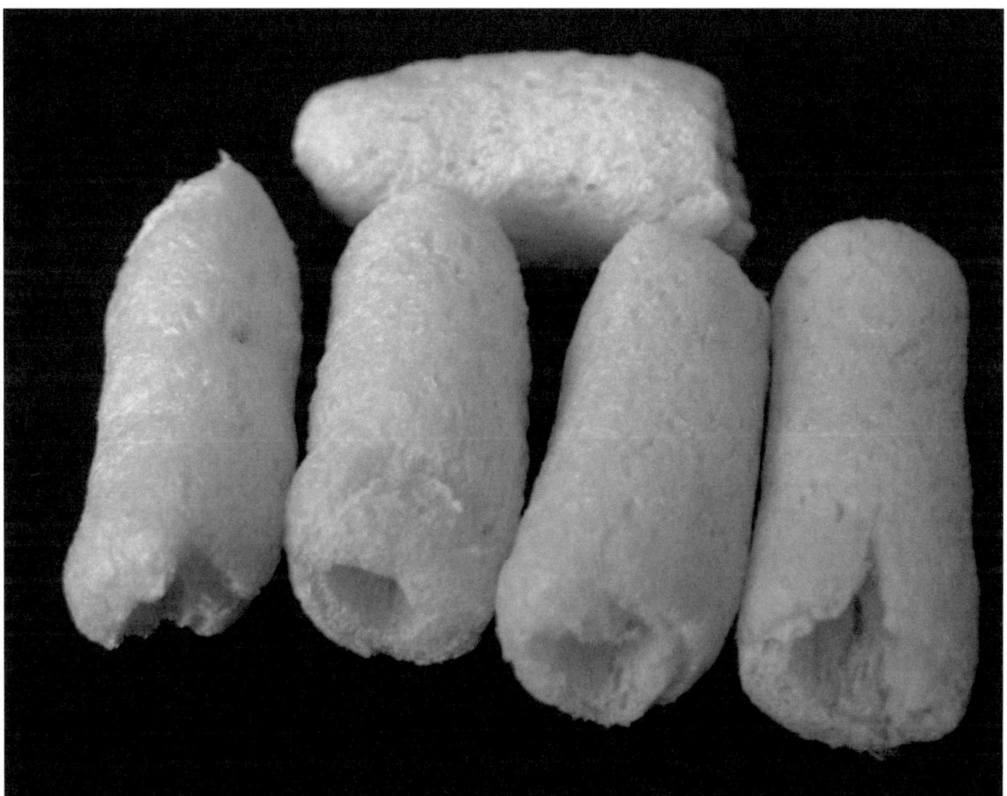

Fig. 1. Extruded starch foams.

As an alternative to very popular polymer foams such as expanded polystyrene (EPS), loose-fill (foamed chips for filling space around goods within a packing box) extruded from starch (fig. 1) is probably the most successful application of starch-based material in cushion packaging. Several patents on the extruded foams based on starch and blends of starch with various additives have been filed (Bastioli et al., 1998a, 1998b; Bellotti et al., 2000; Xu & Doane, 1998) and the material is commercially available. Considerable effort has been made to study the influence of extrusion conditions, moisture content and composition on the physical properties of starch-based foams (Bhatnagar & Hanna, 1995; Tatarka & Cunningham, 1998).

Extruded starch foams are generally water soluble, and their properties are sensitive to moisture content. Greatest expansion and lowest densities are generally achieved through the use of modified high amylase starches. Various synthetic polymers, like poly(vinyl alcohol) or polycaprolactone, have been blended with unmodified starches to produce foams with lower densities and increased water resistance.

2. Thermoplastic starch

Starch biodegrades to carbon dioxide and water in a relatively short time compared with most synthetic polymers. Considering some drawbacks of the existing technologies of biodegradable materials manufacture, in the recent years there have been started large-scale researches to increase amount of starch in starch-plastic composites to the highest possible level. The final objective of these investigations is to obtain commercial items for one-time use, produced from pure starch and to exclude synthetic polymers from the formulation. Thermoplastic starch (TPS) seems to be a perfect solution because it can be processed with conventional technologies used in synthetic plastic manufacture (extrusion, injection moulding) (Shogren et al., 1993; Wiedmann & Strobel, 1991).

To obtain thermoplastic starch, thermal and mechanical processing should disrupt semi crystalline starch granules. As the melting temperature of pure starch is substantially higher than its decomposition temperature there is a necessity to use plasticizers, for example water. Under the influence of temperature and shear forces, disruption of the natural crystalline structure of starch granules and polysaccharides form a continuous polymer phase is reported (Avérous et al., 2001; Nashed et al., 2003; Shogren et al., 1993; Van Soest et al., 1996).

TPS produced from starch plastified only with water becomes very brittle at room temperature. To increase material flexibility and improve processing other plasticizers are also used, e.g. glycerol, propylene glycol, glucose, sorbitol and others (Nashed et al., 2003; Van Soest et al., 1996).

To improve the mechanical properties of TPS based materials also other additives can be applied, like emulsifiers, cellulose, plant fibres, bark, kaolin, pectin and others (Avérous et al., 2001; Ge et al., 2000).

Thermoplastic starch can be obtained by the proper treatment with temperature and pressure in the presence of a plasticizer, like water, glycerol or sorbitol. The plasticizer penetrates into the starch granules and disrupts the initial crystallographic structure. Due to temperature and shear forces the material undergoes a melting process and forms a continuous amorphous mass that does not exhibit diffraction anymore. If the total thermal and mechanical energy provided to the starch is insufficient the product will show unmolten starch granules of clear crystallographic structure. Similarly, an insufficient amount of plasticizer may result in incomplete destruction of the crystallographic structure of starch (Souza & Andrade, 2002; Van Soest & Knooren, 1997).

One of the major drawbacks connected with starchy materials is their brittleness. This is related to a relatively high glass transition temperature T_g. This temperature marks the transition from a highly flexible state to a glassy one. T_g is considered the most important parameter for determining the mechanical properties of amorphous polymers and for the control of their crystallisation process (De Graaf et al., 2003).

The effect of starch plastification with water was repeatedly studied and various techniques for glass transition temperature were compared. The method of differential scanning calorimetry DSC is used most commonly, but the glass transition temperature found by DSC can be 10 - 30°C higher than the T_g value obtained by NMR (nuclear magnetic resonance) or DMTA (dynamic mechanical thermal analysis). The analysis of the influence of water on the T_g of amylose and amylopectin showed that the very branched amylopectin had a slightly lower glass transition temperature than the amylose. On the grounds of published researches and practical observation it can be stated that starchy material containing water is generally in the glassy state and therefore brittle under natural conditions (De Graaf et al., 2003; Moates et al., 2001; Myllarinen et al., 2002).

However, the results of measurements published by various authors are inconsistent to a high degree due to complex changes that occur in starch due to high temperatures and different measurement conditions.

Zeleznak & Hoseney (1997) found that the glass transition temperature of wheat starch with 13 - 18,7% moisture varies between 30 and 90°C and that T_g is likely to be lower than room temperature if the starch humidity increases above 20%. Van Soest et al. (1996) detected a T_g of 5°C for extruded potato starch with 14% moisture content, while at higher moisture T_g could not be determined. Shogren showed that the glass transition temperature for starch with 7 - 18% moisture ranged from 140 - 150°C (Shogren, 1993).

Myllarinen et al. (2002) indicated that T_g of amylose and amylopectin may equal the room temperature when the water content in a blend is 21%, however, at the same glycerol content it goes up to 93°C. This leads to the conclusion that glycerol is a less effective plasticizer than water. Moreover, on the basis of calculations they found out that a glycerol content as high as 35% is required to let T_g drop to the room temperature.

Also for other plasticizers, like sorbitol the T_g of TPS decreases at increasing concentrations. Yu et al. (1998) hold that TP maize starch with 10% moisture and 25-35% glycerol shows a T_g running from 83 - 71°C. Van Soest & Knooren (1997) proved that potato TPS with 11% moisture and 26% glycerol had a T_g = 40 °C, whereas for the materials of higher moisture and glycerol content it fell below 20°C. Lourdin et al. (1997) reported that potato starch of 13% moisture with 15% glycerol content had the T_g around 25°C, while at 25% glycerol the T_g dropped to around 0°C.

The TPS mechanical properties depend on the temperature of starch production, water content as well as quantity and type of added plasticizers and aid materials. The most considerable influence on the changes in mechanical properties proved to be the amount of plasticizer and aid materials.

The most common plasticizers like, glycerol, glycol or sorbitol possess the same hydroxyl groups as those appearing in starch, thus being compatible with starch macrogranules.

Increase of the plasticizer content brings about a decrease in tensile strength of thermoplastic starch, whereas the elongation at break increases. Starch is a natural polymer containing numerous hydrogen bonds between the hydroxyl radicals in its molecules, therefore it manifests substantial tensile strength values. Glycerol, sorbitol or glycol behave like diluents and decrease the interaction between molecules and consequently, they diminish tensile strength. At the same time they act as placticizers that improves macromolecular mobility and leads to a rise in elongation at break (Yu et al., 1998).

The increase of elongation at break at increasing plasticizer content occurs at some ranges of glycerol only. If this content surpasses 35%, a decrease of the elongation at break is noted. This effect is caused by too high percentage of glycerol, and therefore the molecular interactions are so weak that some interactions between starch molecules are replaced by the interactions between glycerol and starch molecules (Liu et al., 2001; Shogren, 1993; Yu et al., 1998).

An increase of water content in the blend induces an decrease of the tensile strength of the TPS and increase of the elongation at break. Yet, if water content exceeds 35% there is detected drop in the elongation at break of the TPS (Shogren, 1993).

Additon of filler materials like cellulose fibres, flax, kaolin or pectin increase the tensile strength but decrease the elongation at break. In turn, urea or boric acid addition improve the elongation at break but decreases the tensile strength (Fishman et al., 2000; You et al., 2003; Yu et al., 1998).

During storage of TPS some recrystallisation of amylose and amylopectin occurs. Together with a longer storage period, and consequently TPS crystallinity, tensile strength increases and elongation at break decreases. The increase of moisture content of the starchy materials storage conditions brings about intensification of their mechanical properties changes (Van Soest & Knooren, 1997).

On the basis of investigations made by Mitrus (2009) it was found out that with glycerol content increase in material mixture, a decrease of maximal stress formed at granulate compression was detected. Besides, blend moisture was also noted to affect on maximum stress generated in a compressed granulate. Along with mixture moisture growth the formed stresses are greater, however a faster drop of stress values was recorded together with glycerol increase. In the case of granulate produced from mixtures with glycerol content over 27% the stresses generated in granulate from more humid blends are lower than those are for more dry ones. The results emphasise that glycerol percentage growth is accompanied by lower tensile strength of thermoplastic starch. Although, water content increase may improve the strength of the obtained material, still it is most probable that some boundary value of total plasticizer content exists and when it has been surpassed the material becomes soften.

The excessive expansion and pore presence abates TPS tensile strength. Investigating of granulate from corn and wheat starch with 20% of glycerol content there were detected very low values of stress. The presence of pores had a significant impact on tensile strength weakening the obtained extrudate. In case of trials with 25% of glycerol the highest tensile strength values were recorded for the materials from corn starch, while the lowest for those containing wheat starch.

The investigations on water absorption by starch without a plasticizer revealed that amylopectin absorbs less water than amylose. The influence of a glycerol addition on water absorption is similar for both starch components (Myllarinen et al., 2002).

At low relative ambient moisture, i.e. below 50%, water content for both, amylopectin and amylose turned out to be lower in the blends with/without glycerol addition. This effect is probably connected with the replacement of strongly structurally immobilised water by glycerol. Similar behaviour was detected for starch plastifyied with sorbitol.

At the relative ambient moisture higher than 50% the mixtures with highest glycerol concentration showed the highest water content. When ambient moisture exceeded 70%, water content in the starch plastifyied by glycerol proved higher than in starch without glycerol.

Although both starch components behave alike at water absorption, in case of high ambient moisture amylose absorbs greater amount of water compared to amylopectin because the crystallisation reduces water absorption of hydrophilic polymers (Nashed et al., 2003, Myllarinen et al., 2002).

3. Methods of production

Starch offers a structural platform to manufacture sustainable, biodegradable foam packaging. As a basic material can be mixed with other polymers or with plasticizers and other additives. This blends can be produced with conventional technologies used in synthetic plastic manufacture.

Starch-based loose-fill can be manufactured, in a one-step process, via an extrusion cooking process. Granular starch and water are fed into an extruder, usually a twin screw, where heat and shear causes the starch to gelatinize. Water, released as steam at the die of the extruder, is the primary blowing agent. Complete expansion or density reduction takes place immediately after the product exits the extruder (Tatarka & Cunningham, 1998).

Application of extrusion-cooking technique to process starch-plasticizer mixtures is one of the most economical and efficient way to produce TPS loose-fill foams. The process conditions are to be very stable and strictly determined according to the expected quality of the extrudates. Also raw materials used (starchy components mixed with the plasticizers) have to be fitted properly. All together need many trials and detailed measurements to find optimal process conditions which can guarantee the best quality of the product. One of the most interesting question connected with extrusion-cooking is energy consumption. Specific Mechanical Energy (SME) consumption is defined as the amount of energy that is consumed per kilogram product. Knowledge of SME is not only important for design purposes, but it is also an indication of the mechanical forces on the material and consequently of degradation an viscous heating during the process. The specific mechanical energy for extrusion cooking of thermoplastic starch are in the order of $2{,}52 \cdot 10^5$ J/kg, which is equivalent to 0,07 kWh/kg. This rather low SME value depends on material composition of the mixtures, temperature and rotation speed (Mitrus & Moscicki, 2009).

Harper (1981) gave a detailed description on the mechanics of starch extrusion. The phenomenon of starch foaming involves the physicochemical properties of starch, which are modified during extrusion (Moscicki, 2011). The rheological properties of the starch plastic are in turn reliant on these physicochemical properties (Mercier et al.,1998), which affect the quality attributes of the foamed product. Extrusion process parameters, such temperature, screw speed, feed rate and moisture have direct influence on density, expansion ratio and other physical properties of extruded foams. Extrusion temperature depends on processed blend composition and ranged from 100°C to 180°C (Bhatnagar & Hanna, 1995; Cha et al., 2001; Shogren et al., 2002).

Moulded foam trays have been developed based on baking technology (Shogren et al., 2002) and are commercially available. Batter is foamed up and dried within heated moulds to foam thin-shelled containers similar to the process for making ice cream cones. The foam structure is featured by a highly porous centre sandwiched by much denser skin layers. The technology is somewhat limited by the slow processing rate necessary to dry off the moisture in the batter, which in turn restricts the maximum wall thickness of the foams.

Technologies for producing bulk starch forms have also been developed. Corrugated foam planks (Lye et al., 1998) made by extrusion foaming of modified cornstarch have been shown to have good cushion performance. The high foam density and cost of the materials, however, have somewhat restricted their widespread applications in packaging. Block foams have been made by combination of extrusion foaming and adhesion technology (Wang et al., 2001). The foams are of lower density and made from low cost wheat flours. When combined with other materials to form lightweight sandwich composites, mechanical properties and resistance to water attack can be drastically enhanced (Song, 2005).

Moulded starch block foams are highly desirable in order to provide biodegradable counterparts to moulded polymeric foams. Recently, a microwave foaming process has been described for making moulded starch foams from extruded pellets (Zhou et al., 2006). This involves converting starch-based raw materials into pellets by extrusion processing and foaming the extruded pellets by microwave heating. The microwaveable starch pellets are compact for transportation and storage and can be expanded using microwave when needed. They may be formulated to produce microwaveable snacks in food industry. In non-food applications, free-flowing foamed balls may be produced for loose-fill packaging. When the pallets are foamed in a mould, lightweight mouldings can be produced in forms of containers, end caps, edge or corner cushion pads for protective packaging, which are difficult to produce with extrusion foaming technology.

4. Properties of the loose-fill foams

Properties of the loose-fill foams highly depend on raw materials and process parameters used in production. Addition of synthetic polymers, plasticisers or other additives, different temperatures of processing are causing changes in products properties.

4.1 Cell structure

Open cells in foams occur if at least part of one wall is missing, creating an opening onto adjacent cells. Tatarka and Cunningham (1998) compared properties of two expanded polystyrene (EPS) based foams (PELASPAN PAC and FLO-PAK S) and six commercially available starch-based foams (CLEAN GREEN, ENVIROFIL, ECO-FOAM, FLO-PACK BIO 8, RENATURE and STAR-KORE), manufactured in extrusion cooking process. They reported that all starch-based foams have higher open-cell content than either EPS-based foam. Considering the manufacturing process used, it is not surprising that starch-based foams have more open cells. The expansion is attributable to the escape of water as steam during the extrusion process, resulting between 96 and 99% open cells. Steam can easily rupture the cell walls because thermoplastic starches have poor melt strength. After exposure to high humidities and temperatures, most foams exhibited a statistically significant, but trivial increase, about 1,0%, in open-cell content. Commercial starch-based foams have an open cellular structure. This differs from patents that claim hydroxypropylated high amylose foams as having a closed-cell structure, but the method used to make this assessment was not disclosed.

Bhatnagar and Hanna (1995) tested commercial polystyrene (PS) loose-fills, commercial starch-based loose-fills and starch-based plastic loose-fills with addition of polystyrene and poly(methyl methacrylate) (PMMA). Tests showed that the PS loose-fills and the starch-based loose-fills had uniform cell size. The commercial starch foams and starch-based PS foams had cells larger than the commercial PS foams. The commercial starch foams also had fewer large-diameter cavities. These cavities can help to reduce the density, but also affect other functional characteristics of the foams.

Willet and Shogren (2002) reported that surface of starch-based foams have many small holes, suggesting that the starch outer wall burst during extrusion foaming. This reflects the low melt strength and elasticity of starch melts and is consistent with previous studies indicating that starch foams have open cells (Tatarka & Cunningham, 1998). Surface of starch foams containing poly(lactic acid) (PLA) or poly(hydroxyester ether) (PHEE) had fewer or no holes. This suggests that these foams have greater melt strength and resistance to bubble rupture. The average cell size is much larger for foams containing addition of polymer, reflecting the higher volume expansion of these foams than the starch-based foams.

According with Zhang and Sung (2007a) foams of PLA/starch were successfully prepared by using water as a blowing agent in the presence of talc, which acted as an effective nucleation agent. Water concentration, foaming temperature, nucleation agent concentration, screw speed, and die nozzle diameter were factors influencing foam forming, cell size, and cell distribution. Water was a good blowing agent for the PLA/starch system. The foam structure was dramatically affected by solid inorganic fillers that acted as nucleation agents. With 0,5% talc the expansion ratio of the foam was dramatically reduced by almost 50%. The expansion ratio was further reduced with talc concentration. At 3% talc, the foam expansion was 11,2, <25% of the expansion ratio of the foam without talc. In contrast, the bulk density remained the same and increased slightly at a talc concentration of 3%. With a reduction in the expansion ratio by talc, the foam cellular structure was also changed dramatically. Compared with foams without talc, the texture of the foam with talc became uniform and fine. Cell size distribution became narrower as talc content increased from 0,5 to 2,0%. Majority cell size (MCS) was also reduced. For example, the MCS was between 0,6 and 1,2 mm for the foam with 0,5% talc, whereas the MCS was about 0,4 mm with 2% talc. At 3% talc, the cell size becomes larger and distribution becomes broader again.

4.2 Foam unit density and bulk density

Foam unit density describes the density of an individual expanded loose-fill foam specimen. Bulk density is more complex than specific density. Bulk density takes into account not only material and foam densities, but also packing efficiency, which depends on size, shape, and uniformity of the loose fill. Packing efficiency describes how well the loose fill fills the voids among adjacent foam specimens and can be measured by the ratio of bulk density to foam density. If this ratio is equal to one, the efficiency is very high because no voids exits among adjacent foam specimens. A low packing ratio can be achieved from irregular shaped foams. A loose-fill product with low packing is most desirable because the end-user reduces material consumption and saves transport costs.

The foam density of starch-based products was much higher than EPS-based ones (Tatarka & Cunningham, 1998). These values ranged between 16,7 and 22,6 kg/m^3. These products are approximately two to three times more dense than EPS-based foams. This difference is attributable to the large difference in density between polystyrene and starch and a lower expansion factor. Dry, unmodified granular starch has a nominal density of 1500 kg/m^3. During the extrusion process, the starch density has been reduced by factors ranging between 60 and 90. Open cells created during expansion will prevent the foam from continuing the expansion.

All commercial starch-based foams have a significantly higher bulk density by a factor of two to three than EPS-based foams or commercial PS loose-fills (Bhatnagar & Hanna, 1995; Tatarka & Cunningham, 1998). Starch-based foams have bulk densities between 8,8 and 11,3 kg/m^3. Foam density of starch-based foams correlated well with bulk density. The correlation coefficient for this relationship is 0,97. The packing ratio of the starch-based foams is between 0,435 and 0,538. Irregular cylindrical shapes impart products with lower packing ratios than does uniform cylindrical or dual cylindrical shapes with two similar dimensions (Tatarka & Cunningham, 1998).

Generally, the bulk density of starch-based foams decreased as extrusion temperature increased, whereas it was not significantly changed as moisture content increased. This might be reasonably explained by changes in cell wall thickness and viscosity caused by extrusion temperature and the changes in the volume and weight of cells after absorbing sufficient water (Cha et al., 2001).

The microwave-foamed extruded pellets have lower densities, ranged between 114 kg/m^3 and 145 kg/m^3 and are significantly denser than commercial extruded starch-based loose-fills. This disparity is attributable to the difference in expansion ratio during the foaming process and the existence of a denser skin layer in the microwave-foamed pellets (Zhou et al., 2006). The formation of this denser skin layer is likely to have resulted from moisture loss from the surface of a pellet during heating and hence a reduced driving force for foaming in this region.

Investigations made by Willet and Shogren (2002) indicated that density of foam prepared from normal corn starch had average density 61,4 kg/m^3, from wheat starch 58,6 kg/m^3 and from potato starch 40,6 kg/m^3. Foam densities for commercial starch foams are approximately 20 kg/m^3. Addition of PLA or PHEE to starch yields foam densities comparable to commercial starch products.

Fang and Hanna (2001) investigated the effect of blending Z class Mater-Bi® (MBI) resin with corn starches on the properties of the starch-based foam materials. They demonstrated that waxy starch had better foaming capabilities than regular starch mainly due to their differences in amylopectin content. Foams made from the waxy starch had slightly lower unit density (83,4 kg/m^3) and lower bulk density (49,2 kg/m^3) than the foams made from regular starch (83,7 and 56,6 kg/m^3 respectively). However, the bulk densities were still much higher than the 8,9 kg/m^3 of commercial EPS foams. The unit density increased as MBI and moisture content increased, indicating less expansion of the extrudates at higher MBI and moisture content. The bulk density data also showed a similar trend as the unit density. The relationship between the unit and bulk densities and the MBI and moisture contents were relatively linear.

Bhatnagar and Hanna (1996) extruded five types of starch (normal corn, wheat, potato, rice and tapioca) with polystyrene (starch to polystyrene ratio 70:30) and addition of three additives: azodicarbonamide, magnesium silicate and polycarbonate. Unit density was dependent on source of starch and type of additive. Foams made from tapioca and corn starches had the unit densities ranging from 36 to 39 kg/m^3 and from 30 to 86 kg/m^3 respectively. Results indicate that corn starch performed better than the wheat starch. The autors reported unit density of 23 to 59 kg/m^3 for commercial starch or starch-based plastic foams, which compares favorably with tapioca and corn starch-based foams in their study. The unit densities of all the foams including commercial starch foams were much higher than commercial polystyrene foams (8,9 kg/m^3)

4.3 Compressive stress

Compressive stress is the maximum force required to compress the foam by 3 mm. High compressive stress implies foams resist compression. Tatarka and Cunningham (1998) reported that the compressive stress of commercial starch-based foams does not significantly differ from EPS-based foams. As-received FLO-PAK BIO-8, STAR-KORE, ECO-FOAM and ENVIROFIL have lower values between 0,0565 and 0,0853 MPa, whereas CLEAN GREEN and RENATURE have higher values of 0,0927 and 0,1051 MPa. Compressive stress of starch-based foams was generally insensitive to changes in relative humidity. At 20% relative humidity (r.h.) and 23°C, only the chemically modified starch-based foams, ECO- FOAM and STAR-KORE, significantly increased compressive stress by 17 and 28%, respectively. At 80% r.h., 23°C, FLO-PAK BIO-8, RENATURE, and CLEAN GREEN significantly decreased compression stress by 22 to 32%. The higher moisture content in these products was sufficient to lower their resistance to compression. Although the chemically modified STAR-KORE and ECO-FOAM and the unmodified ENVIROFIL absorbed 13 to 16 % water at this condition, compressive stress did not significantly change. Chemically modified starches produced foams with good resistance to compression over a broad humidity range (Tatarka & Cunningham, 1998).

Fang and Hanna (2001) found that compressibility of the waxy starch samples were significantly lower than those of the regular starch samples, indicating that the regular starch foams were more rigid than the waxy starch foams. Foam compressibility increased with increases in the MBI content, but compressibility increased more rapidly as the moisture content increased. High moisture content in the processed materials resulted in more rigid products, which was due to low expansion.

Lin et al. (1995) show direct influence of water activity index on a compressive strength of extruded starch foams. A maximum compressive strength occurred at water activity of 0,53 which is close to normal ambient conditions. Further increase in water activity levels result in a decrease in compressive strength. Water acts as a plasticizer, which lubricates the starch chains. As the water activity increased, the starch chain mobility is also increased. This chain softening results in lower resistance to compression and lowers compressive strength values. It was observed that the foams become more flexible at higher water activity levels, probably due to the plasticization of foam walls by water. At lower water activity levels the starch chain mobility is decreased. This chain stiffening might result in higher resistance to compression and cause a higher compressive strength values. Examination of the samples showed that samples contain tiny fissures or fractures and, with the lower water activity came the more fissures. The effect of fissure or fracture formation in compressive strength of extruded foams at lover water activity levels is more important than the effect of the chain mobility of starch in compressive strength. In other words, fissure or fracture formations controls the compressive strength. The maximum compressive strength at water activity level of 0,53. It seems to be caused by the combination effect of starch chain mobility and fissure or fracture formation.

During measurements a power-law relationship was observed between compressive strength and foam density (Lin et al., 1995, Willett & Shogren, 2002; Zhou et al., 2006). Denser foams tend to have thicker cell walls and hence resist deformation better than lower density foams with thinner cell walls. A strong correlation exists between foam density and compressive strength, regardless of the type of polymer blended with the corn starch.

According to Zhou et al. (2006), some of the mechanical properties such as compressive modulus of elasticity, compressive stress and deformation energy at 40% strain of foamed pellets made from the Superfine flour and purified wheat starch were close to that of EPS block. This suggests that by further optimizing the cell structure and flexibility of cell wall material, it is possible to produce foam by a microwave process so that its mechanical properties match that of EPS block. The starch loose-fill produced by extrusion foaming have very low density, and thus they are suitable for cavity filling in packaging for light weight goods under low compressive stress. While the microwave processed starch foams have relatively high density, they are more suitable for packaging under high compressive stress levels or for heavy goods.

Greater compressibility values correspond to a sample which is more difficult to compress, while easily compressible samples have lower values of compressibility. Researches of Bhatnagar and Hanna (1995) indicated, that the starch-based PMMA foam had the lowest value of compressibility, which was consistent with its low resiliency. This was expected because this product also had the largest cell size. The starch-based PS loose-fill had more compressibility than the commercial PS foam. They found that compressibility was significantly affected both by the type of starch and type of additive (Bhatnagar & Hanna, 1996). For foams made from corn and tapioca starches compressibility values were quite comparable to that of commercial polystyrene foams. Compressibility values of potato starch extrudates were dependent on type of additive, with magnesium silicate and polycarbonate giving comparatively better products.

4.4 Resiliency

Resiliency describes the ability of the foam to recover to its original form after deformation. Resiliency less than 100% implies that the polymer was strained beyond its elastic limit, for example, by cell wall rupture, which prevents the foam from recovering to its original state.

The resiliency of comercial starch-based foams with values between 69,5 and 71,2% are, as a group, about 10% lower on a relative basis than EPS-based foams (Tatarka & Cunningham, 1998). After conditioning, the resiliency of all starch-based foams were significantly lower, with values between 60 and 70%. Although starch-based foams absorbed 13 to 16% moisture after conditioning at 80% r.h. and 23°C, the 62 to 67% resiliency retained is sufficient for the product to function.

Fang and Hanna (2001) showed that the resiliency value of the regular starch samples was 89,5% compared to 86,6% for the waxy starch foams. The resiliency increased with increases in MBI and moisture content. The addition of Mater-Bi® resin improved the mechanical properties of the starch foams. However, moisture content had a more significant positive effect than the MBI content on resiliency. The achieved results ranged between 80 and 95% which indicated a good deformation resiliency.

According to Lin et al. (1995) extruded starch foams at water activity levels below 0,23 are brittle and easily crushed, disintegrating upon compression. It was evident that, in these conditions, they are not suitable for packaging use. The resiliency of foams increased with increasing water activity levels (0,33 - 0,53). This was probably due to the increase in starch chain mobility of extruded foams at the higher water content. Further increases in water activity (0,53 - 0,75) caused the foams to soften and thus decreased the resiliency.

Bhatnagar and Hanna (1995) reported that the resiliency of starch-based PS loose-fill is not different from the resiliency of PS loose-fill or starch loose-fill. The ideal elastic body will have a resiliency of 100%, whereas the foams had resiliency of the order of 80 to 90%. The resiliency of starch-based PMMA foam was significantly lower than the PS loose-fill. They found resiliency values of 97% and 96% for commercial polystyrene and commercial starch foams, respectively. They found that, resiliency was significantly affected by the type of starch but not by the type of additive (Bhatnagar & Hanna, 1996). For corn and tapioca starches, the resiliency was quite comparable to that of commercial polystyrene loose-fill. Potato starch extrudates had good resiliency with all the additives.

Nabar et al. (2006) investigated influence of polymer addition into starch on foam resiliency. The control starch foams provided a resiliency of 69,7%. The addition of poly(butylene adipate-co-terephthalate) (PBAT) improved the resiliency considerably, from 69,7% to 85,9% at a PBAT content of 7% of the starch used. Polycaprolactone and cellulose acetate helped increase the spring index up to ~78%, while poly(vinyl alcohol) and methylated pectin barely increased the resiliency to ~71-72%. When glyoxal was added as a crosslinker, it increased the rigidity of the starch foams and, thus, the resiliency of the foams decreased by ~3%.

Zhang and Sung (2007b) tested the properties of PLA/starch foams. The foam samples dried at 135°C for 2 h, then immediately undergoing mechanical compression testing, showed no recovery at all, and all inside structure fractured into pieces, regardless of starch/PLA ratios.

The foam samples conditioned at 50% relative humidity for one week at room temperature and then tested yielded up to 73% recovery after 1 h of force removal. The recovery was reduced with the reduction in starch content. Water can be a good plasticizer for starch, and PLA could reduce water diffusion into the starch phase. After immersing the foams in distilled water at room temperature for one week, however, foams with PLA showed instant recovery to their original shape after one minute of force removal regardless of PLA/starch ratio.

The addition of poly(hydroxyamino ether) (PHAE) to starch improved the foam resiliency considerably from 69,7% (only starch) to 93,5% at a PHAE content of 7% of the starch used (Nabar & Narayan, 2006).

4.5 Friability

Friability is a measure of the fragmentation of foam during handling. Fragmentation of loose-fill during handling and use is an important product quality concern among end-users. Tests made by Tatarka and Cunningham (1998) indicated that the friability of commercial starch-based foams ranged between 0,003 and 2,3%. Although these values are lower than EPS-based foams, they are not significantly different. After conditioning, the friability of these starch-based foams increased significantly when exposed to 80% r.h. and 23°C 50% r.h. and 35°C. Quantitatively, starch- and EPS-based foams fragmented similarly at 2 to 6 % of the total weight, but starch-based foams broke down into a fine dust, whereas virgin EPS-based foams broke into large fragments.

Willet and Shogren (2002) founded that friability of starch/polymer foams is high for all tested formulations and starches at 10% r.h. Of the control starch foams, only the high amylose starch has friabilityof less than 95%. Under these conditions, the starch matrix is well below its glass transition temperature and its brittle. As the relative humidity increases to 50%, the equilibrium moisture content of the starch rises, and the foams are consequently less brittle. The friability decreases significantly, but only the high amylose starch foam displays negligible friability (0,4%). The addition of polymer generally reduces friability at 50% r.h. Corn starch loss-fills with the greatest density also exhibited greater friability, while corn starch foams with low densities had insignificant levels of friability. Friability decreases as polymer surface concentration increases. The presence of a ductile polymer on the foam surface may retard the formation of cracks and fragments under impact by the wooden blocks. One cannot, however, rule out indirect effects of the polymer on foam structure, i.e. lower foam density and thinne cell walls as contributing to the greater flexibility of the foam structure.

5. Conclusions

Although the use of starch in loose-fill products gives advantages in the form of biodegradability and environmental protection, these products have been criticized for their imperfection compared with EPS loose-fill products. EPS- and starch-based foams have differences, but the differences do not compromise performance.

These products differ with respect to composition and method of manufacture. Foam and bulk densities, which are higher by a factor of two to three times than either EPS-based foams, are attributable to the density of starch, which is 50% higher than polystyrene homopolymer and to the direct water-to-steam expansion process, which creates a predominately open cellular structure that stops foam expansion. Starch-based foam loose-fill is very hygroscopic. Foam density of starch-based products is significantly increasing between 10 and 30% after conditioning at high humidity.

The compressive stress of most starch-based foams does not differ significantly from EPS products. Chemically modified starches gives foams with good retention of compressive stress over a broad humidity range.

The resiliency of starch-based foams with values between 69,5 and 71,2% are, as a group, about 10% lower on a relative basis than EPS foams. Although starch-based foams absorbs 13 to 16 wt % moisture after conditioning at 80% r.h. and 23°C, these products are retaining between 62 and a 67% resiliency.

Both starch- and EPS-based foam fragmentation amounts to 2 to 6 wt %, but starch-based breaks down into a fine dust, whereas EPS-based foams breaks into large fragments.

All starch-based foams have a significantly higher foam and bulk density and open cell and moisture content than EPS-based foam. Both product types have similar compressive stress, resiliency, and friability. Starch-based foams are more sensitive to changes in relative humidity and temperature than EPS-based foam, but the higher amount of absorbed moisture does not compromise its mechanical integrity.

Generally, extrusion technique can be successfully employed for starch-based foams production. The physical properties of loose-fills, such as density, porosity, cell structure, water absorption characteristics and mechanical properties are highly dependent on the raw materials and additives. Mechanical behaviour of foamed pellets can be adjusted effectively by controlling the cell structure through using different additives. At room temperature and 50% relative humidity, some mechanical properties, such as compressive strength or compressive modulus of elasticity are comparable to commercial EPS foams.

Starch-based foams can be prepared from different starch sources replacing 70% polystyrene with biopolymer starch. Functional starch-based plastic foams can be prepared from different starch sources depending on their availability.

Starch-based foams with polymer addition (for example: PS, PMMA, PHEE) exerts improved properties in comparison with 100% starch foams. The addition of polymers significantly increases radial expansion and gives low density foams. Compressive strength is depending primarily on foam density, and not on starch type or polymer structure. Friability is reduced when polymer is present in the foam.

Foams of PLA/starch can be successfully prepares by using water as a blowing agent in the presence of talc, which acts as an effective nucleation agent. Water is a good blowing agent for the PLA/starch system. Talc at 2% gives the PLA/starch foam fine foam cell size and uniform cell size distribution.

The addition of Mater-Bi® is affecting the foam expansion characteristic. High levels of MBI are resulting in low radial expansions and high densities. The resiliency is improving as the levels of MBI and moisture contents are increasing. The MBI-starch foams have the potential to be used as an environmentally friendly loose-fill packaging material.

6. References

Avérous, L.; Fringant, C. & Moro, L. (2001). Starch - based biodegradable materials suitable for thermoforming packaging. *Starch*, Vol.53, No.8, (August 2001), pp.368-371, ISSN 1521-379X

Bastioli, C.; Bellotti, V.; Del Tredici, G.; Montino, A. & Ponti, R. (1998). Biodegradable foamed plastic materials. US Patent 5,736,586

Bastioli, C.; Bellotti, V.; Del Tredici, G. & Rallis, A. (1998). Biodegradable foamed articles and process for preparation thereof. US Patent 5,801,207

Bellotti, V.; Bastioli, C.; Rallis, A. & Del Tredici, G. (2000). Expanded articles of biodegradable plastic material and a process for the preparation thereof. Europe Patent EP0989158

Bhatnagar, S. & Hanna, M. A. (1995). Properties of extruded starch-based plastic foam. *Industrial Crops and Products*, Vol.4, No.2, (February 1995), pp. 71–77, ISSN 0926-6690

Bhatnagar, S. & Hanna, M. A. (1996). Starch-Based Plastic Foams From Various Starch Sources. *Cereal Chemistry*, Vol.73, No.5, (May 1996), pp. 601-604, ISSN 0009-0352

Cha, J. Y.; Chung, D. S.; Seib, P. A.; Flores, R. A. & Hanna, M. A. (2001). Physical properties of starch-based foams as affected by extrusion temperature and moisture content. *Industrial Crops and Products*, Vol.14, No.1, (October 2000), pp. 23–30, ISSN 0926-6690

De Graaf, R.A.; Karman, A.P. & Janssen, L.P.B.M. (2003). Material properties and glass transition temperatures of different thermoplastic starches after extrusion processing. *Starch*, Vol.55, No.2, (December 2002), pp. 80-86, ISSN 1521-379X

Fang, Q. & Hanna, M.A. (2001). Characteristics of biodegradable Mater-Bi®-starch based foams as affected by ingredient formulations. *Industrial Crops and Products*, Vol.13, No.3, (August 2000), pp. 219-227, ISSN 0926-6690

Fishman, M.L.; Coffin, D.R.; Konstance, R.P. & Onwulata C.I. (2000). Extrusion of pectin/starch blends plasticized with glycerol. *Carbohydrate Polymers*, Vol.41, No.4, (July 1999), pp. 317-325, ISSN 0144-8617

Ge, J.; Zhong, W.; Guo, Z.; Li, W. & Sakai, K. (2000). Biodegradable polyurethane materials from bark and starch. I. Highly resilient foams. *Journal of Applied Polymer Science*, Vol.77, No.12, (June 2000), pp. 2575-2580, ISSN 1097-4628

Harper, J. M. (1981). *Extrusion of foods*, CRC Press Inc., ISBN 0849352037, Boca Ration, FL., USA

Lin, Y.; Huff, H.E.; Parsons, M.H.; Iannotti, E. & Hsieh, F. (1995). Mechanical properties of extruded high amylase starch for loose-fill packaging material. *Lebensmittel-Wissenschaft und-Technologie*, Vol.28, No.2, (September 1994), pp. 163-168, ISSN 0023-6438

Liu, Z.Q.; Yi, X.S. & Feng, Y. (2001). Effects of glycerin and glycerol monostearate on performance of thermoplastic starch. *Journal of Materials Science*, Vol.36, No.7, (July 2000), pp. 1809-1815, ISSN 0022-2461

Lourdin, D.; Bizot, H. & Colonna, P. (1997). "Antiplasticization" in starch - glycerol films?. *Journal of Applied Polymer Science*, Vol.63, No.8, (April 1996), pp. 1047-1053, ISSN 1097-4628

Lye, S. W.; Lee, S. G. & Chew, B. H. (1999). Characterisation of biodegradable materials for protective packaging. *Plastics Rubber and Composites Processing and Applications*, Vol.27, No.8., (May 1999), pp. 336–383, ISSN 09598111

Mercier, C.; Linko, P. & Harper, J. M. (1998). *Extrusion Cooking* (2nd edition). AACCH Inc., ISBN 0-913250-67-8, St. Paul, Minnesota, USA

Mitrus, M. (2009). TPS and Its Nature. In: *Thermoplastic Starch. A Green Material for Various Industries*, Janssen L.P.B.M., Moscicki L. (Eds.), 77-104, Wiley-VCH Verlag GmbH&Co. KGaA, ISBN 978-3-527-32528-3, Weinheim, Germany

Mitrus, M. & Moscicki, L. (2009). Extrusion-Cooking of TPS. In: *Thermoplastic Starch. A Green Material for Various Industries*, Janssen L.P.B.M., Moscicki L. (Eds.), 77-104, Wiley-VCH Verlag GmbH&Co. KGaA, ISBN 978-3-527-32528-3, Weinheim, Germany

Moscicki, L. (2011). *Extrusion-Cooking Techniques. Applications, Theory and Sustainability*. Wiley-VCH Verlag GmbH&Co. KGaA, ISBN 978-3-527-32888-8, Weinheim, Germany

Myllärinen, P.; Partanen, R.; Seppälä, J. & Forssell P. (2002). Effect of glycerol on behaviour of amylose and amylopectin films. *Carbohydrate Polymers*, Vol.50, No.4, (February 2002), pp. 355-361, ISSN 0144-8617

Nabar, Y.; Narayan, R. & Schindler, M. (2006). Twin-Screw Extrusion Production and Characterization of Starch Foam Products for Use in Cushioning and Insulation Applications. *Polymer Engineering and Science*, Vol.46, No.4, (February 2006), pp. 438–451, ISSN 1548-2634

Nabar, Y.U.; Draybuck, D. & Narayan, R. (2006). Physicomechanical and Hydrophobic Properties of Starch Foams Extruded with Different Biodegradable Polymers. *Journal of Applied Polymer Science*, Vol.102, No.1, (July 2006), pp. 58–68, ISSN 1097-4628

Nashed, G.; Rutgers, R.P.G. & Sopade, P.A. (2003). The plasticisation effect of glycerol and water on the gelatinisation of wheat starch. *Starch*, Vol.55, No.3-4, (March 2003), pp. 131-137, ISSN 1521-379X

Shogren, R.L. (1993). Effects of moisture and various plasticizers on the mechanical properties of extruded starch. In: *Biodegradable polymers and packaging*, Ching Ch., Kaplan D.L., Thomas E.L., (Eds.), 141-149, Technomic Publishing Company Inc., ISBN 1566760089, Lancaster-Basel, USA-Switzeland

Shogren, R.L.; Fanta, G.F. & Doane, W.M. (1993). Development of starch based plastics - a reexamination of selected polymer systems in historical perspective. *Starch*, Vol.45, No.8, (May 1993), pp. 276-280, ISSN 1521-379X

Shogren, R. L.; Lawton, J. W. & Tiefenbacher, K. F. (2002). Baked starch foams, starch modifications and additives improve process parameters, structure and properties. *Industrial Crops and Products*, Vol.16, No.1, (January 2002), pp. 69–79, ISSN 0926-6690

Song, J. H. (2005). Lightweight sandwich panels based on starch foam. *Proceedings of CIMNFC LINK seminar*, Warwick University, Warwick, 19th April 2005

Souza, R.C.R. & Andrade, C.T. (2002). Investigation of the gelatinization and extrusion processes of corn starch. *Advances in Polymer Technology*, Vol.21, No.1, (October 2001), pp.17-24, ISSN 1098-2329

Tatarka, P. D. & Cunningham, R. L. (1998). Properties of Protective Loose-Fill Foams. *Journal of Applied Polymer Science*, Vol.67, No. 7, (February 1998), pp. 1157–1176, ISSN 1097-4628

Van Soest, J.J.G. Benes, K. & de Wit, D. (1996). The influence of starch molecular mass on the properties of extruded thermoplastic starch. *Polymer*, Vol.37, No.16, (December 1995), pp. 3543-3552, ISSN 0032-3861

Van Soest, J.J.G.; Bezemer, R.C.; de Wit, D. & Vliegenthart J.F.G. (1996). Influence of glycerol on melting of potato starch. *Industrial Crops and Products*, Vol.5, No.1, (September 1995), pp. 1-9, ISSN 0926-6690

Van Soest, J.J.G. & Knooren, N. (1997). Influence of glycerol and water content on the structure and properties of extruded starch plastic sheets during aging. *Journal of Applied Polymer Science*, Vol.64, No.7, (November 1996), pp. 1411-1422, ISSN 1097-4628

Wang, B.; Song, J. H. & Kang, Y. G. (2001). Modelling of the mechanical behaviour of biodegradable foams—From physical fundamentals to applications. In: *Engineering plasticity and impact dynamics*, L. C. Zhang, (Ed.), 117–134, World Scientific, ISBN 981-02-4803-2, Singapore

Wiedmann, W. & Strobel, E. (1991). Compounding of thermoplastic starch with twin-screw extruders. *Starch*, Vol.43, No.4, (February 1991), pp. 138-145, ISSN 1521-379X

Willett, J. L. & Shogren, R. L. (2002). Processing and properties of extruded starch/polymer foams. *Polymer*, Vol.43, No.22, (June 2002), pp. 5935–5947, ISSN 0032-3861

Xu, W. & Doane, W. M. (1998). Biodegradable polyester and natural polymer compositions and expanded articles therefrom. US Patent 5,854,345

You, X.; Li, L.; Gao, J.; Yu, J. & Zhao, Z. (2003). Biodegradable extruded starch blends. *Journal of Applied Polymer Science*, Vol.88, No.3, (April 2002), pp. 627-635, ISSN 1097-4628

Yu, J.; Chen, S.; Gao, J.; Zheng, H.; Zhang, J. & Lin, T. (1998). A study on the properties of starch/glycerine blend. *Starch*, Vol.50, No.6, (December 1998), pp. 246-250, ISSN 1521-379X

Zeleznak, H.F. & Hoseney, R.C. (1987). The glass transition in starch, *Cereal Chemistry*, Vol.64, pp. 121-124, ISSN 0009-0352

Zhang, J.-F. & Sun, X. (2007a). Biodegradable Foams of Poly(lactic acid)/Starch. I. Extrusion Condition and Cellular Size Distribution, *Journal of Applied Polymer Science*, Vol.106, No.2, (June 2007), pp. 857–862, ISSN 1097-4628

Zhang, J.-F. & Sun, X. (2007b). Biodegradable Foams of Poly(lactic acid)/Starch. II. Cellular Structure and Water Resistance, *Journal of Applied Polymer Science*, Vol.106, No.5, (August 2007), pp. 3058-3062, ISSN 1097-4628

Zhou, J.; Song, J. & Parker, R. (2006). Structure and properties of starch-based foams prepared by microwave heating from extruded pellets. *Carbohydrate Polymers*, Vol.63, No.4, (November 2005), pp. 466–475, ISSN 0144-8617

6

Thermoplastic Starch

Robert Shanks and Ing Kong
Applied Sciences, RMIT University, Melbourne
Australia

1. Introduction

Thermoplastics are polymers that can flow when heated above a melting or vitrification temperature. They undergo plastic deformation, meaning viscous flow with often-complex rheology due to their large molar mass, entanglements, interactions and chain branches. Starch is a natural polymer with complex levels of structure that impinge upon thermoplastic deformation. Natural polymers are no different to synthetic polymers once some added levels of structural complexity are understood (Wunderlich, 2011). Starch is a semi-crystalline polymer that does not melt in the traditional sense to form a liquid. Starch melting does mean loss of crystallinity due to disruption of hydrogen bonds, however melting occurs in the presence of a moderate (10-30 % w/w) water content. Starch crystals contain about 9-10 % w/w of bound water, where bound water means water that does not freeze when cooled below 0 °C. Additional water is required for melting of starch at convenient temperatures below the boiling temperature of water and the degradation temperature of the starch.

Starches are applicable to thermoplastic processing, unlike other polysaccharides such as cellulose and various gums. Cellulose is a natural structural polymer designed for regularity of packing, chain stiffening and strong cohesion via hydrogen bonding. Cellulose is a structural polymer forming cell walls in plants and it cannot be melted at moderate temperatures when only water is present. Cellulose has interesting solution properties of practical importance since solution processing is the method for transforming cellulose into fibres and films. Starch is an energy storage polymer designed for reversible chain propagation and reversible structure formation to allow access to enzymes for its energy forming degradation. Starch is a biopolymer that is biodegradable, suitable for green packaging materials providing that it can be formulated and readily processed into usable forms. Traditional processing equipment such as extrusion and thermoforming can be used with adaption for the specific characteristics of starch.

Thermal processing of starch has been reviewed as to changes in microstructure, phase transitions and rheology as a consequence of processing technique, conditions and formulations (Liu, Xie, Yu, Chen, Li, 2009). Starch has a unique microstructure that contributes multiphase transitions during thermal processing that provides an illustrative model to demonstrate conceptual approaches to understanding structure-processing-property relationships in all polymers. A unique characteristic of starches is their thermal processing properties that are considerably more complex than those of conventional polymers, because multiple chemical and physical reactions take place during processing. Examples of phenomena taking place during processing are: water diffusion, granular expansion, gelatinization, decomposition, melting and crystallization. Of the phase transitions, gelatinization is of most importance because it is the means of conversion of starch to a thermoplastic. The starch decomposition temperature is higher than its pre-gelatinization melting temperature. Conventional processing techniques such as extrusion, injection molding, compression molding, thermoforming and reactive extrusion, have been adapted for processing thermoplastic starch.

Starch is a versatile biopolymer obtained from renewable plant resources such as maize, wheat and potato harvests. Starch consists of two component polymers, amylose (AM) and amylopectin (AP). Amylose is the linear polysaccharide, poly(α-1,4-glucopyronosyl). Amylopectin is poly(α-1,4-glucopyronosyl)) with many a -1,6-glucopyronosyl branches. The molar mass of AM is large, $>10^6$ g/mol, while AP is $>10^7$ g/mol. There are various crystal forms of starch, due to double helix formation of linear regions of AP. The crystalline and amorphous regions assemble in layered formations to ultimately constitute the starch granules. Starch is economically competitive with polymers derived from petroleum for manufacture of packaging materials. Starch based materials are biodegradable.

Thermoplastic starch (TPS) has attracted much attention due to its thermoplastic-like processability with temperature and shear, though the structures being disrupted are more complex than those of synthetic thermoplastics. However, TPS is no different to any other polymer with respect to linear and branched structures, molar mass, glass transition temperature, plasticiser modification, crystallinity and melting temperature. Starch is a stereo-regular polymer with chirality, directionality of the chains, branching and a high density of hydrogen bonding. Polymerisation of glucose into starch not only builds the primary chains, but forms the secondary regular packing of chains into crystals and the clustering of crystals into tertiary cell structures.

The aim of this review is to provide a framework for the transition from native starch, from its primary molecular structure, secondary structure and tertiary granules to thermoplastic starch with its properties that parallel and contrast with synthesis thermoplastics. Objectives will be to interpret chemical structure changes in forming thermoplastic starch and additives that will assist with native structure dissociation, processing and the properties of materials formed from starch compositions.

2. Structure and chemistry of starch

The structure of starch considered from a top down approach begins with visual features and mines toward chemical bonding: starch consists of granules, and the granules have cell walls separating them. Within each cell are crystal bundles interspersed with amorphous starch, lipids and waxes and then the individual crystalline regions (Shogren, 2009). Several crystal structures are formed depending on the source of the starch: A (cereal starch), B (tuber starch), C (a combination of A and B crystals), V (retrograded starch) type. Starches consist of amylose and amylopectin at the molecular level in helical coils. Figure 1 shows a schematic of various levels of starch morphology and chemical structures for amylopectin, with amylose amorphous regions omitted. The primary structure of starch is the linking of glucose units into a continuous chain of amylose and with branches in amylopectin.

Adjacent amylopectin branches form a double helical secondary structure that is the basis of crystallinity in starch granules. The double helical structures then associate into a tertiary structure of a superhelix of the secondary double helical configuration. Further superstructures form combined with inter-crystalline amylose and associated lipids. Figure 2 shows a starch tetramer to illustrate a stereochemical view of the primary starch structure, representative of an amylose starch chain without branching.

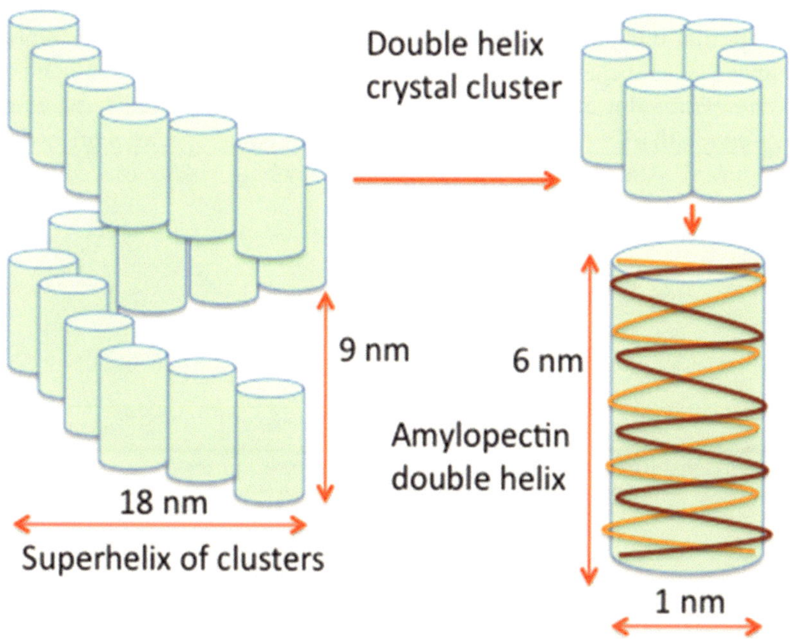

Fig. 1. Diagram of starch amylopectin superstructures

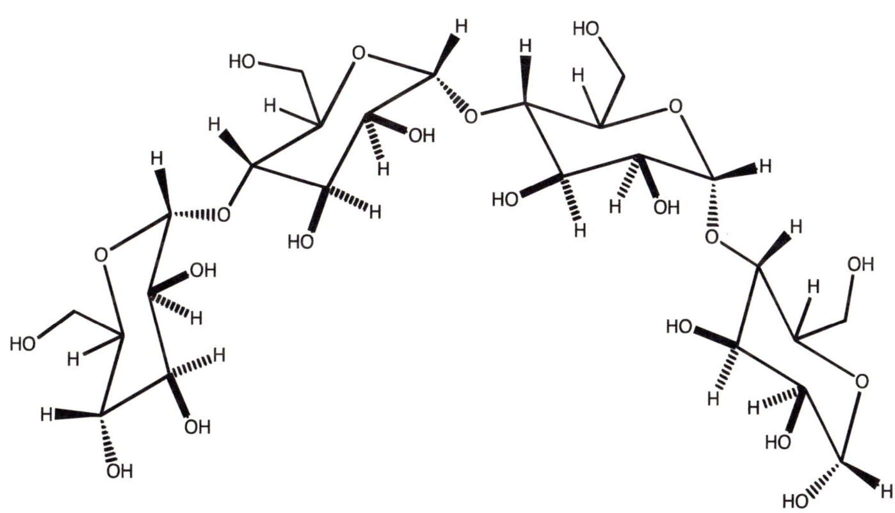

Fig. 2. Chemical structure of a starch tetramer showing conformation of glucose and stereochemistry of hydroxyl substituents and the α-1,4-glycosidic linkages

The glucose monomer units in starch have all hydroxyl groups in the plane of the six-membered ring. Asymmetry is introduced by the out-of-plane alpha-1-link to carbon 4 of an adjacent glucose. The asymmetry of the alpha-1,4-link causes starch to form a helical structure in contrast to the planar beta-1,4-link of cellulose that makes a planar cellulose structure. Biomolecules are characterised by the stereochemistry of their monomers that translates through into the morphology of their secondary and tertiary structures. Thus starch structures are all controlled from a bottom-up approach. Amylopectin is the same as amylose except that it contains alpha-1,6-links at chain branches as well as alpha-1,4 links. The chain links and branches are directional with only one chain end in a hemi-acetal form that is the end hydroxyl group is attached to a C1. All other chain ends, including those of the branches end with a C4 hydroxyl group. An idealized amylopectin branched molecule is shown in Figure 3. After each branch point the branching chains can coil together as a double helix.

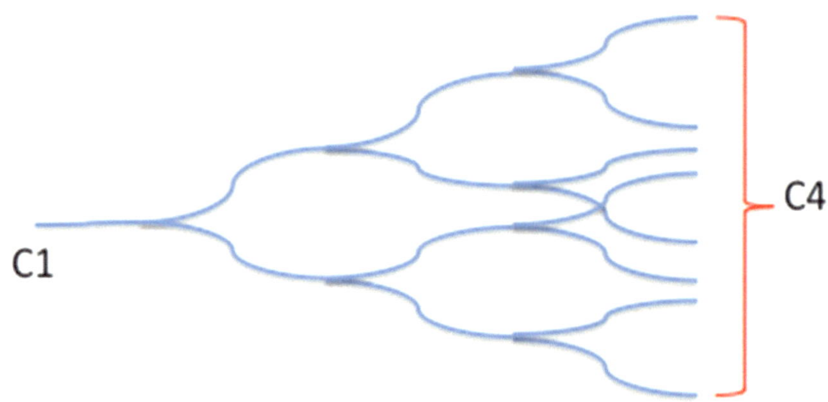

Fig. 3. Schematic of amylopectin with one C1 hemi-acetal end group

In contrast to synthetic thermoplastics such as polyethylenes, it is the branched starch molecules, amylopectin, that crystallise as double helices in the native starch granules. With polyethylene the linear chain or linear chain segments crystallize and branches are excluded from the crystals. The linear amylose resides in amorphous regions between the crystals. Amylose is therefore more water-soluble because it can be extracted from amorphous regions. Similar to synthetic thermoplastics, high amylose starches have more suitable rheology for processing by extrusion. Linear molecules can flow better than branched molecules and amylose has a lower molar mass though it can be in the range of 1000 kg mol^{-1}, while amylopectin is many times larger. Starch is like a polar, hydrogen bonded form of polyethylene, with linear and branched (or low density) type molecular architectures. There are few synthetic polymers that can be considered as analogues of starch. Poly(vinyl alcohol) is a linear synthetic polymer with a hydroxyl group on each monomer residue. Of synthetic polymers poly(vinyl alcohol) is most like amylose, and it is one of the few synthetic polymers to be reasonably rapidly biodegradable. This may be why poly(vinyl alcohol) has often been used as a blending polymer for starch. Poly(vinyl alcohol) is crystallisable, however it does not have branches, at least not to the extent of amylopectin for which there is no synthetic polymer analogue. While in the starch granule it is amylopectin that crystallises, after formation of amorphous starch it is the linear amylose that crystallises. This implies that the linear polymer, amylose, forms the more kinetically and thermodynamically stable crystals.

3. Thermoplastic starch

Formation of soluble starch or thermoplastic starch requires disruption of starch granules and their supramolecular structures, dissociation of complexes with lipids and melting of crystals with the assistance of added water. Figure 4 shows an environmental scanning electron microscope (SEM) image of corn starch granules. Though there is bound water within starch that varies with ambient humidity, water is typically added. A water concentration of 25 % w/w will give a gelatinisation temperature with a range of 60-70 °C. Gelatinisation is assisted by shear. An extruder, either a twin-screw or single-screw, is suitable for continuous shear processing. Alternatively batch mixers with a wiping action can be used.

Fig. 4. Environmental SEM of corn starch granules

A starch solution can be formed by thoroughly mixing starch with cold water to form a uniform suspension. The suspension will be a milky colour with an relatively high viscosity. The suspension can then be heated without coagulation. Gradual heating produces a clear solution of increased viscosity. If the solution is stored for several days it will gradually become opalescent through to milky. Starch solutions are not stable because hydrogen bonds within and between starch molecules are more stable than the hydrogen bonds with water that keep the starch in solution. Formation of starch-starch hydrogen bonds accompany ordering of starch molecules into crystalline structures different from the original starch granules. The ordering can be considered as a lyophilic liquid crystalline behaviour. Excessive heating of a starch in water solution will cause the starch to flocculate. This will be experienced as a continuous increase in viscosity until gelation and then formation of a strengthening gel structure. This gelation is consistent with lower critical solution temperature (LCST) behaviour. LCST behaviour is observed with solutions/solubility of other highly hydrogen bonding polymers in water, such as poly(methacrylic acid).

Conversion of starch into a packaging material requires disruption of the starch granules and their constitutive crystals into a flowable thermoplastic. High AM content is preferred for preparation of thermoformable starch due to the better flow properties of the linear polymer and its decreased tendency to crystallise. High amylose corn starch is the major source of amylose-rich starch and it can have amylose contents of up to 80 %. Typically high AM starch can be processed in an extruder if it is first equilibrated with water and any other plasticisers, normally alcohols such as glycerol. Water must diffuse into the granules and hydrogen bond with the starch. Upon heating the granules undergo destructurisation and the crystals melt to form a uniform viscous starch fluid. Processing, strength and stability of the starch is assisted by addition of a hydrophilic polymer such as poly(vinyl alcohol), poly(oxyethylene), poly(caprolactam) or poly(vinylpyrrolidone). After formation of a starch film the added polymer stabilises the amorphous structure by hydrogen bonding with the starch. This prevents recrystallization of the starch to form V-type crystals in a process called retrogradation. Another way to limit retrogradation is to use substituted starch such as hydroxyethyl, hydroxypropyl or acetoxy starches. The substituents interrupt the structural regularity of the starch and thereby restrict crystallization.

The viscosity increases considerably as the molecules are freed from their ordered superstructures and the gelatinous mass becomes transparent. The gelatinised starch is amorphous and referred to as thermoplastic starch (TPS) since it can be reprocessed so long as water content is retained or replenished. The glass transition temperature of starch is uncertain, but likely to be about -50 to -10 °C with the water content suitable for gelatinisation and subsequent processing. TPS has properties expected of any thermoplastic though water content is required for them to be revealed. TPS has a glass transition temperature, it can flow under shear when heated and it can slowly crystallise. TPS sheets are transparent, though depending on the extent of shear mixing there can be traces of starch granules recognisable as faint outlines. These are called ghost granules and they disturb the clarity of sheets and films of TPS, which is a problem for some packaging applications.

Disruption of the ordered structures within starch is necessary to prepare thermoplastic starch. Heat is required along with shear to disrupt native starch structures to form a homogeneous thermoplastic starch that can flow or form into shapes. Water content, shear degradation by chain cleavage and thermal degradation by dehydration or bond rupture are limitations in the processing of TPS. Residual crystallinity and re-crystallisation formed by the retrogradation process causes loss of film clarity and embrittlement. After processing amylopectin enhances the properties of TPS and retards retrogradation in TPS. Retrogradation involves the more thermodynamically and kinetically favourable crystallisation of the linear amylose to form V-type crystals that are distinct from native starch crystals.

Gelatinisation of cassava and potato starches at low water levels (>20 % w/w) in the presence of sodium chloride caused a decrease in gelatinisation temperature (Farahnaky, Farhat, Mitchell, Hill, 2009). A range of moisture contents was studied and the decrease in gelatinisation temperatures was 30 °C at a water content of 11 % w/w for example.

Retrogradation of waxy corn starch has been studied using differential scanning calorimetry (DSC) and it was found that an endotherm developed after 2-5 h and increased with time (Liu, Yu, Tong, Chen, 2010). Two-phase transitions were identified; the Gr transition was due to helix-helix dissociation while the M1r transition was due to a helix-coil transition. This indicated reformation of mainly the amylopectin molecules due to an original structural memory. The retrogradation in this high amylopectin starch was due to reforming of the amylopectin double helix. The process was inhibited by increased amylose content. Cassava starch retrogradation has been studied using X-ray diffraction and DSC and it was found that with this starch the gelatinisation procedure has no significant effect on the melting enthalpy upon retrogradation, while the storage temperature was a factor (Rodriguez-Sandoval, Fernandez-Quintero, Cuvelier, Relkin, Bello-Perez, 2008).

Starch with about 15-25 % w/w water shows a gelatinisation peak in differential scanning calorimetry (DSC) at 60-70 °C. This endotherm is a double peak with this lower temperature peak being due to disruption of starch–lipid complexes and the high temperature peak being due to melting of starch crystals. The endotherm has a relatively small area, heat of fusion. The term melting is used to describe disruption of crystals in starch, though it does not mean melting in the sense of crystals becoming a flowable melted liquid. The gelatinised starch is a fluid that can flow under shear because of the presence of water, so it could be described as a concentrated starch–water solution.

There is uncertainty in the position of the glass transition temperature. When Tg is plotted against Tm for many synthetic polymers a trend is observed that $Tg \sim 2/3 \cdot Tm$ (temperatures in Kelvin). Therefore Tg of starch should be below 0 °C. Some careful DSC and dynamic mechanical analysis (DMA) measurements have shown Tg to be in a range -10 to 0°C. In this range Tg can be obscured or uncertain due to the large melting endotherm of free water in the processing of starch. The glass transition of waxy maize starch and water (1:2 w/w) has been studied using DSC after different degrees of gelatinisation (Sang, Alavi, Shi, 2009). A three-phase model with an amorphous phase, a rigid amorphous phase and a crystalline phase was proposed to interpret the data. Gelatinisation had an onset of 61.5 °C with a maximum of 70.3 °C and an end temperature of 81.7 °C. The Tg temperature and detection depended upon prior thermal history and gelatinisation, however a distinct subzero Tg of annealed native starch granules was observed after annealing at -7 °C.

3.1 Plasticisers

Water is the main plasticiser, however it is too transient to be used as the sole plasticiser for TPS. Other less volatile plasticisers are glycerol, pentaerythritol, polyols, sugar alcohols, poly(oxyethylene)s, poly(oxypropylene)s, non-ionic and anionic surfactants. Hydrogen bonding polar liquids are most suitable. Plasticisers with a high affinity for water, such as glycerol, can exhibit anti-plasticisation at particular concentrations, typically low concentrations depending on water content. Anti-plasticiser activity has been found, for different reasons in dioctyl phthalate plasticised poly(vinyl chloride). Anti-plasticisation may occur when the plasticiser has a higher affinity for water than does starch, thus the plasticiser-water coupling lowers the total plasticiser available to hydrogen bond with starch. This causes an increase in the *Tg*, gelatinisation temperature and increased brittleness.

Plasticiser activity has been related to water sorption isotherms that give molar sorption enthalpies of starch–plasticiser systems. Competitive plasticisation has been assessed using glass transition temperature models. The study validated the anti-plasticization limit for glycerol to be ~10–15 % w/w, however for xylitol, its anti-plasticization limit did not manifest until 20 % w/w (Liu, Chaudhary, Ingram, John, 2011). A new hydrogen bonding highly polar plasticizer, N,N-bis(2-hydroxyethyl)formamide, has been synthesized and used to prepare TPS from corn starch (Dai, Chang, Yu, Ma, 2008). Hydrogen bonding between the plasticizer and starch was confirmed using Fourier transform infrared (FTIR) spectroscopy and SEM showed granules to be completely disrupted after gelatinization with the new plasticizer. Comparison of the plasticizer with glycerol was made and at low relative humidity, modulus was higher while at high relative humidity extension at break was higher when the new plasticizer was included compared with glycerol.

Wheat starch plasticization in the water–glycerol combination was used to study the time-temperature regime for melt processing of TPS using excess plasticizers (Li, Sarazin, Favis, 2008). Gelatinization occurred between 1 min and 3 min with an excess of plasticizers with a rapid reduction of granule phase size.

Plasticisers for starch function the same as plasticiser in synthetic polymers. An additional function of starch plasticisers is that they form a complex via hydrogen bonding with the starch and prevent retrogradation, which will cause embrittlement due to gradual recystallization. Where the plasticiser is a polymer the correct term is an allomer, which then is a polymer blend as treated in the next section. Retrogradation is more severe in high amylose starch since it is the amylose that forms the V-type crystals. However high amylose starch is the most suitable for processing to form TPS. Additives are required to stabilise TPS against retrogradation to ensure reliable properties over time.

Fatty acids such as stearic acid are mostly hydrophobic, however the hydrophobic chain preferentially resides inside of starch helical coils forming a complex. A similar complex is formed with iodine giving a red to blue colour depending on the length of the coil and the number of iodine molecules aligned inside the coil. The interior of starch coils is more hydrophobic that the exterior due to the stereo-aligned pendant hydroxyl groups. The iodine complexes with glycans have been used to elucidate the structure of dispersed starch molecules, where iodine can bind with corn starch with only 8 % w/w water content. Complexes with iodine and different starches (corn and potato) of varying water content were compared and potato starch where amylose was more closely involved with crystal structures, was found to form more effective complexes based upon formation of a more intense colour (Saibene D, Zobel H F, Thompson D B, Seetharaman K, 2008). Layered films based on lipophilic starch and gelatin were formed with varying amounts of fatty acids (palmitic, lauric, myristic, capric, caproic and caprylic) and sorbitol plasticizer (Fakhouri, Fontes, Innocentini-Mei, Collares-Queiroz,, 2009). The addition of fatty acids decreased opacity and elongation while decreasing tensile strength and water vapor permeability.

Native corn starch was plasticized with water, glycerol and stearic acid and gelatinized in a twin-screw extruder (Pushpadass, Kumar, Jackson, Wehling, Dumais, Hanna, 2009).

Gelatinization temperature range depended on glycerol content and lipid-amylose complexes were formed depending on stearic acid and moisture content. Extrusion caused fragmentation of the starch that was detected by size exclusion chromatography and cross-polarisation magic angle spinning nuclear magnetic resonance spectroscopy.

3.2 Starch–Polymer blends

Substituted starch can be extrusion gelatinised in the presence of added hydrophilic polymer with water as the only plasticizer. Without addition of glycerol, or other alcohol, the gelation and plasticity of the starch is only dependent on water content. The subsequent equilibrium water content controls the flexibility, tensile strength and elongation. The added hydrophilic polymer contributes to the strength, and together with the substituted groups, restricts retrogradation. In addition to the chemical processes contributing to gelation, the extrusion conditions including shear, residence time and temperature physically provide gelation and homogenisation. Polymers that have been blended with TPU are poly(vinyl alcohol) pure or with some residual acetate, biopolyesters such as poly(lactic acid), poly(hydroxyl alkanoate)s; poly(butylene succinate) and poly(butylene adipate). Other polar thermoplastics, such as poly(N-vinylpyrrolidone) and poly(caprolactam), are suitable though biopolymers are preferred to maintain overall biodegradability.

Starch-poly(caprolactone) (PCL) blends have been produced by extrusion with an in-line rheometer slit die (Belard, Dole, Averous, 2009). A multilayer structure could be generated with a PCL outer skin that was formed depending on the molar mass of the PCL. The phase separation between starch and PCL showed that the polymers were not miscible in the blend, yet they were expected to be compatible. Starch-poly(lactic acid) (PLA) blends have been filled with nano-clay (Cloisite 30B) to form nano-composite blends (Lee, Hanna, 2009). The clay was found to be present as tactoids that influenced melting temperature, onset degradation temperature, radial expansion ratio, density, bulk compressibility and bulk spring index.

TPS requires that starch be a continuous phase and exhibit flow and processability. In some starch blends the starch is dispersed like filler in a continuous phase of a second polymer. This is the case with starch-polyethylene blends where starch is dispersed in the polyethylene, usually with compatabilisation via a maleic anhydride grafted polyethylene or use of a substituted starch. Such starch blends enhance biodegradablilty of the matrix polymer through degradation of the starch leaving a porous framework of synthetic polymer that then slowly degrades. Degradation is usually too slow to be classified as biodegradable according to international standard definition.

3.3 Composites

Composites of TPS are prepared for the same advantages as for synthetic thermoplastics. Typical mineral fillers include talc, clays, silica and cellulose fibres particularly microcrystalline cellulose and cellulose nano-fibres. The filler can be dispersed in starch during the gelatinisation extrusion process. It is advantageous if a twin-screw extruder with multiple inlet ports is used so that gelatinisation, plasticiser and filler addition can be separated processing steps within the one extrusion cycle.

Composites are used to reinforce, limit properties dependence on moisture or humidity, as well as restrain retrogradation. Composites tend to be more brittle, which is a problem for TPS sheets. Combinations of composites with plasticiser are used to arrive at optimum properties where inert fillers assist with reduction of problems associated with the starch phase.

A starch film with poly(vinyl alcohol) and nano-silica was prepared (Huali-Tang, Shangwen-Tang, Peng-Zou, 2009). The nano-silica was found to be evenly distributed and the films exhibited improved mechanical properties, water resistance and optical clarity when nano-silica was present. A nano-composite was prepared from starch and unmodified montmorillonite and it was found that water resistance and water vapour permeability decreased when only small amounts of the clay were added. (Maksimov, Lagzdins, Lilichenko, Plume, 2009). The permeability of water determined as a function of orientation of the platelet clay layers calculated by a method of orientation averaging of permeability characteristics of separated structural elements with planar orientation of lamellar filler particles. A comparison of nano-composites prepared from three modified montmorillonite clays (Cloisite Na+, Cloisite 30B and Cloisite 10A) was made under 60 % and 90 % relative humidity (Perez, Alvarez, Mondragun, Vazquez, 2008). The nano-composite matrix consisted of a blend of starch and PCL. Each of the nanocomposites showed decreased water diffusion coefficient considered to be due to a tortuous path for diffusion. Elongation at break increased though in general mechanical properties decreased upon exposure to water.

TPS materials blended with poly(vinyl alcohol) and reinforced with sodium montmorillonite were prepared and found to have highly exfoliated clay structures (Dean K M, Do M D, Petinakis E,Yu, 2008). PVAlc was found to be important for expanding inter-gallery spaces in the clay. Composites containing both PVAlc and clay were found to have superior increases in tensile modulus and strength, due to both enhanced interfacial interactions and retardation of retrogradation.

Layered clays like monmorillonite have been much studied in starch composites. As described above Cloisite clays are featured in the literature. Swelling of clays with water involves migration of the small water molecules into clay galleries where the water solvate intergallery sodium ions and the silicate anions on surfaces. Solvation causes the layer spacing to increase against the ionic attractions, with stabilisation due to the solvation, though with loss of entropy by the water molecules. Swelling occurs similarly with many polar organic molecules such as alcohols, amines, carbonyls and ethers. Swelling occurs with polar polymers such as poly(oxyethylene) (POE). POE ether groups solvate sodium ions and through hydroxy end-groups the surface silicate anions are solvated. Migration of the larger POE into galleries will be slower and solvation will be less strong than with water, but the loss of entropy is less for polymer adsorption onto a surface than it is with water. A special case of POE is crown ethers where the cyclic structure has a cavity surrounded by ether groups that is the correct size to accommodate a sodium ion. Ion-exchange of inter-gallery sodium ions with alkylammonium ions is an efficient way to retain the ionic balance while introducing hydrophobic groups into the galleries. The hydrophobic groups reduce the interlayer adhesion of the clay and when clay layers separate the surface energy of the layers is reduced to that of hydrocarbons. Rheological properties of clays are twofold. The first is due to absorption of water, or other polar molecules, from the liquid phase and binding them by solvation. Solvated clay is suspended as colloidal particles that flocculate to form high viscosity and under shear separate to form reduced viscosity. Clay particles are thus shear thinning and due to the time required for flocculation, they are thixotropic. A second mechanism is where the solvated clay layers exfoliate providing greatly increased surfaces for solvation and associated rheological behavior. Clay with inter-gallery sodium ions, or other alkali metals ions, can exfoliate under high shear force, though the resulting surface energy increase is unfavorable. Ion exchange with alkylammonium ions creates hydrophobic clay gallery surfaces that are exfoliated with lower shear and the lower surface energy of the layers provides a more stable system. After shearing has stopped, the layers can associate through ionic end groups or through hydrophobic interactions between surfaces. The alkylammonium treated clays are shear thinning and thixotropic agents, and can further be describes as hydrophobic thickening agents.

Clay is swelled in water and other polar media by intercalcation due to solvation of the gallery sodium ions and the anionic silicate surfaces. The binding of the clay layers by the gallery sodium ions is overcome by the solvation energy gained minus the entropy lost by the solvating molecules. Exfoliation of the clay increases surface energy extensively due to the huge surface area of the many exfoliated clay layers. Thin layers are an unstable state because their mass is nearly all surface. If the layer were fluid, they would contract into spherical beads. The clay layers are of sub-nanometre thickness making the surface to volume extremely high. This huge surface area must be reduced by solvation if the exfoliated state is to be thermodynamically supported. Shear can provide energy for exfoliation, but after the shearing to process the layers must be stabilized by solvation or recombination. Starch is hydrophilic and able to adsorb to the layers and assist with stabilization with less loss of entropy than required by many water molecules. Thus, exfoliated sodium montmorillonite is unstable even with polar molecules or macromolecules adsorbed onto the surfaces.

An alternative is where the gallery sodium ions have been exchanged by alkylammonium ions. Ionic neutrality is maintained but the clay layers are separated by alkyl chains and bound by dispersive forces from the alkyl chains. The shear force required to separate the layers is low and the exfoliated sheets have relatively low hydrocarbon surface energy. The surface energy required to be supported for stabilization of the clay layers is low and their tendency to recombine is low. Solvation by water or polar molecules can occur along the layer edges where polar silicates are exposed. The clay layers are analogous to layered surfactants with both polar and non-polar regions. Starch has hydrophobic as well as hydrophilic functionality. The starch hydrophobic groups can stabilize the low surface generated by exfoliation of treated clay with minimal loss of entropy compared with solvation by many small molecules. Creation of thin layers of any material requires an accompanying mechanism for reduction of surface energy with minimal entropy loss to the system. Reinforcement, diffusion reduction and rheological action of clays will be enhanced by hydrophobic modification, even in aqueous or polar environments.

All-starch nano-composites have been formed from cassava starch reinforced with waxy starch nano-crystals with glycerol plasticizer (Garcia, Ribba, Dufresne, Aranguren, Goyanes, 2009). A reinforcing effect was found in the composites, along with increased equilibrium water content. The reinforcement was interpreted as due to strong hydrogen bonding interactions between filler and matrix, and the nano-crystals were well dispersed, which provided decreased permeability. Similarly, nano-composites have been formed by reinforcing waxy starch films with starch nanocrystals, using sorbitol as plasticizer by formation from solution by casting and evaporation to form films (Viguie, Molina-Boisseau, Dufresne, 2007). Thermal and mechanical characterization was performed after conditioning in moist atmospheres.

Advanced nano-composites of TPS have been formed with multi-walled carbon nano-tubes (MWCNT) with glycerol plasticizer for potential application as electroactive polymers (Ma, Yu, Wang, 2008). The composites displayed restrained retrogradation, increased modulus and tensile strength, with decreased toughness. The composites were conductive though conductivity was sensitive to water content. The percolation threshold for conductivity occurred at a MWCNT content of 3.8 % w/w.

TPS biocomposites have been formed by blending with PLA and reinforcement with coir cellulose fibres with maleic anhydride as compatabiliser or coupling agent (Lovino, Zullo, Rao, Cassar, Gianfreda, 2008). The composites showed a high degree of biodegradability as determined by carbon dioxide production upon composting. SEM showed crazing patterns on the surface and the growth of bacteria on the surfaces was observed using optical microscopy. Hybrid composites of TPS blends with PVAlc plasticized with glycerol and reinforced with layered clay and jute fabric (Ray, Sengupta, Sengupta, Mohanty, Misra, 2007). The hybrid composites were prepared by a solution casting method. The combined filler of clay and jute are of vastly differing dimension scales. The mechanical properties were superior and fracture surfaces demonstrated strong bonding between the matrix and jute. The clay filled matrix did not display plastic deformation.

3.4 Stabilisers

Starch is readily biodegradable and a natural energy storage substance in plants. It is a nutrient source for microorganisms and this makes starch materials a target for various fungal and bacterial growths. Biodegradability may begin before the required end-of-life. Additives may be included to prevent premature biodegradation though the additive level must not prevent ultimate biodegradation, nor be toxic to packaging contents and hence users.

An important area of investigation is to develop additives that decrease the susceptibility of starch materials to moisture and cyclic humidity conditions. Stabilisation against retrogradation is important, though this is usually a function of the plasticisers and filler. Thermal degradation of starch occurs by loss of water. First free water is lost, followed by weakly bound water, then bound water and the TPS no longer has the plasticizing effect of the water. During degradation water formed from dehydration reactions of the starch is eliminated and the starch degrades forming a carbonaceous residue. Figure 5 shows degradation of starch studied using thermogravimetry with water evolved from the start of the scan until about 120 °C. A sharp degradation mass loss begins at about 300 °C under the scanning conditions of 20 °C min^{-1} under nitrogen with switch to air at 810 °C. After the rapid degradation, a broad temperature range degradation occurred from 350 °C as the residual organic species degraded to form a carbonaceous residue that was oxidized in air after 810 °C. Comparison is made with a starch–calcium carbonate composite that contained less water and degraded at higher temperature. The calcium carbonate decomposed to calcium oxide with elimination of carbon dioxide above 700 °C.

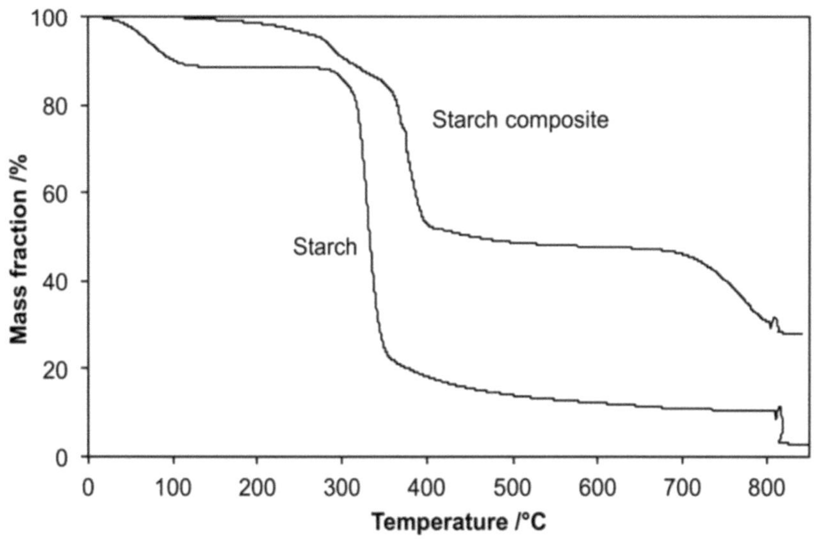

Fig. 5. Thermogravimetry of TPS and a TPS–calcium carbonate composite

3.5 Starch derivatives

Starch is a potentially great biomaterial resource due to its natural abundance and biodegradability. However, some serious problems also exist in starch-based materials, such as poor long-term stability caused by water absorption, high hydrophilicity, poor mechanical properties and poor processability. To improve the properties of starch-based materials, extensive studies have been focused on chemical modification of starch.

Many reactions have been used to derivatize starch due to available hydroxyl groups. The hydroxyl groups on starch are slightly more acidic than typical alcohol hydroxyl groups. This is why base catalysis is effective for most of the reactions summarized in this section. Carboxylic anhydrides, such as maleic anhydride are used for vinyl functionalization. Acyl halide via a Schotten-Bowmann reaction using base catalysis, or sulfonyl chloride. Reaction with aldehyde for acetal formation, such as crosslinking with glutaraldehyde. Vinylsulfone reaction via a Michael reaction is used for adding an ethyl sulfone derivative. Azinyl chloride (cyanuryl trichloride) can be used to link reactive dyes or with an alkyl group. Epoxy ring opening is used to form 2-hydroxypropyl starch or in general 2-hydroxylalkyl starch. Lactone ester ring opening such as caprolactone has been used to form polycaprolactone grafts. Oxazoline (ring opening) has been used to form a convenient linking group. Etherification is used to form methylated starch, such as with iodomethane or dimethyl sulfate. Carboxylation using chloroacetic acid has been used to form carboxymethyl starch. Isocyanate reaction has been to form a urethane link to starch, however isocyanate is toxic though it will be likely react to completion.

Maleated TPS has been used in reactive extrusion melt blending with poly(butylene adipate-*co*-terephthalate) (PBAT) for blown-film application (Raquez, Nabar, Narayan, Dubois, 2008). Maleated TPS was formed by reaction of maleic anhydride and corn starch with glycerol plasticizer in an extruder, followed by addition of PBAT further along the extruder screw to form the complete reaction compatabilitzed blend in a single step. Grafting was via transesterification via the maleic anhydride functionality of the TPS. Confirmation of successful grafting was obtained from soxhlet extraction and infrared spectroscopy. Some blends contained 70 % w/w PBAT therefore it is likely that PBAT was the continuous phase with dispersed TPS, though the interfacial area was high due to a fine phase morphology.

If a functional group reaction is to be used for reactive processing it must be capable of complete reaction within the residence time in the extruder and the by-products must be suited to remaining in the final product. Few organic functional groups can react fast without forming an equilibrium, which is why step growth polymerization is limited to few reactions. The more reactive substances tend to be toxic and not stable in water. The reactions summarized are suitable for batch reaction in solution.

Radical initiated derivatization of starch is another alternative. Ceric ion, hydrogen peroxide to form an alkoxy radical on starch can be used for reaction with an additive monomer. A thermal peroxide reaction such as with t-butyl peroxide is less selective for grafting than a redox system. The reactant needs not be a polymer, grafting can occur between starch and another polymer; if in an emulsion, the other polymer could be hydrophobic. Radical reactions are generally faster than functional group reactions, but residual monomer will remain unless oligomer or polymers are used as the graft. Crosslinking (gel formation) will be a problem and require a low radical concentration.

Sol-gel reactions using tetraethoxysilane (TEOS), or tetrabutyltitanate (TBuTi), provide in-situ formation of the corresponding oxides, silica and titanium dioxide, gives high dispersion and bonding with starch. Borate or boric acid precursors such as borate ester form complexes that can crosslink starch. Any residual sol compound will continue to react with water in the starch until complete conversion. These composites are likely to have interlinked chains making them difficult to process, though some may be thixotropic enabling both processing and subsequent strength development.

Complexation or adsorption of starch onto surfaces can be used to modify starch. For example, alumina with a positive zeta potential, or silica such as precipitated or fumed silica with a negative zeta potential, and surface-active clays or minerals that have hydrophilic edges and hydrophobic faces such as montmorillonite, kaolin and talc can strongly adsorb starch. No reaction is required; dispersion is required and while this may result in high viscosity or gelation; the fluid dispersion is likely to be shear thinning or thixotropic.

3.6 Oxidised starch

Starch is chemically modified in various ways with oxidation being a common process. Starch has been oxidised using hypochlorite resulting in an increase in carboxylic acid and carbonyl groups (Sangseethong, Lertphanich, Sriroth, 2009). Oxidation rate depended on the alkalinity of the reaction medium and this influenced the viscosity of the oxidised starch solution, decreased the gelatinisation temperature though retrogradation was slightly increased. The light transmission was less changed with oxidised starch. Banana starch was oxidised and acetylated, then the product was used to form TPS films (Baruk, Zamudio-Flores, Bautista-Baoos, Salgado-Delgado, Arturo, Bello-Perez, 2009). Oxidation increased solubility while acetylation decreased solubility. The oxidised starch showed a high modulus and lower elongation at break that was not significantly changed by acetylation, though acetylation reduced the barrier properties.

4. Processing

Shear for disruption of super-structures can take place in a batch internal mixer for small-scale preparations. Extrusion is more practical for pilot scale or commercial production. After formation of TPS sheets or films final shaping may be by thermoforming. Extrusion of starch is best performed in a twin-screw extruder where custom combinations of rheological elements can be assembled along the screw. Zone of high shear will assist with disruption of granules while uncoiling of molecules can take place in less shear intensive zones. Formulations often require inlets for plasticiser, filler or other additives along the extruder barrel. Escape of volatiles such as steam will be required, without loss of other materials. At the extruder die the TPS will emerge with a higher moisture content than the equilibrium moisture content of TPS sheet or film. A drying zone will be needed before the product TPS is wound into a coil for storage, transport or prior to further shaping.

Extrusion or high shear is required to disrupt the native starch structure and produce a uniform composition with other components. The extrudate must be a uniform continuous stream with rheology suitable for shaping. The process is more complex than extrusion of typical thermoplastics but the outcome is similar.

Starch foams can be produced using partial vaporisation of entrained water to form a cellular structure. The foams can be in the form of continuous extrudates or popcorn type granules that are used as protective inserts in packaging.

5. Properties

Starch crystal structures are characterised using wide-angle X-ray scattering (WAXS). A Kratzky powder camera is used on starch specimens that have been cryo-ground into powder. Figure 6 shows a WAXS pattern for waxy potato starch with diffraction peaks characteristic of B type crystals; camera type: powder, source 0.154 nm, power 4.0 kW, aperture 200 μm, range 5-35°, count time 10 s^{-1}, interval 0.05 data/°. Figure 7 shows a WAXS pattern for corn starch captured under the same conditions shown above. Figure 8 shows a WAXS pattern for TPS, that is gelatinised starch where the lower curve has been aged, compared with the upper curve, and small peaks indicating onset of retrogradation crystallisation are beginning to emerge.

The modulus of TPS compositions is typically high compared with synthetic thermoplastics. Elastic properties at low strain are measurable however TPS has low, moisture dependant, elongation at break. The high modulus of TPS has dependence upon moisture, other plasticisers, fillers and recrystallization. TPS has high high strength, that is break stress for brittle materials or yield stress for ductile materials, due to inter- and intra-molecular hydrogen bonding, and strong adhesion to blended polymers and fillers. Fracture of starch materials tend to be brittle and increasingly brittle with time due to moisture and recrystallization.

Dynamic mechanical properties (modulated force thermomechanometry (mf-TM)) is used for detection of damping peaks, elastic and loss modulus changes with temperature. Mf-TM was performed using a Perkin-Elmer Diamond DMA in tensile mode. A synthetic frequency consisting of 0.5, 1, 2, 10, 20 Hz was applied with constant amplitude of 10 μm. Fourier analysis deconvoluted the data into the five constituent frequencies. A tan(δ) curve is shown for each heating rate scan since these curves showing the best resolution. Examples of storage and loss curves are shown for heating at 10 °C min^{-1}. The specimen was coated with paraffin oil and a higher heating rate than normally used for mf-TM (typically 2 °C min^{-1} used) to minimise moisture loss during the scan. The damping curves in Figure 6 show a phase change at about 90 °C interpreted to be due to melting of crystals formed by retrogradation.

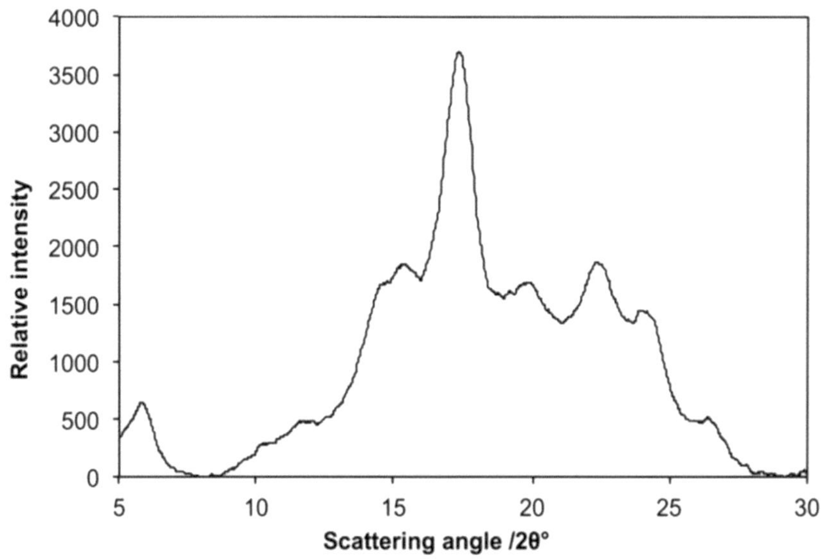

Fig. 6. WAXS pattern for waxy potato starch (B type crystals)

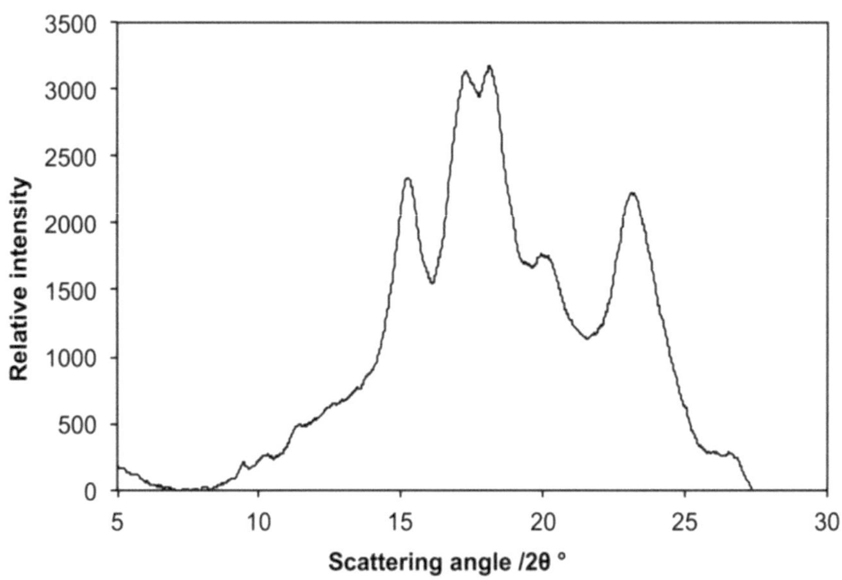

Fig. 7. WAXS pattern for corn starch (A type crystals)

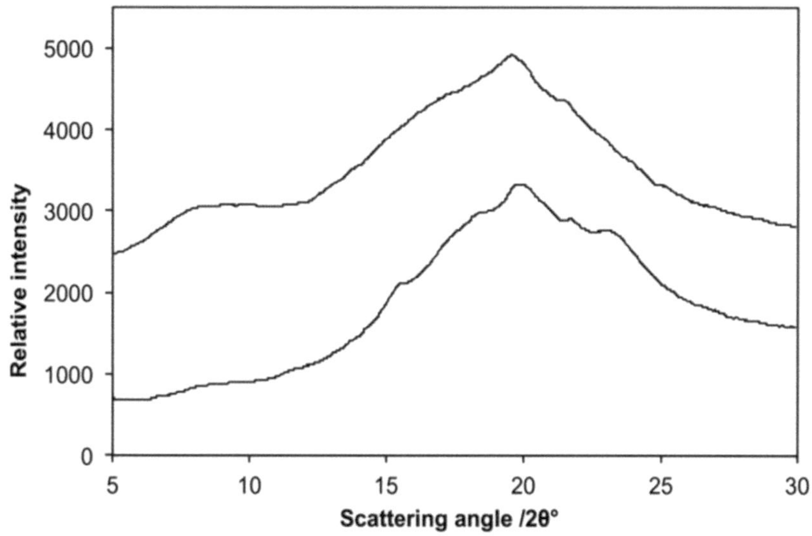

Fig. 8. WAXS patterns for TPS where the lower curve has been aged compared with the upper curve

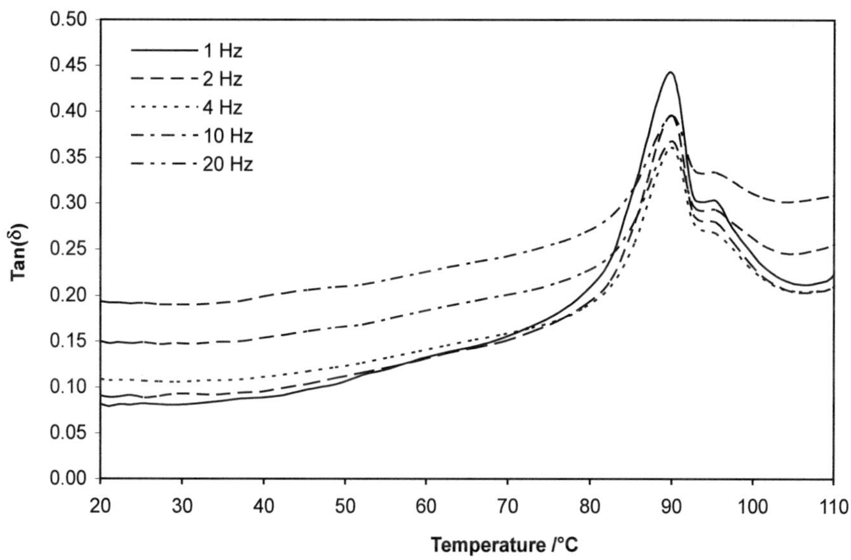

Fig. 9. Thermoplastic starch sheet, mf-TM tensile mode at 10 °C min^{-1}

TPS sheets are often brittle and mf-TM is performed better using single cantilever bend mode instead of tensile mode. Figure 10 shows such an analysis performed using a TA Instruments Q800 DMA with 0.2 % strain, 1 Hz frequency and heating at 2 °C min^{-1}. The specimen was coated with paraffin oil prior to clamping in the instrument to minimize moisture loss during the scan. The glass transition denoted by the peak of loss modulus or damping factor is at a higher temperature than that shown in Figure 9 since at the slower scan rate loss of moisture during the scan made the specimen more brittle with higher glass transition temperature. Scans were performed at 1, 2, 5 and 10 °C min-1 and 10 °C gave the best scan though with highest thermal lag. Thermomechanical analysis requires a compromise between moisture loss and thermal lag of the instrument.

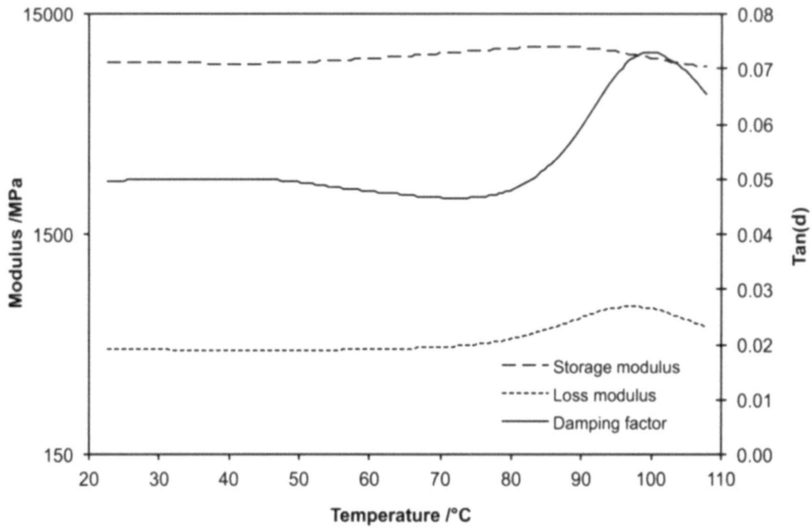

Fig. 10. Thermoplastic starch sheet, mf-TM single cantilever bend mode at 2 °C min-1

Dielectric properties of TPS are a sensitive to changes in structure due to the density of polar groups and water content. Thermally stimulated depolarization current (TSDC) has been used to study the dielectric relaxations in cassava starch on semi-crystalline and amorphous variations (Laredo, Newman, Bello, Muller, 2009). Three secondary relaxation modes were detected and interpreted as due to short-range orientations of polar groups or to main chain restricted motion. Moisture plasticization contributed to relaxation Vogel-Tammann-Fulcher parameters that confirmed cooperative relaxation. A heterogeneous amorphous phase resulted in a bimodal distribution of relaxation times.

Vapour transmission - oxygen, moisture, other volatiles, tends to be low compared with other polymers due to density of hydrogen bonds and polarity to restrain diffusion of small molecules. Vapour transmission is dependant upon moisture content and any crystal structures remaining or formed. Moisture sorption is a problem, however moisture does not readily pass through TPS sheets.

5.1 Property changes with time

Thermoplastic starch structure changes with time, temperature and humidity. These changes are the greatest limitation for adoption of TPS in commercial applications. TPS is hydrophilic and during ambient humidity changes the water content of the starch varies. There is a water content hysteresis depending upon whether the humidity is increasing or decreasing because desorption of hydrogen bonded water is delayed compared with adsorption of water. Moisture and other TPS components exchange with packaging contents, either water is absorbed into the TPS, softening it, while drying the package contents or water is absorbed into the package contents drying and embrittling the TPS package.

Similar to moisture content other plasticisers in TPS can bleed or become leached or extracted. Small molecule plasticisers such as glycerol are relatively involatile though they may migrate to the surface with water transfer or become extracted by adjacent materials such as packaging contents. Other plasticisers may separate depending upon changes in water content or due to their compatibility with TPS. Combinations of plasticisers create complex phase diagrams that may involve crystallinity and mutual attractions.

Humidity responsive starch-poly(methyl acrylate) films were formed as graft copolymers that were plasticized with water and urea (Willett, 2008). The films display significant shrinkage that was composition independent in relative humidities greater than 50 % and in relative humidities above 75 % shrinkage was correlated with the urea:starch weight ratio that determines the equilibrium moisture content of the films. A master curve was constructed by shifting shrinkage data with respect to a reference relative humidity, demonstrating that relaxation processes in the starch phase control film shrinkage that was accompanied by loss of orientation within the starch. While moisture sensitivity is to be avoided in typical TPS applications and it is an undesired property, it can be exploited in materials that are sensors to humidity when controlled shrinkages are designed into the structure.

A severe property change is retrogradation caused by re-crystallisation of amylopectin or crystallisation of amylose to for new V-type crystals. Retrogradation causes embrittlement of TPS and loss of optical clarity often accompanies this. Retrogradation is usually more rapid in high amylose starches, since the linear chains are most mobile to form single helix V-type crystals. Re-crystallisation of amylopectin occur by reforming double helical crystals by inter-coiling of adjacent chain branches.

Further changes in TPS with time can arise from mould growth due to presence of combined moisture and nutrients, however this form of change is specific to the next section.

6. Biodegradation

Starch is a nutrient for many organisms and since water is present in the final structure the starch is readily biodegraded. Water can easily be absorbed by starch resulting in disintegration of the material by partial solubility. Partially solubilised starch is even more readily biodegraded by enzymes principally from microorganisms. Starch–poly(vinyl alcohol) blends were plasticised with glycerol and reinforced with unripe coconut fibres to form fully biodegradable composites (Rosa, Chiou, Medeiros, Wood, Mattoso, Syed, Inam, 2009)). Degradation rate slowed when fibre content increased, though crystallinity was not changed in the composites. The blend and composites degraded slower in composite that the pure TPS.

7. Applications

Commercialisation of thermoplastic starch and starch-based polymers has been progressing. Due to the inherent biodegradability of starch materials they are allied mostly to packaging as films or sheet that can be subsequentially shaped by thermoforming into custom packaging requirements. Starch are most suited for packaging of dry products., otherwise there will be a diffusion transfer and equilibrium established between the TPS package and its contents. TPS foams are suited for damping impacts to protect fragile products. TPS solve the issue of disposal of packaging materials because they are biodegradable into environmentally friendly fragments, depending on the nature of non-starch components of TPS compositions.

8. Future directions

Improved processing techniques due to equipment design (screw element design, feed stages, steam release and plasticiser addition) to accommodate retention of water content. Plasticiser structures that are specific to starch and can be used with starch–filler composites, while limiting moisture susceptibility are required. Where plasticisers are not able to enhance properties, polymer blends may be more suited due to the large size of polymer molecules compared with plasticisers. The primary limitations to TPS are moisture variation and retrogradation causing embrittlement during storage that is not desired for packaging materials. A challenge is for starch with more specific initial properties to be available. The structure of synthetic properties can be controlled better than that of biopolymers causing a major set-back for materials from natural resources where properties may change with source, season or plant variant. This makes preparation of commercial products with precise tolerances difficult.

9. Conclusion

Native starch has complex supra-molecular structures that must be disrupted and dissociated to form amorphous thermoplastic starch. Water, other plasticisers, temperature and shear are important in gelatinising starch to form TPS. TPS is processed using extrusion then formed using injection moulding or thermoforming, the same as synthetic thermoplastics. TPS can be combined with other polymers in blends and fillers to form composites. Problems with TPS are water sorption and retrogradation, causing properties to change over time and under prevailing ambient conditions. There are many varieties of starch with differing composition, processing behaviour and properties. This makes development of TPS materials highly specific to particular starch types during development and application. Starch has interesting theoretical properties as a polymer. There is not theoretical distinction between synthetic and natural polymers, except for the complex intermolecular interactions resulting from polarity and hydrogen bond formation and the multileveled hierarchy of ordered structures that form. Starch presents many characteristics, found in synthetic polymers, in the one polymer: linear chains with a single repeat unit, branching, hydrogen bonding, crystallinity, gelatinisation, melting and glass transition phenomena, liquid crystalline characteristics, retrogradation, upper critical solution temperature behaviour, solubility, gelation all combined to form the technology of thermoplastic starch production.

10. Appendix

10.1 Abbreviations

DMA dynamic mechanical analysis
DSC differential scanning calorimetry
FTIR Fourier transform infrared
LCST lower critical solution temperature
mf-TM modulated force thermomechanometry
MWCNT multi-wall carbon nano-tube
PBAT poly(butylene adipate-*co*-terephthalate)
PCL poly(caprolactone)
PLA poly(lactic acid)

PVAlc poly(vinyl alcohol)
SEM scanning electron microscope
TBuTi tetrabutyltitanate
TEOS tetraethoxysilane
Tg glass transition temperature
Tm melting temperature
TPS thermoplastic starch
TSDC thermally stimulated depolarization current technique
WAXS wide-angle X-ray scattering

11. References

Baruk P, Zamudio-Flores, Bautista-Baoos S, Salgado-Delgado R È, Arturo L, Bello-Perez. (2009). Effect of oxidation level on the dual modification of banana starch: The mechanical and barrier properties of its films. *Journal of Applied Polymer Science*, 112, 822-9.

Belard L, Dole P, Averous L. (2009). Study of pseudo-multilayer structures based on starch-polycaprolactone extruded blends. *Polymer Engineering & Science*, 49, 1177-1186.

Dai H, Chang P R, Yu J, Ma X, (2008). N,N-Bis(2-hydroxyethyl)formamide as a New Plasticizer for Thermoplastic Starch. *Starch - Stärke*, 60, 676-684.

Dean K. M, Do M D, Petinakis E, Yu L, (2008). Key interactions in biodegradable thermoplastic starch/poly(vinyl alcohol)/montmorillonite micro- and nanocomposites. *Composites Science and Technology*, 68, 1453-1462.

Fakhouri, F. M., Fontes, L. C. B., Innocentini-Mei, L. H., and Collares-Queiroz, F. P. (2009). Effect of Fatty Acid Addition on the Properties of Biopolymer Films Based on Lipophilic Maize Starch and Gelatin. *Starch - Stärke*, 61, 528-536.

Farahnaky A, Farhat IA, Mitchell JR, Hill SE. (2009). The effect of sodium chloride on the glass transition of potato and cassava starches at low moisture contents. *Food Hydrocolloids*, 23, 1483-1487.

Garcia N L, Ribba L, Dufresne A, Aranguren M I, Goyanes S. (2009). Physico-Mechanical Properties of Biodegradable Starch Nanocomposites. *Macromolecular Materials and Engineering*, 294, 169-177.

Huali Tang HX, Shangwen Tang, Peng Zou,. (2009). A starch-based biodegradable film modified by nano silicon dioxide. *Journal of Applied Polymer Science*, 113, 34-40.

Laredo E, Newman D, Bello A, Muller A J. (2009). Primary and secondary dielectric relaxations in semi-crystalline and amorphous starch. *European Polymer Journal*, 45, 1506-1515.

Lee S Y, Hanna M A. (2009). Tapioca starch-poly(lactic acid)-Cloisite 30B nanocomposite foams. *Polymer Composites*. 30, 665-672.

Li G, Sarazin P, Favis B D, (2008). The Relationship between Starch Gelatinization and Morphology Control in Melt-Processed Polymer Blends with Thermoplastic Starch. *Macromolecular Chemistry and Physics*, 209, 991–1002.

Liu H, Xie F, Yu L, Chen L, Li L. (2009). Thermal processing of starch-based polymers. *Progress in Polymer Science*, 34, 1348-68.

Liu, H., Chaudhary, D., Ingram, G., and John, J. (2011). Interactions of hydrophilic plasticizer molecules with amorphous starch biopolymer—an investigation into the glass transition and the water activity behavior. *Journal of Polymer Science Part B: Polymer Physics*, 49, 1041-1049.

Liu, H., Yu, L., Tong, Z., and Chen, L. (2010). Retrogradation of waxy cornstarch studied by DSC. *Starch – Stärke*, 62, 524-529.

Maksimov R D, Lagzdins A, Lilichenko N, Plume E. (2009). Mechanical properties and water vapor permeability of starch/montmorillonite nanocomposites. *Polymer Engineering and Science*, 49, 2421-2429.

Perez C J, Alvarez V A, Mondragun I, Vazquez A, (2008). Water uptake behavior of layered silicate/starch-polycaprolactone blend nanocomposites. *Polymer International*, 57, 247-253.

Pushpadass H A, Kumar A, Jackson D S, Wehling R L, Dumais J J, Hanna M A. (2009). Macromolecular Changes in Extruded Starch-Films Plasticized with Glycerol, Water and Stearic Acid. *Starch – Stärke*, 61, 256-266.

Rodriguez-Sandoval E, Fernandez-Quintero A, Cuvelier G, Relkin P, Bello-Perez L A, (2008). Starch Retrogradation in Cassava Flour From Cooked Parenchyma. *Starch – Starke*, 60, 174-180.

Rosa MF, Chiou B-S, Medeiros ES, Wood DF, Mattoso LHC, Syed WJO, Inam H. (2009). Biodegradable composites based on starch/EVOH/glycerol blends and coconut fibers. *Journal of Applied Polymer Science*, 111, 612-618.

Saibene D, Zobel H F, Thompson D B, Seetharaman K, (2008). Iodine-binding in Granular Starch: Different Effects of Moisture Content for Corn and Potato Starch. *Starch – Starke*, 60, 165-173.

Sang Y, Alavi S, Shi Y-C. (2009). Subzero Glass Transition of Waxy Maize Starch Studied by Differential Scanning Calorimetry. *Starch – Stärke*, 61, 687-695.

Sangseethong K, Lertphanich S, Sriroth K. (2009). Physicochemical Properties of Oxidized Cassava Starch Prepared under Various Alkalinity Levels. *Starch - Stärke*. 61, 92-100.

Shogren R L. (2009). Estimation of the Intrinsic Birefringence of the A, B and V Crystalline Forms of Amylose. *Starch - Stärke*, 61, 578-581.

Wunderlich B. (2011). Do biopolymers behave the same as synthetic high polymers? *Journal of Thermal Analysis and Calorimetry*, 106, 81-84.

Raquez J-M, Nabar Y, Narayan R, Dubois P, (2008). In situ compatibilization of maleated thermoplastic starch/polyester melt-blends by reactive extrusion. *Polymer Engineering and Science*, 48, 1747-1754.

Willett J L, (2008). Humidity-Responsive Starch-Poly(methyl acrylate) Films. *Macromolecular Chemistry and Physics*, 209, 764-772.

Lovino R, Zullo R, Rao M A, Cassar L, Gianfreda L, (2008). Biodegradation of poly(lactic acid)/starch/coir biocomposites under controlled composting conditions. *Polymer Degradation and Stability*, 93, 147-157.

Ma X, Yu J, Wang N, (2008). Glycerol plasticized-starch/multiwall carbon nanotube composites for electroactive polymers. *Composites Science and Technology*, 68, 268-273.

Ray D, Sengupta S, Sengupta S P, Mohanty A K, Misra M, (2007). A Study of the Mechanical and Fracture Behavior of Jute-Fabric-Reinforced Clay-Modified Thermoplastic Starch-Matrix Composites. *Macromolecular Materials and Engineering*, 292, 1075-1084.

Viguie J, Molina-Boisseau S, Dufresne A, (2007). Processing and Characterization of Waxy Maize Starch Films Plasticized by Sorbitol and Reinforced with Starch Nanocrystals. *Macromolecular Bioscience*, 7, 1206-1216.

7

Retrogradation and Antiplasticization of Thermoplastic Starch

Yachuan Zhang and Curtis Rempel
Richardson Centre for Functional Foods and Nutraceuticals, University of Manitoba, Winnipeg, MB, R3T 2N2, Canada

1. Introduction

Petrochemical-based plastics are widely used in modern society due to their high effective mechanical and barrier properties (Farris et al., 2009; Siracusa et al., 2008). However, petrochemical-based plastics have become an environmental concern as they are not biodegradable or recyclable. Replacing the petrochemical-based polymers with biopolymers which are renewable has become an attractive idea and necessitates research on bioplastics (Debeaufort et al., 1998). Among the biopolymers, starch is considered as one of the most promising candidates for bioplastics due to its abundant availability, annually renewability, competitive price, and potential performance, including thermoplasticity (Lai & Padua, 1997; Mali et al., 2005). Native starch does not have thermoplastic properties. However, when additional plasticizers, elevated temperatures and shear are present, native starch does exhibit thermoplastic properties. Standard techniques, such as extrusion and injection moulding, used for producing petrochemical-based plastics, can be used in thermoplastic processing of starch (Guilbert et al., 1997). Some of thermoplastic starch (TPS) has been developed into commercial products, like compost bags, packaging materials (loose fillers and films), coatings, mulch films and disposable diapers (Jovanovic et al., 1997; Lai et al., 1997). TPS film and coating are being developed for the meat, poultry, seafood, fruit, vegetable, grains and candies industry sectors (Debeaufort et al., 1998). A drawback for use of starch is that TPS products age with time during storage due to starch retrogradation, which significantly changes quality, acceptability, and shelf-life of the TPS products. This review will summarize the current knowledge of TPS pertaining to its plasticization, retrogradation, and antiplasticization.

2. Basics of starch plasticization

Native starch does not have any thermoplastic properties without addition of plasticizer(s) (e.g., water, glycerol, sorbitol, etc.). The products made from native starch are readily broken into fragments when they are dried in ambient conditions due to strong intermolecular hydrogen bonding in the amylose and amylopectin macromolecular chains (Ma et al., 2007). But in the presence of plasticizers and elevated temperatures and shear, native starch readily melts and flows, allowing for its use as an extruding, injection moulding, or blowing material, similar to most conventional synthetic thermoplastic polymers (Ma et al., 2007). The role of plasticizers is to attract the water molecules around them, reduce the intermolecular interactions between the starch molecules, and then increase the flexibility of native starch (Ke & Sun 2001). This process of overcoming the brittleness in starch by softening the structure and by increasing the mobility of the macromolecular chains, resulting in a lowering of processing temperature, is termed as plasticization of starch. During the plasticization process, starch is transformed from a semicrystalline granular material into a system containing granular remnants, or to an amorphous paste with no structure at all (Smits et al., 2003).

Three theories have been proposed to account for the mechanisms of plasticization. These are lubricity theory, gel theory, and free volume theory. Lubricity theory proposes that the plasticizer acts as a lubricant and lubricates movements of the macromolecules over each other. Gel theory proposes that the plasticizer disrupts the interaction of starch chain bonds. Free volume theory proposes that plasticizer increases free volume between the starch chains and lowers its glass transition temperature (T_g). The commonality of these is that plasticizer is considered to interpose itself between the starch chains and reduce the forces holding the chains together (Gioia & Guilbert, 1999).

Commonly used plasticizers for TPS include water, glycerol, sucrose, fructose, glucose, glycols, urea, formamide, ethanolamine, and ethylene bisformamide, and amino acids (Huang et al., 2006; Ma & Yu, 2004; Pushpadass et al., 2008; Wang et al., 2008; Yang et al., 2006a, 2006 b; Zhang & Han, 2006a, 2006b). These chemicals are small in molecular size and are hydrophilic. Water and glycerol have traditionally been considered as the most effective plasticizers. Urea, formamide, ethanolamine, and ethylene bisformamide, which contain -CO-NH- functional groups, have recently been proven to be good plasticizers, for they are believed to suppress retrogradation and improve mechanical properties of TPS (Ma et al., 2005; Ma et al., 2006). Zullo & Iannace (2009) found that a urea/formamide mixture worked more effectively than glycerol in making homogenous and robust TPS films. Ma & Yu (2004) compared the hydrogen bond energy of the urea-starch, formamide-starch, acetamide-starch, and glycerol-starch, and concluded that urea, formamide and acetamide formed stronger hydrogen bonds with starch. Consequently urea, formamide and acetamide are more effective in plasticizing starch than polyols.

A minimum of 20% glycerol or any other suitable plasticizer is required to plasticize starch successfully (Pushpadass et al., 2008). With increasing plasticizer amount, properties of TPS, like tensile strength, Young's modulus, and glass transition temperature (T_g), decrease, while elongation and gas permeability increase. A TPS film which contains 25% glycerol is reported to exhibit maximum tensile strength and optimum modulus of elasticity (Pushpadass et al., 2008).

For industrial manufacturing of TPS, extrusion processing is a realistic approach. A typical extruder consists of a hopper, barrel, feed screw, thermocouples, and die (Fig. 1). Starch pellets or beads are fed from the hopper along the feed screw through the barrel chamber. As starch pellets or beads travels along the barrel, it is subject to friction, compression, and heated zones. The result is that the starch pellets homogeneously melt and mix as they traveled through the feed screw to the die. The die is precisely machined with a pattern opening such that the extruded starch mix takes the die pattern for its cross sectional area. TPS extrudates from the die solidify quickly. Before solidifying, TPS extrudates can be blown into films, sheets, or be moulded into desired shapes (Thunwall et al., 2006).

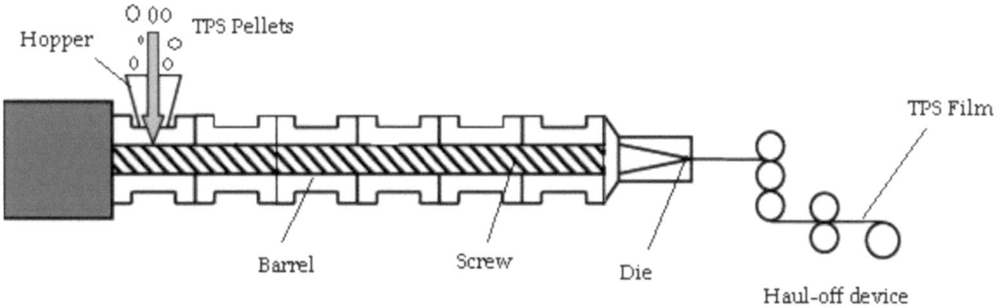

Fig. 1. A schedule of a typical extruder (Adapted from Li et al., 2011)

3. Mechanism of TPS retrogradation

Zhang & Han (2010) showed that gelatinized and plasticized TPS contains about 10% crystallinity, meaning 90% of TPS is amorphous region. Starch polymers in the amorphous region are metastable and consequently they automatically retrogradate over time during storage. Mechanism of starch retrogradation has been intensively investigated. Liu & Han (2005) observed starch crystal formation of amylose and amylopectin film respectively under an inverted phase-contrast microscope. They found amylose film featured a layer of dendrites with 90-degree branching (Fig. 2a), while amylopectin film showed a network of interlinked clusters with an amorphous background (Fig. 2b). Lian et al. (2011) observed the starch retrogradation under a microscope. They found starch nuclei in 2 h and crystals in 48 h (Fig. 3a & b). Russell (1987) postulated amylose formed nuclei by congregating, and amylopectin formed crystalline lamellae by prolongating the rod-like growth of crystals. Delville et al. (2003) proposed a mechanism for formation of starch crystalline lamellae shown in Fig.4, presenting the crystalline cluster formation of amylopectin. The cluster formation begins with the formation of crystalline lamellae composed of double helices of amylopectin short chains (symbolized with rectangular boxes). Then, the packing of double helices forms crystalline clusters (Delville et al., 2003).

Tako & Hizukuri (2002) proposed starch retrogradation mechanisms at a molecular level, in which starch retrogradation occurs as a result of intermolecular hydrogen bonding between O-6 of D-glucosyl residues of amylose molecules and OH-2 of D-glucosyl residues of short side-chains of amylopectin molecules (Fig. 5). The starch retrogradation was also attributed to intermolecular hydrogen bonding between OH-2 of D-glucosyl residues of amylose molecules and O-6 of D-glucosyl residues of short side-chains of amylopectin molecules (Fig. 6). In addition to intermolecular hydrogen bonding between amylose and amylopectin, hydrogen bonding between O-3 and OH-3 of D-glucosyl residues on different amylopectin molecules also occurs (Fig. 7). The intramolecular hydrogen bonding might occur between OH-6 and adjacent hemiacetal oxygen atom of the D-glucosyl residues within an amylose molecule (Fig. 5 and 6), while the intramolecular association within an amylopectin molecule was not suggested to exist (Tako & Hizukuri, 2002).

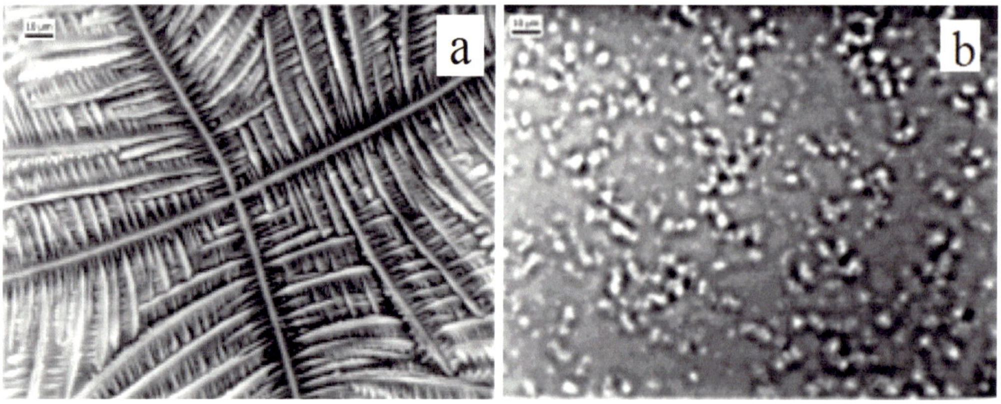

Fig. 2. Photomicrographs of amylose film (a) and amylopectin film (b). Amylose film shows a dendrite feature, while amylopectin film shows a cluster network with an amorphous background. (From Liu & Han, 2005)

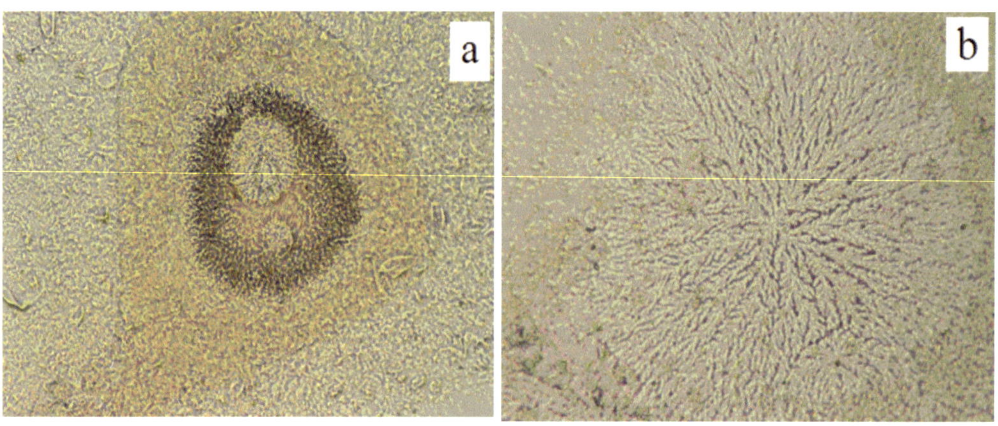

Fig. 3. Optical micrographs of retrograded starch of sweet potato aging 2 h (a) and aging 48 h (b). (From Lian et al., 2011)

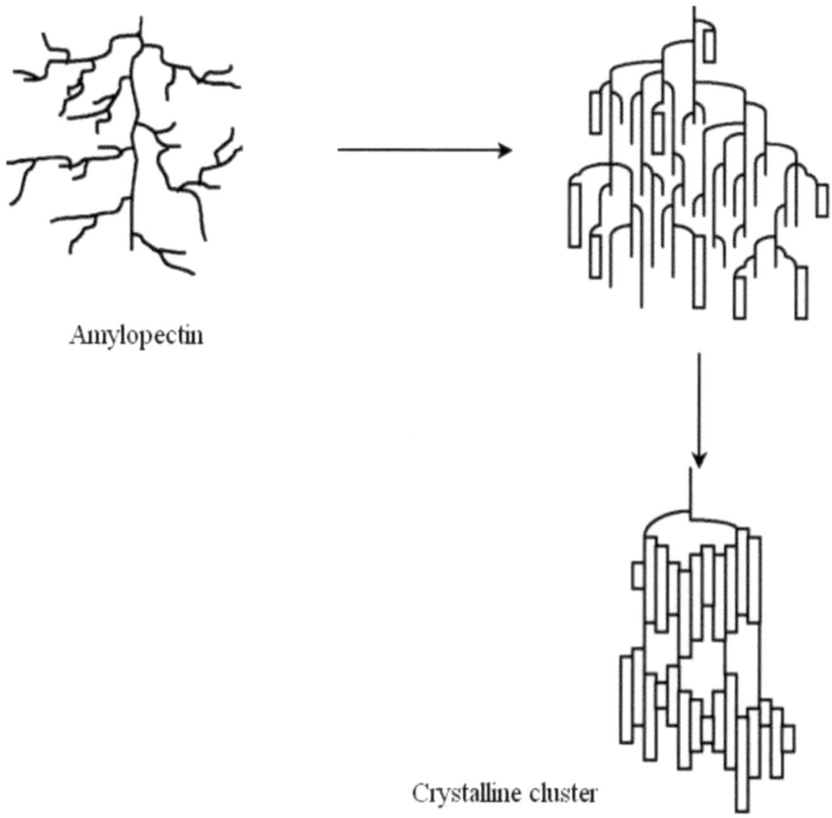

Fig. 4. Schematics of amylopectin retrogradation at the rubbery state (amylopectin double helices are represented as rectangles). (From Delville et al., 2003)

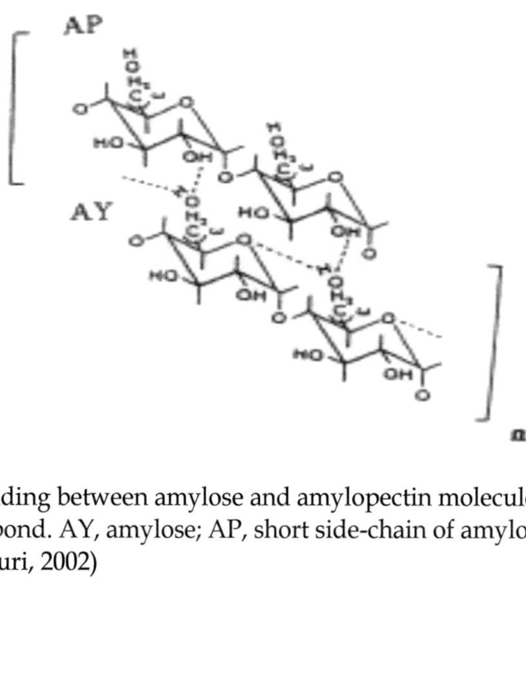

Fig. 5. Hydrogen bonding between amylose and amylopectin molecules (Dotted lines represent hydrogen bond. AY, amylose; AP, short side-chain of amylopectin molecules) (From Tako & Hizukuri, 2002)

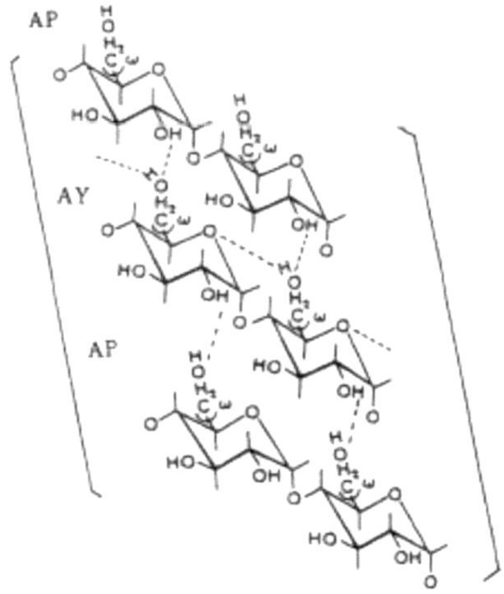

Fig. 6. Retrogradation mechanism of starch (Dotted lines represent hydrogen bond. AY, amylose; AP, short side-chain of maylopectin molecules) (From Tako & Hizukuri, 2002)

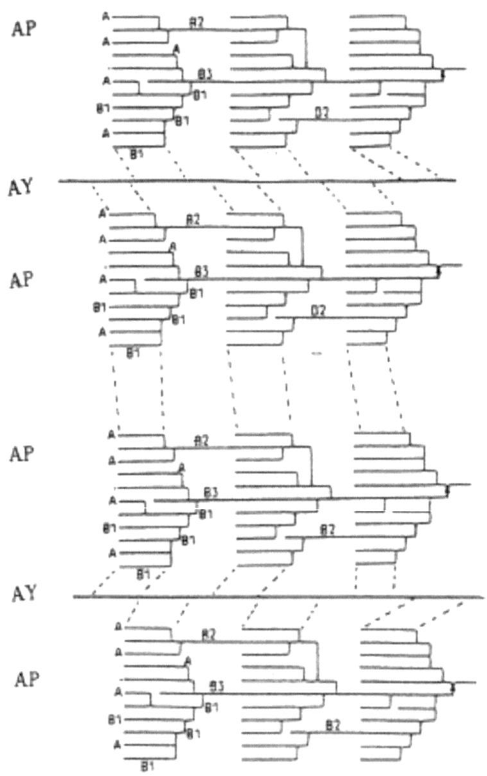

Fig. 7. Association between amylose and amylopectin molecules (Dotted lines, represent the hydrogen bonding sites). Two or more short-chains of amylopectin molecules may interact with one amylose molecule. Self-association within amylopectin molecules may also take place (From Tako & Hizukuri, 2002)

4. Effect of retrogradation on the property of thermoplastic starch

Properties of TPS products depend upon its microstructure. Zhang & Han (2010) studied the effect of crystallinity on the properties of the starch film, including moisture content, gas permeability, and tensile properties, etc. When crystallinity of glycerol-starch film increased from 6.0% to 8.0%, moisture content decreased from 11.0% to 8.0%. Similar phenomena were observed in sorbitol-, fructose-, and mannose-films. During the retrogradation process, starch chains aligned to form crystalline lamellas. Water molecules, as well as plasticizer molecules, are pushed out and evaporate resulting in reduced moisture content. Retrogradation of starch was also observed to have negative effect on gas permeability, including oxygen permeability (OP) and water vapor permeability (WVP) (Zhang and Han, 2010). Sorbitol-film had a reduction in OP from 20×10^{-7} to 0.3×10^{-7} cc mm h^{-1} kPa^{-1} m^{-2} when its crystallinity increased from 7.0% to 12.0%, while OP in glycerol-film was reduced from 6×10^{-7} to 4×10^{-7} cc mm h^{-1} kPa^{-1} m^{-2} when its crystallinity increased from 5.0% to 9.0%. A similar trend was also found in WVP. When crystallinity of glycerol-film increased from 6.0% to 9.0%, its WVP decreased from 1.2 to 0.9 g mm m^{-2} h^{-1} kPa^{-1}. Pushpadass & Hanna (2009) explained that the increase in crystallinity with time decreased the free volume in the TPS network and resulted in decrease in WVP. Zhang & Han (2010) proposed that crystallite in the films results in difficulty for gas molecules to diffuse within the starch films, leading to low permeability. Crystallinity also resulted in decreased elongation of the film. Elongation of mannose-film decreased from 4.0% to 0.2% when its crystallinity increased from 6.0% to 19.0%, elongation of fructose-film from 4.0% to 1.5% when its crystallinity increased fom 6.0% to 18.0%. Pushpadass & Hanna (2009) reported similar findings. They reported the relative crystalline content of TPS samples increased from 3 % (after 4 h extrusion) to 7%, 14%, and 17% after 3, 30, and 120 d of storage. As a result, tensile strength of TPS samples increased by 39.3 - 134.1%, elongation decreased by 48.0 - 81.1%, and WVP decreased by 6.1 % - 19.3%. A lower E value means that TPS becomes stiffer, less flexible and more difficult to handle. Crystallites may act as physical crosslinking points which generate internal stresses of TPS, leading to the increase in tensile strength and decrease in elongation (Delville et al. 2003)

5. Technologies to study the starch retrogradation

Many analytical techniques have been developed to monitor the starch retrogradation based on the TPS property changes. These methods include differential scanning calorimetry (DSC), differential thermal analysis (DTA), X-ray diffraction (XRD), and enzymatic susceptibility, and others. Karin et al. (2000) summarized some of the methods. For TPS crystallinity study, DSC and X-ray diffraction have proven to be extremely sensitive and therefore valuable tools to quantify retrograded starches.

5.1 Differential scanning calorimetry

Differential scanning calorimetry (DSC) is a thermo-analytical technique. It measures temperatures and heat flows associated with thermal transitions in a material sample. A reference with a well-defined heat capacity over the range of temperature is identically scanned. The difference in the heat flow between the sample and reference is detected and recorded. A plot of the differential heat flow between the reference and sample as a function of temperature is developed. When the sample undergoes a physical transformation, such as glass transition, melting, crystallization, more (or less) heat will need to flow to it than the reference to maintain both at the same temperature, resulting in a significant deviation in the difference between the two heat flows. When using DSC for starch retrogradation, the deviation between the heat flows results in a peak in the DSC curve. An example of this is illustrated in Fig. 8, using wheat starch.

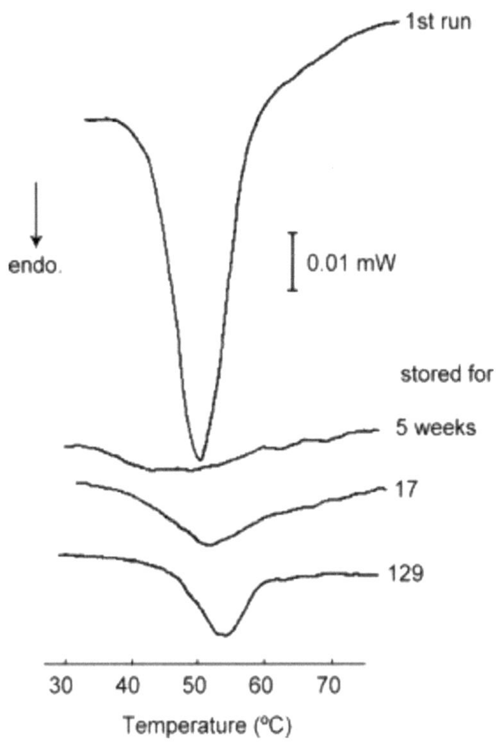

Fig. 8. Observed DSC curves for gelatinization of native wheat starch (upper) and retrograded wheat starch after storage time of 5, 17, and 129 weeks. (From Kohyama et al., 2004)

The first run in Fig. 8 represents gelatinization of wheat starch. The peaks after 5, 17, and 129 weeks were broader and shallower than that of the first run, and became deeper and more pronounced with increasing storage time.

Tian et al. (2011) used DSC to determine the retrogradation degree of rice starch by equation 1.

$$DR(t) = {\Delta H_t - \Delta H_0} \Big/ {\Delta H_\infty - \Delta H_0} \times 100\% \tag{1}$$

where $DR(t)$ represents the crystalline fraction (%) developed at storage time t, ΔH_0 is the enthalpy change (J/g) at storage time zero, ΔH_t is the enthalpy change at storage time t, and ΔH_∞ is the enthalpy change at unlimited time (35 days in the study). From 0 to 21 days, rice starch retrogradation went from 10% to 50%.

Application of DSC on the starch retrogradation study has a limitation. As mentioned by Biliaderis (1990), starch retrogradation consists of two processes: the rapid gelation of amylose solubilized during gelatinization and the slower recrystallization of amylopectin. DSC is only used to determine the latter phenomenon, since the re-organized region in amylopectin melts while the re-organized region in amylose does not (Gidley & Bulpin, 1987; Russell, 1987).

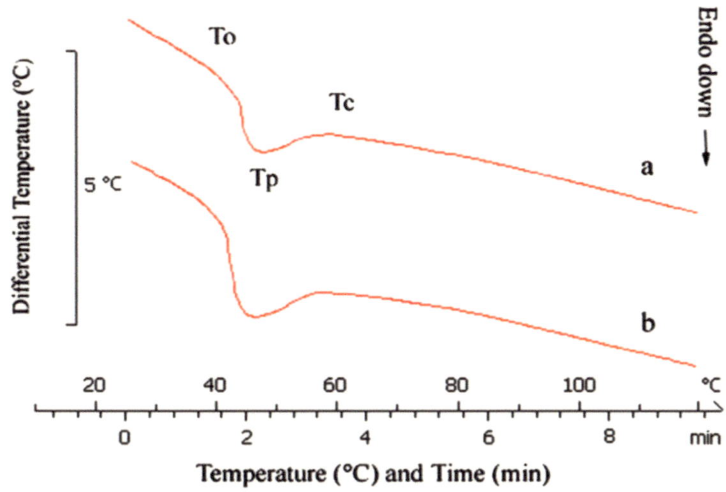

Fig. 9. DTA curves for (a) the gelatinized and (b) the 35-day retrograded rice starch containing 1.9% amylose. T_o is the onset temperature, T_p the peak temperature, and T_c the conclusion temperature, respectively. (From Tian et al., 2011)

5.2 Differential thermal analysis

An alternative technique, which shares much in common with DSC, is differential thermal analysis (DTA). In this technique it is the heat flow to the sample and reference that remains the same rather than the temperature. When the sample and reference are heated identically, phase changes and other thermal processes cause a difference in temperature between the sample and reference. The differential temperature is then plotted against temperature or against time as a DTA curve that provides data on the physical and chemical transformations, such as crystallization, gelatinization (Wada et al., 1979), melting, and sublimation (Morita, 1956). Fig. 9 shows a DTA curve for retrogradation of rice starch. Tian et al. (2011) used DAT technology to study rice starch retrogradation, which contained 29.3%, 13.5%, and 1.9% amylose. They found the storage time from 0 to 35 days slightly decreased the onset temperature (T_0), the peak temperature (T_p), and the conclusion temperature (T_c) for the retrograded starch. This was attributed to the fact that the number of the perfect crystal nuclei was reduced with storage time and the imperfect crystalline nuclei were increased during the retrogradation. By using equation 2, crystalline fraction of the retrograded starch was calculated.

$$DR(t) = \left({\Delta T_t - \Delta T_0} \big/ {\Delta T_\infty - \Delta T_0} \right) \times 100\% \qquad (2)$$

where *DR(t)* is the crystalline fraction (%) formed at storage time t, ΔT_t is the maximum differential temperature for the *t* day retrograded starch, ΔT_0 is the maximum differential temperature for the gelatinized starch, and ΔT_∞ is the maximum differential temperature for the long-term, for example 35 days, retrograded starch. With equation 2, Tian et al. (2011) calculated the rice starch crystallized from around 10% to 50~60% with increasing of the storage time from 0 to 21 days. The *DR* data obtained from DTA was slightly higher than that evaluated from the DSC, but no significant difference between the two data was found.

5.3 X-ray diffraction

The application of X-ray diffraction (XRD) on the study of starch crystallinity has shown that three X-ray diffraction crystal patterns exist in native starch granules, namely A-type, B-type, and C-type. The typical XRD patterns of TPS are characterized by sharp peaks associated with the crystalline portion area and an amorphous area (Fig. 10). The amorphous fraction of the sample can be estimated by the area between the smooth curve drawn following the scattering hump and the baseline joining the background within the low and high-angle points. The crystalline fraction can be estimated by the upper region above the smooth curve (Mali et al., 2006). Equation 3 was used to estimate crystallinity of the TPS.

$$DR = \frac{I_c}{I_c + I_a} \times 100\% \tag{3}$$

where *DR* is the crystalline fraction (%), I_c is the crystalline area on the X-ray diffractogram and I_a is the amorphous area on the X-ray diffractogram (Kalichevsky et al., 1993; Yoo & Jane, 2002). With the X-ray method, Zhang & Han (2010) studied the anti-plasticization of pea starch film. They announced that native pea starch film contained about 6.0% crystalline structure, and the crystallinity increased with plasticizer content increasing until 25%. The X-ray method is one of most popular methods for studying TPS structure and its application has been widely studied (Garcia et al., 2000; Smith et al., 2003; Zhang & Han, 2010).

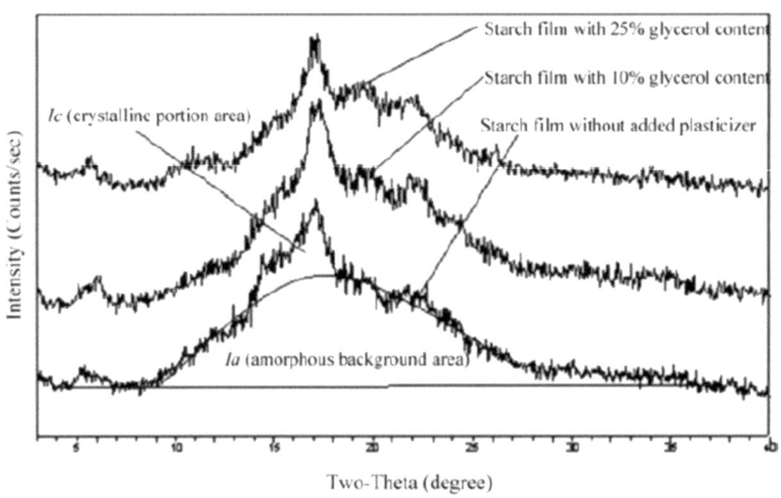

Fig. 10. X-ray diffraction patterns of peas starch films plasticized by 0, 10, and 25 % glycerol, respectively. (From Zhang & Han, 2010)

5.4 Enzymatic susceptibility

As a result of retrogradation, starch forms resistant starch which is resistant to the enzyme hydrolysis. Enzymatic method has been developed which is bases on measuring the resistant starch level to determine the starch retrogradation degree. Tsuge et al. (1990) gave a detailed procedure on how to conduct enzymatic measurement on the retrogradated starch. Tian et al. (2010) repeated it with rice starch and α-amylase from *Bacillus subtilis*. Equation 4 was used to estimate the retrogradation degree.

$$DR = {100 \times (b-c)}/{a-c} \tag{4}$$

where DR is the crystalline fraction (%), a represents absorbance read of total starch fraction in a spectrophotometer, b the absorbance of starch fraction to be tested, and c the absorbance of the complete digestion of starch. Wave length of 625 nm was selected for the absorbance measurement. Tian et al. (2010) measured rice starch retrogradation and found 7 - 28% crystallinity during storage of 1 - 14 hours.

5.5 Size-exclusion high performance liquid chromatography

Size-exclusion high performance liquid chromatography (SE-HPLC) is usually used to determine the molecular weight (M_W) of starch components, according to the different retention time (T_R) for its components. Tian et al. (2010) used it to estimate amylose retrogradation in rice starch and compared the results with that of enzymatic measurement and DSC. Equation 5 was used to calculate the amylose recrystalline fraction in the rice starch.

$$DR_{(t)} = {A \times (T_t - T_0)}/{(T_\infty - T_0)} \times 100\% \tag{5}$$

Where $DR(t)$ represents the recrystalline fraction of amylose (%) developed at time t, T_t is retention time of the sample stored for t hours, T_0 is the retention time of the gelatinized sample, T_∞ is limiting retention time of 24 h storage sample, and A is the amylose area (%) occupied of the peak (Tain et al., 2010).

Tian et al. (2010) found two peaks in SE-HPLC plot with retention time 13.2 min and 16.0 min respectively. The first peak was ascribed to the amylopectin retrogradation and the second one amylose retrogradation. Analysis on retrogradation degree of rice starch showed 7 - 25% crystallinity during storage of 1 - 14 hours. This data corroborated that obtained by enzymatic measurement, but not with DSC data. DSC is reported to suitable for amylopectin retrogradation study, not for amylose crystallization. Tian et al. (2010) further concluded that SE-PHLC is an effective method to measure the amylose retrogradation during short-term storage time, i.e. 0 - 24 h.

6. Factors which can affect the retrogradation of starch

While many factors may affect the retrogradation of starch, the botanical source is the most important one. Different botanical sources produce starches which are different with respect to properties such as amylose/amylopectin ratio (Gudmundsson & Eliasson, 1990; Klucinec

& Thompson, 2002), lipid content (Gudmundsson & Eliasson, 1990), and amylopectin fine structure (Kalichevsky et al., 1990). These properties will strongly influence the starch retrogradation kinetics. Ottenhof et al. (2005) stated that potato TPS (34 ± 1% water content, w/w) showed the highest rate of retrogradation (~0.17 h^{-1}) followed by waxy maize (~0.12 h^{-1}), while wheat TPS was the slowest (~0.05 h^{-1}). In addition of the botanical source, factors such as storage time, environmental temperature, and TPS moisture and plasticizer content, influence starch retrogradation also (Mali et al., 2002).

6.1 Storage temperature

At a temperature which is higher than the T_g, TPS is in the rubbery state and its starch molecule retrogradation (or recrystallization) in the amorphous phase occurs easily. The rate of starch retrogradation depends on difference between the storage temperature and T_g (Mali et al., 2006), with increasing retrogradation rate for higher temperature, especially under the conditions when TPS being stored at high relative humidity or high plasticizer contents (Delville et al., 2003). Conversely, when TPS is stored at temperature below the T_g, starch polymers are in a stable glassy state, and retrogradation does not occur or is extremely slow (Baik et al., 1997). Williams-Ferry-Landel (WLF) equation was commonly used to study the kinetic of the retrogradation process (Baik et al., 1997) at different temperature value.

$$log_{10}\left(\frac{t}{t_g}\right)=\frac{-C_1(T-T_g)}{C_2+(T-T_g)} \tag{6}$$

where t is the time of crystallisation, t_g the time to crystallise at T_g, C_1 and C_2 are constants (17.44 and 51.6K, respectively), and T is the temperature (K). Currently, there are numerous studies available on native starch, but limited studies are available on TPS retrogradation rate results at different storage temperatures.

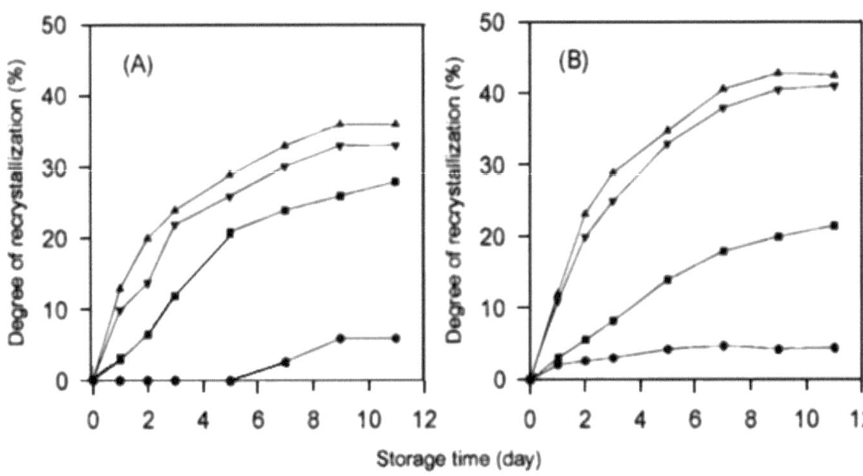

Fig. 11. Changes in degree of retrogradation of nonwaxy (A) and waxy (B) rice starch gels during storage at various temperatures; ●, 30; ■, 20; ▲, 4; ▼, 0 °C. (Adapted from Baik et al., 1997)

6.2 Storage time

Storage time increases starch retrogradation. Fig. 11 shows the relationship between degree of retrogradation of waxy rice starch gels with storage time. Rice starch becomes increasingly retrogradated with storage time, until the 9th day when it levels off (Baik et al., 1997). Mali et al. (2002) found as storage time increased, the width of the X-ray diffraction peak of starch samples decreased but its peak intensity increased, showing an increase in crystallinity of starch.

6.3 Storage relative humility and plasticizer content

In the rubbery state, high relative humidity (RH) favors starch macromolecular mobility which in turn facilitates the development of retrogradation (Delville et al., 2003). Glycerol content slows the crystallization kinetics in starch (Delville et al., 2003). According to Mali et al. (2006), glycerol limited the crystal growth and recrystallization by interacting with the polymeric chains and interfering with polymer chain alignment due to steric hindrances. Controversially, Garcia et al. (2000) reported that plasticizers (including glycerol and water) favored polymer chain mobility and allowed the development of a more stable crystalline structure during shorter periods of storage. Similarly, Smits et al. (2003) found that starch films without plasticizers formed less recrystallinity than the plasticized starch films. They attributed this phenomenon to the mobility of starch polymer chains, because plasticized starch polymers could easily vibrate and align up to form crystallites, while the unplasticized starch polymers interact with each other strongly and lose their mobility. Zhang & Han (2010) found that plasticizer concentration plays a critical role in starch retrogradation. When plasticizer content is greater than 25%, the plasticizer limits the starch polymer retrogradation. Otherwise, the plasticizer will favour the crystallization of the starch chains. Fig. 12 shows the relationship of starch film crystallinity and the plasticizer concentration.

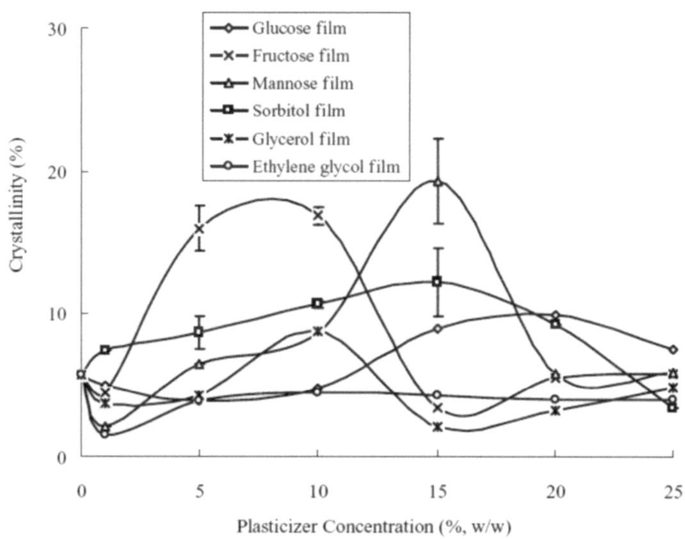

Fig. 12.Relationship of crystallinity of pea starch film and the plasticizer concentration. Crystallinity of the starch films increases with plasticizer concentration increasing until 25%. Bars indicate mean ± standard deviation. (From Zhang & Han, 2010)

7. TPS antiplasticization and its explanation

Antiplasticization is a well-known phenomenon for synthetic polymers. This phenomenon also exists in TPS and has been receiving increasing attention recently (Chang et al., 2006). Contrary to the orthodox behaviour of plasticization, plasticizer can reduce elongation and gas permeability, and increase tensile strength and modulus of elasticity of TPS when it is introduced in low amount under a critical level (generally <25%) (Godbillot et al., 2006; Lourdin et al., 1997; Zhang & Han, 2010). Fig. 13 shows strain-stress curves of pea starch films plasticized by glycerol and mannose respectively, indicating the starch films plasticized by glycerol and mannose had less stress and strain than the film without any plasticizers.

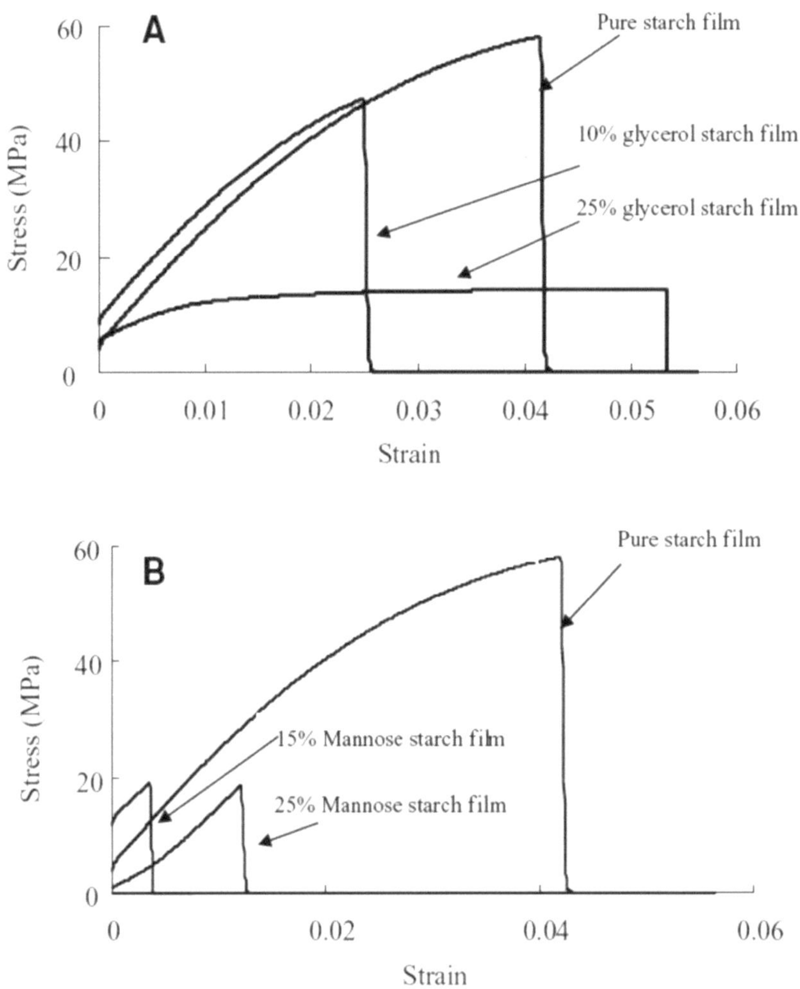

Fig. 13. Strain-stress curves of pea starch films plasticized by glycerol and mannose respectively, indicating the starch films plasticized by glycerol and mannose had less stress and strain than the film without any plasticizers. (From Zhang & Han, 2010)

Mechanism of anti-plasticization effect of plasticizer on TPS has not been clearly understood, but some efforts have been made to elucidate its mechanism. Zhang & Han (2010) found that starch retrogradation results in antiplasticization phenomena. The crystallinity of the pea starch films increased with plasticizer content increasing from 1% to 20%, leading to decrease in MC, OP, WVP, and E, and increase in EM. Plasticizer performs plasticization or antiplasticization depending on its concentration. Addition of plasticizers at the range of low to intermediate concentration level (1% to 25%) facilitates the formation of crystallites in the starch films, leading to the antiplasticization. Zhang & Han (2010) further proposed an anti-plasticization model. Due to the movement or vibration of the starch polymer chains, water and plasticizer molecules were pushed aside gradually from starch polymers. D-glucosyl residues of the amylose or amylopectin, which used to be separated by water or plasticizers molecules, interacted to form strong hydrogen bonds causing retrogradation or recrystallization. However, when starch polymer was plasticized by high content plasticizer (>25%), the plasticizer molecules could not be pushed aside completely from the starch polymers. Then, the plasticizers performed their plasticization effect, which was to interrupt interaction between the hydrogen bondings of starch polymers, increase the TPS elongation, reduce its T_g, and prevent retrogradation of starch chains.

8. Summary

Thermoplastic starch (TPS) has attained more attention for its potential to replace the conventional polymers. Retrogradation occurs in TPS with time and affects its properties and applications. TPS contains normally 10% crystallinity, but this value changes with storage time, temperature, atmosphere relative humidity, and plasticizer content. Starch retrogradation mechanisms are discussed at molecular level. Methods to measure the retrogradation degree, such as differential scanning calorimetry, differential thermal analysis, X-ray, etc. are also reviewed. Changes in TPS property, such as tensile strength, elongation, gas permeability, are due to the retrogradation of starch polymers and these are described.

9. References

Baik, M.Y., Kim, K.J., Cheon, K.C., Ha, Y.C., Kim, W.S. (1997). Recrystallization kinetics and glass transition of rice starch gel system. *Journal of Agricultural and Food Chemistry*, 45, pp. 4242-4248.

Biliaderis, C. G. (1990). Thermal analysis of food carbohydrates. In: *Thermal Analysis of Foods*, Harwalkar, V.R. and Ma, C.Y. pp. 120-134. Elsevier Applied Science, ISBN: 9781851664368, London.

Chang, Y.P., Abd Karim, A., Seow, C.C. (2006). Interactive plasticizing-antiplasticizing effects of water and glycerol on the tensile properties of tapioca starch films. *Starch*, 55, pp.304-312.

Debeaufort, F., Quezada-Gallo, J.A., Voilley, A. (1998). Edible films and coatings: tomorrow's packagings: a review. *Critical Reviews in Food Science and Nutrition*. 38, pp.299-313.

Delville, J., Joly, C., Dole, P., Bliard, C. (2003). Influence of photocrosslinking on the retrogradation of wheat starch based films. *Carbohydrate Polymers*, 53, pp.373-381.

Farris, S., Schaich, K.M., Liu, L.S., Piergiovanni, L., Yam, K.L. (2009). Development of polyion-complex hydrogels as an alternative approach for the production of biobased polymers for food packaging applications: a review. *Trends in Food Science & Technology*, 20, pp.316-332.

Garcia, M.A., Martino, M.N., Zaritzky, N.E., Plata, L. (2000). Microstructural characterization of plasticized starch based films. *Starch*, 52, pp.118-124.

Gidley, M.J., Bulpin, P.V. (1987). Crystallization of malto-oligosaccharides as models of the crystalline of starch: minimum chain-length requirement for the formation of double helices. *Carbohydrate Research*, 161, pp.291-300.

Gioia, L.D., Guilbert, S. (1999). Corn Protein-Based Thermoplastic resins: effect of some polar and amphiphilic plasticizers. *Journal of Agricultural and Food Chemistry*, 47, pp.1254-1261.

Godbillot, L., Dole, P., Joly, C., Roge, B., Mathlouthi, M. (2006). Analysis of water binding in starch plasticized films. *Food Chemistry*. 96, pp.380-386.

Gudmundsson, M.; Eliasson, A.C. (1990). Retrogradation of amylopectin and the effects of amylose and added surfactants/emulsifiers. *Carbohydrate Polymers*, 13, pp.295-315.

Guilbert, S.; Cuq, B.; Gontard, N. (1997). Recent innovations in edible and/or biodegradable packaging materials. *Food Additives and Contaminants*, 14, 6, pp.741-751.

Huang, M., Yu, J., Ma, X. (2006). High mechanical performance MMT-urea and formamide-plasticized thermoplastic cornstarch biodegradable nanocomposites. *Carbohydrate Polymers*, 63, pp. 393-399.

Jovanovic, S., Jeremic, K., Jovanonic, R., Donlagic, J., Dunjic, B. (1997). Preparation of thermoplastic starch. *Journal of the Serbian Chemical Society*, 62, pp. 623-629.

Kalichevsky, M.T., Blanshard, J.M. V., and Marsh, R.D.L. (1993). Applications of mechanical spectroscopy to the study of glassy biopolymers and related systems. In: *The glassy state in foods*. Blanshard, J. and Lillford, P.J., pp. 133-156, Nottingham University Press, ISBN: 9781897676202, Leicestershire, UK.

Kalichevsky, M.T.; Orford, P.D.; Ring, S.G. (1990). The retrogradation and gelation of amylopectin from various botanical sources. *Carbohydrate. Research*. 198, pp. 49-55.

Karin, A.A., Norziah, M.H., Seow, C.C. (2000). Methods for the study of starch retrogradation. *Food Chemistry*, 71, pp. 9–36.

Ke, T.; Sun, X. (2001). Thermal and mechanical properties of poly (lactic acid) and starch blends with various plasticizers. *Transactions of the American Society*, 44, pp. 945-953.

Klucinec, J.D.; Thompson, D.B. (2002). Amylopectin nature and amylose-to-amylopectin ratio as influences on the behavior of gels of dispersed starch. *Cereal Chemistry*, 79, pp. 24-35.

Kohyama, K., Matsuki, J., Yasui, T., Sasaki, T. (2004). A different thermal analysis of the gelatinization and retrogradation of wheat starches with different amylose chain lengths. *Carbohydrate Polymers*, 58, pp. 71-77.

Lai, H.M., Padua, G.W. (1997). Properties and microstructure of plasticized zein films. *Cereal Chemistry*, 74, pp. 771-775.

Lai, H.M., Padua, G.W., Wei, L.S. (1997). Properties and microstructure of zein sheets plasticized with palmitic and stearic acids. *Cereal Chemistry*, 74, pp. 83-90.

Li, M., Liu, P., Zou, W., Yu, L., Xie, F., Pu, H., Liu, H., Chen, L. (2011). Extrusion processing and characterization of edible starch films with different amylose contents. *Journal of Food Engineering*, 106, pp. 95-101.

Lian, X., Zhao, S., Liu, Q., Zhang, X. (2011). A photographic approach to the possible mechanism of retrogradation of sweet potato starch. *International journal of biological macromolecules*. 48, pp. 125-128.

Liu, Z., Han, J.H. (2005). Film-forming characteristics of starches. Journal of Food Science, 70, pp. E31-E36.

Lourdin, D., Bizot, H., Colonna, P. (1997). "Antiplasticization" in starch-glycerol films? *Journal of Applied Polymer Science*, 63, pp. 1047-1053.

Ma, X.F., Yu, J.G. (2004). The plasticized containing amide groups for thermoplastic starch. *Carbohydrate Polymers*, 57, pp. 197-203.

Ma, X.F., Yu, J.G., Ma, Y.B. (2005). Urea and formamide as a mixed plasticizer for thermoplastic wheat flour. *Carbohydrate Polymers*, 60, pp. 111-116.

Ma, X.F., Yu, J.G., Wan, J.J. (2006). Urea and ethanolamine as a mixed plasticizer for thermoplastic starch. *Carbohydrate Polymers*, 64, pp. 267-273.

Ma, X.F., Yu, J.G., Wang, N. (2007). Fly ash reinforced thermoplastic starch composites. *Carbohydrate Polymers*, 67, pp. 32-39.

Mali, S., Grossmann, M.V.E., Garcia, M.A., Martino, M.N., Zaritzky, N.E. (2002). Microstructural characterization of yam starch films. *Carbohydrate Polymers*, 50, pp. 379-386.

Mali, S., Grossmann, M.V.E., Garcia, M.A., Martino, M.N., Zaritzky, N.Z. (2005). Mechanical and thermal properties of yam starch films. *Food Hydrocolloids*, 19, pp. 157- 164.

Mali, S., Grossmann, M.V.E., Garcia, M.A., Martino, M.N., Zaritzky, N.E. (2006). Effects of controlled storage on thermal, mechanical and barrier properties of plasticized films from different starch sources. *J Food Engineering*, 75, pp. 453-460.

Morita, H. (1956). Characterization of starch and related polysaccharides by differential thermal analysis. *Analytical Chemistry*, 28, pp. 64-67.

Ottenhof, M.A., Hill, S.E., Farhat, I.A. (2005). Comparative study of the retrogradation of intermediate water content waxy maize, wheat, and potato starches. *Journal of Agricultural and Food Chemistry*, 53, pp. 631-638.

Pushpadass, H.A., Hanna, M.A. (2009). Age-induced changes in the microstructure and selected properties of extruded starch films plasticized with glycerol and stearic acid. *Industrial & Engineering Chemistry Research*, 48, pp. 8457-8463.

Pushpadass, H.A., Marx, D.B., Hanna, M.A. (2008). Effect of extrusion temperature and plasticizers on the physical and functional properties of starch films. *Starch*. 60, pp. 527-538.

Ressull, P.L. (1987). The aging of gels from starches of different amylose/amylopectin content studied by differential scanning calorimetry. *Journal of Cereal Science*, 6, pp. 147-158.

Siracusa, V., Rocculi, P., Romani, S., Rosa, M.D. (2008). Biodegradable polymers for food packaging: a review. *Trends on Food Science & Technology*, 19, pp. 634-643.

Smits, A.L.M., Kruiskamp, P.H., van Soes, J.J.G. Vliegenthart J.F.G. (2003). The influence of various small plasticizers and malto-oligosaccharides on the retrogradation of (partly) gelatinized starch. *Carbohydrate Polymers*, 51, pp. 417-427.

Tako, M., Hizukuri, S. (2002). Gelatinization mechanism of potato starch. *Carbohydrate Polymers*. 48, pp. 397-401.

Tian, Y., Li, Y., Xu, X., Jin, Z. (2011). Starch retrogradation studies by thermogravimetric analysis (TGA). *Carbohydrate Polymers*, 84, pp. 1165-1168.

Tian, Y., Li, Y., Xu, X., Jin, Z., Jiao, A., Wang, J., and Yu, B. (2010). A novel size-exclusion high performance liquid chromatography (SE-HPLC) method for measuring degree of amylose retrogradation in rice starch. *Food Chemistry*, 118, pp. 445-448.

Thunwall, M., Boldizar, A., Rigdahl, M. (2006). Extrusion processing of high amylose potato starch materials. *Carbohydrate Polymers*. 65, pp. 441-446.

Tsuge, H., Hishida, M., Iwasaki, H., Watanabe, S., Goshima, G. (1990). Enzymatic evaluation for the degree of starch retrogradation in foods and foodstuffs. *Starch*, 42, pp. 213-216.

Wada, K., Takahashi, K., Shirai, K., Kawamura, A. (1979). Differential thermal analysis (DTA) applied to examining gelatinization of starches in foods. *Journal of Food Science*, 44, pp. 1366-1372.

Wang, N., Yu, J., Chang, P.R., Ma, X. (2008). Influence of formamide and water on the properties of thermoplastic starch/poly(lactic acid) blends. *Carbohydrate Polymers*, 71, pp. 109-118.

Yang, J.H., Yu, J.G., Ma, X.F. (2006a). Study on the properties of ethylenebisformamide and sorbitol plasticized corn starch (ESPTPS). *Carbohydrate Polymers*, 66, pp. 110-116.

Yang, J.H., Yu, J.G., Ma, X.F. (2006b). Preparation and properties of ethylenebisformamide plasticized potato starch (EPTPS). *Carbohydrate Polymers*, 63, pp. 218-223.

Yoo, S.H., Jane, J.L. (2002). Structural and physical characteristics of waxy and other wheat starches. *Carbohydrate Polymers*, 49, pp. 297-305.

Zhang, Y., Han, J.H. (2006a). Mechanical and thermal characteristics of pea starch films plasticized with monosaccharides and polyols. *Journal of Food Science*, 71, pp. E109-E118.

Zhang, Y., Han, J.H. (2006b). Plasticization of pea starch films with monosaccharides and polyols. *Journal of Food Science*. 71, pp. E253-E261.

Zhang, Y., Han, J.H. (2010). Crystallization of high-amylose starch by the addition of plasticizers at low and intermediate concentrations. *Journal of Food Science*, 75, pp. N8-N16.

Zullo, R., Iannace, S. (2009). The effects of different starch sources and plasticizers on film blowing of thermoplastic starch: Correlation among process, elongational properties and macromolecular structure. *Carbohydrate Polymers*. 77, pp. 376-383.

Part 2

Modifications of Thermoplastic Elastomers

8

Thermoplastic Elastomers

Robert Shanks and Ing Kong
Applied Sciences, RMIT University, Melbourne
Australia

1. Introduction

An elastomer is defined by mechanical response not by chemical structure. Elastomers comprise a diverse range of chemical structures although they are characterized as having weak intermolecular forces. An elastomer will undergo an immediate, linear and reversible response to high strain to an applied force. This response has a mechanical analogy with a spring according to Hooke's Law. Non-linear, time dependent mechanical response is distinguished as viscoelasticity according to the parallel spring and dashpot model. Time dependent irreversible response is a viscous response according to a dashpot model. An ideal elastomer will only exhibit an elastic response. Real elastomers exhibit a predominantly elastic response, however they also exhibit viscoelastic and elastic responses especially at higher strains.

The chemical structure and molecular architecture of elastomers is tightly related to elastomeric mechanical response. High strain requires a polymer with high molar mass preferred. Many materials can exhibit an elastic response, that is immediate, reversible and linear strain with stress, however only a polymer can exhibit additionally high strain. High strain is due to uncoiling of random molecular coils into more linear conformations. The limit to elastic response is when molecules are in fully extended conformations. This mechanism is due to uncoiling of chain segments. Molecules do not move relative to each other, there are reversible random coiling not translational motions.

Reversibility and immediate response is obtained with macromolecules that have flexible chains with weak intermolecular forces. Rigid groups such a benzene, bulky side-chains such as isopropyl, polar groups such as ester and hydrogen bonding groups such as hydroxy are not desirable if a polymer is to be an elastomer. This description supposes elastomeric properties at ambient temperatures, since at elevated temperatures above the glass transition temperature many polymers become elastomers.

At high extensions and when under strain for longer times viscous flow occurs, known as creep when over longer times. Chemical cross-linking prevents viscous flow, the movement of molecules relative to each other. Elastomers are cross-linked after moulding or shaping to fix molecules into their relative positions. Once cross-linked the unstrained shape of an elastomer cannot be altered and the elastomer cannot be reprocessed or recycled. The permanence brought about by cross-linking and the need to perform a cross-linking reaction on elastomers are disadvantages for their applications.

2. Thermoplastic elastomer

A thermoplastic elastomer has all the same features as described for an elastomer except that chemical cross-linking is replaced by a network of physical cross-links. The ability to form physical cross-links is the opposite to the chemical and structural requirements of an elastomer just described. The answer to this dilemma is that thermoplastic elastomers must be two-phase materials, and each molecule must consist of two opposite types of structure, one the elastomeric part and the second the restraining, physical cross-linking part. Thermoplastic elastomers are typically block copolymers.

The elastic block should have high molar mass and possess all of the others characteristic required of an elastomer. The restraining block should resist viscous flow and creep. One restraining block can be used per macromolecule, giving a diblock copolymer (AB), or one restraint block at each of the elastomer can be used giving a triblock copolymer (ABA). Specific polymers will be described in the context of these general principles in the following sections. To provide an example of thermoplastic elastomer block copolymer structures the monomers butadiene and styrene are chosen.

Elastomer Type	Soft phase, T_g (°C)	Hard phase, T_g or T_m (°C)
SBS	-90	95 (T_g)
SIS	-60	95 (T_g)
SEBS	-55	95 (T_g) and 165 (T_m)[a]
SIBS	-60	95 (T_g) and 165 (T_m)
Polyurethane elastomers	-40 to -60[b]	190 (T_m)
Polyester elastomers	-40	185 to 220 (T_m)
Polyamide elastomers	-40 to -60[b]	220 to 275 (T_m)
Polyethylene-poly(-olefin)	-50	70 (T_m)[c]
Polypropylene/poly(ethylene-propylene)	-50	50 to 70 (T_m)[c]
Poly(etherimide)-polysiloxane	-60	225 (T_g)
Polypropylene/hydrocarbon rubber[d]	-60	165 (T_m)
Polypropylene/nitrile rubber	-40	165 (T_m)
PVC-(nitrile rubber+DOP)	-30	80 (T_g) and 210 (T_m)
Polypropylene/poly(butylacrylate)	-50	165 (T_m)
Polyamide or polyester/silicone rubber	-85	225 to 250 (T_m)

Notes:
[a]In blends containing polypropylene
[b]The values are for polyester and polyether respectively
[c]The values are presumably the result of the short length of polyethylene and polypropylene segments
[d]EPDM, EPR, butyl rubber and natural rubber
Source: Holden, 2011; Drobny, 2007

Table 1. Glass transition and crystalline melt temperatures of major TPEs

Polymerisation of butadiene via 1,4-addition gives the elastomer poly(1,4-butadiene). This polymer is a hydrocarbon with low intermolecular forces, no rigid or bulky groups and a relatively flexible chain, except for the double bond between carbons 2-3. The cis stereoisomer of the double bond is preferred over trans since this decreases chain regularity.

The transform is more regular and crystallinity can occur, which will prevent elastomeric response. Poly(butadiene) would need to be cross-linked to be a useful elastomer. Polystyrene is a glassy polymer with glass transition temperature =100 °C so it will resist flow and creep at ambient temperatures, but it can flow and be moulded at temperatures above T_g. A diblock copolymer of butadiene and styrene will provide the combination of properties required for a thermoplastic elastomer when the butadiene content is higher. Poly(butadiene-b-styrene) (BS) has two separate phases, a continuous polybutadiene phase with dispersed poly(styrene) phase. The matrix phase gives the overall elastomeric response while the dispersed islands are the restraining physical cross-links. Glass transition and crystalline melt temperatures for major TPEs are given in Table 1.

3. Thermodynamics of elasticity

Elastomers extend and contract by conformational change from a compact random coil to extended chain. The random coil can have many possible conformations resulting in a high entropy. A fully extended chain can only have one conformation resulting in low entropy. The extended chain will spontaneously contract into a random coil since the entropy of the process is favourable. Enthalpy is not a contributor for an ideal elastomer since intermolecular forces are minimal. Entropy is overcome by a mechanical force deforming the elastomer. Thermodynamic equations are applied to elastomeric deformation and recovery. By analogy with an ideal gas, elastomers that conform to the thermodynamics are called ideal elastomers. As in the case of real gases, real elastomers deviate from ideality.

Deviations of elastomers are the result of:

- At high extensions elastomer chains become fully extended between cross-links, chemical or physical, and as the distribution of chains become fully extended the stress-strain response becomes non-linear.
- At high extensions the extended chains can pack closely forming crystals that cannot be further extended. Crystallisation appears as stress whitening and it is only maintained with the deforming force since the intermolecular interactions are too weak to prevent entropy-controlled recovery.
- Molecular entanglements prevent free molecular uncoiling and therefore cause deviation from a linear elastic response. Often entanglement caused deviations are more pronounced in a first extension-recovery cycle that differs from subsequent cycles when performed together.
- In compression deviations from ideality are caused by the finite size of the molecules and the limited free volume available for molecules to occupy.

The aim of this review is to compile the updated knowledge on thermoplastic elastomers in general, practically their structures, syntheses, processing methods, mechanical properties and applications.

4. Structure of thermoplastic elastomers

Thermoplastic elastomers (TPEs) are defined as a group of polymers that exhibit instantaneous reversible deformation (to be an elastomer). Most of the TPEs consist of continuous phase that exhibit elastic behaviour and dispersed phase that represents the physical crosslinks. If the dispersed phase is elastic then the polymer is a toughened thermoplastic, not an elastomer. Elastomer reversibility must have physical cross-links, therefore these crosslinks must be reversible. Physical crosslinks do not exist permanently and may disappear with the increase of temperature.

Generally, thermoplastic elastomers can be categorized into two groups: multi-block copolymers and blends. The first group is copolymers consist of soft elastomers and hard thermoplastic blocks, such as styrenic block copolymers (SBCs), polyamide/elastomer block copolymers (COPAs), polyether ester/elastomer block copolymers (COPEs) and polyurethane/elastomer block copolymers (TPUs). TPE blends can be divided into polyolefin blends (TPOs) and dynamically vulcanized blends (TPVs).

Thermoplastic elastomers are known as two-phase system consisting of rubbery elastomeric (soft) component and rigid (hard) component. The soft phase can be polybutadiene, poly(ethylene-*co*-alkene), polyisobutylene, poly(oxyethylene), poly(ester), polysiloxane or any of the typical elastomers while the hard phase are polystyrene, poly(methyl methacrylate), urethane, ionomer - poly(ethylene-*co*-acrylic acid) (sodium, Mg, Zn salt), ethylene propylene diene monomer, and fluropolymers. The structure representing a styrenic TPE is shown schematically in Figure 1.

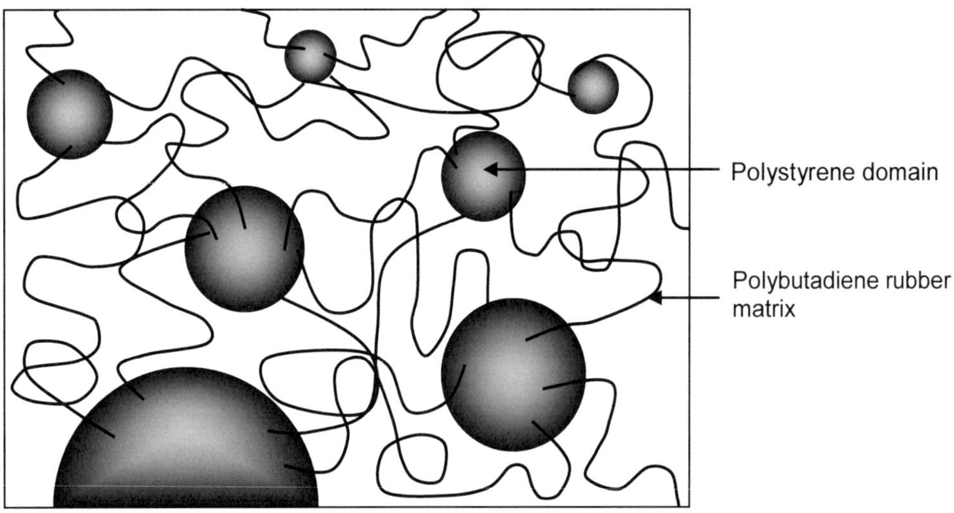

Fig. 1. Schematic of a styrene-butadiene-styrene block copolymer

5. Synthesis of thermoplastic elastomers

TPEs are two phase polymers, however they can be synthesised in one reaction step or in two or three steps to create each phase separately.

5.1 One-step methods

One-shot method is a commonly used industrial technique to prepare polyurethanes. The urethane reaction involves a diisocyanate (hard segment) and a diol (soft segment) (see Figure 2). Generally two diols are required, a chain extender or short chain diol and an elastomeric hydroxyl terminated polymer. Examples of a diisocyanate are methane 4,4'-diphenyl diisocyanate (MDI), 2,4- and 2,6-toluene diisocyanate (TDI) and 1,6-hexane diisocyanate (HDI). A chain extender may be 1,4-butanediol. When MDI and butanediol react they form a polyurethane with alternating monomer units connected by urethane groups, though other functional groups also form as by products of the reaction. This polyurethane is not elastomeric and it constitutes the hard phase of a typical TPU. Hydroxyl terminated elastomers include polyethers: poly(oxyethylene), poly(oxybutylene), polyesters: poly(ethylene succinate), poly(butylene succinate), poly(ethylene adipate), poly(butylene), hydroxyl terminated polybutadiene and hydroxyl terminated poly(butadiene-*co*-acrylonitrile).

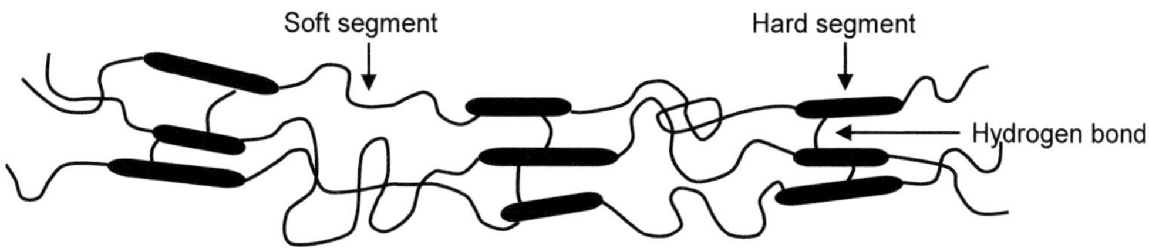

Fig. 2. Schematic representation of TPUs composed of alternating hard segment and soft segment structures

These polymeric diols react with isocyanate and are linked into the TPU as a complete elastomer block. The hard and soft (elastomeric) chain segments phase separate with the hard segments as a dispersed minor phase, since the soft segments must form a continuous phase if elastomeric properties are to be displayed. The reactivity difference between the -OH groups of the polyol and the chain extender with different isocyanate groups affect the sequence of hard segments in the polymer chain. Thus, polyurethanes obtained by using this method have a more random sequence. However, the polymer is highly crystalline due to the favoured reaction between polyol and diisocyanate before extended polymer growth has occurred.

A chain growth polymerisation can be used to form a TPE in one step. An example is a poly(ethylene-*co*-butene) with a high butane content, polymerised using a single-site metallocene initiator. This polyethylene can undergo phase separation due to crystallisation, crystals are the physical crosslinks, and the highly branched structure will exhibit elastomeric properties. It may need to be blended with a less branched polyethylene to increase the physical cross-links. Alternatively it can be partially cross-linked by dynamically vulcanising by extrusion with a peroxide initiator. While chemical cross-links are formed this type of polyethylene can still be processed as a thermoplastic. Dynamic Vulcanization can be applied to poly(ethylene-*co*-propylene) rubber (EPR) that may be blended with a thermoplastic polyethylene to provide a binding crystalline phase.

5.2 Two-step methods

TPU can be synthesised by a two-step method, which is known as prepolymer method. The reaction may be carried out in two steps where excess diisocyanate is added to the polymeric diol to form an isocycanate terminated pre-polymer with excess diisocyanate monomer that is then reacted or chain extended with the monomeric diol to form the segmented TPU structure. The polymeric diol can be a biopolymer or biodegradable polymer such as castor oil. Comparison has been made between poly(butylene succinate) and poly(butylene adipate) as the soft phase in TPU (Sonnenschein et al., 2010). The succinate derived TPU exhibited higher soft phase glass transitions and more hard phase to soft phase interactions than the adipate derived TPU, due to higher carbonyl content and hence polar interactions in the succinates. Abrasion resistance was a function of overall hard phase volume fraction.

Sequences are found to be more regular in the polymer obtained via prepolymer method compared to the one shot method. The structural regularity leads to a better packing of hard segments where physical cross-linking points are easier to form. Hence, a two-step method gives a product of better mechanical properties than a one-step method does (Table 2). Again, the solubility of these two products is different. The polyurethanes obtained from one-method are soluble in some of the common solvents, but the polyurethanes from the prepolymer process could not be dissolved in any common solvents (Figure 3).

Properties	**One-shot**	**Prepolymer**
Elongationat break (%)	550	559
Hardness Shore A	80	82
Modulus at 100% (MPa)	5.6	6.6
Tensile strength at break (MPa)	14.0	11.9
T_g (°C)	-35	-34

Table 2. Comparison of properties of TPUs made from one-shot and prepolymer methods

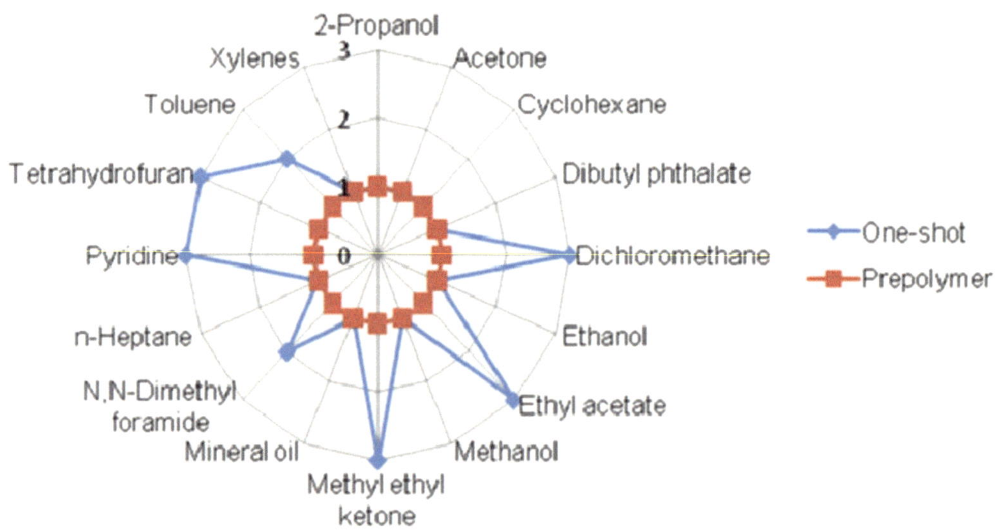

Source: Herbert & Nan, 2010

Fig. 3. Solubility results of TPUs made from one-shot and prepolymer methods in various solvents at room temperature (1: not soluble; 2: soluble after a few days; 3: soluble)

Dynamic vulcanization is a widely used method to prepare thermoplastic elastomers comprising partially or fully cross-linked elastomer particles in melt-processable thermoplastic matrix. Thermoplastic vulcanizates are prepared by melt-mixing the elastomer and thermoplastic in an internal mixer or in a twin-screw extruder. After a well-mixed blend has formed, in the second step, vulcanizing agents such as cross-linkers or curatives are added. Vulcanization of the rubber polymer takes place during the continuation of the mixing process under conditions of high temperature and high shear. According to the earlier investigation (Aubert et al., 2004) the best combinations of elastomer and thermoplastic are those in which the surface energies of the two components are matched, the entanglement molecular weight of the elastomer is low and the thermoplastic is at least 15% crystalline. The most common used compositions are based on dynamically vulcanized ethylene propylene diene monomer (EPDM) and polyolefins. Others blends include butyl and halobutyl rubbers and polyolefin resins, polyacrylate rubber and polyolefins (Soares et al., 2008) and butadiene-acrylonitrile rubber and poly(vinyl chloride) (Passador et al., 2008).

Anionic polymerization remains as an important technique to for the preparation of well-defined styrene butadiene triblock copolymer. Poly(styrene-b-butadiene-b-styrene) (SBS) is an example of a tri-block copolymer, though di-block copolymers are also formed from the same monomers. Styrene is first initiated with butyllithium and polymerised until all of the styrene has reacted. The polystyrene has an anionic end group with lithium counter-ion. Butadiene is added and the polymerisation continues forming a butadiene block. After all of the butadiene has reacted, more styrene is added and the polymerisation continues until all styrene has reacted. Then the polymerisation is terminated by addition of a protic substance such as methanol or water. Termination may be carried out after the second polymerisation step to give a di-block copolymer. The size of blocks is determined by the concentration of initiator and the amounts of monomers added at each step. Molar mass distribution is characteristically low for anionic polymerisation so the macromolecular architecture is accurately controlled.

Carbocationic polymerization has a more complex system than the anionic polymerization described above. It has been used to produce block copolymers with polyisobutylene mid-segments, or poly(styrene-b-isobutylene-b-styrene) (SIBS). This polymerization involves a three-step progression: (i) controlled initiation, (ii) reversible termination (quasi-living systems), and (iii) controlled transfer. The initiators have two or more functionalities. The polymer segments are produced sequentially from monomers as in anionic polymerization. The initiator is reacted with isobutylene at the first stage. The product is a difunctional living polymer. It can initiate further polymerization when more styrene monomers are added. After termination, this gives the block polymer SIBS. Polyisobutylene is the only mid segment that can be produced by this method while there are many aromatic polymers that can form the end segments.

6. Processing methods

Thermoplastic elastomers are technologically very attractive because they can be processed as thermoplastics, this is their main advantage compared with cross-linkable elastomers. They can be re-melted or devitrified and shaped again. Hence, they are generally processed by extrusion and injection moulding, which are the most common processing methods used by thermoplastics. A disadvantage is that TPE have an operating temperature below that at which the hard phase becomes dimensionally unstable.

Several factors need to be taken into account during the processing of TPEs, including viscosity or rheology of the two-phase polymer, temperature at which the hard phase can be processed, thermal stability since the complex structures will potential have several weak chemical links, thermal conductivity since the hard phase is surrounded by soft phase, crystallinity in the hard phase that must be melted with excess enthalpy, and moisture that may cause hydrolysis at processing temperatures.

6.1 Extrusion

Extrusion is a high volume manufacturing process for fabricating parts from thermoplastic elastomers. This processing technique is essential in the melting of raw materials and shaping them into different continuous profiles. The most common extrusion methods are film and sheet extrusion, blow film extrusion, cast film extrusion, coextrusion, tubing extrusion and extrusion coating. The end products made by extrusion are pipe/tubing, wire insulation, film, sheets, adhesive tapes and window frames.

Basically, the extrusion process involves heating a thermoplastic above its melting temperature and forcing it through the die. The extruder is a heating and pressurizing device involves one or more screws operating in a heated barrel. The key determinant of an extruder's performance is the screw. It has three main functions to perform: feeding and conveying the raw material feed; melting, compressing and homogenizing the material; and metering and pumping it through the extrusion die at a constant rate. Raw thermoplastic elastomer material is fed into the barrel of the extruder and comes into contact with the screw. As a melt delivery device, the rotating screw forces the polymer forward into barrel which is heated at a desired temperature. After leaving the screw, the molten travels through a screen pack/plate breaker, where the contaminants in the melt are removed. Breaker plate also creates back pressure in the barrel which is needed for uniform melting and proper mixing of polymer. After that, the molten enters the die, where the cross section of the extruded product is determined.

6.2 Injection molding

Injection molding is by far the most used processing technique of producing parts from thermoplastic elastomers due to its high productivity. Injection molding machines and molds are very expensive because of the high pressures required and complexity of the process control. However, the shortcoming of this technique is balanced by its ability to produce a complex finished part in a single and rapid operation.

The principle of injection molding is very simple. The plastic material is fed into the injection barrel by gravity through hopper. Upon entrance into the barrel, the polymer is heated to the melting temperature. It is then forced into a closed mold that defined the shape of the article to be produced. The mold is cooled constantly to a temperature that allows the molten to solidify and the mold is opened, the finished product is ejected and the process continues.

The injection molding process is capable in producing a variety of parts, from the smallest components to entire body panels of car in a single molding operation. Other part designs obtained from injection molding include threads, springs, storage containers, mechanical parts and automotive dashboards.

6.3 Compression molding

Compression molding was among the first method of molding to be used to produce plastic parts. However, it is by far less used than injection molding. Generally, this method involves four steps. First of all, the raw polymer materials in pellets or powder form are placed in a heated and open mold cavity. The mold is closed with another half of the mold and at the same time, pressure is applied to force the materials into contact with all mold areas. The materials soften under high pressure and temperature, flowing to fill the mold. The part is hardened under pressure by cooling the mold before removal so the part maintains its shape.

There are six important considerations that should be bear in mind, they are the proper amount of material, the minimum energy required to heat the material, the minimum time required to heat the material, the proper heating technique, the force needed to ensure that shots attains the proper shape, the design of the mold for rapid cooling. Compression molding of TPEs usually requires longer heating and cooling time due to their high melting points. Separate platens can be used to solve this problem where one is hot press that is electrically heated and another one is cold press that is water cooled. The part is hot pressed under pressure and then transferred immediately to the cold press to chill it under pressure. The hot press is usually preheated to reduce the total cycle time.

6.4 Transfer molding

Transfer molding is a process which the polymer is melted in a separate chamber known as pot then forced into a preheated mold through a sprue, taking a shape of the mold cavity. The mold is cooled down before opening. Thermoplastic elastomers usually have high viscosity and longer transfer time is needed. The temperature of the mold should be maintained at above melting temperature of the polymer to avoid premature cooling or freezing before the completion of transfer. The important variables during the process of transfer molding are the type of polymer, melting point of the polymer, pot hold time, transfer pressure, transfer rate and the mold cooling time.

6.5 Blow molding

Blow molding is a manufacturing process that is used to produce hollow plastic parts. There is a wide variety of materials can be used in this process, including but not limited to high density polyethylene (HDPE), low density polyethylene (LDPE), polypropylene (PP), Poly(vinyl chloride) (PVC) and Poly(ethylene terephtalate) (PET). The basic process begins with the melting of thermoplastic and extruding it through a die head to form a hollow tube called a parison. The parison is then clamped between two mold halves, which close around it and the parsion is inflated by pressurized air until it conforms to the inner shape of the mold cavity. Lastly, the molds open and the finished part is removed.

Basically, there are three types of blow molding used to form the parison. In extrusion blow molding, plastic is melted and extruded using a rotating screw to force the molten through a die head that forms the parsion. Injection blow molding is part injection molding and part blow molding where the molten plastic is injection molded around the core pin and then the core pin is transferred to a blow molding station to be inflated. There are two stretch molding techniques. In one-stage process, the preform is injection molded which is then transferred to the blow mold where it is blown and ejected from the machine. In the two-stage process, preform is injection-molded, stored for a short period of time, and blown into container using a reheat blow machine.

6.6 Thermoforming

Thermoforming is a process which uses heat and pressure or vacuum to transform thermoplastic flat sheet into a desired three-dimensional parts. The sheet is drawn from large rolls or from an extruder and then transferred to an oven for heating to its softening temperature. The heated sheet is then transferred to a preheated, temperature-controlled mold. Vacuum is applied to remove the trapped air and deform the sheet into the mold cavity, where it is cooled to retain the formed shape. After that, a burst of reverse air pressure is applied to break the vacuum and assist the formed part out of the mold. The principal factors in this process include the forming force, mold type, sheet prestretching, the material input form and the process phase condition. These factors have a critical effect on the quality and properties of the final products.

6.7 Calendaring

Calendaring is a process where a large amount of molten plastic is fashioned into sheets by passing the polymer between a set of rollers. The rollers are hot and keep the polymer in its semi-molten state. This allows the molten to be rolled many times until the desired thickness is reached. The sheet is then rolled through cold rollers to enable it to go hard and then wound up into rolls. Calendar for thermoplastics generally operates in four-roll units made up of three banks, each bank being wider than the preceding one. The advantages of calendar over extruder are the possibility of calendar to produce embossed films, sheets and laminates and the higher output than extruder. Examples of the final products are cling film, shrink film, clear, translucent rigid sheets for blister packaging and opaque flexible film.

7. Plasticisers

Plasticisers are the additives added to polymeric materials to improve their flexibility and durability by spacing them apart. A plasticiser-polymer mix contains more free volume than a pure polymer, thus, the plasticised polymer need to be cooled to a lower temperature to reduce its free volume which defines glass transition temperature of the polymer. There are many types of plasticisers can be used to modify the thermoplastic elastomers and enhancing their utilities. Each of the plasticisers has a compatibility with a specific type of polymer. Among them, ester plasticisers have a well known function in the TPEs due to its exceptional ability to provide improved low-temperature by plasticising the soft phase while allow the hard phase to stay intact for strength and high temperature properties. Some examples of ester plasticisers are phthalate esters which are used in situations where good resistance to water and oil is required, adipic esters which are used for low temperature or resistance to UV light and trimellitic esters which are used in automobile interiors and where resistance to high temperature is required. TPU tend to be internally plasticised by choice of monomer combination. For polyolefin elastomers, it is important to identify plasticisers which have low glass transition temperature and high boiling temperature and which are compatible over a broad temperature range with both rubber and polyolefin plastic components.

8. Additives

Common additives in TPEs include those materials added during or after polymerization to prevent their degradation, during monomer recovery, drying and compounding, and also storage.

8.1 Antioxidants

Among the additives, antioxidants are used to prevent oxidation and degradation. Primary or free radical scavenging antioxidants, which have reactive hydroxyl and amine groups, inhibit oxidation via chain terminating reactions. Secondary antioxidants inhibit oxidation of polymers by decomposing hydroperoxides.

8.2 Nucleating agents

Nucleating agents are generally used to enhance the formation of nuclei for the growth of crystal in the polymer melt. A higher degree of crystallinity and a more uniform crystalline structure in the hard phase can be obtained by adding a nucleating agent in the polymer. Nucleating agents can be classified as inorganic additives (talc, silica), organic compounds (salts of mono- or polycarboxylic acids) and polymer. Nucleating agents may be used to enhance crystallinity of a hard phase segment.

8.3 Colorants

Colorants are often referred to as dyes and pigments. Generally, dyes are soluble in water while pigments are not. The colours from dyes are produced from the light absorption and they are transparent. Pigments produce colours from the dispersion of fine particles throughout the resin. Inorganic pigments are thermally stable than organic pigments. They are less transparent and resistant to migration, chemicals and fading. Some examples of inorganic pigments are oxides, sulfides, hydroxides, and other complexes based on metal.

8.4 Flame retardants

Most thermoplastics are flammable, burning easily when heated to high temperature. Flame retardants are added to polymer to delay the ignition and burning of polymer. Char-formers form a foamy porous protective barrier on the polymeric material to shield it for further combustion. Flame retardants acting in the condensed phase deposit a layer on the surface of polymer to prevent it from the heat source while flame retardants acting in gas phase interrupt the combustion chemistry of the fire.

9. Composites

Various fillers and reinforcements have been introduced into thermoplastic elastomers to enhance their processability and mechanical properties, as well as to reduce material costs.

Most common fillers used in TPEs include cubic and spheroidal fillers (calcium carbonate, silica, carbon black), fibrous fillers (glass fibers, aramid fibers), platy fillers (kaolin, mica, talc) and nanofillers (carbon nanotubes, nanoclays, nanosilica). Reinforcing TPEs with fillers such as silica, clay, carbon black, carbon nanotubes, natural fiber results in better thermal and mechanical properties of the composites.

Carbon black composites with polyether polyurethane exhibited a percolation threshold of 1.25 % v/v and significant conductivity at 2 % v/v carbon black content (Wongtimnoi et al., 2011). Electric field induced strain was observed due to an increase in dielectric constant. Polyester thermoplastic elastomers reinforced by mica showed significant increment in the flexural, thermal and electrical properties with an increase in the filler concentration. The improved thermal properties are attributed to the small and uniform crystallite size distribution with the addition of mica (Sreekanth et al., 2009). Composites containing silica and poly(styrene-b-ethylene-co-butylene-b-styrene) (SEBS) block copolymer-based thermoplastic elastomer showed an improvement in the mechanical properties such as tear strength due to the strong interaction between the fillers and polymer matrices where the silica particles are wetted by the polymer (Veli et al., 2009) . Polypropylene/natural rubber (PP/NR) and poly(propylene-ethylene-propylene-diene-monomer) (PP/EPDM) reinforced by kenaf natural fibre with maleic anhydride polypropylene (MAPP) as a compatibilizer agent has significantly increased the tensile strength, flexural properties and impact strength as compared to unreinforced thermoplastic elastomer. The improvement achieved in mechanical properties was due to the interaction both matrix system and kenaf fibre (Anuar & Zuraida, 2011).

10. Morphology

The disperse hard-phase in TPE self-agglomerates after processing to form reinforcing domains within the elastomeric matrix. The morphology contribution to elastomeric properties has been investigated using a semi- phenomenological approach (Baeurle et al., 2005). The authors describe an extended domain model for the size and distribution of hard phase within the elastomer and the contribution to stress relaxation times. Relaxation was modelled using a stretched exponential function to correlate stress decay due to multiple length-scales and time-scales. Behaviour under stress for long times resulted from plastic flow, chain pull-out from hard domains and finally disruption of the hard domains.

Transmission electron microscopy (TEM) is one of the most powerful equipment to characterize the structure and morphology of thermoplastic elastomers. It is often used to interpenetrate polymer networks (IPNs), morphology and crystallinity of hard blocks, structural evolution of segmented copolymers under strain and blends. The morphology of the IPNs transformed from SBS by using γ-radiation as shown by TEM is a homogeneous and sponge-like network (Robert et al., 2003). Sometimes, TEM is associated with other techniques, particularly small-angle X-ray scattering (SAXS), wide-angle X-ray scattering (WAXS), small-angle neutron scattering (SANS), and atomic force microscopy (AFM) to provide more information on the microstructure of TPEs. TEM and SANS have been used to examine the morphology of tough-semi- and full- IPNs with SIS crosslinkings and PS at different content (Bhavna & Robert, 2001). TEM showed similar structure for both samples which are different in total of PS content, while for SANS, the plot of intensity against the scattering vector q showed that both structures are significantly different with the tough

IPNs (lower PS content) having two types of domains. The morphology of natural rubber (NR)/high density polyethylene (HDPE) reinforced with carbon black composite was examined using TEM, SAXS and SANS (Kazuhiro et al., 2005). TEM image showed that NR and HDPE were phase-separated in the blends and carbon black nanoparticles were located in the NR domains. This phenomenon can be explained by the chemical groups on the surface of carbon blacks which chemically absorb olefin well.

11. Elastomer polymer blends

Elastomers are often used in blends with other polymers. When the elastomer is the minor component it will constitute a disperse phase. A disperse phase elastomer will be a toughening additive for the matrix phase that could be a thermoplastic or a thermoset polymer.

When the elastomer is the major component it will be a matrix phase and the overall blend will be an elastomer. The disperse phase blended polymer will contribute to physical cross-links that will prevent creep and assist with reversibility of elastomer deformation.

12. Mechanical properties

12.1 Stress-strain

When tested under a tensile stress-strain condition, TPE behave as elastomers until the yield stress after which they can undergo plastic flow, part of which may be viscoelastic (time-dependant recovery) and part will be permanent set. The pre-yield region represents an elastomeric response where the physical cross-links are not deformed. The stress-strain curve of SBS is shown in Figure 4.

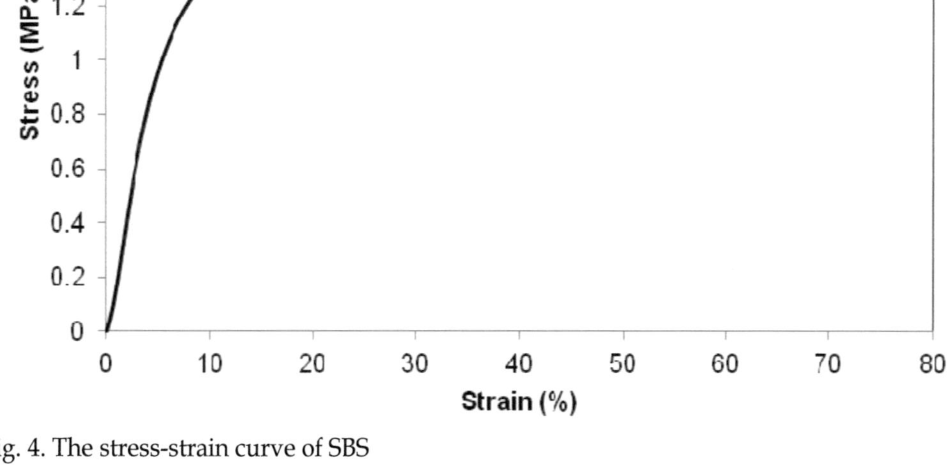

Fig. 4. The stress-strain curve of SBS

12.2 Dynamic mechanical analysis

Dynamic mechanical analysis (DMA), uniaxial tensile and microscopic properties were used to observe properties and transitions at lower temperatures to the glass transition of the elastomer phase (Adhikari et al., 2003). Two distinct transitions were detected at the lower temperatures evaluated, an elastomeric to ductile thermoplastic transition, and a thermoplastic to brittle transition. A typical DMA curve of SBS is shown in Figure 5. Scanning electron microscopy (SEM) revealed an associated change in deformation mechanism.

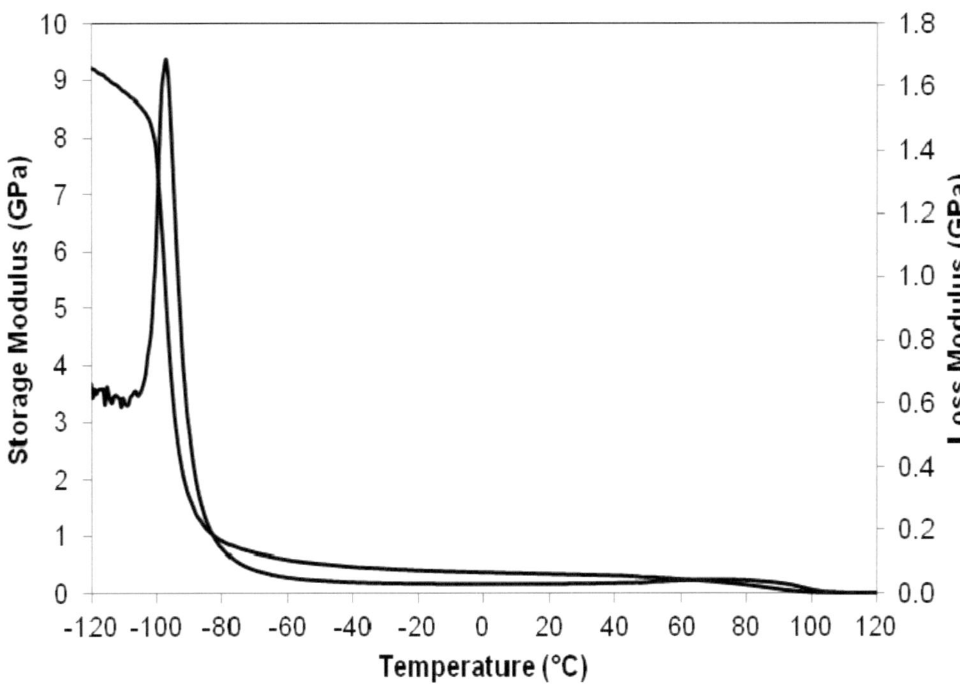

Fig. 5. The storage and loss modulus curves of SBS

12.3 Creep and recovery

Creep is a problem for TPE because there are only physical crosslinks that can dissociate and flow, chemical crosslinks are permanent and creep will be low. TPEs soften and melt with increasing temperature, showing creep on extended use. Creep resistance and tensile strength are generally directly related. A softer TPE will creep more and have less tensile strength than a harder TPE. Recovery should be elastic with some viscoelasticity, permanent set is not suitable in TPE. The deformation behaviour of TPEs during creep flow can be divided into three strain regimes: linear regime at low strains where the recovery from creep is complete; the transient regime, where the viscosity shows a maximum; flow regime where steady state morphology is formed. Creep and recovery curve of SBS is shown in Figure 6.

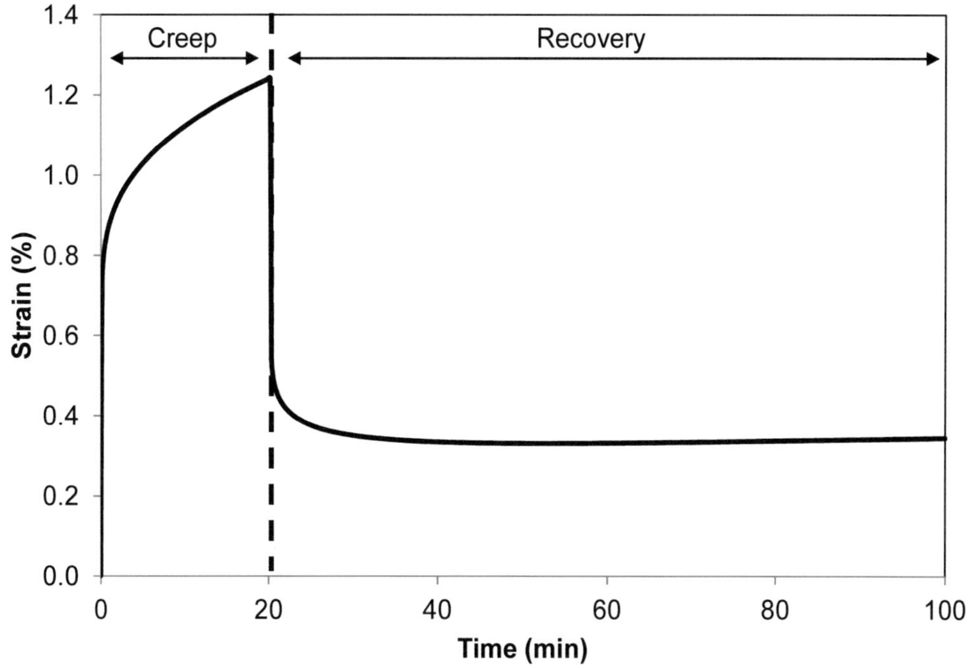

Fig. 6. The creep and recovery curve of SBS

12.4 Stress relaxation

Stress relaxation is a technique can be used as a physical method to determine the domain structure of the TPEs by studying the time-dependent deformation behaviour of the segmented block copolymer. There is no stress relaxation occurs in ideal elastomer.

12.5 Strain hardening

At high strain, strain hardening takes place and converts TPE from elastic behaviour to leathery or stiffer characteristic so that elasticity is lost while tensile strength and modulus will be increased. In time, the recovery of the deformation occurs suggesting that the part where deformation occurs is viscoelastic.

12.6 Tear strength

Tear strength which describes how well the elastomer resists tearing. TPE is stretched and the amount of force required is recorded. Peel strength is a measure of how well a TPE has bonded to a rigid substrate.

13. Applications

Thermoplastic elastomers have been widely used in automotive sector, medical devices, mobile electronics, household appliance sector and construction to replace conventional vulcanized rubber.

- Automotive: windshield seal (SEBS), wire/cable (SEBS, TPU), fibre reinforced soft touch surface for interior (TPO), gaskets (TPV) and spoiler (SEBS)
- Medical devices: syringe (TPV), medical tubing (TPO), medical wrapping and packaging (TPU)
- Mobile electronics: wire/cable (SEBS), earplugs (TPV), cell phone (TPV).
- Household appliance sector: sporting goods (TPU), footwear soles (SBS), toys (SEBS), adhesives (SIS)
- Construction: gaskets (SEBS)

The only constraint of TPEs is the physical reversible crosslinks need to be disrupted by heat to mould, but they maybe disrupted during use.

14. Future directions

Thermoplastic elastomer structure, performance and specialty applications are interconnected with block copolymer molecular characteristics. Block continuity, molar mass, molar mass distribution and stereochemistry must be controlled. New polymerisation techniques and initiators are expanding the choice of monomers that can be polymerised under controlled conditions. This allows monomer selection and molecular architecture to provide elastomers with chemical resistance, self-healing, abrasion resistance and unique mechanical performance to be prepared for special applications.

Further thermoplastic elastomers can be prepared by creating blends of an elastic polymer with a dimensional stabilising polymer. Inclusion of ionomers as a physically cross-linking phase can be extended with carboxylates, sulfonates and phosphates with various metal ions. In conjunction with polyfluorocarbon elastomers chemical resistance can potentially be improved.

Nanocomposites are being formed with elastomers where the nanoparticles form self-assembled clusters or bridges to provide physical cross-links. Nano-particles have traditionally been used to modify elastomeric properties. Such modification will be expanded through increasing knowledge of nano-composite preparation and morphologies. Carbon black is a much used nano-fillers that is now being assisted or replaced by carbon nanotubes, graphenes and silicas with a diversity of surface modifications. Carbon blacks are known to form reversible clusters and to binder elastomer molecules within the clusters. Nano-silicas remain in multi-particles aggregates while forming reversible agglomerates that enhance absorption of elastic energy. Now carbon nanotubes and graphenes have been found to create reversible networks at low volume fraction.

15. Conclusions

The characteristics of an elastomer require that there is a mechanism to provide reversible deformation. Only the elastic component of the three-component viscoelastic model must be active. The viscous contribution resulting from molecules sliding past each other, this results in irreversible flow, is eliminated by cross-linking in a thermoset elastomer. Physical crosslinks are present in a thermoplastic elastomer as a second vitrified or crystalline phase. The viscoelastic component can be reduced by minimising chain stiffness and intermolecular interactions in the continuous elastic phase. Thermoplastic elastomers offer ease of processing the same as thermoplastics without the need for a separate curing reaction. Waste material can be reprocessed and production rates will be fast consistent with a thermoplastic. Upper application temperature limitations exist dependent upon the glass transition or melting temperature of the hard phase. Stress resistance is limited to the yield stress of the hard phase since permanent deformation will follow distortion or flow of the hard phase. Thermoplastic elastomers are enhanced by fillers, with nano-fillers having particular relevance when small amounts can support the hard phase. In the soft phase fillers will modify the elastic response. Structural diversity is found in thermoplastic elastomers with many chemical structures such as polyurethanes and polyolefins available as both thermoset and thermoplastic elastomers.

16. References

Adhikari, R., Godehardt, R., Huy, T. A., & Michler, G. H. (2003). Low temperature tensile deformation behavior of styrene/butadiene based thermoplastic elastomer. *KGK, Kautschuk Gummi Kunststoffe*, Vol. 56, No. 11, pp. 573-577, ISSN 0948-3276

Anuar, H., & Zuraida, A. (2011). Improvement in mechanical properties of reinforced thermoplastic elastomer composite with kenaf bast fibre. *Composites Part: B*, Vol. 43, pp. 462-465, ISSN 1359-8368.

Aubert, Y., Coran & Raman, P. P. (2004). Thermoplastic Elastomers Based on Dynamically Vulcanized Elastomer-Thermoplastic Blends, In: Thermoplastic Elastomers, Holden, G., Kricheldorf, H. R., and Quirk, R. P., pp. 143-182, Hanser Publishers, ISBN 1-56990-364-6, Munich.

Baeurle, S. A., Hotta, A., & Gusev, A. A. (2005). A new semi-phenomenological approach to predict the stress relaxation behavior of thermoplastic elastomers. *Polymer*, Vol. 46, No. 12, pp. 4344-4354, ISSN 0032-3861.

Bhavna, B. H., & Robert, P. B. (2001). Interpenetrating polymer networks based on a thermoplastic elastomer, using radiation techniques. *Radiation Physics and Chemistry*, Vol. 62, pp. 99-105, ISSN 0969-806X.

Drobny, J. G. (2007). *Handbook of Thermoplastic Elastomer*, William Andrew Publishing, ISBN 978-0-8155-1549-4, New York, USA.

Herbert, C., & Nan, T. (2010). Property difference of polybutadiene-derived thermoplastic polyurethanes based on preparative methods. *Cray Valley Technical Bulletin*, 5706

Holden, G. (2011). Thermplastic Elastomers, In: *Applied Plastics Engineering Handbook: Processing and Materials*, Myer, K., pp. 77-92, Elsevier, ISBN 978-1-4377-3514-7, Oxford, UK.

Kazuhiro, Y., Satoshi, A., Hirokazu, H., Satoshi, K., Chudej, D., Pasaree, L., Jipawat, C., & Areeratt, K. (2005). Structural study of natural rubber thermoplastic elastomers and their composites with carbon black by small-angle neutron scattering and transmission electron microscopy, *Composites Part: A*, Vol. 36, pp. 423-429, ISSN 1359-835X.

Passador, F. R., Rodolfo Jr. A. & Pessan, L. A. (2008). Dynamic vulcanization of PVC/NBR blends, *Proceedings of the Polymer Processing Society 24th Annual Meeting*

Robert, P. B., Martin, G. M., & Robert, B. K. (2003). Small angle neutron scattering and transmission electron microscopy studies of interpenetrating polymer networks from thermoplastic elastomers. *Nuclear Instruments and Methods in Physics Research B*, Vol. 208, pp. 58-65, ISSN 0168-583X.

Soares, B. G., Santos, D. M. & Sirqueira, A. S. (2008). A novel thermoplastic elastomer based on dynamically vulcanized polypropylene/acrylic rubber blends. *Express Polymer Letters*, Vol. 2, pp. 602-613, ISSN 1788-618X.

Sonnenschein, M. F., Guillaudeu, S. J., Landes, B. G., & Wendt, B. L. (2010). Comparison of adipate and succinate polymers in thermoplastic polyurethanes. *Polymer*, Vol. 51, No. 16, pp. 3685-3692, ISSN 0032-3861.

Sreekanth, M. S., Bambole, V. A., Mhaske, S. T., & Mahanwar, P. A. (2009). Effect of concentration of mica on properties of polyester thermoplastic elastomer composites. *Journal of Minerals & Materials Characterization & Engineering*, Vol.8 , No. 4, pp. 271-282, ISSN 1539-2511.

Veli, D., Nursel, K. & Osman, G. E. (2009). Effects of fillers on the properties of thermoplastic elastomers. *SPE Plastic Research Online*, 10.1002/spepro.002518

Wongtimnoi, K., Guiffard, B., Bogner-Van de Moortele, A., Seveyrat, L., Gauthier, C., & Cavaille, J. Y. (2011). Improvement of electrostrictive properties of a polyether-based polyurethane elastomer filled with conductive carbon black. *Composites Science and Technology*, Vol. 71, No. 6, pp. 885-888, ISSN 0266-3538.

9

Modification of Thermoplastics with Reactive Silanes and Siloxanes

Jerzy J. Chruściel* and Elżbieta Leśniak
Technical University of Łódź, Faculty of Chemistry,
Institute of Polymer and Dye Technology,
Łódź
Poland

1. Introduction

In a contemporary world goods made from plastics and other polymeric materials are applied in many areas of our life. Growing practical applications are mainly stimulated by better properties of modified polymers, in a comparison with the polymeric materials used so far. On a world polymer market a biggest production concerns thermoplastics, thus modification of their properties has become one of a most important research challenges in a field of a polymer chemistry and technology, and materials science as well.

Silicones (polysiloxanes) are a large and most important group of various inorganic-organic (hybrid) compounds and materials, composed of silicon and oxygen atoms in their main chains and organic substituents bound to silicon. Silicones play an important role among polymers with special properties, because they possess many unusual features. Even an addition of a very small amount of silicones causes a crucial improvement of properties of modified materials. Most importantly: silicones increase hydrophobicity and improve water resistance and thermal stability of many materials. Silicones exhibit excellent chemical, physical, and electrical properties. Most popular organosilicon polymers are polydimethyl-siloxanes (PDMS). Silicones are mainly applied as silicone oils, rubbers, and resins (Noll, 1968; Rościszewski & Zielecka, 2002). Similar positive effects on properties of polymers and other materials can be reached by the addition of reactive silanes, siloxanes, and silicates, which are also used very often in practice for the modification of polymeric and inorganic materials. An important practical meaning have also other organosilicon polymers (polysilanes, polycarbosilanes, polysilazanes, etc.), and many functional silanes with different chemical structures, containing reactive groups, mostly bound to silicon atom, but also quite often attached to carbon.

We observe a continuously growing interest in applications of reactive silanes and polysiloxanes in many different fields of science (with focus on materials science) and the chemical technology, and this is a subject of our review.

---

*Corresponding Author

2. Reactive silanes and siloxanes

2.1 Functional groups and reactive silanes in silicon chemistry

In organosilicon chemistry most important are following functional groups: ≡Si-Cl, ≡Si-OR, ≡SiOCOR (R - usually alkyl group), ≡Si-H, and ≡Si-CH=CH_2. They are present both in all kinds of organosilicon compounds and polymers and can be grafted on silica, silicates, and some inorganic fillers. Functional organic silanes include first of all: chlorosilanes, alkoxysilanes, and acetoxysilanes (e.g. dimethyldichlorosilane, methyltriethoxysilane, methyltriacetoxysilane, etc.), and different functional and carbofunctional silanes (Brook, 2000, Eaborn, 1960; Kricheldorf, 1996). Moreover very often are used inorganic chlorosilanes: silicon tetrachloride $SiCl_4$, trichlorosilane $HSiCl_3$, and dichlorosilane H_2SiCl_2, and silicate compounds: esters of silicic acid, and especially tetraethoxysilane $Si(OEt)_4$, tetramethoxysilane $Si(OMe)_4$, and tetracetoxysilane $Si(OAc)_4$. Chlorosilanes are produced on an industrial scale by Rochow-Miller process, and their fast reactions with alcohols (or metal alkoxides) and carboxylic acids lead to appropriate alkoxysilanes or acyloxysilanes, respectively.

2.2 Functional siloxanes and polysiloxanes

Siloxane oligomers (linear and cyclic) and polysiloxanes (silicones) are mainly prepared by so called a multistep "hydrolytic polycondensation" of chlorosilane monomers of different functionality. This step-growth process consists of elementary hydrolysis and condensation reactions:

$$\equiv\text{Si-Cl} + H_2O \longrightarrow \equiv\text{Si-OH} + HCl \qquad (1)$$

$$\equiv\text{Si-Cl} + \text{HO-Si}\equiv \longrightarrow \equiv\text{Si-O-Si}\equiv + HCl \qquad (2)$$

$$\equiv\text{Si-OH} + \text{HO-Si}\equiv \longrightarrow \equiv\text{Si-O-Si}\equiv + H_2O \qquad (3)$$

Linear oligo- and polysiloxanes with diferent functional (α,ω-terminal or pendant) groups and functional cyclosiloxanes find many applications in further syntheses. These functionalities include most often silanol, chlorosilyl, hydrosilyl, vinyl, and allyl groups.

2.3 Carbofunctional Silanes (CFS)

Carbofunctional silanes (CFS), which are used for a long time in a processing of plastics may be considered as precursors of carbofunctional polysiloxanes. The carbofunctional silanes are compounds with a general formula $X_nSi(R'Y)_{4-n}$, where: R' - alkylene chain, Y - functional group, e.g. Cl, NH_2, NR_2, OH, OCOR, NCO, CH_2=CH, SH, and X - a functional group sensitive to hydrolysis (Cl, OR, OCOR). Alkylene chain R' is usually built of three methylene groups (Marciniec et al., 1990, 1992a, 1993; Chruściel et al., 2008a).

The carbofunctional silanes are usually prepared in a three step synthesis. In a first step a catalytic hydrosilylation reaction of allyl chloride with hydro(alkyl)chlorosilanes is carried out, in a second step - alcoholysis of addition products, and in a last step - substitution of chlorine atom in (chloropropyl)trialkoxysilane $Cl(CH_2)_3SiMe_n(OR)_{3-n}$ by a catalytic nucleophilic substitution - take place (Guliński et al., 2003).

2.4 Carbofunctional Polysiloxanes (CFPS)

Weak interactions between polysiloxane chains (through van der Waals forces) are responsible for poor mechanical properties of these polymers at room temperature, even in the case of PDMS with a very high molecular weights (Abe & Gunji, 2004). An improvement of mechanical properties of PDMS can be reached by addition of fillers (most often silica is used), by a dense crosslinking in the presence of peroxides, or by synthesis of siloxane-organic copolymers of different structures (block, segmented, or graft). Since both systems are thermodynamically non-miscible it is impossible to prepare siloxane-organic copolymers from (α,ω-dihydroxy) polydimethylsiloxanes, and it was necessary to use for this purpose α,ω-dihydro- or α,ω-divinylpolysiloxanes, but first of all to use the carbofunctional polysiloxanes (CFPS) (Chruściel et al., 2008b).

The general structure of carbofunctional polysiloxanes is presented on Fig. 1.

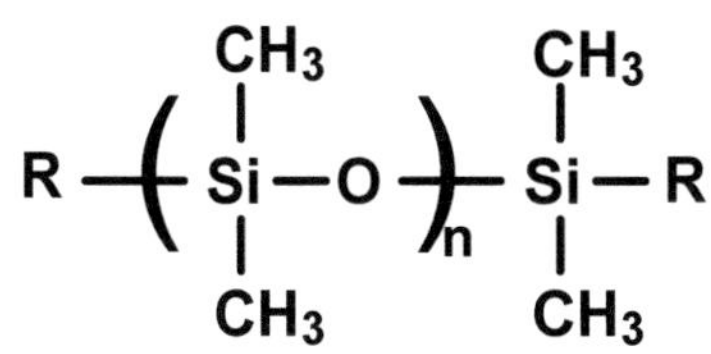

Fig. 1. The chemical structure of α,ω-dihydro-, α,ω-divinyl-, and the carbofunctional polysiloxanes (R = H, CH = CH_2, aminoalkyl, hydroxyalkyl, chloroalkyl groups, etc.)

2.4.1 Synthesis of CFPS

Macromolecules of the carbofunctional polydimethylsiloxanes (CFPS) are terminated with different functional alkylene groups. Most often on both ends of CFPS chains exist hydroxypropyl, aminopropyl, glicydoxypropyl, or methacryloxypropyl groups. The carbofunctional polysiloxanes may also be terminated with alkene groups (e.g. allyl groups) or with arylamine end groups: -$C_6H_4NH_2$ (Kawakami & Abe, 1992). Quite often for synthesis of the carbofunctional polydimethylsiloxanes are used α,ω-dihydrosiloxanes, which are synthesized in condensation reactions of polysiloxane-α,ω-diols with chloro(hydro)-dimethylsilane. The CFPS containing terminal carbinol groups (C-OH) are products of hydrosilylation reaction of different α,ω-dihydropolysiloxanes with allyl derivatives (Marciniec et al., 1992b), e.g. allyloxytrimethylsilane (Greber & Jäger, 1962), followed by hydrolysis reaction of alkoxysilane end groups (Scheme 1).

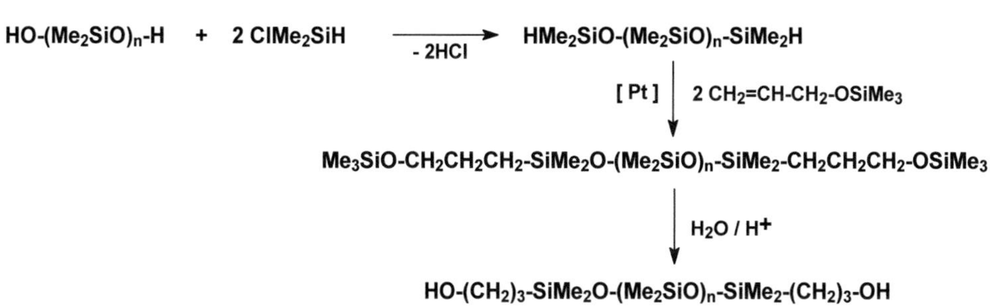

Scheme 1. The preparation of α,ω-di(hydroxypropyl)polydimethylsiloxanes.

The hydrosilylation is the fundamental reaction used for the preparation of the carbo-functional polysiloxanes (Brook, 2000; Marciniec et al., 1992b), and platinum Speier's and Karstedt's catalysts are not enough effective in the case of addition of allylamine to tetra-methylcyclotetrasiloxane D_4^H (D^H = MeHSiO), giving low yield and selectivity of products (Wu & Feng, 2001). Instead of a very useful catalyst of this reaction is platinum oxide PtO_2, resistant to "poisoning" by amine groups, giving the product with almost 100 % yield and the very good selectivity, determined by a ratio of isomers γ to β 93:7 (Zhou et al., 2004).

A very common method of the preparation of the carbofunctional polysiloxanes are catalytic equilibration reactions of cyclic siloxanes, e.g. octamethylcyclotetrasiloxane $(Me_2SiO)_4$ (D_4) with carbofunctional disiloxanes, which are carried out in acidic or basic media (Scheme 2) (Yilgor & McGrath, 1988; Harabagiu et al., 1996).

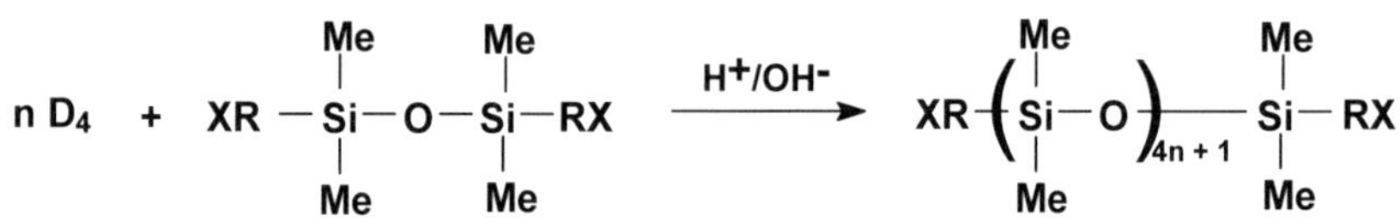

Scheme 2. The preparation of CFPS by the catalytic equilibration of the carbofunctional disiloxanes with D_4.

For example, by the equilibration of D_4 (D = Me_2SiO) with (3-aminopropyl)methyldiethoxy-silane and hexamethyldisiloxane $Me_3SiOSiMe_3$, polydimethyl-*co*-[(aminopropyl)(methyl)]-siloxane oils (containing side aminopropyl groups) were prepared (Yang et al., 2007).

By anionic copolymerization of D_4 with $(Ph_4SiO)_4$ and $(ViMeSiO)_4$ towards 1,3-bis(3-amino-propyl)tetramethyldisiloxane (as a chain terminating agent) poly(dimethyl-*co*-diphenyl)-siloxane containing terminal aminopropyl groups was prepared. This product was utilized for a synthesis of a segmented poly(imide-siloxane) copolymer, and subsequently for the preparation of a hybrid nanocomposite, reinforced with silica (Liaw, 2007a) or titania TiO_2 (Liaw, 2007b).

The carbofunctional polysiloxanes, in which alkylfunctional groups are attached to the silicon atoms inside of the polymer chain were also prepared by the catalytic equilibration of cyclotetrasiloxane, containing aminopropyl functional groups (D_3D^{NH2}), with decamethyl-tetrasiloxane MD_2M (D = Me_2SiO, M = $Me_3SiO_{0.5}$) in the presence of $Me_4NOSiMe_3$ (TMAS) (Zhou et al., 2004; Yang et al., 2007), as it is presented on a Scheme 3:

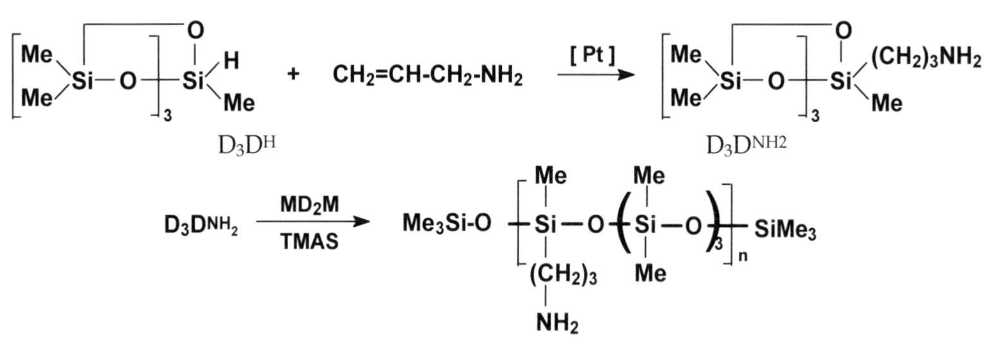

Scheme 3. The preparation of CFPS with pendant aminopropyl functional groups.

By a controlled hydrolytic copolycondensation of (3-aminopropyl)(methyl)diethoxysilane and [(3-aminoethylamino)propyl](methyl)dimethoxysilane - a mixture of linear and cyclosiloxane oligomers containing 3-aminopropyl and [(3-aminoethylamino)propyl substituents was prepared (Kichler et al., 2003). They were applied as new generation polycationic supports of DNA (*"gene transfer reagents"*).

Carbofunctional PDMS with different photoactive linkages, i.e. benzoin (Yagci et al., 1994), glicydoxyl (Harabagiu et al., 1995), or benzylacryl groups (Iojoiu et al., 2000) were also prepared.

The synthetic methods, mentioned in this Chapter, do not cover many other methods of the preparation of the carbofunctional polysiloxanes.

2.5 Polyhedral Silsesquioxanes (POSS)

Silsesquioxanes (SSO) are important hybrid materials. Silsesquioxane or T unit of general formula $RSiO_{1.5}$ can exist in several structural types such as random, ladder, cage or semi-cage structures. SSO with a cage structure is also called polyhedral oligomeric silsesquioxane (POSS). POSS are organic-inorganic molecules, approximately 1-3 nm in size, with a general formula $(RSiO_{1.5})_n$ or T_n, where R is mostly organic group, which can be suitable for polymerization or grafting (Baney et al., 1995). Among various oligosilsesquioxanes, aromatic POSS are the most interesting compounds because of their high temperature stability (> 500 °C for octaphenylsilsesquioxane).

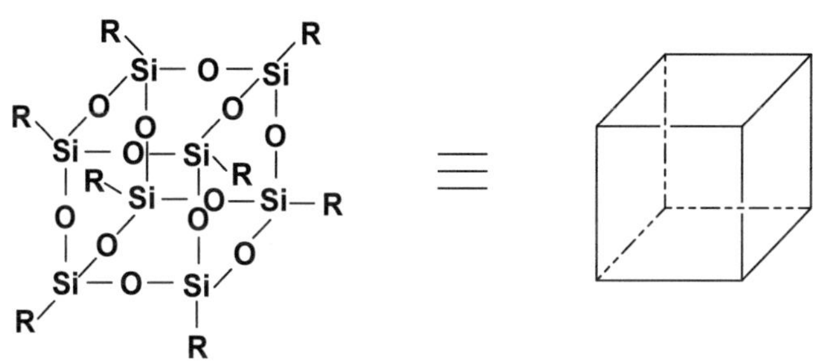

Fig. 2. Chemical structure of T_8 (Williams et al., 2011).

3. Technical applications of reactive silanes and siloxanes

An interest concerning the carbofunctional silanes is still growing and broadens a scale of their numerous applications. A great practical meaning have also carbofunctional polysiloxanes.

3.1 Applications of carbofunctional silanes

The carbofunctional silanes $X_nSi(R'Y)_{4-n}$ have found misceleneous practical applications thank to the presence of two kinds of the functional groups X and Y. In practice formation of chemical bonds with an inorganic material is possible, as a result of reactions of functional groups X, which most often are alkoxy groups. On the other hand the functional groups Y enable reactions with organic polymers. Most popular, available on market, are compounds of a structure $(RO)_3Si(CH_2)_3Y$, in which as alkoxy substituents are present methoxy or ethoxy groups, and as organic functional groups Y: amine -NH_2, glicydyloxy -O-CH_2-(CH-CH_2O) or methacryloxy -O-CO-C(CH_3)=CH_2 groups.

3.1.1 Processes sol-gel

The carbofunctional silanes are often used in sol–gel processes for a preparation of hybrid organic–inorganic materials (Brinker, 1990; Schmidt, 1984, 1994). In sol–gel systems take place hydrolysis and condensation reactions of alkoxysilanes, described by equilibrium elementary reactions (4)-(6), which are catalyzed both by acids and bases:

$$\equiv Si\text{-}OR + H_2O \Leftrightarrow \equiv Si\text{-}OH + ROH \quad (4)$$

$$\equiv Si\text{-}OR + HO\text{-}Si\equiv \Leftrightarrow \equiv Si\text{-}O\text{-}Si\equiv + ROH \quad (5)$$

$$\equiv Si\text{-}OH + HO\text{-}Si\equiv \Leftrightarrow \equiv Si\text{-}O\text{-}Si\equiv + H_2O \quad (6)$$

In an acidic medium hydrolysis reactions undergo faster than condensation reactions, but in a basic medium condensation reactions are faster. Products of the sol–gel process of tetra-ethoxysilane are: silicagel, branched poly(ethoxysilicates) or silsesquioxanes, as it is presented on a Scheme 4 (Abe & Gunji, 2004).

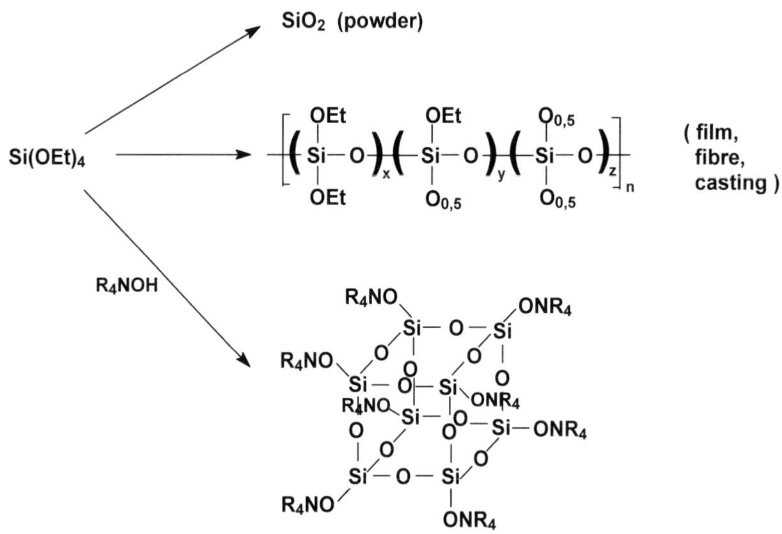

Scheme 4. Products formed in the sol–gel process of tetraethoxysilane.

New 2,6-di-tert-butylphenol antioxidants and their derivatives containing trialkoxysilane linkages were prepared by the sol-gel method. They were applied for a stabilization of isotactic polypropylene (Nedelcev et al., 2007). The sol-gel technique is also applied in order to get organic materials with better properties, through an introduction of metal alkoxides.

3.1.2 Modification of polymers properties by CFS

The carbofunctional silanes have found wide applications in modifications of many organic polymers. For example, unsaturated polyesters modified with (methacryloxypropyl)silanes are utilized in a manufacture of glass-reinforced laminates. Silanes containing oxirane groups are applied for surface treatment of fillers which are used in epoxide composites. They are also utilized for the modification of novolac resins through reaction of hydroxyl groups of phenol with epoxy groups. Polycarbonates (PC) and polysulphones (PSU) form amide or sulphonamide bonds, and polyurethanes - urea bonds, respectively. This way modified polyamides, PC and PSU can be easier reinforced with the fillers, giving composite materials showing very good mechanical properties (Arkles, 1977).

For the modification of polymers most often are used (aminoalkyl)silanes. Chlorinated polymers, e.g. poly(vinyl chloride) (PVC), easily form quaternary salts with (aminoalkyl)-silanes (Scheme 5).

Silane agents, terminating polymer chains (*"endcappers"*), were prepared by addition Michael reactions of (3-aminopropyl)trimethoxysilane to acrylates. They were used for an introduction of trimethoxysilane groups on chain ends of polyurethanes, enabling their cross-linking under action of atmospheric moisture. It established a base for elaboration of a technology of newer one-component, solventless adhesion materials (*"solvent-free"*) (Nomura et al., 2007).

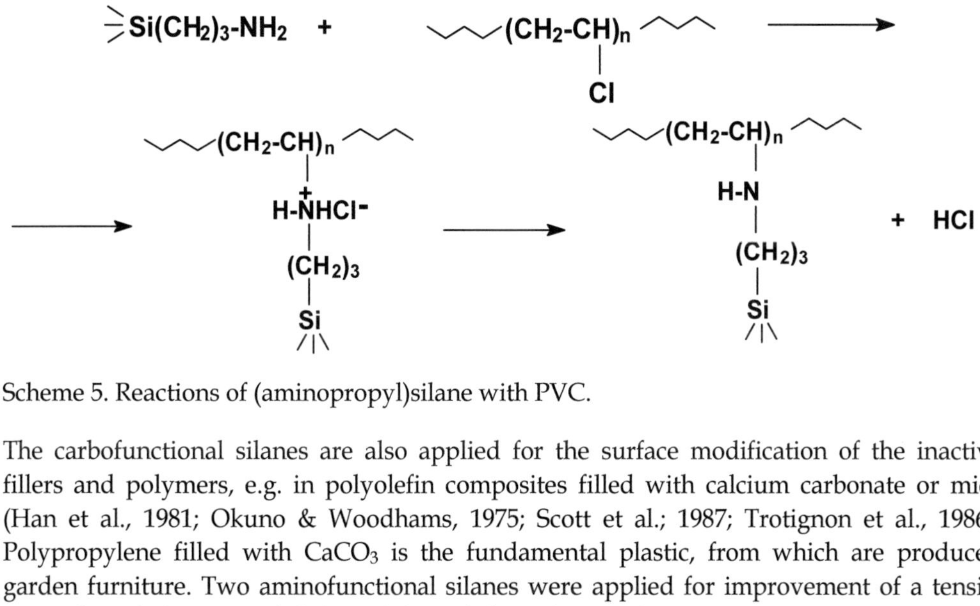

Scheme 5. Reactions of (aminopropyl)silane with PVC.

The carbofunctional silanes are also applied for the surface modification of the inactive fillers and polymers, e.g. in polyolefin composites filled with calcium carbonate or mica (Han et al., 1981; Okuno & Woodhams, 1975; Scott et al.; 1987; Trotignon et al., 1986). Polypropylene filled with $CaCO_3$ is the fundamental plastic, from which are produced garden furniture. Two aminofunctional silanes were applied for improvement of a tensile strength and decrease of deformability of the polypropylene composites containing non-modified $CaCO_3$ and modified with silanes (Demjen et al., 1997, 1998, 1999; Pukanszky, 2005). The polycondensation of triethoxysilyl groups under process conditions takes place quickly as well and leads to a reinforcement of interactions between ingredients of the composite PP/$CaCO_3$ (Demjen et al., 1997).

3.1.3 Modification of polyolefins

Polyolefins are the most widely used group of commodity thermoplastics. A cross-linking is one of modification methods of polyolefins (PO) properties, such as high temperature stability, chemical and stress cracking resistance, and also as shape memory. It can be done by different methods: irradiation, peroxide process, azo process, and moisture curing of silane-functionalized polyolefins (Munteanu, 1997; Morshedian & Hoseinpour, 2009). The last technique (SIOPLAS technology) was developed in the late 1960s, and modified in the late 1980s. Both irradiation and peroxide cross-linking processes have some disadvantages such as high costs for irradiation equipment, the possibility of pre-curing in peroxide cross-linking processes, the inability to crosslink PP through tertiary-bonded carbon atoms and the formation of voids in cable insulations.

3.1.3.1 Silane-crosslinked polyolefins

The new cross-linking SIOPLAS technology, patented by Dow Corning, is a two-step process, based on the chemistry of organofunctional silanes, which are first grafted on polyolefins chains. In a second post-reactor step the siloxane cross-linkages (Si-O-Si) are formed. Alkoxysilanes, most often vinyltrimethoxysilane (VTMS), are melt grafted on polyolefins in processing equipment, usually extruders, in the presence of small amounts of peroxides as radical initiators. Vinyltriethoxysilane (VTES), and especially 3-methacryloxypropylsilane have limited application. Usually, about 2 % of silanes are used, and dicumyl peroxide as the initiator is used (5-15 wt. %). The silane grafting reaction is very fast and it allows the choice of reactive processing to be the industrial synthetic method. Within a few minutes of the residence time of all components in the extruder, very high grafting yields are obtained (at least 80 % of the added silane is grafted onto PE) in an optimized industrial process. The possibility of silane homopolymerization may be almost excluded under the conditions of reactive processing. The reactive processing techniques enable silane grafting onto any kind of olefin homopolymer (LDPE, linear LLDPE, HDPE, very low density VLDPE, PE waxes, all kinds of PP), PO blends or copolymers (EVA, EP and EPDM elastomers), and polyisobutene (Munteanu, 1997).

The silane-grafted POs are still thermoplastic cross-linkable products, which are processed in the usual way in finished goods subsequently subjected to cross-linking in the presence of moisture. The alkoxysilyl groups hydrolyse to form silanol groups Si-OH, which condense to generate the siloxane bonds responsible for the formation of the three dimensional network. Before the moisture cross-linking step a master batch containing 0.05-0.15 wt. % (and even 2.0-3.6 wt. %) of an organotin catalyst is melt mix with the silane-grafted PO in order to achieve cross-linking rates of practical importance. Cross-linking takes place beside the production line in hot water tanks or steam chambers, most often at 60-90 °C. Under these conditions the curing time is in a range from a few hours or days.

The silane cross-linking has many advantages among curing technologies. Most of those advantages are the consequence of separating the cross-linking step from the shaping step and the specific structure of the network as well, the latter feature is responsible for better thermomechanical properties in comparison with peroxide cross-linked networks.

A modified one-step version of this technology, called as MONOSIL process, appeared few years later. All the components, i.e. polyethylene + silane + grafting initiator + cross-linking catalyst, were added in a high-shear mixing extruder wit a longer screw (Thomas & Bowrey, 1977; Swarbrick et al., 1974).

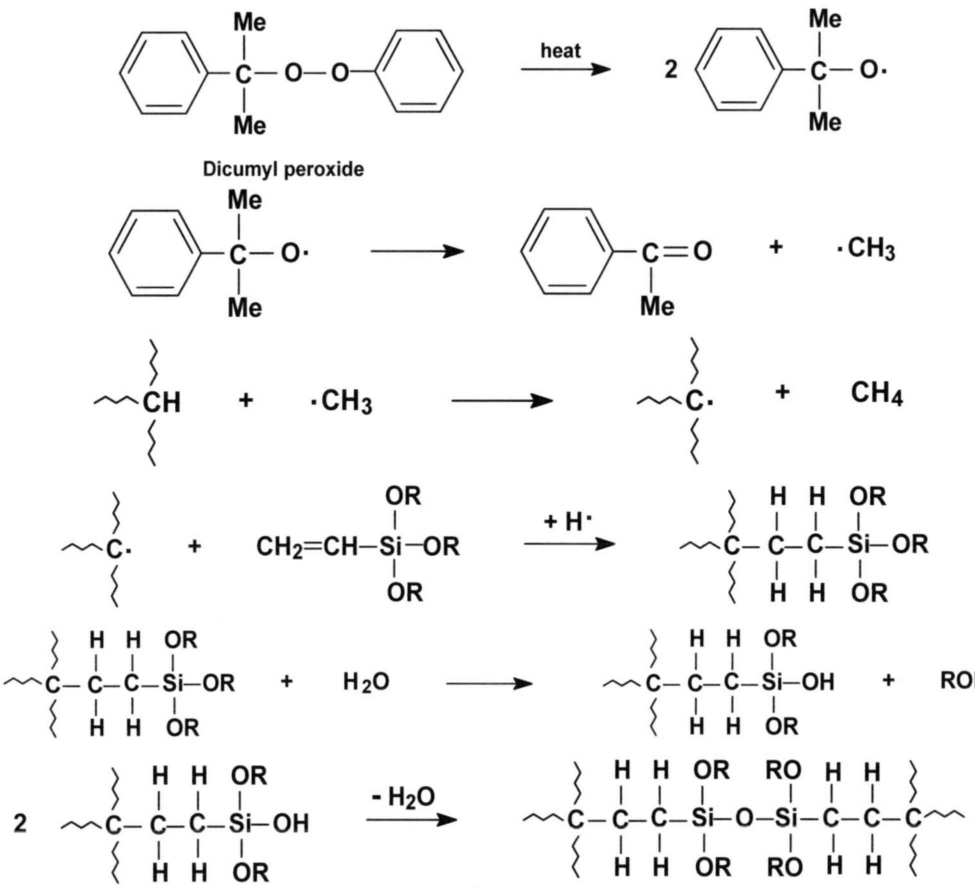

Scheme 6. Basic reactions involved in silane grafting onto PE and further cross-linking of PE (Morshedian & Hoseinpour, 2009).

An alternative method to the silane grafting is a one-step random copolymerization of ethylene with vinyltrialkoxysilanes at high pressure (150-350 MPa) and temperature (180-290 °C) in the presence of free radical initiators (usually peroxides), which was developed by Union Carbide and Mitsubishi Petrochemicals. This process has some advantages, since the silane units (< 5 wt. %) can be randomly and homogeneously distributed in the polymer chain, so it was possible to achieve a certain degree of cross-linking with a small amount of the silane incorporated in the polymer. There is also no limitation in the choice of antioxidants as in the case of peroxide and silane grafting cross-linking process (Eagles, 1990; Munteanu, 1997).

A copolymer of ethylene with 1,3-divinyl-1,1,3,3-tetramethyldisiloxane, a poly(ethylene-*co*-divinylsiloxane), was synthesized in a high pressure reactors at 180-220 °C and p ≈ 1300 bar (Smedberg et al., 2004).

3.1.3.2 Hydrosilylation of polyolefins

A hydrosilylation of terminal double bonds in polyolefins through reactive processing is a next important modification method of these thermoplastics:

$$\equiv Si\text{-}CH{=}CH_2 + H\text{-}Si\equiv \longrightarrow \equiv Si\text{-}CH_2\text{-}CH_2\text{-}Si\equiv$$

Polypropylene containing terminal double bonds was modified with a hydride-terminated polydimethylsiloxane (PDMS) at three different temperatures through a hydrosilylation reaction, catalyzed by Karstedt's platinum catalyst in the melt phase (Malz & Tzoganakis, 1998; Long et al., 2003, 2004). The double bonds were formed by peroxide initiated degradation of PP in an extruder or a batch mixer. An organic peroxide, Lupersol 101, was used in concentrations of 0.5-5 wt. %. A hydride-terminated polydimethylsiloxane was used as a substrate hydrosilylating the terminal double bonds of the degraded polypropylene (DPP). Reactive processing experiments were carried out in a hot press, a batch mixer, and a single screw extruder. The reaction time of the hydrosilylation reaction was short enough to be completed in a screw extruder. This made possible the simultaneous extrusion and modification of polypropylene.

Two different reaction mechanisms were used to initiate the hydrosilylation reaction: a radical chain addition mechanism and a platinum catalyzed mechanism using platinum Karstedt's catalyst. It was found that both reactions, degradation and hydrosilylation, could be performed simultaneously, by using high peroxide concentrations, while previously degraded polypropylene could be hydrosilylated with catalytic amounts of a peroxide. For the catalytic mechanism, the required stabilization of the platinum colloid formed in this mechanism was accomplished by adding *t*-butylhydroperoxide as a co-catalyst.

The hydrosilylation of polybutadiene (PB) with hydrosilanes is catalyzed by rhodium and platinum catalysts, and platinum-nanoclusters and it is a source of silyl derivatives of PB, many of them have functional group and can find useful applications (Guo et al., 1990; Iraqi et al., 1992; Chauhan & Balagam, 2006; Chauhan et al., 2008).

A comprehensive study on the surface characteristics of hydrosilylated polypropylene was conducted by combining macroscopic thermodynamics, microstructure, and chemical composition measurements (Long et al., 2003, 2004). A positive effect of a poly(dimethylsiloxane) modified polyolefin additive on the processing and surface properties of linear low density polyethylene (LLDPE) was observed (Zhu et al., 2007). A polydimethylsiloxane (PDMS) modified polyolefin, obtained by reacting an ethylene-ethyl acrylate-maleic anhydride (EEAMA) copolymer with an amine terminated PDMS in the melt phase, was used as a processing aid to facilitate the extrusion of LLDPE. Surface properties of hydrosilylated polyolefins could be further modified by annealing in supercritical carbon dioxide (Zhu & Tzoganakis, 2008). Two hydrosilylated polypropylenes (PP) and polyethylenes (PE) were obtained by reacting with di- and multi-functional hydride terminated poly(dimethylsiloxane). Processing properties of polypropylene were improved by the addition of a small amount (~2 wt. %) of silicone oil (Zhang et al., 2010).

A PDMS grafted copolymer, which can be potentially applied as a processing agent or a surface property modifier, was synthesized via reactive melt mixing of ethylene–ethyl acrylate–maleic anhydride terpolymer (EEAMA) and aminopropyl terminated PDMS (McManus et al., 2006).

Crosslinked styrene-isoprene-siloxane copolymers were applied for a preparation of membranes useful for pervaporative removal of volatile organic compounds from water.

The membranes were prepared from the crosslinked copolymer composed of elastomeric glassy styrene-isoprene prepolymer and oligo- or polyhydrosiloxanes (Kerres & Strathmann, 1993; Kujawski et al., 2003). The pervaporation properties of the membranes, used for pervaporation of water–methyl acetate and water–methyl *tert*-butyl ether mixtures, depend on the siloxane content in the membrane and were much better than properties of the commercially available hydrophobic membranes.

High molar mass copolymers of ethylene or propylene with α-olefin macromonomer, monovinyl-functional silsesquioxane cage 1-(9-decenyl)-3,5,7,9,11,13,15-heptaethylpenta-cyclo-[9.5.1.13,9.15,15.17,13]octasiloxane, were obtained in the presence of metallocene catalysts, activated with alumoxanes. These copolymers, contained up to 25 wt. % (1.2 mol. %) of spherosilicate pendant units, which accounted for a decrease of the melting temperature by 18 K with respect to PE. A thermostability under air was improved in the polyethylene copolymer in comparison to PE (Tsuchida et al., 1997).

3.2 Applications of carbofunctional polysiloxanes

The carbofunctional polysiloxanes are most often used for syntheses of hybrid silicone-organic copolymers and for the modification of many polymers and plastics. This way, numerous new polymeric materials with profitable properties were obtained.

3.2.1 Modification of vinyl and acryl polymers by CFPS

The graft copolymers of polysiloxane with styrene, methyl metacrylate, and chloroprene were synthesized by a radical polymerization from the commercial carbofunctional polysiloxanes containing side and terminal mercaptopropyl groups (Fawcett et al., 2001).

3.2.1.1 Polystyrene

Block copolymers comprised of chemically distinct polymers covalently joined together self-organize in bulk into well-defined nanoscopic structures including lamellar, cylindrical, and spherical morphologies. There are a lot of block copolymers based on styrene monomers with reactive alkoxysilyl groups. A series of well-defined poly(3-methacryloxypropyl-triethoxysilane)-*b*–polystyrene (PMAPTES-*b*-PS) diblock copolymers was prepared via sequential RAFT polymerization (Zhang et al., 2007). Poly(acryloxypropyltriethoxysilane)-*b*-polystyrene (PATPES-*b*-PS) diblock copolymers were prepared by the nitroxide-mediated polymerization (NMP) using alkoxyamine initiators (Gamys et al. 2010). The core-shell particles prepared by emulsion polymerization of styrene and subsequent addition of MPTMS have been used to produce hybrid nanocapsules (Ni et al., 2008). A series of well-defined rod-coil block copolymers poly(dimethylsiloxane-*b*-poly{2,5-bis[(4-methoxyphenyl)-oxycarbonyl]styrene} (PDMS-*b*-PMPCS) containing a CO_2-philic block PDMS was synthesized through ATRP (Shi et al., 2011). The nanostructure of these copolymers changed from lamellae (LAM) to hexagonally packed cylinders (HEX) according to the volume fraction of PDMS in block copolymers. The microemulsion photocopolymerization of styrene and butyl acrylate in the presence of silane coupling agent, such as MAPTMS, using 2-hydroxy-2-methylpropiophenone as photoinitiator was carried out as well (Wan et al., 2009). The copolymer particles were nearly spherical and were very uniform with the number average particle size of 25.5 nm and M_w/M_n of 1.11.

Particles with structural and compositional anisotropy are great interest for potential applications such as optical switches and filters. Anisotropic particles are composed of various materials such as functional polymers and inorganic materials with a high refractive index. Nagao et al. (2008) demonstrated monodisperse, anisotropic particles composed of organic and inorganic materials, prepared in a two step soap-free emulsion polymerization. In the first step, seed polymer particles were prepared in the presence of MPTMS in water. In the second step, another polymerization of styrene and/or methyl methacrylate was conducted in the presence of seed particles, which induced anisotropic protrusion of polymer from the seed particles. This method is applicable to the preparation of anisotropic polymer particles containing inorganic particles such as silica.

The transparent hybrid materials were obtained by a bulk polymerization with styrene, ethylene glycol dimethacrylate (EGDMA), and silicone macromonomer containing PDMS chains with urethane and methacrylate functionalities (SiUMA) (Daimatsu et al., 2008). The prepared P(S-*co*-SiUMA-*co*-EDGMA) copolymers were transparent, similarly to pure PS, while P(S-*co*-SiUMA) copolymers were opaque.

The simple method for preparing block copolymers of PDMS and vinyl monomers was reported (Oz & Akar, 2006). PDMS-*b*-PAN and PDMS-*b*-PS copolymers were produced in one step with redox system of α,ω-dihydroxypropyl terminated PDMS/ceric ammonium nitrate at ambient temperature in water. The copolymer products had much higher contact angle values than corresponding homopolymers, although their silicone content was as low as 1-2 wt. %. These copolymers may act as compatibilizers in the blending application of vinyl polymers with polysiloxanes.

3.2.1.2 PVAL

Silicone polyoxyalkylene copolymers are the most known silicone surfactants because of their biodegradability, biotolerance and water solubility, e.g. poly(vinyl alcohol)-*graft*-PDMS (Pouget et al., 2008). These copolymers have been synthesized in aqueous dispersion of a mono- or diepoxy terminated PDMS with pendant alcohols of P(VAL-*co*-VAC). The last copolymer was used as a dispersant to stabilize the PDMS microsuspension and the reaction took place at the interface of the silicone droplets. The preparation of PVAL-PDMS copolymers is of great interest, especially for the preparation of PVAL-PDMS amphiphilic copolymers. The incorporation of PDMS block in the PVAL structure would enhance its flexibility and resistance to moisture. Lacroix-Desmazes et al. (2008) reported synthesis of triblock copolymer PVAL-*b*-PDMS-*b*-PVAL.

3.2.1.3 PVDF

Poly(vinylidene fluoride) (PVDF), semicrystal membrane materials, were recently obtained. They showed high strength and good chemical resistance. However their relative hydrophobicity and inactivity have limited their use in materials for membrane separation of oil and biological molecules. For these applications the increase of PVDF hydrophilicity was necessary. PVDF as a macroinitiator was grafted with 3-methacrylate propyltrimethoxysilane (MAPTMS) via ATRP of MAPTMS. The flux of the membrane prepared from silylfuctionalized PVDF was much higher than that of the membrane prepared from parent material (Chen & Kim, 2008). Another limitations in application of PVDF are its poor solubility and difficulty in cross-linking. Low-molecular-weight copolymers obtained by copolymerization of fluorinated monomers bearing an ω-trialkoxysilane function with vinylidene fluoride (VDF), were successfully cross-linked in the presence of moisture and these cross-linked materials showed interesting repellent properties with respect to boiling acetone, water, acids, and oil (Guiot et al., 2006).

3.2.1.4 Acryl polymers

Some silyl derivatives of polymers such as acrylates, methacrylates show membrane properties with the selectivity with reference to oxygen. Thus silyl methacrylate may be used in contact lenses. The synthesis of 2-hydroxyethyl methacrylate (HEMA) polymers modified with some silyl groups was carried out. The increase of polymethacrylates solubilities with increase of incorporation of silyl groups was observed (Assadi et al., 2005).

Polymerization strategies to overcome limiting monomer conversion in silicone-acrylic miniemulsion polymerization were presented (Rodriguez et al., 2008). The waterborne silicone-modified acrylic materials were obtained from divinyl terminated PDMS and methyl methacrylate (MMA), butyl acrylate (BA), and acrylic acid (AA).

Thermoresponsive shape memory polymers (SMPs) are the most studied materials because of their potential applications in biomedical fields. Poly(N-isopropylacrylamide) (PNIPAAm) is a typical example that exhibits a lower critical solution temperature (LCST) at ca. 32 °C in water solution. Below this temperature PNIPAAm chains are soluble due to the hydrogen bonding interactions between polymer chains and water molecules, whereas the phase separation occurs and the polymers precipitate from the solution when the temperature is above the LCST. A series of monodisperse microgel with different particle sizes and cross-linking density were prepared by precipitation copolymerization of NIPAAm with 3-methacryloxypropyltriethoxysilane (MAPTES). The experimental results revealed that the microgel exhibited temperature sensitivity and the phase transition temperature approximately at 31 °C. A decrease of MAPTES content resulted in a reduction of the final hydrodynamic diameters of the microgel (Zhang et al., 2009).

A self-assembly silane monolayer was formed by 3-aminopropyltriethoxysilane (APTES), followed by the graft polymerization of NIPAAm on the glass surface. These kind of glasses bonded with NIPAAm films have potential applications as environmentally switchable materials (Wang et al., 2005).

PDMS have been widely used in a variety of biomedical applications, but their hydrophobicity is often a problem. This can be overcome with the introduction of hydrophilic polymers, such as poly(N,N-dimethylacrylamide) (PDMAAm). The synthesis of a new family of amphiphilic multiblock and triblock copolymers (PDMAAm-*b*-PDMS-b-PDMAAm) via RAFT polymerization was proposed (Pavlovic et al., 2008). A novel ABC triblock copolymer with very low surface energy: PDMS-*b*-PMMA-*b*-PHFBMA (2,2,3,3,4,4,4-heptafluorobutyl methacrylate) was successfully synthesized via ATRP by Luo et al. (2008). The lowest surface energy of these triblock copolymers reached 3.03 mN/m, thus the PDMS-*b*-PMMA-*b*-PHFBMA could be applied as an antifouling coating.

A next interesting application of the CFPS is their utilization in synthesis of interpenetrating polymer networks (IPN) with fluorinated acrylates. These siloxane-fluoroacryl IPN networks were prepared *in situ* from of α,ω-(3-hydroxypropyl)polydimethylsiloxanes, isocyanurate derivatives of hexamethylene diisocyanates $R(NCO)_x$, 3,3,4,4,5,5,6,6,7,7,8,8,8-tridecafluoro-1-octanol, ethylene glycol dimethacrylate (EGDMA), dicyclohexyl acrylate, and dipercarbonate (DCPD), towards dibutyltin dilaurate (DBTDL). A transparency of the obtained materials gave an evidence that no phase separation took place. The fluorinated polymers exhibit good hydro- and oleofobicity (Darras et al., 2007).

3.2.1.5 Radical polymerization of dissymmetric fumarates

Polymerization of fumarate monomers provide polymers with a substituent in every methylene unit. These polymers are applied as high performance materials, as bio-degradable polymers, low dielectric materials, oxygen permeability membranes, liquid crystal, contact lens (CL) materials, etc. Among them, CL is known well as the lens equipped on a cornea foe vision correction, and the CL materials are required both for the oxygen permeable property for breathing of corneal cell and the hydrophilic property for not repelling a tear and to protect a lipid deposition. Polymerizations of amphiphilic fumarate (containing hydrophilic and hydrophobic substituents) monomers give polymers with high oxygen permeability, hydrophilicity and transparency required for CL materials. Novel dissymmetric fumarate monomers having both alkoxyethyl group [2-(2-methoxyethoxy)-ethyl, 2-(2-(2-methoxy-ethoxy)ethoxy)ethyl] and a bulky 3-[tris(trimethylsiloxy)silyl]propyl group were radically polymerized with styrene and N-vinylpyrrolidone (NVP). Various membranes, prepared by copolymerization of fumarate monomers with NVP or NVP and HEMA, showed much better transparency, as compared to the membranes containing 3-[tris-(trimethylsilyloxy)silyl]propyl methacrylate (Ohnishi et al. 2009).

3.2.2 Modification of condensation thermoplastics by CFPS

3.2.2.1 Polyesters

Enzymatic polymerization is an environmentally friendly approach to polymer synthesis and in the contrast to traditional chemical methods it does not need harsh reaction conditions and metallic catalysts. Lipase-catalyzed synthesis of aromatic silicone polyesters (SAPEs) and polyamides (SAPAs) have been reported (Poojari & Clarson, 2010). The SAPEs were synthesized using α,ω-dihydroxyalkyl-terminated PDMS, and the SAPAs were prepared with α,ω-diaminoalkyl-terminated PDMS via transesterification reaction with dimethyl terephthalate in toluene at a temperature in the range 80–90 °C. Both types of polymer were liquids at room temperature. Polysiloxane-polyester copolymers were synthesized by polycondensation of a series of diacids and α,ω-bis(3-hydroxypropyl)-PDMS with Novozyme-435. The use of lipase as the catalyst led to carry out the polycondensation reaction under mild conditions (Guo et al., 2008).

A series of novel thermoplastic elastomers, poly(ester-siloxane)s, have been synthesized (Antic et al., 2010). These materials were based on poly(butylenes terephthalate) (PBT) as the hard segments and PDMS-containing prepolymers as the soft segments. In order to increase the compatibility of PDMS copolymers of siloxanes with terminal aliphatic polyester such a triblock copolymer polycaprolactone-*block*-polydimethylsiloxane-*block*-polycaprolactone (PCL-*b*-PDMS-*b*-PCL) was used. The thermal stability of the poly(ester-siloxane)s was higher than that of PBT homopolymer, and the reduced crystallinity of the hard segments have been observed.

Thermoresponsive shape memory polymers (SMPs) are a class of materials that change shape upon exposure to heat. They are lightweight, easy to fabricate and may be biodegradable. Poly(ε-caprolactone) (PCL)-based SMPs have received much attention due to the biocompatibility, biodegrability, and elasticity of PCL. PCL is useful as a crystalline switching segment for SMPs as its T_{trans} is a well-defined T_m in the range of 45-60 °C with increasing M_n. Such T_{trans} values are useful for *in vivo* deployment as well as other applications which require low heating. To modify properties of PCL-based SMPs, AB polymer networks have been prepared through incorporation of organic hard or soft segments. PDMS with extremely low T_g (-123 °C) is a particularly effective soft segment candidate. Schoener et al. (2010) prepared organic-inorganic SMPs comprised of inorganic PDMS segments, terminated by aminopropyl groups, and organic PCL segments. The resulting SMP network exhibited excellent shape fixity and recovery. By changing the PDMS length, the thermal, mechanical, and surface properties were systematically alternated.

Carbon nanotubes (CNTs) are the ideal reinforcing agents for high-strength polymer composites because of their tremendous mechanical strength, their Young's modulus is ca. 1000 GPa and it is much higher than that of the conventional carbon fibres (200-800 GPa). For good reinforcement effect the surface of CNTs must be strongly bonded to the polymer matrix. Sidewall functionalization of CNTs, introduction mainly carboxylic groups, makes possible to chemically attach these nanofillers with polymers. Chen and Shimizu (2008) prepared the nanocomposites of poly(L-lactide) (PLLA) reinforced by aminopropylisooctyl POSS modified multi-wall nanotubes (MWNTs) (MWNT-*g*-POSS). They have observed the homogeneous dispersion of MWNTs throughout the PLLA MWNT-*g*-POSS composites without any aggregation. The fractial surface of the composites showed not only a uniform dispersion of MWNTs but also a strong interfacial adhesion with the matrix.

3.2.2.2 Polyamides

Alongi et al. (2009) have reported the preparation of inorganic-organic hybrids made of POSS and polyamide 6 (PA 6) produced by melt mixing of the two components or by the *in situ* polymerization of ε-caprolactam in the presence of POSS. The last method did not result in any POSS self-aggregation in the polymer matrix, achieving a very fine cubic dispersion of nanometric dimensions. The nearly nano-sized aromatic polyamide particles with amino groups, known as poly(amino-amide) PAMAM, were prepared by Yoshioka (2009). They were modified with silane coupling agent, i.e. 3-glycidoxypropyltriethoxysilane (GPTES), and further complexed with ZnO particles. The finally ZnO-based materials have specific optical properties: they show excellent optical absorption properties in UV region and high transmittance in the visible region.

3.2.2.3 Polyimides

Polyimides (PIs) are well-known engineering plastics with excellent thermal, mechanical, dielectric, and optical properties. They also have good chemical resistance and high dimensional stability. However the main deficiency of aromatic polyimides is their insolubility in organic solvents and infusibility or extremely high glass transition temperature (T_g) which makes their processing very difficult. One approach to increase the solubility and processability of polyimides is the introduction of flexible linkages or bulky units into the polymer chain. The incorporation of PDMS sequences in PIs has afforded new copolymers with good processability, low water absorption, atomic oxygen resistance, and excellent adhesion. The unique properties of the imide-siloxane copolymeric materials (PSIs) make them especially attractive for applications in microelectronics and as structural adhesives.

A series of amorphous poly(imide siloxane)s (PIS) with different PDMS contents and segmental lengths were synthesized via condensation reaction by Ku & Lee (2007). A variety of morphologies of PIS films, including unilamellar vesicle, multilamellar vesicle, sea-island and others, were found as the function of the content and the segmental length of PDMS, as well as the coexistence of large-scale phase separations and nano-scale phase separation of approximately 20 nm.

Lu et al. (2006) prepared, using sol-gel method, polyimide/polydiphenylsiloxane (PI/PDPS) composite films with high thermal stability near pure PI. Polysiloxanes well dispersed in polyimide matrix, without macroscopic separation for the composite films was observed with low content of DPS, while large domain of polysiloxane was formed in films with high DPS content. The introduction of DPS into PI improved the elongation at break but the composite films still remained with higher modulus and tensile strength.

Aromatic polyimides (PIs) have been of great interest in gas separation membranes because of their gas selectivity and excellent thermal and mechanical properties. Recently, novel triamine-based hyperbranched PIs are applied for high-performance gas-separation materials. Hybridization of PIs with inorganic compounds has been focused also to improve the gas transport properties. Hyperbranched PIs prepared by polycondensation of a triamine 1,3,5-tris-(4-aminophenoxy)benzene and 4,4'-(heksafluoroisopropylidene) diphthalic dianhydryde (6FDA) were modified by sol-gel reaction using tetramethoxysilane (TMOS), methyltrimethoxysilane (MTMS), and 3-aminopropyltrimethoxysilane (Suzuki et al., 2008).

New copolymers, polysiloxane-imides (PSIs), have been prepared from α,ω-(bisamino-propyl)dimethylsiloxane oligomers (ODMS) and aromatic dianhydrides: 1,2,4,5-benzene-tetracarboxylic dianhydride (PMDA), and 6FDA, (Krea et al., 2004). Membranes of these new PSIs have been used to remove polar organics from water by pervaporation.

Bismaleimide (BMI) has the unique combination of a high service temperature, good toughness and epoxy-like processing. Composites of BMI with surface-modified SiO_2 nanoparticles by amino-functionalized silane coupling agent has been studied (Yan et al., 2008). The nanocomposites with surface-modified SiO_2 showed better wear resistance and lower frictional coefficient than that with unmodified fillers.

3.2.2.4 Modification of polycarbonates by CFPS

Polycarbonates (PCs) belong to an extremely useful class of engineering thermoplastics known for their toughness and optical properties. Most commercialized PCs are based on bisphenol A and these polyesters have the advantageous properties of superior heat resistance, transparency, dimensional stability and self-extinction, exceptional impact resistance and ductility at or below room temperature. Other properties such as the modulus, dielectric strength and tensile strength, are comparable to those of other amorphous thermoplastics. The main limitations of PCs are their poor scratch, ultraviolet (UV), and solvent resistance, and their sharply decreased impact strength at low temperature.

PC/silicone oil blends, based on two kind of silicone oils: PDMS and polydimethyldiphenyl-siloxane PDMDPS, have been studied (Kim & Kim, 2008). A transition from brittle to ductile failure was observed with increasing PDMDPS content. At temperature of -30 °C the impact strength increased from 8 kg_fcm/cm to 52 kg_fcm/cm for neat PC and PC/PDMDPS blends respectively.

An interesting group of thermoplastic elastomers (TPE), characterized by a great stiffness at low temperature, are block copolymers prepared from polycarbonates (PC) and PDMS (PC/PDMS). Copolymers PC/PDMS are usually prepared in reaction of phosgene with bisphenol A in the presence of PDMS containing terminal bisphenol groups (van Aert et al., 2001). Block PC/PDMS copolymers find applications in a manufacture of membranes, which are used for selective separation of gases (LeGrand, 1972).

Multi-walled carbon nanotubes/polycarbonate nanocomposites (MWNT/PC) have been prepared (Wang et al., 2010). Functionalized MWNTs-COOH with α,ω-3-aminopropyl-PDMS, were used to produce MWNT/PC nanocomposites. The results showed that siloxane-modified carbon nanotubes were dispersed well in the PC matrix, and the tensile strength, flexural strength, flexural modulus, and flame retardancy of MWNT/PC composites were better than these of MWNT-COOH/PC. Siloxane-modified MWNTs can improve the electrical properties of the nanocomposites at low loading in PC.

3.2.3 Modification of polyurethanes and polyureas by CFPS

Thermoplastic polyurethanes (TPUs) are obtained by a polyaddition reaction of diisocyanates to long, hydroxyl-terminated oligomers (polyols) and short diols, which are often named as chain extenders. The polyurethanes are multi-block copolymers consisting of hard segments (HS) and soft segments (SS). The thermodynamic incompatibility between alternating segments usually induces a two-phase structure with soft domains containing mostly the polyol moieties, and hard domains made up of diisocyanate-chain extender sequences. If incompatibility exists between the two block components, a microphase separation will occur and the hard domains will provide a reinforcement to the system. The type of diisocyanate has a marked effect on a strength of a final material. The most often applied isocyanates are: 4,4'-methylenediphenyl diisocyanate (MDI), tolylene diisocyanate (TDI), bis(4-isocyanatocyclohexyl) methane (HMDI), isophorone diisocyanate (IPDI). TPUs are widely used for high-performance applications, as: medical implants (Arkles, 1983), membranes, adhesives and coatings, especially when a high tear and a tensile strength or good wear and abrasion resistance are required. Another special features of TPUs are their low-temperature elasticity, a smoothness for the touch, an electrical insulation and a good chemical resistance.

Over the past three decades, considerable attention has been directed to utilization of the PDMS as a soft segment component in polyurethanes and polyurethaneureas. Polysiloxane SS are introduced by PDMS-diols or hydroxyalkenyl or aminopropyl or ethyl-piperazine terminated PDMS. However, the tensile strength and elongation at break of these materials compared to those based on polyether (PET) or polyester (PES) SS are visibly poorer. Thus, various co-SS components, mainly based on poly(alkenyl oxide) have been utilized. Sheth et al. (2005) selected poly(propylene oxide) (PPO) as a second SS component, because his solubility parameter (23.5 $J^{1/2}/cm^{3/2}$) is in between that of PDMS (15.6 $J^{1/2}/cm^{3/2}$) and urea (45.6 $J^{1/2}/cm^{3/2}$). The inter-segmental mixing between PPO and urea segments could modify the nature of interphase between the soft matrix and the hard urea microdomains.

In order to reduce emission of volatile organic compounds, water borne polyurethanes (WPUs) have attracted much attention. However, the WPUs have shortcomings, such as lack of resistance to water, a surface hydrophilicity and the low strength. These disadvantages can be eliminated by using a different mixing composition of soft segments. Chen et al. (2005) prepared the anionic WPUs from hydroxyl-terminated PDMS, poly(tetramethylene oxide) (PTMO) as the soft segments, IPDI as the hard segment, and 2,2-bis(hydroxymethyl)-propionic acid (DMPA) as the ionic centre. DMPA was used to form a water-dispersible urethane prepolymer. The tensile strength of WPU-PDMS films decreased with an increasing content of PDMS and a water repellency reached 80 %, which was equal to a capability of the silicone rubber. These waterborne siloxane containing PUs are useful as textile coatings, leather finishing, sealants, plastic coatings, and a glass-fibre sizing.

A POSS-modified ionomeric polyurethane aqueous dispersion have been also reported for their improved reaction to fire and resistance to atomic oxygen as well as to the possibility to build nanostructured surfaces. As with other polymer systems, many of the POSS hybrids do yield elastomeric PU with improvements in physical and thermal properties at relatively low levels of POSS incorporation. The waterborne PU hybrid dispersions were synthesized by incorporating amino- and hydroxyl-terminated POSS macromers into the PU ionomeric backbones, using a homogeneous solution polymerization, followed by a solvent exchange with water (i.e., the "acetone process") (Nanda & Wicks 2006). The absence of crystalline domains evidenced the incorporation of POSS macromer in the PU hard segments.

Various researchers have shown that the gas permeability of PU membranes increases with the decrease of the hard segment content and the increase of the soft segment molecular weight. In addition, a correlation has been established between the gas permeability and the chemical nature of the polyols and chain extenders. Queirez & de Pinho (2005) investigated structural characteristics and gas permeation properties of PDMS-PPO-urethane/urea bi-soft segment membranes. The permeability of the membranes to CO_2, O_2 and N_2 increased with the increase of PDMS content. The $P(CO_2)/P(N_2)$ permeability ratio was higher for the bi-soft segment membranes in comparison with PDMS/PU membranes.

Some properties of PUs (high tear, tensile strength, good wear, abrasion resistance) are strongly jeopardized as soon as the service temperature of the material exceeds 70 to 80°C, mostly because of a creep phenomena. Dassin et al. (2002) tried to improve the working temperature of commercial TPU by creating a well-controlled chemical curing. The idea was to have an easily processible material with an improved thermomechanical behaviour. A commercial polyester-type TPU at first was reacted with a diisocyanate to create allophanate branching units, which at second step was grafted with 3-aminopropyltrimethoxysilane. Such materials could be crosslinked by a hydrolytic condensation mechanism, resulting in increased thermomechanical properties.

3.2.4 Modifications of other polymers and polymeric materials by CFPS

The CFPS were also used for synthesis of conductive siloxane-pyrrole block copolymers. An incorporation of flexible PDMS chains into the rigid polymer chain improves not only mechanical properties, but also solubility in different non-polar solvents. (N-pyrrole)-polysiloxane precursor, prepared in reaction of N-glicydylpyrrole with aminopropyl groups of PDMS was copolymerized with pyrrole (PY) by electrochemical methods, giving block copolymer PY-PDMS. Classical conductive polymers are insoluble and infusible. A common way to facilitate their processing are syntheses of block or graft copolymers containing conductive segments and conventional segments, increasing solubility of copolymer, e.g. block copolymers of styrene and acetylene, styrene and thiophene, graft copolymers of pyrrole on polystyrene, and on poly(methyl methacrylate). Recently a new multifunctional polysiloxanes, containing mercapto-, thiocarbamide, and (hydroxyalkyl)ether linkages were applied for the modification of silica fillers used for the production of rubber tires (Kalaycioglu et al., 1998).

Copolymers with siloxane moieties serve as connectors with organic polymers (compatibilizers). Siloxane segments of the copolymers migrate to the surface, while organic segments act as anchors of siloxane blocks in the polymeric material.

3.3 Composites and nanocomposites from thermoplastic polymers

Composite materials consist of a polymer matrix and various inorganic fillers. They have been widely used in order to improve the mechanical, thermal, barrier, and other properties of the polymer. However it is known that the improvement of one property can result the worsening of another. It is anticipated that such problems can be overcome if the inorganic additive exists in the form of a fine dispersion within the polymer matrix producing a nano-composite. Silane modified fillers and nanofillers are used for these purposes.

3.3.1 Modification of fillers properties by CFS

The carbofunctional silanes are most often applied as so called adhesion promoters, which are also called as *"silane coupling agents"* (Pleuddeman, 1991). The silane adhesion promoters are directly added to polymer compositions (usually in a quantity 0.2–2 % wt. %) or in a form of water and water-alcoholic dispersions. The substituent X (e.g. alkoxy group) may undergo condensation reactions with hydroxyl groups of the filler:

or hydrolysis reactions:

$$\equiv Si-OR + H_2O \longrightarrow \equiv Si-OH + ROH$$

Thus alkoxy groups are replaced with silanol groups Si-OH, which are able to form chemical bonds with inorganic materials, e.g. glass, metals or fillers (Scheme 7) (Haas & Wolter, 1999):

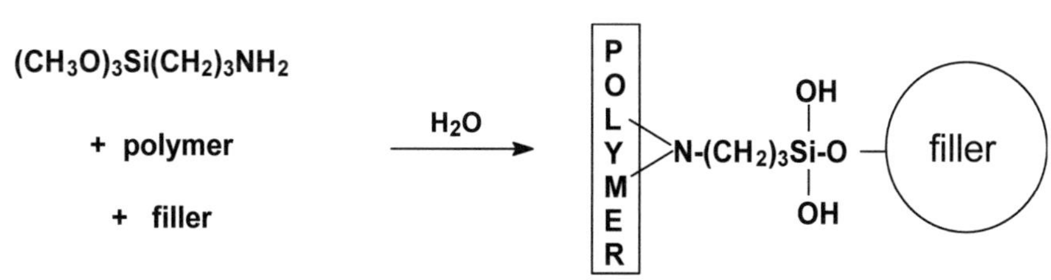

Scheme 7. An adhesion model in a system: polymer–silane–filler.

The silanol functional groups Si-OH also undergo homocondensation reactions, with a formation of thermodynamically more stable siloxane systems Si–O–Si. Alternatively, Si-OH groups may react with OH groups of a surface of the filler. This way on the surface of the filler a layer of crosslinked polysiloxane is formed. An universality of an application of the fillers in plastics results from necessity of an improvement of mechanical, thermal, electrical, and magnetic properties of the polymers. The most often used fillers are: silica, calcium carbonate, calcium sulfate (gypsum), barium sulfate, calcium, aluminum (kaolin), and magnesium (talc and sepiolite) silicates, aluminum oxide and hydroxide, metal oxides (e.g. ZnO, MgO, CaO, TiO_2, Fe_2O_3, Pb_3O_4). An addition of nanofillers of particie size 10–50 nm profitably affects an increase of abrasion resistance of transparent polymer coatings and enables preparation of homogeneous composite of nanoparticles in a polymer matrix. The addition of 20–40 wt. % of the silica and alumina can improve abrasion resistance of material. The non-modified silicon and aluminum oxides have a large number of hydroxyl groups on the surface and thus they are hydrophilic materials. In the presence of non-polar liquids, e.g. acrylates, an aggregation of nanoparticles of these fillers takes place, which causes "thickening" of their mixtures with acrylates. The modification of the fillers surfaces with the carbofunctional silanes substantially improves their dispersibility in organic media. The use of trimethoxysilanes with methacryloxypropyl or vinyl groups for the modification of nanoparticles of SiO_2 and Al_2O_3 allows to prepare transparent abrasion resistant coatings on the surface. Acrylic laquers containing nanoparticles of the fillers are utilized as top, decorative layers of furniture, as decorative layers of aluminum foils and as varnishes for floor and coatings.

Nanosphers of a mesoporous silica, functionalized with (3-aminopropyl)trimethoxysilane and a sol of gold form nanocomposites with a specific surface area around 1000 m^2/g (Botella et al., 2007) were prepared. Instead of (trialkoxy)propylsilanes containing different carbofunctional groups: amine, aldehyde, acrylate, isocyanate, thiol, ether, triacetoxy-diamine, and trialkoxysilanes containing aldehyde and nitrile groups attached to a longer alkylene "connector" were grafted on the surface of cobalt ferrite $CoFe_2O_4$. They were utilized for the preparation of water dispersions of nanoparticles with supermagnetic properties (de Palma et al., 2007).

3.3.2 Applications of silane-modified nano-fillers in thermoplastic composites

Depending on chemical structure and geometry three classes of nanoparticles were widely investigated:

- layered materials such as nanoclays (e.g. montmorillonite MMT) which were characterized by one nanometric dimension, referred to as 2D nanoparticles;
- fibrous materials, such as carbon nanotubes, which were characterized by elongated structures with two nanometric dimentions and referred to as 1D nanoparticles;
- particulate materials, such as polyhedral oligosilsesquioxane (POSS) and spherical silica nanoparticles, which were characterized by three nonometric dimension and sometimes referred to as 0D nanoparticles.

For good properties of these polymer nanocomposites it is required the homogeneous dispersion of nanoparticles in the polymer matrices. The best way to avoid aggregation is to graft polymer chains onto the particles covalently, forming organic-inorganic hybrid materials. Grafting techniques include the "grafting to", "grafting through", and "grafting from" methods. "Grafting to" method involves the reaction of appropriate end-capped functional groups or side pendant groups in the polymers with the particles. This method is simple, but the graft density is fairly low because the diffusion of polymer chains to the surface of the particles is hindered sterically. "Grafting to" method gives relatively low grafting densities. "Grafting from" technique, in which a monomer polymerization is initiated from groups bound to particle surface is expected to lead to higher surface grafting densities, because monomer can more easily diffuse towards the reactive centre. In the "grafting through" method the surface of the particle is modified with a polymerizable group. The graft density may be relatively high as compared with the "grafting to" method, but since the monomer-modified particle is multifunctional cross-linking between polymer chains may be occur.

A variety of polymerization techniques, including conventional free radical, cationic, anionic, ring-opening, and controlled radical polymerizations (CRPs) or living radical polymerization (LRP) have been used to graft the polymer onto the particle surfaces. In comparison with the conventional radical polymerization CRPs can control the architecture, molecular weight, and molecular weight distribution, and they are simple on operation procedure and versatile on monomers in comparison with ionic polymerizations. Out of the three most common LPR methods, nitroxide mediated polymerization (NMP), atom transfer radical polymerization (ATRP), reversible addition-fragmentation chain transfer (RAFT) polymerization, RAFT is arguably the most applicably technique (Chrissopoulou & Anastasiadis, 2011).

3.3.2.1 Composites based on polyacrylates

Silica is the most applied filler in polymers based acrylate moieties. Polyacrylate-SiO_2 particle are usually produced by emulsion or microemulsion polymerization. The investigations on developing emulsifier-free latexes have been made. An emulsifier-free emulsion polymerization and sol-gel process were used for preparation of polyacrylate-silica hybrids in which tetraethoxysilane (TEOS) and poly(vinyl alcohol) (PVAL) were used as a silica precoursor and colloid stabilizer, respectively (Shen et al., 2009).

Surfactant-free emulsion polymerization, in which neither surface treatment for nanosilica particles nor additional surfactant or stabilizer is required is one of the methods for preparation of inorganic-organic composites. SiO_2-PMMA composite particles were prepared via conventional emulsion polymerization by the aid of acid-base interaction between the silanol groups of unmodified silica surface and amino groups of 4-vinylpyridine. The morphology of the composite particles could be multicore-shell, rapsberry-like or normal core-shell, depending upon the emulsifier content, monomer/silica ratio, silica particles size, and monomer feed method (Cheng et al., 2006).

Sol-gel process is often applied to prepare polymer-silica organic-inorganic hybrids, in which a nanometer size silica component is dispersed in a polymer matrix. In polymer-silica hybrids films silica component works as the hard segment in soft polymer matrix. Silane coupling agents are commonly used to achieve miscibility of the polymer and silica. Transparent polyacrylic-silica hybrid thin films were obtained from MMA, trimethylolpropanetrimathacrylate, water based monodispersed colloidal silica and coupling agent: MAPTMS. A mixture of ethylene glycol and ethyl ether was used for preventing the phase separation of the prepared hybrid materials (Yu & Chen, 2003). PMMA-silica composites were synthesized via sol-gel reaction. The base acrylic polymers were prepared by copolymerization of MMA with MPTMS or vinyltrimethoxysilane (VTMS) and obtained copolymers were following condensed with TEOS (Brostow et al., 2008).

Covalent linking of polymer chains onto silica surface can be achieved by grafting methods. PMMA-grafted silica microparticles were synthesized via radical photopolymerization of MMA, initiated from N,N-diethyldithiocarbamate (DEDT) groups previously bound to the silica surface ("grafting from") (Derouet & Thuc, 2008). The same technique of grafting was used for preparation of poly(octadecyl acrylate)(PODA)-grafted silicas (Mallik et al., 2009). They synthesized two kinds of ATRP initiators, immobilized them on silica, and then ATRP was carried out from these initiator-grafted silicas to obtain comb-shaped polymer-grafted silica. Polymers possessing long side groups, i.e. PODA are often termed comb-shape polymers. These PODA-SiO_2 composites showed unique separation behaviour with ordered-disordered phase transition of long alkyl chains and high selectivity towards polycyclic aromatic hydrocarbons (PAHs) in the ordered (crystalline) state.

Mesoporous silicas are worthwhile fillers for polymers. The pores can be occupied by polymer chains, especially during melt blending. To realize the polymerization inside the silica pores, the pores should be empty, so the most suitable method is bulk polymerization. Perez et al. (2007) used mesoporous silica, with the interconnected porous structure where the monomer can be trapped, as a reinforcement agent for PMMA. Some polymeric chains grew into the silica and the interaction of polymer with filler was increased.

Silicone nitride (Si_3N_4) is an important ceramic material used for various applications because of its high strength, high thermal-shock resistance, low coefficient of thermal expansion, and good wear resistance. It has been proved to be an effective filler for the improvement of mechanical properties, and in particular the wear resistance of polymeric materials. To limit the nanoparticles agglomerates and improving their dispersivity in polymer matrix the surface modification of fillers is necessary. A novel macromolecular coupling agent tercopolymer BA-MMA-VTES (useful for surface modification of silicone nitride nanopowder) was synthesized. The macromolecular coupling agent bonded covalently on the surface of nano-sized Si_3N_4 particles was used for preparing of an organic coating layer (Xia et al., 2008).

3.3.2.2 Composites based on polyolefins

Isotactic polypropylene PP is a commodity polymer with widespread applications. To improve its ductility and toughness, impact modifiers like various polyolefinic elastomers (POEs) are commonly used. Inorganic fillers such as talc, silica, $CaCO_3$, $BaSO_4$, $Mg(OH)_2$ and $Al(OH)_3$ are used as reinforcing agents in these PP-blends. During the last decade PP-based nanocomposite blends, containing finely dispersed nanofillers, have received great interest. Due to the nonpolar character of the PP, functionalization of the polymer matrix and/or treatment of the nanoparticles are needed to achieve a good dispersion of the rigid nanoparticles and satisfactory mechanical properties. PP was functionalized by peroxide-initiated grafting of vinyltriethoxysilane (VTES), to give a PP-*g*-VTES derivative, which resulted in improved compatibility between the matrix and the filler. The mechanical properties of the composites were greatly enhanced in terms of tensile and flexural strength, while impact strength was preserved when the silane-treated nanosilica was used (Bailly & Kontopoulou, 2009).

In order to promote dispersion of nano-silica in PP grafted polymerizable foaming agent p-vinylphenylsulfonylhydrazide (VPSH) onto nanoparticles was prepared via free-radical polymerization. The grafted VPSH played double role when it was melt mixed with PP. The side sulfonylhydrazide groups were decomposed with volatile gas products, which blowed up the surrounding matrix that pulled apart the agglomerated nanoparticles, while the remaining backbone of the grafted polymer helped to improve the filler-matrix interaction (Cai et al., 2006).

Aluminium anhydride is an effective flame retardant for thermoplastics for two decades. Surface treatment of mineral fillers can ease their incorporation into thermoplastic melts, such as PP. Whilst most surface modifiers can act as effective dispersant for fillers, only the silane and functionalised polymer based systems can also act as effective coupling agents in thermoplastics based composites. Liauw et al. (1995) compared three surface modifiers including a dicarboxylic acid anhydride (DAA), a silane based system and a maleanised polybutadiene (MPBD) in PP/$Al(OH)_3$ composites. At filler levels below these required for effective flame retardation, the DAA treated filler gave the composite with the best mechanical properties. At higher filler levels, i.e. above 59 % w/w, the composites based on $Al(OH)_3$ coated with the strongly coupling silane afforded the best properties. The latter treatment proved to be most effective when the polymer matrix was an impact modified PP.

A special case of nanocomposites is obtained by mixing polymers with layered silicates (nanoclays). The best known of layered silicates is montmorillonite (MMT). Layered silicates composed of sheet-like platelets that are about 1 nm in thickness and 100-1000 nm in width and length, so they possess high aspect ratios and large surface area. The high aspect ratio of the clay platelets permits significant reinforcement at relatively low loadings if the high degree of exfoliation or intercalation is achieved. It is well known that the dispersion of clay tactoids in a polymer matrix can result in the formation of three types of composites: (1) conventional composites that contain clay tactoids dispersed simply as a segregated phase; (2) intercalated polymer-clay nanocomposites, which are formed by the infiltration of one or more molecular layers of polymer into the clay host galleries; (3) exfoliated polymer-clay nanocomposites as characterized by low clay content, a monolithic structure, and a separation between clay layers that depend on the polymer content of the composite. Exfoliation is particularly desirable for improving specific properties that are affected by the degree of dispersion and the resulting interfacial area between polymer and clay nanolayers.

The layered silicate particles are usually hydrophilic and their interaction with non-polar polymers are unfavourable. There are two main objectives of surface modification on MMT: (1) to expand the interlayer space, allowing large polymer molecules to enter into the clay galleries, and (2) to improve the miscibility of MMT with the polymer to achieve a good dispersion of layered structure within the polymer matrix.

The maleic-anhydride functionalized polypropylene or polyethylene are the most commonly used to improve the interfacial bonding between the clay and respective polymers. Maleated polyethylene (PEgMA)/aminosilane compatibilizer has been very effective for preparing polyethylene-clay nanocomposites (Sanchez-Valdes et al., 2009). Silane grafting is one of methods to functionalize polyolefins, particularly for the preparation of silane-grafted water-cross-linked polyethylene. Lu et al. (2005) prepared methacryloxypropyltrimethoxysilane (MAPTMS)-grafted PE (PE-*g*-MAPTMS) by melt grafting reaction, and then blend it with organically modified montmorillonite to make PE-*g*-MAPTMS/MMT nanocomposite.

3.3.3 Applications of POSS-modified thermoplastic polymers

POSS are a class of versatile building blocks used for the preparation of organic-inorganic hybrid polymers with designed properties. The POSS molecules can be modified into monomers, initiators, or chain transfer agents (CTA) for a living polymerization. The POSS molecule was used as monomer to prepare the POSS-containing homopolymers or copolymers (Lichtenhan et al., 1995). In recent years the preparation of the POSS-containing hybrid polymers with the novel architecture has been focused on a living and controlled-living polymerization techniques, e.g. the synthesis of the homopolymers and triblock copolymers from a POSS-based methacrylate monomer using atom transfer radical polymerization (ATRP) was carried out (Pyun et al., 2003). The star-shaped hybrid polymers were prepared via ATRP of methyl methacrylate using the octafunctional POSS molecule as an initiator (Costa et al., 2001).

Reversible addition-fragmentation chain transfer (RAFT) polymerization has a wider selectivity of the monomers containing carboxyl, amino and ionic groups and it can be used under a broad range of experimental conditions including aqueous solutions. The POSS-containing RAFT agent, prepared of minopropylisobuthyl polyhedral oligomeric silsesquioxane was used in the RAFT polymerization of N-isopropylacrylamide (NIPAM) to produce tadpole-shaped organic-inorganic hybrid poly(N-isopropylacrylamide) (PNIPAM). The same POSS-containing RAFT agent was also successively applied in the RAFT polymerization of styrene to produce the organic-inorganic hybrid homopolymers and block copolymers (Zhang et al., 2009).

Recently the living/controlled polymerization technique is combined with "click chemistry" which is based on the copper-catalyzed Huisgen 1,3-dipolar cycloaddition between azides and alkynes. This reaction can be performed under mild conditions and it has a good tolerance of functional groups. Click chemistry has been mostly used to functionalize the end-groups or pendant groups of polymers or to built polymers with well-defined structures. The click reaction occurs between a polymer and small organic molecules or between two different polymer chains. The copper-catalyzed click reaction between alkyne-functional POSS molecules and mono-, di-, and pentafunctional azido-terminal polymers made by ATRP proceeds smoothly to form monochelic (tadpole-shaped), and ditelechelic (dumbbell-shaped) linear hybrid polymers as well as pentatelehcelic, star-shaped hybrids. The inversion of the procedure is also possible, i.e. the reaction of azido-functional POSS molecule was "clicked" to alkyne-terminated polymers (Müller et al., 2010).

It should be noted that most of the POSS molecules are intrinsically hydrophobic, while many commercial monomers are polar. Attachment of the polar group to POSS molecules make these compounds compatible with many polymers. For example oligosilsesquioxanes with poly(ethylene glycol)s, POSS-PEG monomers, because of their hydrophilic chemical structure are soluble in water and readily dispersed in polar monomers. The hydroxyl containing POSS molecules were used as an initiator in the ring-opening polymerization of lactide in the presence of a stannous catalyst.

Optical transparent films of a single POSS compound are hardly formed without cross-linking reagents due to their high symmetry and crystallinity. It could be speculated that lower the symmetries of the POSS derivatives decrease their crystallinity and would provide optical transparent film forming properties. Such kinds of materials are regarded as thermo-plastic hybrids possessing low-k or refractive index.

Dumbbell-shaped trifluoropropyl substituted POSS derivatives linked by simple aliphatic chains (ethane, propane, hexane) to reduce their symmetries were synthesized. These derivatives formed optical transparent films depending on their aliphatic linkages under the low temperature, which would open the way to apply to coating on various thermally unstable materials (Araki & Naka, 2011).

3.4 Silicones and silica as flame retardant additives

Silicone-based compounds (silicones, silsesquioxanes, silicas, and silicates) usually are used not only as fillers, incorporated in the polymer matrix but also as the flame retardant additives. They are endowed with excellent thermal stability and high heat resistance, with very limited release of toxic gases during thermal decomposition. Several types of silicone polymers were applied as flame retardants in polycarbonates (Iji & Serizawa, 1998). For the PC materials modified with branched methyl- and phenylsiloxanes and end-capped by methyl groups, limiting oxygen index (LOI) increased over 35 %, whereas LOI was about 27 % for pure PC. The effect of content and block size of PDMS on LOI values was investiga-ted for PC-*b*-PDMS copolymers. The PDMS block size influenced the dispersibility of the PDMS in the PC and the moderate PDMS dispersion (i.e. 50 nm mean inclusion size) resulted in high flame retardancy (Nodera & Kanai, 2006). Silica particles "*in situ*" formed by thermal degradation of PDMS mostly remained within the char layer. This highly oxidized char layer has the structure which prevented volatile fuel production and served as an additional thermal insulator.

The effect of silica gel structure on the flammability properties of thermoplastics has been investigated by many workers. The effect of pore volume, particle size and surface silanol concentration of silica gel in polypropylene (PP) was studied (Gilman et al., 1999). The performances of various types of silica, silica gel, fumed silica and fused silica as flame retardants in PP and polyethylene oxide were investigated. The fumed silica and silica gel accumulating near the surface acted as a thermal insulation layer and reduced the polymer concentration near the surface in contact with flame. The fused silica did not accumulate near the surface and it is mainly present in the polymer melt layer. The accumulation of silica on the surface of the burned polymer has been also observed in PMMA composites (Kashiwagi et al., 2003).

Nanometric particles in polymer matrices are known to enhance the fire resistance. POSS with general formula $(RSiO_{1.5})_8$ is an inorganic silica-like nanocage. Eight organic groups located at the corners give possibility for compatibility POSS compounds with organic polymers. These inorganic nanocages are referred to as preceramic compounds. On combustion of such polymer composite, POSS acts as a precursor forming thermally stable ceramic materials at high temperature. The incorporation of POSS in polymers modifies both the viscosity and the mechanical properties of the molten polymer. It also affects the thermal stability and fire performances by reducing the quantity of heat released upon combustion. Metal-bearing POSS nanoparticles at tiny concentrations (ca. 1 wt. %) can markedly enhance the char yield in PP (Fina et al., 2005).

Halogen free siloxane modifiers can significantly reduce heat release characteristics of thermoplastic polyurethanes (TPUs). Segmented PUs derived from PTMO, MDI, 1,4-butanodiol were modified with secondary aminoalkyl functional PDMS via solution polymerization (Wang et al., 2000). Addition of 15 % of the PDMS allowed to reduce the cone calorimetry heat release rate by a factor of about 2/3 and hence improved fire resistance, while maintaining mechanical behaviour. It was suggested that the low surface energy characteristics of PDMS promoted migration to the air-polymer interface to form a predominately PDMS enriched surface, which was oxidized at elevated temperatures in air to a silicate-like material and this served as a protective layer, which further reduced burning of the underlaying polyurethane.

Water repellent of silicone-based materials, forming in the fire protective coatings, appear to be attractive flame-retardants, which have many advantages such as low smokes, low toxicity and halogen-free. It is known that intumescent flame retardant (IFR) system is composed of an acid source, a carbonization compound and a blowing agent. The main disadvantages of IFR system are its moisture sensitivity and poor compatibility with polymer matrix. Flammable properties of polyolefins (e.g. LOI for polypropylene is only 17 %) restrict their application in electric and electrical devices, wire and cable, public chairs, packaging films and transportations. One of the most effective methods for obtaining the halogen-free flame-retardant polyolefin is use of IFR. The effect of polydimethylsiloxane (PDMS) in IFR polypropylene containing melamine phosphate (MP) was studied (Lu et al., 2009). It was found that the values of LOI increase gradually with the increase of siloxane content in PP/MP/PDMS composites. Studies of the water resistance of these materials showed improved water leaching of MP. A polysiloxane was found to be a very effective additive for the flame retardancy and water resistance of the IFR-PP materials. An improved thermal stability for PP/MP/PDMS composites was also observed. IFR system containing polysiloxane and silane-modified SiO_2, ammonium polyphosphate (APP) in PP matrix was investigated as well (Gao et al., 2010). It was showed that both polysiloxane and silica effectively enhance the flame retardancy of the IFR-PP, and LOI values exceeded 36 %.

Silane-functionalized PP and maleic anhydride modified PP were efficient as a coupling agents for PP-aluminium hydroxide (AH) composites, in order improve the interaction between the phases. However, PP modified with vinyltriethoxysilane showed better effect on the mechanical properties (Bohrz et al., 2006).

3.5 Applications of silane-modified thermoplastics in medicine

Synthetic polymers with pendant sugar moieties are of great interest not only as simplified model of biopolymers bearing oligosaccharides, but also artificial glycoconjugates in biochemistry and medicine. A new polysiloxane copolymer with pendant glucosylthioureylene groups was prepared in reaction of amino-functionalized PDMS with glucosyl isothiocyanate. They kept the optical activity and could be used not only as biomedical and biotechnological materials, but also as a model of amphiphilic polymers. Another examples of these materials are saccharide-based polysiloxanes, i.e. PDMS grafted amylase or PDMS grafted saccharose (Zhou et al., 2004).

The CFS have also found applications for production of contact lenses. This type of membranes must to pass oxygen and this ability is determined on a base of so called *"equivalent oxygen percentage"* (EPO). A minimal value of EPO equals 5-7 %. For the soft lenses made of PDMS, with thickeness 0.2 mm, EPO equals ~20 %. Hydrophobic properties of this material can lead to a damage of a cornea, and lenses made of pure PDMS are not used. They are produced of silicone-methacrylic copolymers or the carbofunctional silanes, i.e. (methacryloxypropyl)trimethylsiloxysilane or (methacryloxypropyl)pentamethyl-disiloxane. Lenses made from these materials reach EPO value ~10 % (Arkles, 1983).

Hydrogels have been extremely useful in biomedical and pharmaceutical applications for a long time. Poly(2-hydroxyethyl methacrylate) PHEMA, a biocompatible hydrogel, can absorb a large amount of water and is used to make ophthalmic prostheses (contact or intraocular lenses), vascular prostheses, drug delivery and soft-tissue replacement. Bioactive PHEMA-silica hybrids can be produced either by addition of silica nanoparticles to HEMA monomer-PHEMA solution, or using HEMA solution and tetraethoxysilane (TEOS) (as a silica precursor) through *in situ* sol-gel processes. A hybrid monomer was prepared of HEMA and 3-aminopropyltriethoxysilane (APTS) and its polymers, and blend materials were synthesized through basic catalyzed hydrolysis and condensation of TEOS, followed by *in situ* radical polymerization of HEMA. Obtained hybrid materials were used to make bioactive scaffolds for bone engineering (Luciani et al., 2008).

Composites prepared from a silicone rubber and a hydrogel are often used as drug release materials. A silicone rubber-polyacrylamide (PAAm) composite hydrogel exhibits pH-independent swelling degree, high specific surface area and non-ionogenic character (Mashak, 2008).

Polylactides, due to their biocompatibility and biodegradability, have also been used for biomedical applications for decades. Poly(lactide-*b*-methylvinylsiloxane-*b*-lactide) triblock copolymers, in which the vinyl groups were functionalized with carboxylic acids were synthesized. At neutral pH, the carboxylate functional polysiloxane central block binds to the surface of magnetic nanoparticles, while the polylactides serve as tail blocks to provide dispersibility in polylactide solvents through interparticle steric repulsive forces. Potential application for these magnetic materials include magnetic field-directed drug delivery, magnetic cell separations and magnetic hypertherma therapy for a treatment of tumors (Ragheb & Riffle, 2008).

The permanent fungicidal and bactericidal properties of polysiloxanes bearing quaternary ammonium salt (QAS) groups have been known for five decades. Novel low surface energy antimicrobial coating, based on hybrid siloxane epoxy coatings containing QAS moieties, was capable of self-contaminating in a variety of environments. These coatings have demonstrated the ability to eliminate up to 99.9 % of pathogenic bacteria on the surface (Pant et al., 2008).

The attachment of α-helical polypeptides to the colloidal silica is interesting as the core-shell composite particles call to mind naturally existing, protein-caged materials like viruses. These hydrophobic, polypeptide-functionalized particles could be used in investigation designed to model enzyme activation or the properties of hydrophobic proteins in cell membranes.

Polyurethane elastomers since 1970s have found a wide range of biomedical applications due to a combination of good biocompatibility, good hydrolytic and oxidative biostability, excellent mechanical properties and good processability. Commercial functionalized copolymers contain ca. 10 wt. % PDMS and are either segmented polyurethanes or polyureas that utilize polycarbonate or polyether as soft segments. They can be utilized in the manufacturing of intra-aortic balloons The anti-thrombogenic property of PDMS based polyurethanes is also utilized in artificial heart grafts, vascular grafts, implants and other such medical applications (Sheth et al., 2004).

4. Concluding remarks

Although carbofunctional silanes and polysiloxanes are known about 40 years they are still often used, and in a literature continually appear new examples of their applications for the modification of properties of polymers and other materials. The CFS have found many different applications for a preparation of polyurethane coating materials, having very good and useful properties. Bis-(3-triethoxysilyl)propyltetrasulphide) is used for the modification of natural and synthetic rubber (Sombatsompop et al., 2007), other CFS are applied for the preparation of silica in elastomers (*in situ*), and for the modification of calcium carbonate, which is used as the filler for elastomers (Zaborski, 2003).

An interest of research centers concerning the CFS and CFPS is still growing and broadens a scale of their numerous applications for modification of polymers. They have a great technological meaning. So far the carbofunctional silanes and polysiloxanes have found many practical applications: these first - mainly as adhesion promoters, polymer modifiers, and crosslinking agents for polyolefins, these latter - in the synthesis of different silicone-organic copolymers with very profitable properties. Perspectives of their further applications seem to be very promising, and especially for the modification of properties of different organic polymers, not only thermoplastics, but also elastomers, chemically hardenable polymers and thermosetting plastics. The carbofunctional silanes are more often used in practice than CFPS. A main barrier in applications of carbofunctional polysiloxanes on a bigger scale is a high cost of their production. However the chemical modification of polymers with CFPS has a very profitable influence on properties of the modified polymers.

In this Chapter we have been able to describe only a part of results concerning applications of silanes, siloxanes and silicate modified thermoplastic polymers.

5. Acknowledgment

Financial support from the European Regional Development Fund in the Operational Programme - Innovative Economy, Project No. UDA-POIG.01.03.01-173/09: "*Silsesquioxanes as polymer nanofillers and modifiers in polymer composites*" is gratefully acknowledged.

6. References

Abe, Y. & Gunji, T. (2004). Oligo- and polysiloxanes, *Progress in Polymer Science,* Vol.29, No.3, pp. 149-182, ISSN 0079-6700

Alongi, J; Monticelli, O.; Russo, S.; Galy, J. & Gerard, J-F. (2009). Morphology of Inorganic-Organic Systems Consisting of POSS and Polyamide 6. *e-Polymers,* No. 028, ISSN 1618-7229

Antic, V.V.; Pergal, M.V.; Govedarica, M.N.; Antic, M.P. & Djonlagic, J. (2010). Copolymers Based on Poly(butylene terephthalate) and Polycaprolactone-*b*-PDMS-*b*-Polycaprolactone. *Polymer International,* Vol.59, No.6, pp. 796-807, ISSN 1097-0126

Araki, H. & Naka, K. (2011). Synthesis of Dumbbell-Shaped Trifluoropropyl-Substituted POSS derivatives Linked by Simple Aliphatic Chains and Their Optical Transparent Thermoplastic Films. *Macromolecules,* Vol.45, No.15, pp. 6039-6045, ISSN 0024-9297

Arkles, B. (1977). Tailoring Surfaces with Silanes. *Chemtech,* Vol. 7, No.12, pp. 766-780, ISSN 0009-2703

Arkles, B. (1983). Look What You Can Make Out of Silicones (Biomedical Applications of Silicones. *Chemtech,* Vol.13, No. 9, pp. 542-555, ISSN 0009-2703

Assadi, M. G.; Makham M. & Tajrezaiy, Z. (2005). Synthesis and Characterization of Some Organosilicon Derivatives of Poly 2-Hydroxyethyl methacrylate with Cubane as A Cross-Linking Agent. *Journal of Organometallic Chemistry,* Vol.690, No.21-22, pp. 4755-4760, ISSN 0022-328x

Bailly, M. & Kontopoulou, M. (2009). Preparation and Characterization of Thermoplastic Olefin/Nanosilica Composites Using a Silane-Grafted PP matrix. *Polymer,* Vol.50, No.11, pp. 2472-2480, ISSN 0032-3861

Baney, R.H., Itoh, M., Sakakibara, A. & Suzuki T. (1995). Silsesquioxanes. *Chemical Reviews,* Vol.95, No.5, pp. 1409-1430, ISSN 0009-2665

Bohrz, N.S. M.; Schneider, M.M.; Edila, E. & Santos, M.R. (2006). Micromolecular Coupling Agents for Flame Retardant Materials. *European Polymer Journal,* Vol.42, No.5, pp. 990-999, ISSN 0014-3057

Botella, P.; Corma, A. & Navarro, M.T. (2007). Single Gold Nanoparticles Encapsulated in Monodispersed Regular Spheres of Mesostructured Silica Produced by Pseudomorphic Transformation. *Chemistry of Materials,* Vol.19, No.8, pp. 1979-1983, ISSN 0897-4756

Brinker, C.J. & Scherer, G.W. (1990). *Sol-Gel Science: the Physics and Chemistry of Sol-Gel Processing,* Academic Press

Brook, M.A. (2000). *Silicon in Organic, Organometallic and Polymer Chemistry,* J. Wiley & Sons, New York - Chichester - Weinheim - Brisbane - Singapor - Toronto; ISBN: 0-471-10658-4

Brostow, W.; Datashvili, T. & Hackenberg, K. Synthesis and Characterization of Poly(methyl acrylate) + SiO_2 Hybrids. *e-Polymers,* 2008, No. 054, ISSN 1618-7229

Cai, L.F.; Huang, X.B.; Rong, M.Z.; W.H. Ruan, W.H. & Zhang, M.Q. (2006). Effect of Grafted Polymeric Foaming Agent on the Structure and Properties of Nanosilica/PP composites. *Polymer,* Vol.47, No.20, pp. 7043-7050, ISSN 0032-3861

De Palma, R.; Peeters, S.; van Bael, M.J.; van den Rul, H.; Bonroy, K.; Laureyn, W.; Mullens, J.; Borghs G. & Maes, G. (2007). Silane Ligand Exchange to Make Hydrophobic Superparamagnetic Nanoparticles Water-Dispersible. *Chemistry of Materials,* Vol.19, No.7, pp. 1821-1831, ISSN 0897-4756

Cheng, X.; Chen M.; Zhou, S. & Wu, L. (2006). Preparation of SiO_2/PMMA composite particles via conventional emulsion polymerization. *Journal of Polymer Science: Part A: Polymer Chemistry,* Vol.44, No.12, pp. 3807-3816, ISSN 1099-0518

Chrissopoulou, K. &. Anastasiadis, S.H. (2011). Polyolefin/Layered Silicate Nanocomposites with Functional compatibilizers. *European Polymer Journal*, Vol.47, No.4, pp. 600-613, ISSN 0014-3057

Chauhan, B.P.S. & Balagam, B. (2006). Silyl Functionalization of Polyolefins. *Macromolecules*, Vol.39, No.6, pp. 2010–2012, ISSN 0024-9297

Chauhan, B.P.S.; Balagam, B.; Raghunath, M.; Sarkar, A. & Cinar, E. (2008). Pt-Nanocluster catalysts and their practical utility in generation of functionalized organosilicon polymers. *Polymer Preprints*, Vol.49, No.2, pp. 897–898, ISSN 0032-3934

Chen, G-X. & Shimizu, H. (2008). Multiwalled Carbon Nanotubes Grafted with POSS and Its Dispersion in Poly(L-lactide) Matrix. *Polymer*, Vol.49, No.4, pp. 943-951, ISSN 0032-3861

Chen, R-S.; Chang, C-J. & Chang, Y-H. (2005). Study on Siloxane-Modified Polyurethane Dispersions from Various PDMS. *Journal of Polymer Science: Part A: Polymer Chemistry*, Vol.43, No.17, pp. 3482-3490, ISSN 1099-0518

Chen, Y. & Kim, H. (2008) Poly(Vinylidene Fluoride) Grafted with 3-trimethoxysilylpropylmethacrylate for Silyl Functional Membranes. *Reactive & Functional Polymers*, Vol 68, No.11, pp. 1499-1506, ISSN 1381-5148

Chruściel, J.; Leśniak, E. & Fejdyś-Kaczmarek, M. (2008a). Carbofunctional silanes and polysiloxanes. Part 1. Applications of carbofunctional silanes. *Polimery*, Vol.53, No.10, pp. 709-716, ISSN 0032-2725

Chruściel, J.; Leśniak, E. & Fejdyś-Kaczmarek, M. (2008b). Carbofunctional silanes and polysiloxanes. Part 2. Applications of carbofunctional polysiloxanes. *Polimery*, Vol.53, No.11-12, pp. 817-829, ISSN 0032-2725

Costa, R.O.R.; Vasconcelos, W.L.; Tamali, R. & Laine, R.M. (2001). Organic/Inorganic Nanocomposite Star Polymers via ATRP of Mathyl Mathacrylate Using Octafunctional Silsesquioxane Cores. *Macromolecules*, Vol.34, No.16, pp. 5398-5407, ISSN 0024-9297

Daimatsu, K.; Anno, Y.; Sugimoto, H.; Nakanishi, E.; Inomata, K.; Ikeda, T. & Yokoi, K. (2008). Preparation, Morphology, and Physical Properties of Transparent PSt Hybrid Materials Containing Silicone Macromonomer. *Journal of Applied Polymer Science*, Vol.108, No.1, pp. 362-369, ISSN 1097-4628

Darras, W.; Fichet, O.; Perrot, F.; Boileau, S. & Tyssie, D. (2007). Polysiloxane-Poly(Fluorinated Acrylate) Interpenetrating Polymer Networks: Synthesis and Characterization. *Polymer*, Vol.48, No.3, pp. 687-695, ISSN 0032-3861

Dassin, S.; Dumon, M.; Mechin, F. & Pascault, J-P. (2002). Thermoplastic Polyurethanes (TPUs) with Grafted Organosilane Moieties: A new way of Improving Thermomechanical Behavior. *Polymer Engineering and Science*, Vol 42, No.8, pp. 1724-1739, ISSN 1548-2634

Demjen, Z.; Pukanszky, B. & Nagy, J. Jr. (1999). Possible coupling reactions of functional silanes and polypropylene. *Polymer*, Vol.40, No.7, pp. 1763-1773, ISSN 0032-3861

Demjen, Z. & Pukanszky, B. (1997). Effect of Surface Coverage of Silane Treated CaCO, on the Tensile Properties of Polypropylene Composites. *Polymer Composites*, Vol.18, No 6, pp. 741-746, ISSN 1548-0569

Demjen, Z., Pukanszky B. & Nagy J. (1998). Evaluation of interfacial interaction in polypropylene/surface treated $CaCO_3$ composites. *Composites, Part A*, Vol.29, No.3, pp. 323-329, ISSN 1359-835x

Derouet, D. & Thuc, C. N. H. (2008). Synthesis and Characterization of Poly(methyl mathacrylate)-Grafted Silica Microparticles. *Journal Applied Polymer Science,* Vol. 109, No.4, pp. 2113-2127, ISNN1097-4628

Eaborn, C., *Organosilicon Compounds,* Butterworth Sci. Publ., London, 1960

Eagles, D.C. (1990). Water crosslinkable copolymer. *Wire Technology International,* Vol. 18, p. 15, ISSN

Fawcett, A.H., Foster A.B., Hania M., Hohn M., McCaffery G.O., Mazebedi J.L. & Mullen E. (2001). *Polymer Preprints,* Vol. 42, No.1, p. 213, ISSN 0032-3934

Fina, A.; Tabuani, D.; Frache, A. & Camino, G. (2005). PP-POSS Nanocomposites. *Polymer,* Vol.46, No.19, pp. 7855-7866, ISSN 0032-3861

Gamys, C.G.; Beyou, E. & Bourgeat-Lami, E. (2010). Micellar Behavior of Well-Defined Polystyrene-Based Block Copolymers with Triethoxysilyl Reactive Groups and Their Hydrolysis-Condensation. *Journal of Polymer Science: Part A: Polymer Chemistry,* Vol 48, No.4, pp. 784-793, ISSN 1099-0518

Gao, S.; Li, B.; Bai, P. & Zhang. S. (2010). Effect of Polysiloxane and Silane-Modified SiO_2 on a Novel Intumescent Flame Retardant PP System. *Polymers for Advanced Technologies,* DOI: 10.1002/pat.1810, ISSN 1099-1581

Gilman, J. W.; Kashiwagi, T.; Harris R.H.; Lomalin, S.; Lichtenhan, J.D.; Jones, P. & Bolf, A. (1999), in: Al-Malaika, S.; Wilkie, C. & Golovoy, C. A. (Eds), *Chemistry and Technology of Polymer Additives,* Blackwell Science, London, UK

Greber, G. & Jäger, S. (1962). Über oligomere Siliciumverbindungen mit funktionellen Gruppen. 12. Mitt. Über Herstellung von oligomeren siliciumorganischen Diolen und Diaminen und ihre Umsetzungen mit organischen Diisocyanaten. *Die Makromolekulare Chem.* Vol.57, No.1, pp. 150-191, ISSN 1022-1352

Guliński J., Marciniec B., Maciejewski H., Pancer K., Dąbek I. & Fiedorow R. (2003). Synthesis of carbo-functional propyltri(alkoxy, methyl)silanes, *Przemysł Chemiczny,* Vol. 82, No.8-9, pp. 605-607, ISSN 0033-2496

Guiot, J.; Ameduri, B.; Boutevin, B. & Lannuzel, T. (2006). Original crosslinking of Poly(Vinylidene Fluoride) via Trialkoxysilane Containing Cure-Site Monomers. *Journal of Polymer Science: Part A: Polymer Chemistry,* Vol.44, No.12, pp. 3896-3910, ISSN 1099-0518

Guo, X.; Farwaha, R.; Rempel & G.L. (1990). Catalytic Hydrosilylation of Diene-Based Polymers. 1. Hydrosilylation of Polybutadiene, *Macromolecules,* Vol.23, No.24, pp. 5047-5054, ISSN 0024-9297

Guo, L.; Zhang, Z.; Zhu, Y.; Li, J. & Xie, Z. (2008). Synthesis Of Polysiloxane-Polyester Copolymer by Lipase-Catalyzed Polycondensation. *Journal of Applied Polymer Science,* Vol.108, No.3, pp. 1901-1907, ISSN 1097-4628

Han, C.D., Van de Weghe, T., Shete P. & Haw J.R. (1981). Effects of coupling agents on the rheological properties, processability, and mechanical properties of filled polypropylene. *Polymer Engineering & Science,* Vol.21, No.4, pp. 196-204, ISSN 1548-2634

Harabagiu, V.; Pinteala, M.; Cotzur C.; Holerca, M.N. & Ropot, M. (1995). Functional polysiloxanes. III: Reaction of 1,3-bis(3-glycidoxypropyl)1,1,3,3-tetramethyldisiloxane with amino compounds. *Journal of Macromolecular Science - Pure & Applied Chemistry,* Vol. A32, No.8-9, pp. 1641-1648, ISSN

Harabagiu, V.; Pinteala, M.; Cotzur, C. & Simionescu B.C.: *The Polymeric Materials Encyclopedia; Synthesis, Properties and Applications*, Salamone J.C. (Ed.), CRC Press, Boca Raton, FL, 1996, Vol. 4, p. 2661

Haas, K.H. & Wolter H. (1999). Synthesis, properties and applications of inorganic–organic copolymers (ORMOCER®s), *Current Opinion in Solid State & Materials Science*, Vol.4, No.6, pp. 571-580, ISSN 1359-0286

Ibarboure, E.; Papon, E. & Rodriguez-Hernandez, J. (2007). Nanostructured Thermotropic PBLG-PDMS-PBLG block copolymers. *Polymer,* Vol.48, No.13, pp. 3717-3725, ISSN 0032-3861

Iji, M. & Serizawa, S. (1998). Silicone Derivatives as New Flame Retardants for Aromatic Thermoplastic Used in Electronic Devices. *Polymers for Advanced Technologies,* Vol. 9, No.10-11, pp. 593-600, ISSN 1099-1581

Iojoiu, C.; Abadie, M.J.M.; Harabagiu, V.; Pinteala, M. & Simionescu, B.C. (2000). Synthesis and photocrosslinking of benzyl acrylate substituted polydimethylsiloxanes. *European Polymer Journal*, Vol.36, No.10, pp. 2115-2123, ISSN 0014-3057

Iraqi, A.; Seth, S.; Vincent, C.A.; Cole-Hamilton, D. J.; Watkinson, M.D.; Graham, I.M. & Jeffrey, D. (1992). Catalytic Hydrosilation of Polybutadienes as a Route to Functional Polymers. *Journal of Materials Chemistry*, Vol.2, No.10, pp. 1057-1064, ISSN 0959-9428

Kalaycioglu, E.; Toppare, L.; Yagci, Y.; Harabagiu, V.; Pintela, M.; Ardelean, R. & Simionescu, B. C. (1998). Synthesis of conducting H-Type Polisiloxane-Polypyrrole Block Copolymers. *Synthetic Metals,* Vol.97, No.1, pp. 7-12, ISSN 0379-6779

Kashio, M.; Sugizaki, T.; Yamamoto, S.; Matsuoka, T. & Moriya, O. (2008). Ring-Opening Polymerization of ε-Caprolactone by Base Catalyst for Synthesis of Grafted Poly-siloxane. *Polymer,* Vol.49, No.13, pp. 3250-3255, ISSN 0032-3861

Kashiwagi, T.; Shields, J.R.; Harris Jr., R.H. & Davis, R.D. (2003). Flame Retardant mechanism of Silica: Effect on Resin Molecular Weight. *Journal of Applied Polymer Science,* Vol.87, No.9, pp. 1541-1953, ISSN 1097-4628

Kawakami, Y.; Yu, S-P. & Abe T. (1992). Synthesis of aromatic diamino-functionalized dimethylsiloxane oligomers and macromonomers, *Polymer Bulletin*, Vol.28, No., pp. 525-530, ISSN 0170-0839

Kichler, A., Sabourault, N.; Décor, R.; Leborgne, C.; Schmutz, M.; Valleix, A.; Danos, O.; Wagner, A. & Mioskowski, C. (1994). Preparation and evaluation of a new class of gene transfer reagents: poly(alkylaminosiloxanes). *Journal of Controlled Release*, 2003, Vol.93, No.3, pp. 403-414, ISSN 0168-3659

Kerres, J. & Strathmann H. (1993). Synthesis and properties of a new AB-cross-linked copolymer membrane system. *Journal of Applied Polymer Science,* Vol.50, No.8, pp. 1405–1421, ISSN 1097-4628

Kim, S-R. & Kim, D-J. (2008). Effect of Silicone Oil on the Morphology and Properties of Polycarbonate. *Journal of Applied Polymer Science*, Vol.109, No.6, pp. 3439-3446, ISSN 1097-4628

Krea, M.; Roizard, D.; Moulai-Mostefa, N. & Sacco, D. (2004). New Copolyimide membranes with High Siloxane Content designed to Remove Polar Organics from Water by pervaporation. *Journal of Membrane Science,* Vol.241, No.1, pp. 55-64, ISSN 0376-7388

Kricheldorf H.R. (1996). *Silicon in Polymer Synthesis,* Springer, ISBN: 3-540-58294-0

Ku, C-K. & Lee, Y-D. (2007). Microphase Separation in Amorphous Poly(imide siloxane) Segmented Copolymers. *Polymer,* Vol.48, No.12, pp. 3565-3573, ISSN 0032-3861

Kujawski W., Kermes J. & Roszak R. (2003). Application of AB-crosslinked polymers composed of styrene/isoprene–siloxane copolymers to pervaporative removal of volatile organic compounds from water. *Journal of Membrane Science*, Vol.218, No.2, pp. 211–218, ISSN 0376-7388

LeGrand, D.G. (1972). *US Pat.* 3 679 774

Liauw, C. M.; Lees, G. C.; Hurst, S. J.; Rothon, R. N. & Dobson, D. C. (1995). *Plastics, Rubber, Composites Processing & Applications,* Vol. 24, No.5, pp. 249-260, ISSN 0959-8111

Liaw W.C. & Chen K.P. (2007). Preparation and Properties of Poly(imide siloxane) Segmented Copolymer/Silica Hybrid Nanocomposites. *Journal of Applied Polymer Science,* Vol.105, No.2, pp. 809-820, ISSN 1097-4628

Liaw W.C. & Chen K.P. (2007). Synthesis and characterization of imide ring and siloxane-containing cycloaliphatic epoxy resins. *European Polymer Journal,* Vol.43, No.4, pp. 1470-1479, ISSN 0014-3057

Lichtenhan, J.D.; Otonari, Y.A. & Carr, M.J. (1995). Linear hybrid Polymer Building Blocks: Methacrylate-Functionalized Polyhedral Oligomeric Silsesquioxane Monomers and Polymers. *Macromolecules,* Vol.28, No.24, pp. 8435-8450, ISSN 0024-9297

Lu H.; Hu Y.; Yang L.; Wang Z.; Chen Z. & Fan W. (2005). Preparation and thermal characteristics of silane-grafted polyethylene/montmorillonite nanocomposites. *Journal of Materials Science,* Vol.40, No.1, pp. 43–46, ISSN 0022-2461

Luo, Z.; He, T.; Yu, H. & Dai, L. (2008). A Novel ABC Triblock Copolymer with Very Low Surface Energy. *Macromolecular Reaction Engineering,* Vol.2, No.5, pp. 398-406, ISSN 1862-8338

Long, J.; Tzoganakis, C. & Chen P. (2003). Surface Characteristics of Hydrosilylated Polypropylene, *Journal of Applied Polymer Science,* Vol.88, No.14, pp. 3117–3131, ISSN 1097-4628

Long, J.; Tzoganakis, C. & Chen P. ((2004). Surface Characteristics of Hydrosilylated Polypropylenes: Effect of Co-catalyst and Reaction Temperature. *Polymer Engineering & Science,* Vol.44, No.1, pp. 56-71, ISSN 1548-2634

Luciani, G.; Constantini, A.; Silvestri, B.; Tescione, F.; Branda, F. & Pezzella, A. (2008). Synthesis, Structure and Bioactivity of pHEMA/SiO_2 Hybrids Derived through in situ Sol-gel Process. *Journal of Sol-Gel Science and Technology,* Vol.46, No.2, pp. 166-175, ISSN 0928-0707

Lv, P; Wang, Z.; Hu, Y. & Yu, M. (2009). Study of Effect Of PDMS in Intumescent Flame retardant PP. Journal of Polymer Research, Vol.16, No.2, pp. 81-89, ISSN 1022-9760

Lu, C.; Wang, Z.; Liu, F.; Yan, J. & Gao, L. (2006). Microstructure and Properties of New Polyimide/siloxane Composite Films. *Journal of Applied Polymer Science,* Vol.100, No.1, pp. 124-132, ISSN 1097-4628

Malz, H. & Tzoganakis C. ((1998). Hydrosilylation of Terminal Double Bonds in Polypropylene Through Reactive Processing, *Polymer Engineering & Science,* Vol.38, No.12, pp. 1976-1984, ISSN 1548-2634,

Marciniec, B.; Guliński, J.; Mirecki, J. & Fołtynowicz, Z. (1990). Silanowe pośredniki adhezji. Cz.1. 3-Aminopropylotrietoksysilan. *Polimery,* Vol.35, No.3, pp. 213-217, ISSN 0032-2725

Marciniec, B. & Guliński J. (1992a). Silanowe środki wiążące. Cz. II. 3-Chloropropylotrialkoksysilany. *Polimery*, Vol.37, No.1, pp. 73-77, ISSN 0032-2725

Marciniec, B.; Guliński, J.; Urbaniak W. & Kornetka, Z.W. (1992b). *Comprehensive Handbook of Hydrosilylation*, Marciniec (Ed.), Pergamon Press, Oxford, ISBN 0080402720

Marciniec, B.; Urbaniak, W. & Maciejewski, H. (1993). Silanowe środki wiążące. Cz. III. 3-Metakryloksypropylotrialkoksysilany. *Polimery*, Vol.38, No.1, pp. 53-58, ISSN 0032-2725

Mashak, A. (2008). In vitro Drug Release from Silicone Rubber-Polyacrylamide composite. *Silicone Chemistry*, Vol.3, pp. 295-301, ISSN 1572-8994

McManus, N.T.; Zhu, S.-H.; Tzoganakis C. & Penlidis A. (2006). Grafting of Ethylene–Ethyl Acrylate–Maleic Anhydride Terpolymer with Amino-Terminated Polydimethylsiloxane during Reactive Processing. *Journal of Applied Polymer Science*, Vol.101, No.5, pp. 4230–4237, ISSN 1097-4628

Morshedian, J. & Hoseinpour, P.M. (2009). Polyethylene Cross-linking by Two-step Silane Method: A Review. *Iranian Polymer Journal*, Vol.18, No.2, pp. 103-128; http://journal.ippi.ac.ir

Munteanu, D. (1997). *Chapter 5: Moisture cross-linkable silane-modified polyolefins*, pp. 196-265. In: S. Al-Malaika (Ed.), *Reactive Modifiers for Polymers*, Blackie Academic & Professional, London-New York, ISBN: 0 7514 0265 6

Nachtigall, S.M.B.; Stedile, F.C.;, Felix, A.H.O. & Mauler, R.S. (1999). Polypropylene Functionalization with Vinyltriethoxysilane. *Journal of Applied Polymer Science*, Vol.72, No.10, pp. 1313–1319, ISSN 1097-4628

Nanda, A.K.; Wicks, D.A.; Madbouly, S. A. & Otaigbe, J.U. (2006). Nanostructured Polyurethane/POSS Hybrid Aqueous Dispersions Prepared by Homegeneous Solution Polymerization. *Macromolecules*, Vol.30, No.20, pp. 7037-7043, ISSN 0024-9297

Nagao, D.; Hashimoto, M.; Hayasaka, K. & Konno, M. (2008). Synthesis of Anisotropic Polymer Particles with Soap-Free Emulsion Polymerization in the Presence of Reactive Silane Coupling Agent. *Macromolecular Rapid Communications*, Vol.29, No.17, pp. 1484-1488, ISSN 1521-3927

Nedelčev T.; Krupa, I.; Csomorová, K.; Janigová, I. & Rychlý, J. (2007). Synthesis and characterization of the new silane-based antioxidant containing 2,6-di-*tert*-butylphenolic stabilizing moiety. *Polymers for Advanced Technologies*, Vol.18, No.2, pp. 157-164, ISSN 1099-1581

Ni, K.; Bourgeat-Lami, E.; Sheibat-Othman, N.; Shan, G. & Fevotte, G. (2008). Preparation of Hybrid Nanocapsules. *Macromol Symp*, Vol.271, No.1, pp. 120-128, ISSN 1521-3900

Nodera, A. & Kanai, T. (2006). Flame Retardancy of a Polycarbonate-PDMS block copolymers: The effect of the dimethylsiloxane block size. *Journal of Applied Polymer Science*, Vol.100, No.1, pp. 565-571, ISSN 1097-4628

Noll, W. (1968). *Chemistry and Technology of Silicones*, New York, Academic Press

Nomura, Y.; Sato, A.; Sato, S.; Mori, H. & Endo, T. (2007). Synthesis of Novel Moisture-Curable Polyurethanes End-Capped with Trialkoxysilane and Their Application to One-Component Adhesives. *Journal of Polymer Science, Part A: Polymer Chemistry*, Vol.45, No.13, pp. 2689-2704, ISSN 1099-0518

Ohnishi, M.; Uno, T.; Kubo, M. & Itoh, T. (2009). Synthesis and Radical Polymerization of Dissymmetric Fumarates with Alkoxyethyl and Bulky Siloxy Groups. *Journal of Polymer Science: Part A; Polymer Chemistry*, Vol.47, No.2, pp. 420-432, ISSN 1099-0518

Okuno K. & Woodhams R.T. (1975). Mica Reinforced PP. *Polymer Engineering & Science*, Vol.15, No.4, pp. 308-315, ISSN 1548-2634

Oz, N. & Akar, A. (2006). Redox Initiation System of Ceric Salt and α, ω-Dihydroxy-Poly(dimethylsiloxane)s for Vinyl Polymerization. *Journal of Applied Polymer Science*, Vol. 102, No.3, pp. 2112-2116, ISSN 1097-4628

Pant, R.R.; Buckley, J. L.; Fulmer, P. A.; Wynne, J.H.; McCluskey, D.M. & Philips, J.P. (2008). Hybrid Siloxane Epoxy Coatings Containing Ouaternary Ammonium Moieties. *Journal of Applied Polymer Science*, Vol.110, No.5, pp. 3080-3086, ISSN 1097-4628

Pavlovic, D.; Linhardt, J. G.; Kunzler, J.F. & Shipp, D.A. (2008). Synthesis of Amphiphilic Multiblock and Triblock Copolymers of PDMS and Poly(N,N-dimethacrylamide). *Journal of Polymer Science: Part A: Polymer Chemistry*, Vol.46, No.21, pp. 7033-7048, ISSN 1099-0518

Pleuddeman, E.P. (1991). *Silane Coupling Agents* Plenum Press, New York, USA

Poojari, Y. & Clarson, S.J. (2010). Lipase-Catalyzed Synthesis and Properties of Silicone Aromatic Polyesters and Silicone Aromatic Polyamides. *Macromolecules*, Vol.43, No.10, pp. 4616-4622, ISSN 0024-9297

Pouget, E.; Garcia, E.H. & Ganachaud, F. (2008). Direct Synthesis of PVA-g-PDMS in Microsuspension. *Macromolecular Rapid Communications*, Vol.29, No.5, pp. 425-430, ISSN 1521-3927

Pyun, J.; Matyjaszewski, K.; Wu, J.; Kim, G.M.; Chun, S.B. & Mather P.T. (2003). ABA Triblock Copolymers Containing Polyhedral Oligomeric Silsesquioxane Pendant Groups: Synthesis and Unique Properties. *Polymer*, Vol.44, No.9, pp. 2739-2750, ISSN 0032-3861

Pukanszky B. (2005). Interfaces and interphases in multicomponent materials: past, present, future. *European Polymer Journal*, Vol.41, No.4, pp. 645-662, ISSN 0014-3057

Queiroz, D. P. & de Pinho, M. N. (2005). Structural Characteristics and Gas Permeation of PDMS/Poly(Propylene Oxide) Urethane/Urea Bi-Soft Segment Membranes. *Polymer*, Vol.46, No.7, pp. 2346-2353, ISSN 0032-3861

Ragheb, R.T. & Riffle, J.S. (2008). Synthesis and Characterization of Poly(lactide-*b*-siloxane-*b*-lactide) copolymers as Magnetite nanoparticle dispersants. *Polymer*, Vol.49, No.17, pp. 5397-5404, ISSN 0032-3861

Rodriguez, R.; Barandiaran, M.J. & Asua, J.M. (2008). Polymerization Strategies to Overcome Limiting Monomer Concentration in Silicone-Acrylic Miniemulsion Polymerization. *Polymer*, Vol.49, No.3, pp. 691-696, ISSN 0032-3861

Rościszowski, P. & Zielecka, M. (2002). *Silikony - właściwości i zastosowanie*, WNT, Warszawa; ISBN: 83-204-2612-X

Sanchez-Valdes, S.; Mendez-Nonell, J.; Medellin-Rodriguez, F.J.; Ramirez-Vargas, E.; Martinez-Colunga, J.G.; Soto-Valdez, H.; Munoz-Jimenez, L. & Neira-Velazquez, G. (2009). Effect of PEgMA/Amine Silane Compatibilizer on Clay Dispersion of PE-Clay Nanocomposites. *Polymer Bulletin*, Vol.63, No.6, pp. 921-933, ISSN 0170-0839

Schmidt, H. (1984). Organicaly modified silicates by sol-gel processes. *Materials Research Society Symposia Proceedings*, Vol. 32, p. 327

Schmidt, H. (1994). Inorganic-Organic Composites by Sol-Gel Techniques. *Journal of Sol-Gel Science and Technology*, Vol.1, No.3, pp. 217-231, ISSN 0928-0707

Schoener, C.A.; Weyand, C.B.; Murthy, R. & Grunlan, M.A. (2010). Shape Memory Polymers with Silicon-Containing Segments. *Journal of Materials Chemistry*, Vol.20, No.9, pp. 1787-1793, ISSN 0959-9428

Scott, C.; Ishida, H. & Maurer F.H.J. (1987). Infrared analysis and Izod impact testing multicomponent polymer composites: polyethylene/EPDM/filler systems. *Journal of Materials Science*, Vol.22, No.11, 3963-3973, ISSN 0022-2461

Shen, Y-D.; Zhao, Y-N.; Li, X-R. (2009). Polyacrylate/Silica Hybrids Prapared by Emulsifier-Rree Emulsion Polymerization and the Sol-Gel Process. *Polymer Bulletin*, Vol 63, No.6, pp. 687-698, ISSN 0170-0839

Sheth, J.P.; Aneja, A.; Wilkes, G.L.; Yilgor, E.; Atilla, G.E.; Yilgor, I. & Beyer, F.L. (2004). Influence of System Variables on the Morphological and Dynamic Mechanical Behavior of PDMS Based Segmented Polyurethane and Polyurea Copolymers: A comparative Perspective. *Polymer*, Vol.45, No.20, pp. 6919-6932, ISSN 0032-3861

Sheth, J.P.; Yilgor, E.; Erenturk, B.; Ozhalici, H.; Yilgor, I. & Wilkes, G. L. (2005). Structure-property Behavior of PDMS Based Segmented Polyurea Copolymers Modified with Poly(propylene Oxide). *Polymer*, Vol.46,No.19, pp. 8185-8193, ISSN 0032-3861

Shi, L-Y.; Shen, Z. & Fan, X-H. (2011). Order-Order Transition in a Rod-Coil Diblock Copolymer Induced by Supercritical CO_2. *Macromolecules*, Vol.44, No.8, pp. 2900-2907, ISSN 0024-9297

Smedberg, A.; Hjeftberg, T. & Gustafsson, B. (2004). Characterisation and crosslinking properties of a poly(ethylene-*co*-divinylsiloxane), *Polymer*, Vol.45, No.14, pp. 4845-4855, ISSN 0032-3861

Sombatsompop, N.; Wimolmala, E. & Markpin T. (2007). Fly-Ash Particles and Precipitated Silica as Fillers in Rubbers. II. Effects of Silica Content and Si69-Treatment in Natural Rubber/Styrene–Butadiene Rubber Vulcanizates. *Journal of Applied Polymer Science*, Vol.104, No.5, pp. 3396-3405, ISSN 1097-4628

Suzuki, T.; Yamada, Y. & Itahashi, K. (2008). 6FDA-TAPOB Hyperbranched Polyimide-Silica Hybrids For Gas Separation Membranes. *Journal of Applied Polymer Science*, Vol.109, No.4, pp. 813-819, ISSN 1097-4628

Thomas, B. & Bowrey M. (1977). *Wire Journal*, Vol. 10, No. 5, p. 88

Tonnar, J.; Pouget E.; Lacroix-Desmazes, P. & Boutevin B. (2008). Synthesis of Poly(Vinyl Acetate)-b-PDMS-b-Poly(Vinyl Acatate) Triblock Copolymers By Iodine Transfer Polymerization. *European Polymer Journal*, Vol.44, No.2, pp. 318-328, ISSN 0014-3057

Trotignon, J.P.; Verdu, J.; Boissard R. & Vallois A. (1986). *Polymer Composites*, Walter de Gruyter, Berlin, p. 191

Tsuchida, A.; Bolin, C.; Sernetz, F.G.; Frey, H. & Mülhaupt R. (1997). Ethene and Propene Copolymers Containing Silsesquioxane Side Groups. *Macromolecules*, Vol.30, No.10, pp. 2818-2824, ISSN 0024-9297

Van Aert, H.A.M.; Nelissen, L.; Lemstra, P.J. & Brunelle, D.J. (2001). Poly(Bisphenol A Carbonate)-PDMS Multiblock Copolymers. *Polymer*, Vol.42, No.5, pp. 1781-1788, ISSN 0032-3861

Wan, T.; Wu, C.; Ma, X-L.; Wang, L.; Yao J. & Lu, K. (2009). Microstructure and Properties of Silane Monomer-Modified Styrene-Acrylate Nanocoatings. *Polymer Bulletin*, Vol.62, No.6, pp. 801-811, ISSN 0170-0839

Wang, L.F.; Ji, Q.; Glass, T.E.; Ward, T.C.; McGrath, J.E.; Muggli, M.; Burns, G. & Sorathia, U. (2000). Synthesis and Characterization of Organosiloxane Modified Segmented Polyether Poluurethanes. *Polymer*, Vol.41, No.13, pp.5083-5093, ISSN 0032-3861

Wang, M.; Li, B.; Wang, J. & Bai, P. (2010). Preparation and Properties of Polysiloxane Grafting Multi-Walled Carbon Nanotubes/Polycarbonate Nanocomposites. *Polymers for Advanced Technologies*, Vol.21, No.6, DOI: 10.1002/pat.1665 , ISSN 1099-1581

Wang, Y-P.; Yuan, K.; Li Q-L.; Wang, L-P.; Gu, S-J. & Pei X-W., Preparation and characterization of poly(N-isopropylacrylamide) films on a modified glass surface via surface initiated redox polymerization. (2005). *Materials Letters*, Vol.59, No.14-15, pp. 1736-1740, ISSN 0167-577x

Williams, K.G.; Gido, S.P. & Coughlin E.B. (2011). *Chapter 4: Polymers and Copolymers Containing Covalently Bonded Polyhedral Oligomeric Silsesquioxanes Moieties*, pp. 167-207. In: C. Hartmann-Thompson (Ed.), *Applications of Polyhedral Oligomeric Silsesquioxanes, Advances in Silicon Science*, Vol. 3, DOI 10.1007/978-90-481-3787-9_4, © Springer Science + Business Media B.V.

Wu, Y.Z. & Feng S.Y. (2001). Viscosity–Molecular Weight Relationship for Aminopropyl-Terminated Poly(dimethylsiloxane). *Journal of Applied Polymer Science*, Vol.80, No.7, pp. 975-978, ISSN 1097-4628

Yagci, Y.; Onen, A.; Harabagiu, V.; Pinteala, M.; Cotzur, C. & Simionescu, B.C. (1994). Photoactive Polysiloxanes. 1. Synthesis of Polydimethylsiloxane with Benzoin Terminal Groups. *Turkish Journal of Chemistry*, Vol.18, pp. 101-104, ISSN 1300-0527

Yan, H.; Li, P.; Ning, R.; Ma, X. & Zhang, Z. (2008). Tribological Properties of Bismaleimide Composites with Surface-Modified SiO_2 Nanoparticles. *Journal of Applied Polymer Science*, Vol.110, No.3, pp. 1375-1381, ISSN 1097-4628

Yang, C.; Gu, A.; Song, H.; Xu, Z.; Fang, Z. & Tong, L. (2007). Novel Modification of Cyanate Ester by Epoxidized Polysiloxane. *Journal of Applied Polymer Science*, Vol.105, No.4, pp. 2020-2026, ISSN 1097-4628

Yilgor, I. & McGrath, J.E. (1988). Advances in Organosiloxane Copolymers, in: *Advances in Polymer Science*, Vol.89, No.1, pp. 1-86, ISSN 0065 3195

Yoshioka, Y. (2009). Preparation of Poly(amino-amide) Particles Complexed with ZnO Particles Using Silane Coupling Agents. *Journal of Polymer Science: Part A: Polymer Chemistry*, Vol.47, No.19, pp. 4908-4918, ISSN 1099-0518

Yu, Y-Y.; Chen, W-C. (2003). Transparent Organic-Inorganic Hybrid Thin Films Prepared from Acrylic Polymer and Aqueous Monodispersed Colloidal Silica. *Materials Chemistry and Physics*, Vol.82, No.2, pp. 388-395, ISSN 0254-0584

Zaborski, M. (2003. Nanotechnology and polymeric nano-composites, *Przemysł Chemiczny*, Vol.82, No.8-9, pp. 544-547, ISSN 0033-2496

Zhang, W.; Deodhar, S. & Yao, D. (2010). Processing Properties of Polypropylene With a Minor Addition of Silicone Oil. *Polymer Engineering & Science*, Vol.50, No.7, pp. 1340–1349, ISSN 1548-2634

Zhang, K; Gao, L. & Chen, Y. (2007). Organic-Inorganic Hybrid Materials by Self-Gelation of Block Copolymer Assembly and Nanoobjects with Controlled Shapes Thereof. *Macromolecules*, Vol. 40, No.16, pp. 5916-5922

Zhang, J-T.; Pan, C-J.; Keller, T.; Bhat, R.; Gottschaldt, M.; Schubert, U. S. & Jandt, K. D. (2009). Monodisperse, Temperature Sensitive Microgels Crosslinked by Si-O-Si Bonds. *Macromolecular Materials & Engineering,* Vol.294, No.6-7, pp. 396-404, ISSN 1439-2054

Zhang, W. & Muller, A.H.E. (2010). Synthesis of Tadpole-Shaped POSS-containing Hybrid Polymers via "Click Chemistry". *Polymer,* Vol.51, No.10, pp. 2133-2139, ISSN 0032-3861

Zhang, W.; Zhung, X.; Li, X.; Iin, Y.; Bai, J. & Chen, Y. (2009). Preparation and Characterization of Organic-inorganic Hybrid Polymers Containing Polyhedral Oligomeric Silsesquioxane via RAFT Polymerization. *Reactive and Functional Polymers,* Vol.69, No.2, pp. 124-129, ISSN 1381-5148

Zhou, C. J.; Guan, R. F. & Feng, S. Y. (2004). The Preparation of a New Polysiloxane Copolymer with Glucosylthioureylene Groups on the Side Chains. *European Polymer Journal,* Vol.40, No.1, pp. 165-170, ISSN 0014-3057

Zhu, S.-H., McManus, N.T., Tzoganakis, C. & Penlidis A. (2007). Effect of a Poly(dimethyl-siloxane) Modified Polyolefin Additive on the Processing and Surface Properties of LLDPE. *Polymer Engineering & Science,* Vol.47, No.9, pp. 1309–1316, 1548-2634

Zhu, S.-H. & Tzoganakis, C. (2008). Surface properties of hydrosilylated polyolefins annealed in supercritical carbon dioxide. *Polymers for Advanced Technologies,* Vol.19, No.4, pp. 258–269, ISSN 1099-1581

10

Advantages of Low Energy Adhesion PP for Ballistics

J. Gonzalo Carrillo-Baeza[1],
W.J. Cantwell[2] and Ricardo A. Gamboa-Castellanos[1]
[1]*Centro de Investigacion Cientifica de Yucatan,*
[2]*The University of Liverpool*
[1]*Mexico*
[2]*UK*

1. Introduction

This chapter shows the advantages of low interfacial adhesion thermoplastic matrix, such as polypropylene (PP), and how it can result in improvements to an aramid fabric for ballistics. In addition to the excellent properties for which it is already well known, the purpose here is to highlight a different aspect, i.e. the advantage of poor adhesion on a fabric arrangement for high velocity impacts, where the basis is related to energy dissipation by matrix debonding, thereby generating load redistribution between primary and secondary yarns in the composite. Another advantage is the weatherproofing aspect conferred to the fabric by the matrix, which protects it from harsh environments. Resistance to frontal impact with high speed projectiles was compared in two arrangements; woven aramid/PP on a consolidated configuration and woven aramid/PP on an independent configuration. The methodology for this comparison consisted in establishing impact energies tested on both configurations; before, during and after penetration, identifying the perforation threshold, V_{50}, and measuring back deformation as well as residual energy when the sample was perforated.

2. Polypropylene and aramid

Polypropylene (PP), one of the most widely used polymers all over the world, has now become the commodity polymer par excellence. One of the most important properties placing it above others is a versatility which allows it to be modified and designed for specific applications, along with an excellent balance of physical, chemical and mechanical properties. Due to the rheological and thermal behavior of molten materials, the materials based on this resin offer an extended period of processability, from injection molding to blown film extrusion. Whichever is the case, thermoplastics offer greater advantages over thermosets given the dexterity with which these materials can be molded, simply by the application of pressure, temperature and a short consolidation period [1].

All these properties conferring such versatility to PP are highly dependent on the degree of crystallinity, which in turn depends on the stereoregularity of the polymeric chain, a term known as tacticity. PP can present three types of tacticity; isotactic, syndiotactic and atactic, where each term represents the balance between three-dimensionally ordered regions and regions with no discernable order (amorphous regions); polymers with high degrees of crystallinity are denominated isotactic and the absence of any degree of crystallinity is known as atactic, while syndiotactic is an intermediate degree of crystallinity [1]. Crystallinity has a very marked effect on mechanical properties, since polymers with high percentages of crystallinity (between 60% and 70%) present increased resistance to traction at the yield point, as well as an increase in rigidity and resistance to flexure; however, they also present a reduction in tenacity and resistance to impact. In contrast, polymers with low percentages of crystallinity show low resistance to traction and flexure but high tenacity and good resistance to impact, as well as excellent transparency, making them very appropriate for use in the production of transparent films with high refraction rates up to 1.49 [1].

With the manipulation of tacticity, it is possible to generate a type of polypropylene with an increased capacity to resist impact; copolymers in blocks, which are part of a new family of elastomers, namely thermoplastic elastomers. This material has the properties of an elastomer, the main difference being that the polymeric chains are not joined by covalent links but by secondary links, in other words, they are elastomers which are not crosslinked. This polypropylene, also called PP-impact (Fig. 1), is composed of polymeric chains with isotactic and atactic blocks, which create highly ordered and amorphous areas. These highly-ordered areas function as covalent links, the main difference being that they can be separated only with temperature.

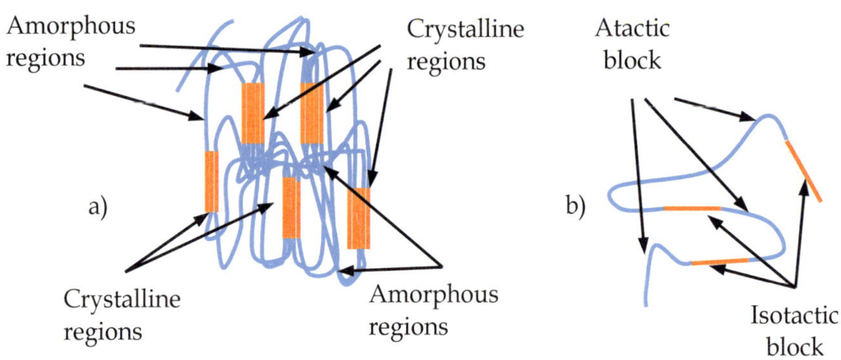

Fig. 1. Block copolymer, a) elastomeric PP, and b) PP chain by blocks

Another way to improve the mechanical properties of PP is by adding a second stage, which may be continuous or discontinuous fibers, thereby forming a composite fiber-reinforced material. This second stage is usually characterized by higher mechanical, thermal, electrical and chemical properties in comparison with those observed in PP bulk, which are subsequently transformed when combine together in a composite. From a mechanical point of view, resistance increases in response to a more optimal distribution of loads in the material, resulting in a better weight-resistance relationship, a property known as specific resistance; defined as the resistance of material per unit of weight, which is greatly appreciated in the transport industry since weight reduction means increased efficiency in any vehicle.

The most common reinforcement used in composite materials is organic fibers, consisting of thousands of filaments with diameters from 5 to 15 μm, obtained from fabric production lines. During production, the material can be obtained in two forms; the first is non-woven fiber, on continuous spools, while the second format is by means of a fabric which can be bi-dimensional or tridimensional, in which there are subcategories depending on the format. This work focuses on a particular study of a bi-dimensional aramid fabric.

Synthetic fiber fabrics are formed by fibers placed lengthwise in two perpendicular directions: where one direction is the warp and the other the weft. The warp usually marks the direction parallel to the fabric production line. Differences between different types of fabric depend on the way the weft is intercalated with the warp, some fabric configurations are shown in Fig. 2. Fig. 2a shows a plain weave configuration, Fig. 2b a satin weave configuration and Fig. 2c a twill weave configuration; these are the most common fabric types used on the market [2]. Each configuration has a preferential application depending on the requirements. The woven configuration is most commonly used in ballistics, since it presents the highest factor of impenetrability in comparison with other fabric configurations. The impenetrability factor of a fabric is a measurement of weave impenetrability as a function of the coverage factor [3].

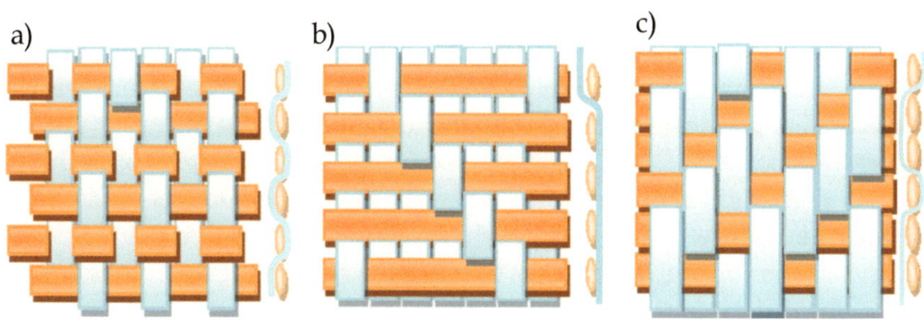

Fig. 2. Configurations of fabrics commonly used in composite materials, a) plain weave, b) satin weave and c) twill weave

One of the most commonly used fibers in composite materials and ballistics is aramid; these fibers were discovered in the early 60's and were put on the market by DuPont in the early 70's under the trademark of Kevlar®, they are well known for their mechanical properties and are currently one of the three most important fibers for applications in composite materials together with fiberglass and carbon fiber. The first structure was conceived by DuPont (Fig. 3) and has since, undergone changes which have improved its properties according to the required applications, this is also the case for Teijin Company with their fiber trademark Twaron® [4].

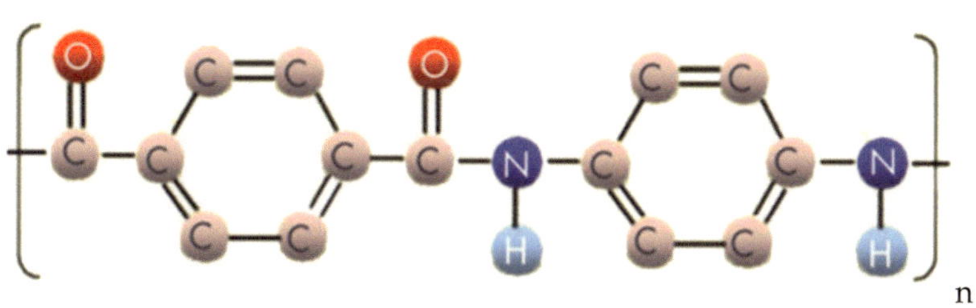

Fig. 3. Chemical structure of Poly(p-phenylene tereftalamide) or PPTA

The word aramid comes from the hybrid aromatic polyamide, where the main difference between the polyamides resides in the fact that 85% of the amide groups are linked to two aromatic rings [5]. Aramid fibers are produced by extracting an acid solution with an appropriate precursor (a polycondensation produced between terephthaloyl chloride and p-phenylenediamine) through a plate with small perforations. During this process, the aramid molecules acquire a high degree of orientation along the fiber, resulting in excellent tensile properties [6].

Due to the excellent specific resistance of this product, its most important application is in personal protection where its capacity to absorb impact energy from a projectile is extraordinary. The types of aramid most commonly used in ballistics are Kevlar 29, Kevlar 49, Kevlar 129 and Kevlar KM2, with important applications, such as the famous PASGT (Personal Armor System for Ground Troops) which have been the bulletproof vests used by US military from the 80's up to 2005, based on Kevlar 29. More recently, the bulletproof vests used during the intervention of troops in Iraq and Afghanistan were made from Kevlar KM2, an improved aramid for ballistics [4].

The ballistic resistance of armor vests made with highly flexible polymeric fibers is based on an energy absorption mechanism in which the impact load is transferred to a network of fibers (fabric) which are in contact with the projectile, making penetration resistance highly related to exclusive parameters of the fabric [7], thus, ballistic resistance is mainly dependent on the interaction between fabric nodes. It is necessary, therefore, to protect the fibers from environments that might degrade their properties. The fabric must be isolated as much as possible from humidity which is particularly harmful as it can affect the ballistic resistance of the fabric, either by degrading the fibers, lubricating them in excess or facilitating separation of the threads at the moment of contact with the projectile.

Friction between the nodes of armor plating promotes better load transfer between fiber bundles as it restricts their mobility. This is further increased by the polymeric matrix, thereby conferring the degree of armor fabric to the composite material (armor-grade composite). Restricting the movement of fibers increases interaction between them and generates other failure mechanisms (interlaminar and intralaminar delamination), which contribute to the process of energy absorption [8]; however, the degree of movement restriction should not reach the point where it might generate fragility. Fig. 4 shows the difference between armor fabric with a polymeric matrix (Fig. 4a) and without a polymeric matrix (Fig. 4b). This figure clearly shows how the matrix limits the movement of the fibers, thereby making more of them interact with each other during an impact at medium speed (300 m/s approximately).

Armor-grade composite materials present several disadvantages, such as the sensitivity of their resistance to manufacturing procedures, the high cost of production processes, the sensitivity and difficulty in locating damage generated at low impact and, most importantly, the difficulty in modeling their mechanical behavior due to the number of parameters involved deriving from the large number of failure mechanisms that intervene in the energy absorption process during a ballistic event. Another disadvantage is the viscoelastic nature of polymeric materials which make them highly dependent on impact velocity [9].

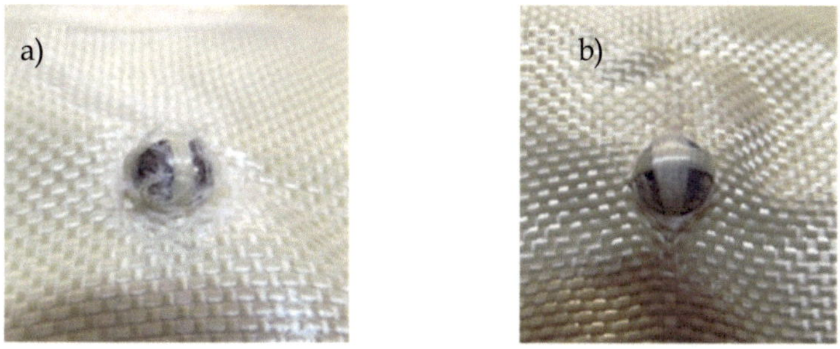

Fig. 4. Aramid samples impacted at different velocities, a) with polypropylene matrix and b) without polypropylene matrix

One particularly important parameter in this type of materials, and in composites in general, is the role played by fiber-matrix interfacial adhesion within mechanical resistance; this parameter which has become extremely important in the design of composite materials now represents a third entity in these materials. The following can give us an idea of the importance of this parameter; it is estimated that in 1 kg of PP with a CaC_3 load at 50% weight fraction, with a nominal particle size of 5 µm, can give an interfacial adhesion area between materials equal to three football fields [10].

The interface is defined as a bi-dimensional surface which divides two phases or components in a system; this is characterized by an abrupt change in properties and chemical composition. This surface does not possess a physical property in itself, since it only exists mathematically (see Fig. 5). The interphase, in contrast with the interface, is a layer in three dimensions surrounding the fiber with properties that are different from those of the fiber and the matrix (Fig. 5); this third entity is commonly used to improve load transfer from the matrix to the fiber, given the incapacity of the interface to carry out this task. Both interface and interphase fulfill the same purpose, to try to achieve load transfer as efficiently as possible from the matrix to the fiber; therefore, this parameter must be controlled in order to determine the behavior of the material.

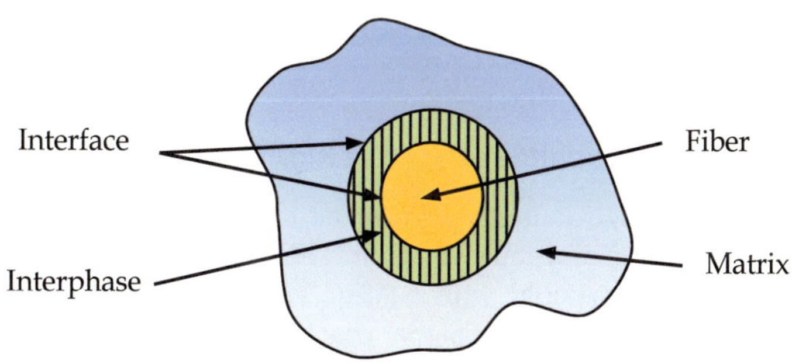

Fig. 5. Constituents commonly found in a fiber-matrix composite

Since load transfer is carried out via the interface, it is of vital importance to understand this parameter as it has considerable influence on the physical constants of the material, such as the elastic modulus, the Poisson ratio and tenacity of fracture. Due to the complexity of the mathematical analysis required for this parameter, no studies were carried out until the last decade; in those studies the only indications of this value are reported with techniques such as pull-out test, which has resulted in considerable controversy due to the lack of a standard which can establish a specific methodology.

The methods for determining the interfacial shear stress can be direct or indirect. Direct methods are those which are analyzed from a micromechanics perspective, where a small representative sample of the unit is used. Some methods we can mention are fiber pull-out, fragmentation, single fiber micro-indentation and single fiber compression; however, due to the close relationship this parameter has with the mechanical properties of the composite, there are also indirect methods to determine interfacial shear stress, which are analyzed from a micromechanical perspective and where the behavior of the whole unit is analyzed in order to determine the levels of interfacial adhesion in the composite. These methods include the variable curvature method, slice compression, ball compression and bundle pullout. There are also some methods which are analyzed at a macromechanical level and by conventional tests; these are able to relate values of tension, compression or flexure at three or four points with interfacial adhesion values in the composite [11].

As it has already established, interfacial adhesion plays an important role in the properties of a material; however, what role does it play in high impact properties? Studies focusing on improved interfacial adhesion in composite materials at tension, compression and shear abound in the literature; however, in high impact it does not appear to be particularly sought after. This is demonstrated in a study carried out in Rohchoon Park, where an aramid/vynilester composite material is characterized at high impact. The material received a superficial treatment to improve adhesion and it was possible to observe a reduction in the ballistic limit when this property is increased. This is precisely where PP can play a very particular role. In addition to all the properties found in PP, it also presents poor interfacial adhesion with practically any material due to its incapacity to generate covalent links. The aim of this work, therefore, is to demonstrate how a polymer with poor interfacial adhesion can be used in applications with high levels of energy absorption by taking advantage of precisely its inert character to dissipate the energy through other mechanisms which are more efficient at high impact, such as back cone formation and load transmission from primary threads to secondary threads.

3. Characterization of the composite material at high impact

To characterize the material at high impacts, the standard STANAG 2920 established by NATO (North Atlantic Treaty Organization) was used. The purpose of this standard is to characterize at high impact any material with ballistic applications, a bulletproof vest, ballistic or combat helmets, or any kind of material produced with this purpose in mind. The projectile used may be a bullet, from which protection is required. For this, a non deformable, spherical, steel projectile (1.11 g) is commonly employed as it offers the highest ballistic limit in high impact tests in comparison to other shapes, compared to ogival, blunt or pointed projectiles [8]. The parameter defined in these tests is the ballistic limit (V_{50}), which is defined as the velocity at which a material fails 50% of the times it is impacted; this parameter, which is calculated with Equation 1, has a statistical origin due to the stochastic nature controlling ballistic events. The V_{50} is determined using the average velocity of six impacts, three which have totally perforated the armored plate and three which have partially perforated it with an interval not greater than 60 m/s between the six impact velocities. Other very important parameters determined in this type of tests are the relationship between impact velocity, absorbed energy and trauma depth.

$$V_{50} = \frac{\sum_{i=1}^{6} V_i}{6} \tag{1}$$

The relationship between impact velocity and absorbed energy registers the amount of energy absorbed by the material at the moment of impact by a projectile at velocities equal or superior to its ballistic limit. The energy absorbed by the material (E_{abs}) is obtained based on the velocity at which the projectile impacts the sample and the velocity at which it exits at the back; these velocities are substituted in Equation 2, where m represents the mass of the projectile, V_{imp} the velocity at which the projectile impacts the sample and V_{res} the velocity at which the projectile exits the back of the material. The velocity at which the material is impacted is obtained with a Chrony chronograph which is capable of registering speeds between 10 and 2 134 m/s with a precision of 99.5%, conferring reliability to the readings. Residual velocity was obtained with the aid of a ballistic gelatine.

$$E_{abs} = \frac{mV_{imp}^2}{2} - \frac{mV_{res}^2}{2} \tag{2}$$

Ballistic gelatine is widely used in criminalistics due to the similarity of this material to the human body during high velocity impact. Due to the behavior presented by this material in response to a high velocity impact, it is possible to calculate the residual velocity with which the projectile impacted the material based on the depth of penetration. The velocity is obtained by characterizing the material during direct frontal impacts, where it is possible to generate an equation relating penetration length of the projectile with the velocity on entering the material. In some studies, such as the one carried out by Jorma Jussila [12] a more detailed procedure for this methodology is presented.

Once the relationship between impact velocity and absorbed energy is obtained, a phenomenon quite particular to this impact regimen emerges; a fall in energy absorption at velocities slightly higher than V_{50} with a subsequent recovery in absorption levels. A logical deduction could be that when a material presents a particular absorption of energy at perforation threshold during high velocity impacts with a 1.11 g projectile, for example 100 Joules, when this projectile impacts at 120 J one might assume that the material will absorb its corresponding part (100 J) and will allow passage of the projectile with a residual energy of 20 J. However, in reality, this does not happen. With projectile impacts at velocities slightly higher than the ballistic limit, what we find is a noticeable reduction in the capacity of the material to absorb energy. A study carried out by Paul Wambua *et al.* [13], shows a composite of natural fiber with a PP matrix, where it can be observe this phenomenon at velocities close to 250 m/s. Fig. 6 shows the curve obtained with this particular behavior.

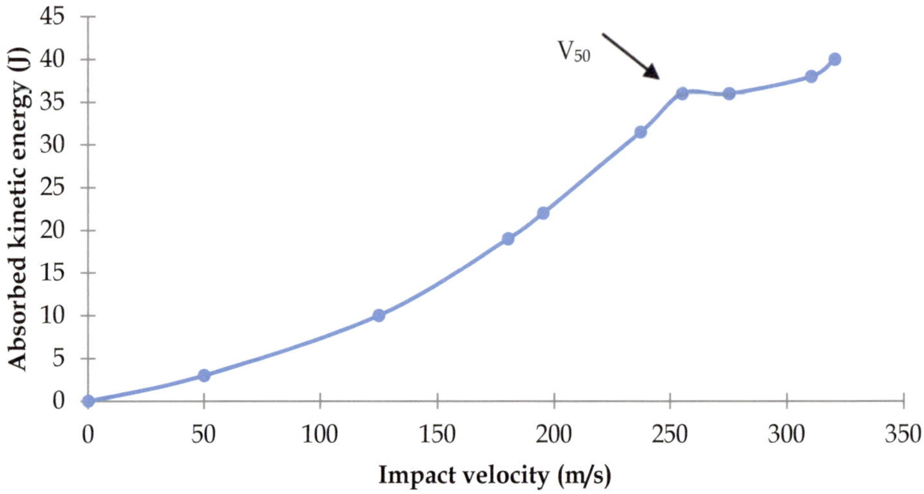

Fig. 6. Energy absorbed by the composite material hemp/PP subjected to impact [13]

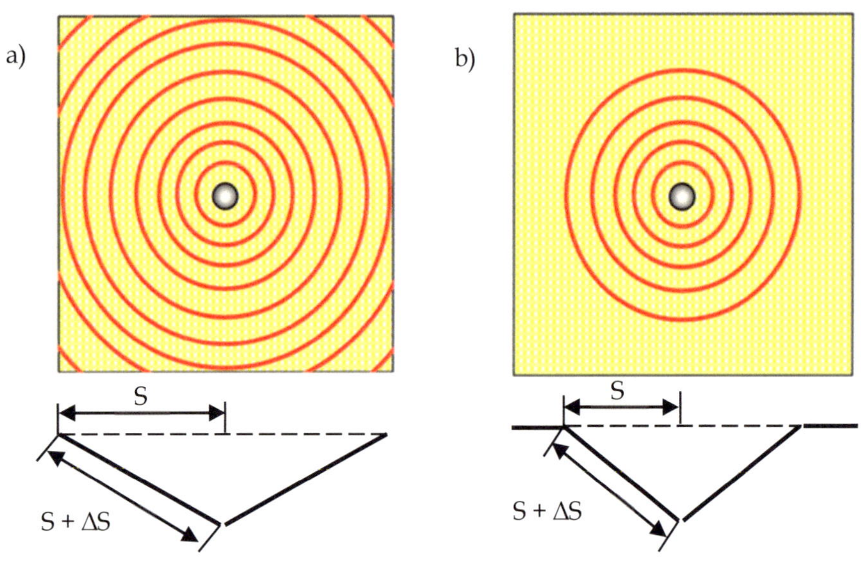

$\varepsilon = \frac{\Delta S}{S} = x \geq 2.5\ \%$, $\varepsilon = \frac{\Delta S}{S} = x \leq 2.5\ \%$

Fig. 7. Unitary deformation (ε) undergone by a laminate at the moment of absorbing impact energy, a) at the ballistic limit, b) above the ballistic limit.

This reduction in energy absorption can be explained, according to some authors, as a reduction in time of residence of the projectile in the material. One of the characteristics that define the ballistic limit of a material is the velocity at which sound can pass through it, the higher this value is, the greater capacity the material has to dissipate energy. Ideally, at the ballistic limit, maximum deformation of the fiber has been reached and this has dissipated the greatest amount of energy, as can be seen in Fig. 7a; with an impact velocity higher than the ballistic limit, the energy is not able to dissipate throughout the material, which generates an area of less deformation (Fig. 7b). At this point, it is important to mention that the sensitivity to deformation of aramid falls below 2% when it is subjected to deformation velocities of 10^3 s^{-1}.

It is important to note that when body armor is in full contact with the user, the material does not have to be perforated to cause fatal injuries in the wearer, since the depth that this material can reach on impact without actual failing, is such that it can cause damage to internal organs. This phenomenon is known as trauma, and refers to the maximum depth of deformation undergone by a material on impact without reaching perforation. The National Institute of Justice Standard (NIJ standard 0101.04) [14] establishes a methodology which determines trauma depth by means of a material denominated witness material, this is a homogenous block of plastilene type ROMA 1 which is placed on the back of the material to be impacted (Fig. 8). The armor plate being tested is impacted at its ballistic limit as this is the point presenting maximum deformation. On impact, the projectile will produce a deformation (δ) which is measured from the unaltered surface of the plastilene block to the lowest point of the depression; the maximum depth permitted by the standard is 40 mm.

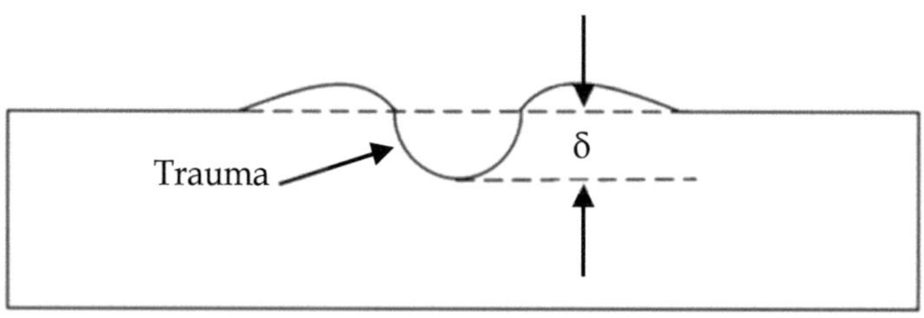

Fig. 8. Measurement of deformation in the witness material

In order to carry out this process of high impact characterization, a test gun is required, for example, a laboratory gas gun. This equipment is based on the generation of pressure in a closed chamber; when the pressure is released, it channels the kinetic energy of the gas to accelerate the projectile towards impact on the target. Fig. 9 shows a general diagram of a high impact test on a sample; the storage tank contains pressurized gas, usually nitrogen or helium, which is released abruptly towards the gas gun containing the projectile, this is then accelerated along the trajectory of the barrel and passes through the chronograph which registers the velocity just before impacting the sample. A steel vice holds the sample in place and the witness material or ballistic gelatine is situated behind it.

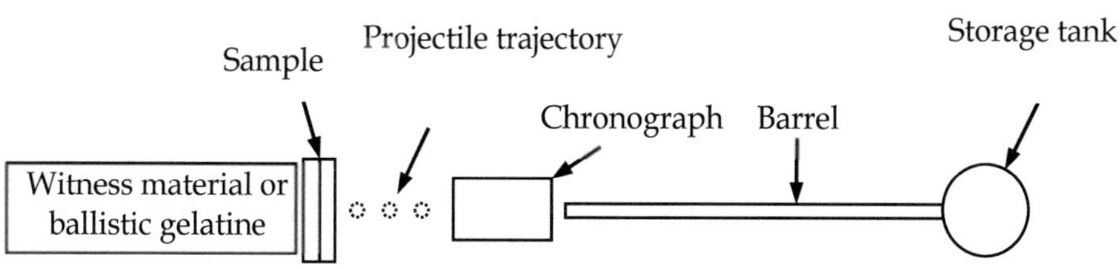

Fig. 9. General diagram of a gas gun

4. Processing of aramid/PP samples tested at high impact

For the processing of samples, an atactic PP in film form was used; the mechanical properties of this polymer are shown in Table 1. The aramid fabric used to reinforce the PP is a balanced woven used for personal body armor which was donated by the fabric Company Carolina Protect Ballistic, the properties of this fiber are included here.

Kevlar® Fabric724 [15]		**Polypropylene film form** [16]	
Yarn type	Kevlar® 129	Elastic modulus	680.6 MPa
Yarn count	1000 denier	Maximum stress	35.83 MPa
Weave	Plain	Maximum deformation	18.89%
Weight	207 g/m²	Glass transition	-10 °C a -18 °C
Count	24 yarns/inch	Melting point	175 °C

Table 1. Properties of the materials used in the aramid/PP composite material

During processing, the samples were divided into two main groups; the first comprising plates in which the aramid and the PP matrix are molded in a single stage, forming a multilayer fiber-reinforced composite (Fig. 10a), and the second comprising arrangements of composite materials where each layer of aramid was independently molded keeping the same volume fraction ratio (Fig. 10b).

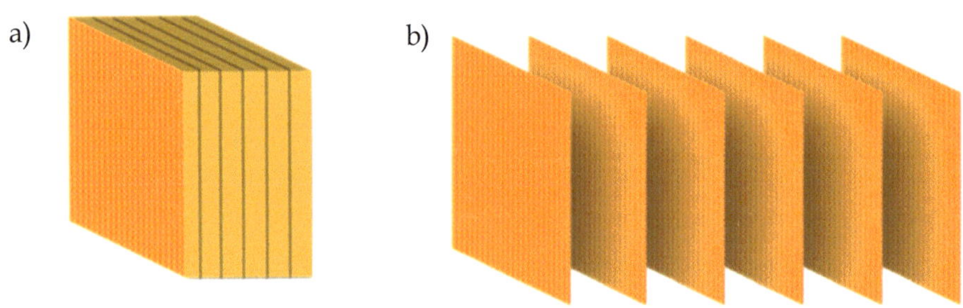

Fig. 10. Configuration of laminates, a) consolidated, b) independent

All these laminates, both consolidated and independent were formed by thermo-molding in a 25 ton press with automatic control of pressure and temperature, which guarantees less variation between the properties of samples within a particular batch. The molding conditions are shown in Table 2. Stacking of material in the molding process is another important factor as it generates a good distribution of the matrix in the fabric, Fig.11 shows this configuration, which grows in accordance with the number of layers required in the material.

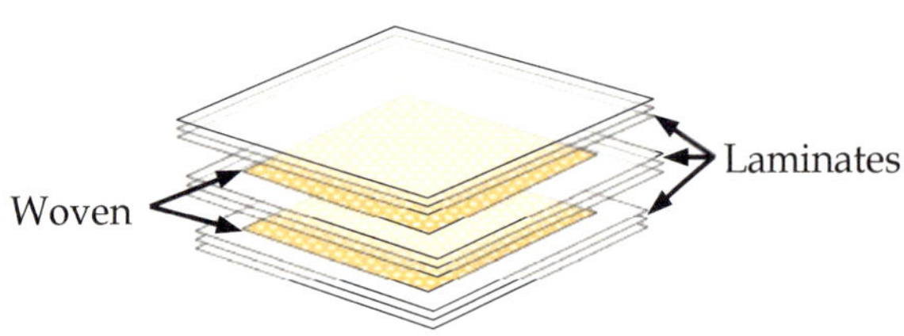

Fig. 11. Stacking of material in the molding process

Molding temperature	185 °C
Molding pressure	878 PSI
Processing period	20 min at 185 °C

Table 2. Laminate molding conditions

Both molding conditions and stacking configurations generate 64% fiber volume fraction in the composite material. This value was obtained thanks to previous studies in which the percentage contained in the composite was varied, it also coincides with values reported in the literature, where the fiber volume fraction recommended is 60% to 70% [17].

The aramid/PP composite material was subjected to the following high impact tests; first, consolidated and independent laminates with two to six layers were characterized at high impact. Each batch per layer consists of six samples in order to determine the ballistic limit in each point and thus create a comparative curve between the ballistic limit and the number of layers between both laminate configurations, independent and consolidated. Subsequently, an intermediate point of four layers was used to carry out the following tests; trauma in four-layer consolidated laminates, trauma in four-layer independent laminates, trauma in aramid fabric arrangements without polymeric matrix, and the velocity curve of residual impact-energy in four layer laminates using ballistic gelatine. Fig. 12 shows a general representation of how a high impact test is carried out with the use of witness material and ballistic gelatine. One important difference in these two tests is that the witness material must be in direct contact with the sample being tested, while the ballistic gelatine may or may not be in contact, in this case it is not in contact with the sample.

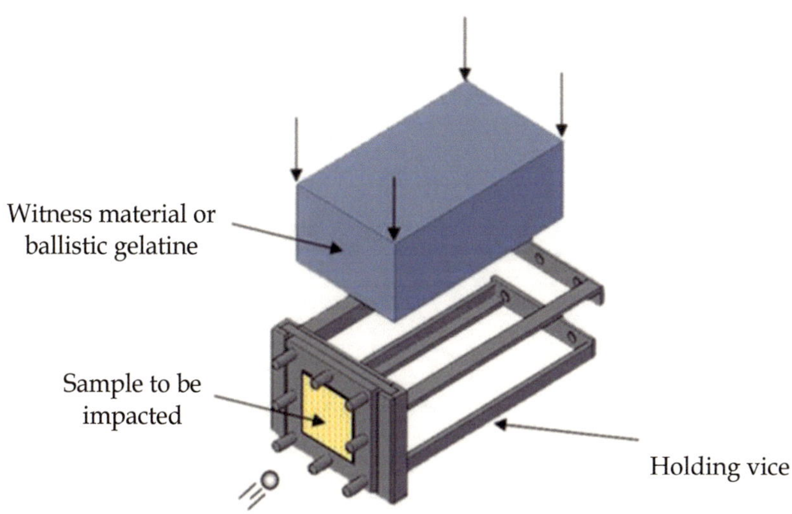

Fig. 12. Representation of an impact test on a sample with material placed behind it.

5. Measurement of interfacial shear stress

Interfacial adhesion was studied using the microdrop technique, this technique was implemented in 1987 by Piggott [18] and since then it has proven its effectiveness in many studies focusing on the determination of interfacial quality in composite materials [11,19]. Basically, the method involves applying a droplet of the matrix to a fiber, this drop must surround the fiber symmetrically so that it may be sustained and separated from the fiber which is normally subjected to tension. The mechanical properties of the fiber-matrix interface were determined in this way (Fig. 13).

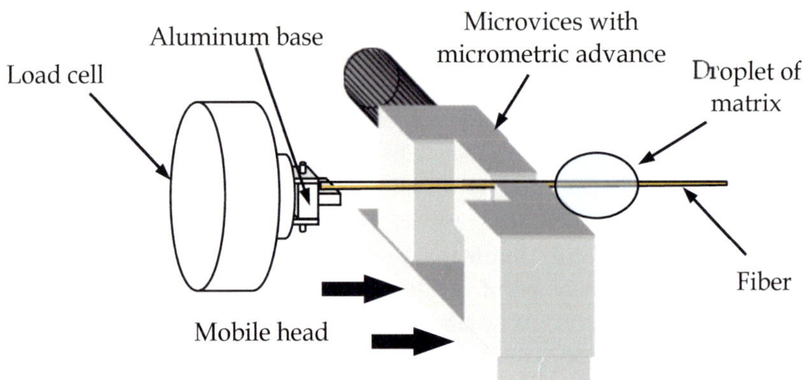

Fig. 13. Representation of a fiber pull-out by microdrop

In order to use this technique in the determination of interfacial shear stress in the aramid/PP arrangement, it was necessary to generate small drops of PP on the fiber (at an interval 80 μm to 200 μm diameter), the drops must surround the fiber completely. This is achieved by grinding the matrix and separating it into different particle sizes. The PP powder obtained is deposited on the fibers arranged in aluminum frames (Fig. 14), where it is subsequently heated in a convection stove to melt the thermoplastic matrix.

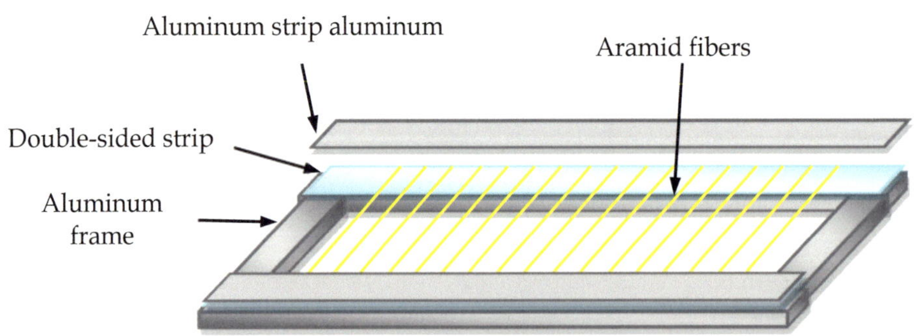

Fig. 14. Aluminum frame with aramid fibers

The samples obtained were separated according to their diameter; those with diameters between 80 μm and 200 μm were used for the interfacial adhesion tests. This property was determined with a microtensometer equipped with a Newport brand mobile head, which is capable of moving with great precision. The equipment consists of a load cell which registers the force required to separate the drop from the fiber by mechanical extraction. The microvises sustaining the drop move lengthwise along the fiber and drag the drop while the fiber is held by the load cell. The unit comprising the mobile head and microvises moves at velocities defined by the user (in this case 0.5 mm/min).

The value of interfacial shear stress is obtained based on the last load supported by the sample, the diameter of the resin drop deposited on the fiber and the length taken up by the drop. These values are substituted in Equation 3, determining the interfacial shear stress.

$$\tau_b = \frac{F_{Max}}{\pi d L} \tag{3}$$

where:

τ_b = Maximum interfacial shear stress (Pa).

F_{Max} = Maximum force reached in the test (gf).

d = Fiber diameter (μm).

L = Length of fiber taken up by the drop (μm)

6. Results of the high impact tests on aramid/PP composite material

A comparison of ballistic behavior was carried out between aramid/PP consolidated and independent laminates with 2 to 6 layers, both with the same fiber volume fraction (64%). Energy absorption is calculated with the projectile mass and at the ballistic limit of the laminate, which determines the kinetic energy that the composite material is capable of absorbing before failure. An increase in the number of consolidated laminates led to a non-linear increase in absorbed energy (Fig. 15). This non-linearity is a consequence of the rigidization undergone by the material as it becomes thicker, thereby causing a restriction in its transversal deformation.

The independent laminates showed an approximate increase of 13% in energy absorption in comparison with their counterpart, consolidated laminates (Fig. 15). The amount of energy absorbed by each laminate is dominated by the rigidity of the material, which increases with the number of layers, resulting in a reduction in the slope of the curve. Composite materials based on epoxy resin show an abrupt change in the slope at four layers, in contrast with the thermoplastic composites in this study which did not register this behavior, even at six layers [20].

Fig. 16 shows energy absorption normalized as a function of the number of layers, from this it can seen clearly that an increase in the number of layers in an arrangement reduces its capacity to absorb energy; however, the reduction was lower for independent aramid/PP arrangements.

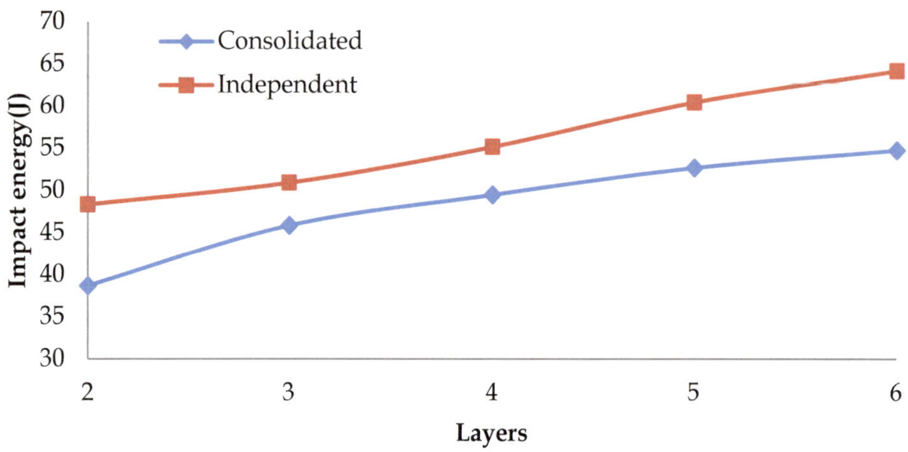

Fig. 15. Energy absorption in consolidated and independent laminates subjected to high impact

This phenomenon of rigidization in laminates was explained by Rohchoon [20], who noted that in composite material subjected to high impact, flexibility is a crucial factor, since this deformation undergone by the material increases the period of contact with the projectile, thus giving the fibers more time to dissipate impact energy. At this point, the velocity of sound propagation in the fiber must be elevated in order to dissipate the highest amount of energy in the least possible time [21]. In the case of a laminate with many independent layers, the adjacent layers impede posterior deformation, restricting energy absorption and consequently reducing the ballistic limit of the laminate. Deformation in a consolidated laminate is restricted only to its thickness, this means that the flexure forces present in a flexible laminate become localized tension forces, concentrating impact energy in small areas, and increasing the pressure experienced in these areas [20]. Fig. 17 shows an example of this situation with a one-layer laminate.

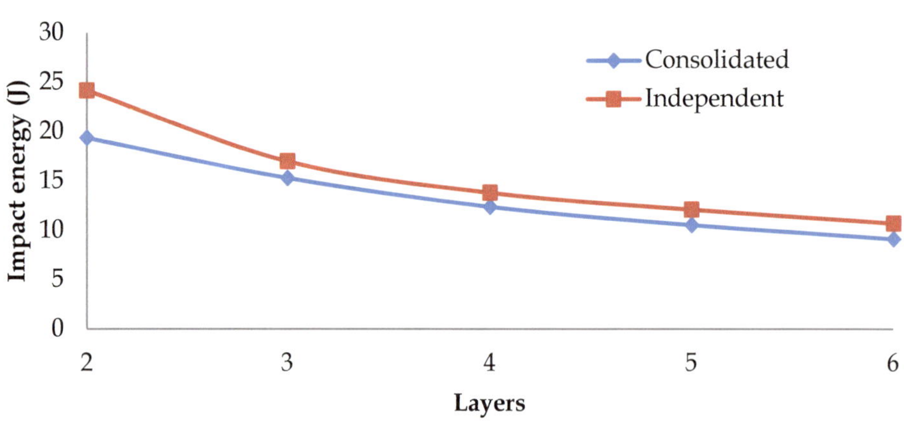

Fig. 16. Energy absorption as a function of the number of layers in consolidated and independent aramid/PP laminates

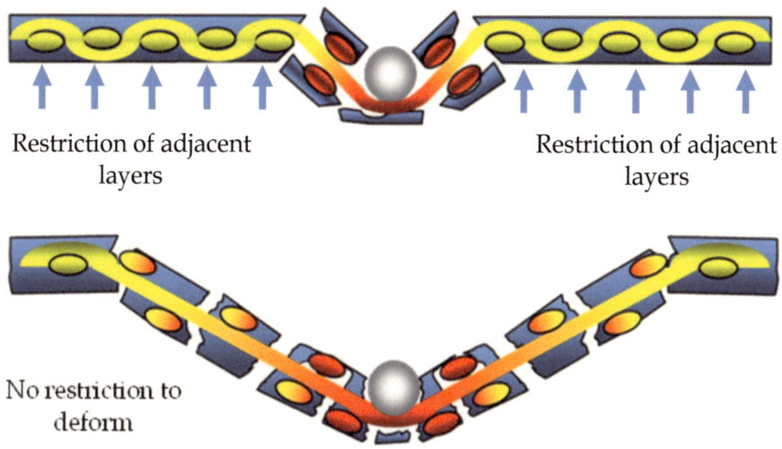

Fig. 17. Laminates subjected to impact, a) with restriction of deformation, b) without restriction of deformation

The results demonstrate how important rigidity of fiber-reinforced composite material is in the process of energy absorption during high impact; the mechanism by which a flexible laminate absorbs energy is completely different, the higher the flexibility of the laminate, the better the dissipation of energy. A laminate with restrictions of deformation concentrates this energy, making the process of energy absorption less efficient.

In order to carry out the tests with ballistic gelatine, it was first necessary to generate a calibration curve for the gelatine in relation to impact velocity; this curve is shown in Fig. 18, with the projectile velocity *vs.* length penetrated by the projectile. Linear regression of experimental data corresponds to penetration velocity of the projectile as a function of distance penetrated in the ballistic gelatine subjected to direct impacts.

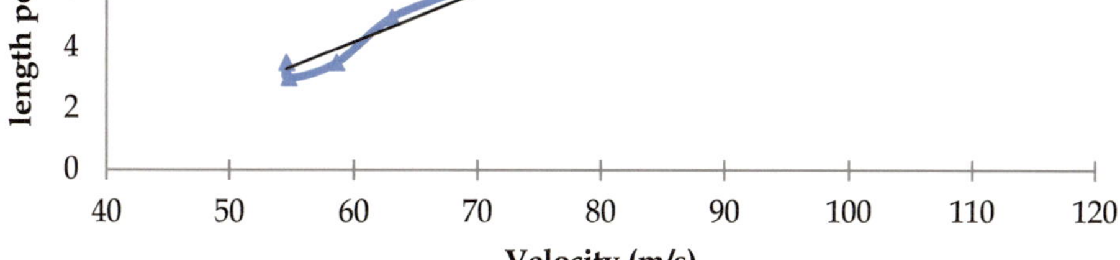

Fig. 18. Calibration curve for ballistic gelatine Bloom 250 at 10% weight

Fig. 19 shows energy absorbed (E_{abs}) by a compact laminate of four layers in relation to impact velocity of the projectile. E_{abs} was calculated with Equation 2, from impact energy (E_{imp}) and residual energy (E_{res}). Where E_{imp} was calculated with the impact velocity registered in the chronograph and the mass of the projectile, and E_{res} was calculated as a function of the length penetrated in the ballistic gelatine.

Energy absorption was calculated in tests on four-layer consolidated laminates, with velocities below the V_{50}, where composite material absorbs all the energy; however, on reaching velocities slightly higher than the ballistic level, a change occurs in energy absorption; this has been mentioned in other studies [13,22].

A linear behavior is clearly observed before the ballistic limit, due to the fact that the composite material has not failed. However, at velocities above the V_{50} a pronounced reduction in energy absorption is registered. Many theories have been proposed to explain this phenomenon, such as thermal effects, others mention a phenomenon called dishing (an indentation in the form of a dish). One of the most widely accepted theories is that of a reduction in the absorption period of impact energy.

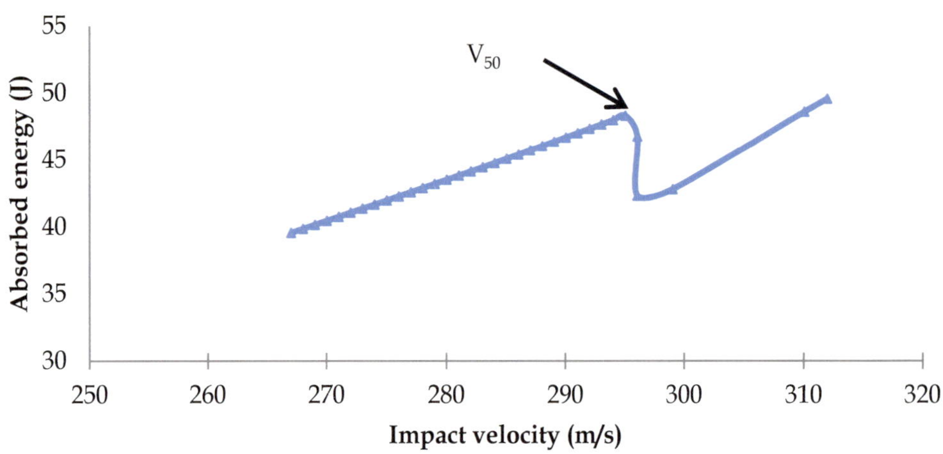

Fig. 19. Energy absorption curve with respect to impact velocity in four-layer samples

Fig. 20 shows an image of plastilene, commercial brand Modelina, after the impact of a steel weight calibrated to determine the corresponding force-deformation. The NIJ standard 0101.04 used to characterize bullet-proof vests, requires the use of a steel weight (1.043 Kg) to measure trauma at high impact in body armor, this weight should form an indentation in the witness material to an average depth (δ) no greater than 20 mm ± 3 mm, calculated from five impacts. Results showed an impact depth of 19.97 mm, indicating that the commercial plastilene Modelina is an adequate witness material for use in determining trauma caused by an impact on an aramid/PP composite material.

One important difference in the test procedures for ballistic gelatine and witness material is that the latter must be in contact with the back of the laminate being impacted and must not be perforated. The fact that it is in contact with the armor plating means that an important mechanism of energy absorption is limited, i.e. posterior deformation. Consequently, during high impact tests on aramid/PP consolidated laminates with four layers, energy absorption falls from 48 J in laminates tested without witness material to 30 J when the witness material is included, a 20% reduction. Fig. 21 shows a laminate which was tested under these conditions.

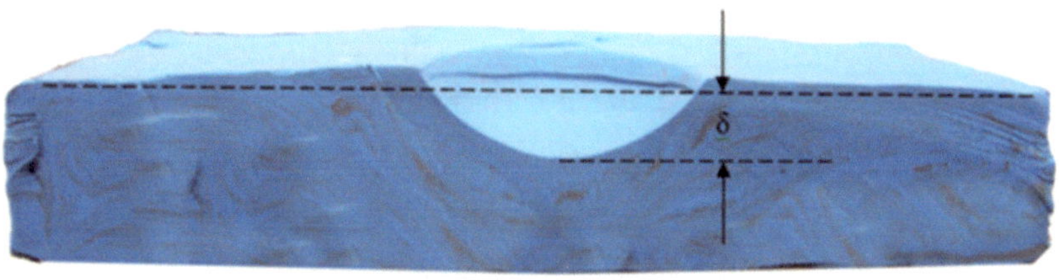

Fig. 20. Impact depth of witness material: commercial plastilene Modelina

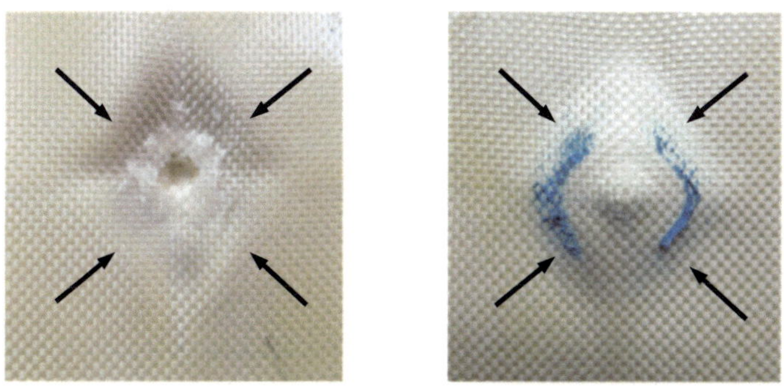

Fig. 21. Four-layer consolidated laminate impacted in presence of witness material

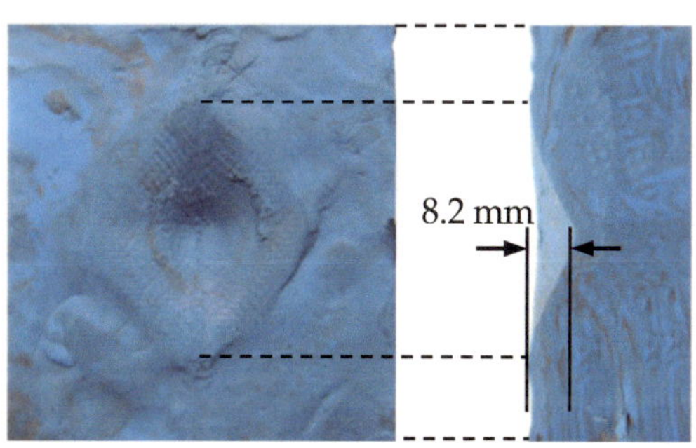

Fig. 22. Witness material used to test impact on an aramid/PP sample

Trauma depth in witness material for the aramid/PP consolidated samples was 8.2 mm, this measurement was carried out from the unaltered surface of the witness material to the deepest point of impact. Fig. 22 shows a frontal image and cross-section of the witness material.

The aramid/PP independent laminates with four layers were tested under the same impact as the consolidated samples, obtaining a trauma value of 8.13 mm. In presence of witness material, the independent laminates showed the same trauma values as the consolidated laminates. However, in tests under the same conditions on body armor without polymeric matrix (four layers of aramid fabric), a trauma depth of 11 mm was observed, which is 27% greater than that of the laminates with polymeric matrix. Images of an impacted armor-plating without a matrix are shown in Fig. 23, where it can observe more clearly the diamond-shaped impact formation, as well as a more pronounced impression of the projectile.

Table 3 presents the values of impact energy and trauma depth generated in the materials. It is important to remember that the maximum trauma depth permitted by the standard is 40 mm. These results demonstrate that the presence of the PP matrix improves posterior deformation of a material even though its configuration allows flexibility, as in the case of independent laminates. In contrast, absence of the matrix increases posterior deformation by 27%. It is also important to take into consideration that the presence of a witness material in contact with the back of the laminate reduces the ballistic limit of the aramid/PP composite material by 20%.

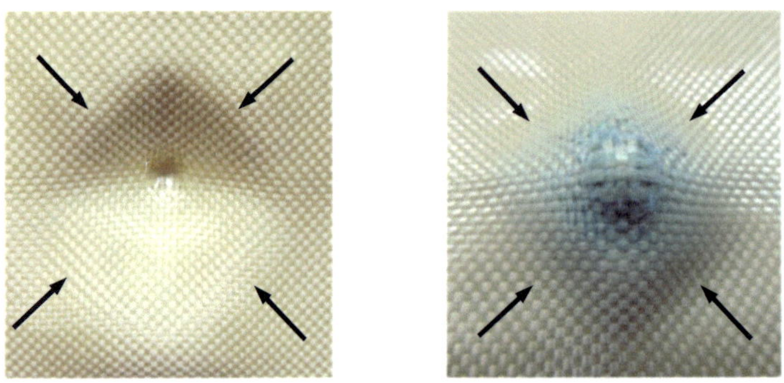

Fig. 23. Four layers of aramid fabric without PP tested with witness material

Configuration	Impact energy (J)	Trauma depth (mm)
aramida/PP consolidated	30.6	8.2
aramida/PP independent	30.6	8.13
aramida without matrix	30.6	11

Table 3. Ballistic properties of armor plating tested in presence of witness material

7. Conclusions

A comparative study of two arrangements tested at high velocity impact was carried out to highlight improvements conferred by a polypropylene matrix to aramid fibers. Two arrangements were made of plain woven aramid fibers/PP, with consolidated and independent configurations. As observed in this work, the presence of the PP matrix generated advantages, such as a reduction in trauma depth in both cases in comparison to the sample without the PP matrix, where the independent laminates presented more back deformation but higher energy absorption than the consolidated laminates. Another advantage that needs to be evaluated, is the protection the matrix confers to the aramid against humidity and substances that may degrade it, such as UV light, thereby extending its lifespan.

8. Acknowledgement

The authors would like to express their gratitude to the Consejo Nacional de Ciencia y Tecnologia, CONACyT, for their support in the funding of this project, CB-2008-01-101680.

9. References

Instituto de Ciencia y Tecnología de Polímeros, Ciencia y tecnología de materiales poliméricos. Instituto de ciencia y tecnología de polímeros, Madrid; 2004, 41.

Gay, Daniel,. Hoa, Suong,. & Tsai, Stephen. (2003). Composite materials, design and applications, CRC Press LLC

Chiou, Minshon,. Ren, Jianrong,. Van, Zijl,. & A, Nicolas. (2001), Artículos Balísticos resistentes a la penetración, Oficina española de patentes y marcas, España

Craig, Benjamin. (2005), High performance fibers for lightweight armor, Available from <http://ammtiac.alionscience.com/pdf/AMPQ9_2ART02.pdf>

Mark, Herman. (2004). Polyamides Aromatic, Encyclopedia of Polymer Science and Technology, Wiley, pp. 558-584, Retrieved from < http://www.wiley.com/WileyCDA/Wiley Title/productCd-0471275077.html>

Anand, Kulshreshtha,. & Vasile, Cornelia. (2002), Handbook of polymer blends and composites, Rapra Technology Limited, Retrieved from < http://www.knovel.com /web/portal/browse/display?_EXT_KNOVEL_DISPLAY_bookid=2219>

Registro Nacional de Armas. (2001), Chalecos Antibalas, Available from: < http://www.aicacyp.com.ar/disposiciones_legales/MA_01_seg.pdf>

Bryan, Cheeseman,. & Travis, Bogetti. (2003), Ballistic impact into fabric and compliant composite laminates, Compos Structure, Vol. 61, pp. 161-173

Donald, Carlucci,. & Sidney, Jacobson. (2008), Ballistics: theory and design of guns and ammunition, CRC Press

Haruntun, Karian. (2003), Handbook of Polypropylene and Polypropylene Composite, Marcel Dekker

Lawrence, Drzal,. Pedro, Herrera-Franco,. & Hence, Ho. (2000), Fiber-Matrix Interface Tests. *Comprehensive Composite Materials, Anthony Kelly and Carl Zweben*, pp. 71-111, Oxford: Pergamon, Retrieved from <http://www.sciencedirect.com/science/book/9780080429939>

Jorma, Jorma. (2004), Preparing ballistic gelatine - review and proposal for a standard method, Forensic Science International, Vol. 141, pp. 91-98

Paul, Wambua,. Bart, Vangrimde,. Stepan, Lomov,. & Ignaas, Verpoest. (2007), The response of natural fibre composites to ballistic impact by fragment simulating projectiles, Composite Structures, Vol. 77, pp. 232-240.

Andrew, Merkle,. Emily, Ward,. James, O'Connor,. & Jack, Roberts. (2008), Assessing Behind Armor Blunt Trauma (BABT) Under NIJ Standard-0101.04 Conditions Using Human Torso Models, Lippincott Williams & Wilkins, Hagerstown

Carolina Protect Ballistic. (2010), Ficha Técnica Comercial, 2010

Gamboa, Ricardo. (2009), Estudio de la Absorción de Energía al Impacto en Laminados Fibroreforzados a Base de Fibras de Aramida y Matriz de Polipropileno, Instituto Tecnologico Superior de Motul, Motul Yucatan

S. Reid,. R, Zhou. (2000), Impact Behaviour of Fibre-Reinforced Composite Materials and Structures, Woodhead Publishing

Michael, Piggott. (2002), Load Bearing Fibre Composite, Kluwer Academic

D, Biro,. G, Pleizier,. & Y, Deslandes. (1992), Application of the microbond technique Part III Effects of plasma treatment on the ultra-high modulus polyethylene fibre-expoxy interface, Journal of Materials Science Letters Vol. 11, pp. 698-701

Rohchoon, Park,. & Jyongsik, Jang. (2003), Effect of laminate thickness on impact behavior of aramid fiber/vinylester composites, Polymer Testing, Vol. 22, pp. 939-946

Ming, Cheng,. Clint, Hedge,. Jean-Pholippe, Dionne,. & Aris, Makris. (2010), Ballistic resistance enhancement by adjusting stress wave propagation paths, Proceedings of 25th international symposium on ballistic, Beijing China

Philip, Cunniff. (1992), An analysis of the effects in woven fabrics under ballistic impact, Textile Research Journal. Vol. 62, pp. 1-15

11

Microinjection Molding of Enhanced Thermoplastics

Mónica Oliveira, Victor Neto,
Maria Fonseca, Tatiana Zhiltsova and José Grácio
Department of Mechanical Engineering, University of Aveiro
Portugal

1. Introduction

Injection molding is, nowadays, a well known and wildly used manufacturing process to produce both thermoplastic and thermosetting polymeric material components, in a large scale, with accuracy and at low prices. Even though the electronics industry provides an economy of scale for the silicon industry, polymer devices can be produced in huge volumes maintaining the requested features and quality, with a great variety of material characteristics, a fact that has considerably open the market to injection molding of micro-components.

A simple miniaturization/rescale of the conventional injection molding process is not valid due to problems related with the rheology of the polymer flow in the micro cavity/channel and requires that all the process layout, as well as adjacent technologies, should be reconsidered, enhanced and properly adapted. The dimensional reduction of the components requires a higher control of the overall accuracy of the devices. The molding blocks of this type of objects are bounded to an amplified wear due to the fact that the surface roughness is dimensionally very close to the dimensions being controlled. The wear of the molding surfaces and the adhesion forces involved in the part extraction stage are greatly influenced by the nature of the material being injected and are enlarged when the molded object shrinks in the core or in the pins of the molding blocks. This is even more critical due to the flow behavior of enhanced polymer materials, developed for highly demanding applications in what concerns material mechanical or chemical properties.

Tailored special polymers are of fundamental importance in the supplement of micro-components. Complex polymeric materials engineered to detain both the mechanical properties and the memory shape suitable for applications such as active control, impose enormous challenges in what concerns both the process and the molding tools.

In conventional injection molding tools, surface engineering is used to improve the molding block performance to obtain parts with superior mechanical quality. In micromolding, surface engineering has a more important role due to the above-sited requisites, in order to reduce the deterioration of the molding impressions, increase its durability and reduce the need for corrective intervention on the tool. Different thin film coatings may be used, but diamond films seem to assume an extreme importance in these types of applications due to its superior properties. The latter are not problem free coatings and not typically used on conventional molding tools. Nevertheless, its use may constitute the premises to solve the problem posed by highly abrasive and shear intensive polymer flows on cavities with geometrical detail and poor access, due to the micro-scale developed with the solely purpose of obtaining parts with high quality requisites.

In this chapter, the challenges that involve the micro-injection of enhanced thermoplastics will be discussed. Special attention will be given to the microinjection technology and tooling, but also to the injection materials, which impose further challenges, the so called enhanced thermoplastics.

2. Microinjection molding

In this section the topic microinjection molding will be developed. The injection molding machine, molding tools and inserts, insert fabrication techniques, rapid heating/cooling process, mould evacuation, demolding, factors affecting the replication quality of micro parts and process monitoring and control will, therefore, be discussed.

2.1 Adaptation of the molding tools and equipment for the micromolding of thermoplastics

2.1.1 Injection molding machines

There are a variety of motives why the conventional injection molding machines fail to satisfy the requirements of the microinjection molding. So far, as typical macroinjection molding machines are concerned, the most important issue becomes its minimum metering size, which apparently is too big in comparison to the tiny quantity of material required to produce a microcomponent. The latter turns precise metering very complicated along with a considerable increase of the time of melt residence in the barrel and eventually may lead to material degradation. In addition, clamping force also has to be reduced in order to guaranty free damage release of the micropart (Giboz et al., 2007; Attia et al., 2009). To accomplish precise, free defect molding of the micro components, every functional system of the macro molding machine has to undergo a series of profound modifications (Giboz et al., 2007). So far, several commercially available and home-made machines are utilized for production of the micromolded components and may be divided in two main groups. In the first one, modifications are accomplished by simple rescale/miniaturization of the metering and injection units addressed to precise dosage of polymer melt in every shot (Giboz et al., 2007). Another approach consists of separation of the plasticization from injection unit, where the plasticization is being performed in the extrusion screw or hot cylinder, mounted at angle to the inject axis. Next, polymer melt enters into the injection unit, where the mini plunger pushes the prepared shot to fill the mould cavity. The above mentioned classification includes both hydraulically and electrically driven microinjection molding machines. However, for the sake of accuracy and repeatability of the process, higher precision of plastic metering and clamping may be achieved by the servo mechanisms of the electrically driven machine allowing for more accurate process control along with low noise level and energy efficiency. (Whiteside et al., 2003; Chang et al., 2007)

2.1.2 Mould insert fabrication

In the conventional injection molding process, cavity is usually machined directly in the mould plate. The procedure, however, has undergone substantial alterations for the tools tailored for the microinjection molding applications. It turns to be more practical, in terms of energy saving and versatility of the mould tool, to machine the impression in the interchangeable mould insert. This way all necessary mould tool transformations can be applied locally and the microparts of different configuration could be produced using the same injection mould. The choice of the technique for the insert fabrication generally depends on three main factors: insert material used, surface finishing (roughness) and the aspect ratio demanded by the application. A number of techniques currently in use for microfabrication include: LIGA-process; silicon etching; laser ablation, micro electrical discharge machining (µEDM) and mechanical micro machining (diamond turning, micro milling) (Rötting et al., 2002).

The LIGA process is a technology developed in the Forschungszentrum Karlsruhe in Germany in 1980s. Since the start of the micro technology development, is has been referred as a suitable technique for fabrication of the high aspect ratio microstructures with surface roughness down to 30 nm and such a low lateral resolution as 200 nm (Despa et al., 1999; Hormes et al., 2003; Munnik et al. 2003). Being a multi stage process, LIGA may be divided in three main steps: lithography, electroforming and plastic molding (Fig. 1). At first step (Lithography), the CAD information for the micro features is stored on the mask membrane (a very thin metal foil) covered with a layer of absorber generally Cu or Au. Synchrotron radiation passes through the transparent part of a lithographic mask and penetrates several hundred microns into a layer of sensitive X ray resist polymer (PMMA). The plastic modified by radiation is removed by solvent, leaving the template of resist structure. On the second step (Electroforming), the space generated by removed plastic is filled by electro deposition with metal normally Ni or Ni based alloys and negative replication of the resist structure is obtained. On the final step of LIGA process, the obtained metal structure is used as a mould insert for micro plastic parts production (Hormes et al., 2003). Among the variety of factors affecting the quality of X-ray LIGA generated patterns, the latter to a great extent is correlated with radiation intensity, mask composition and substrate type.

Novel approaches for improvement of X-ray LIGA process for micro insert fabrication are in permanent research in academy (Kim et al., 2006; Meyer et al., 2008). Although with deep ray X LIGA it is possible to obtain very accurate patterns of micro features with high aspect ratios up to 100 and micro structures with size less than 250 nm, it is still not a widespread commercial technique for micro replication, being time consuming and costly (Despa et al., 1999; Meyer et al., 2008). UV–LIGA and IB (Ion Beam) LIGA technologies are less complex and costly comparing to the X-ray LIGA process. In the former application, the ultraviolet source instead of the X-ray is used to expose the resists, while in the latter impressions is obtained by irradiating of photoresist materials with light ions (Munnik et al., 2003; Yang et al., 2006). It, however, should be pointed out that despite the fact that size of the micro structures obtained by this technique may reach 100 nm, the electrons are very light and may cause the loss of resolution and poor surface finishing at depth (Munnik et al. 2003). In case of low volume production, silicon microstructured inserts may be a suitable alternative to the expensive LIGA process. The impressions in silicon are usually obtained by wet etching and in spite of the inherent fragility of the material low roughness surface finishing makes this technique quite satisfactory for molding of the microparts in limited quantities (Heckele & Schomburg, 2004).

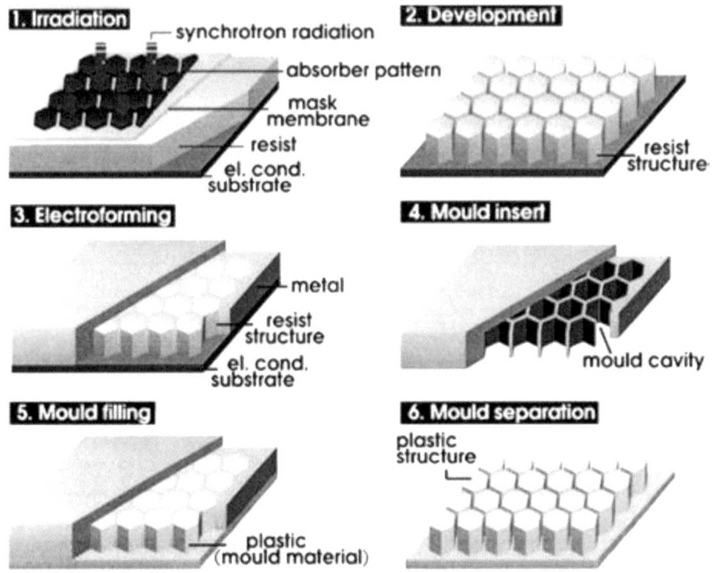

Fig. 1. Principle process steps for the fabrication of microstructures by the LiGA technique (Hormes et al., 2003).

The direct structuring techniques are far more attractive economically and can be easily applied for the micro inserts fabrication from hard metals and alloys. When the complex geometries of the micro features are concerned, a variety of micro cutting techniques are currently available and may be enumerated as follows: micro milling (for microgrooves and micro 3D shapes); micro turning (for micro pins); and micro drilling (for micro holes) (Franssila, 2004). To achieve the required cavity surface finishing, diamond and cemented carbide cutting miniaturized tools are currently in use whether their applications will eventually depend on the properties of processed material (Franssila, 2004). Although with diamond cutting tool such fine resolution as of 48 nm of roughness could be achieved, its applicability is limited to the soft non ferrous metals as brass, aluminium, copper and nickel (Davim & Jackson, 2009). More resilient materials like steel are processed with the cemented carbide tool of diameters down to 50 µm yet with less surface resolution (0.1-0.3 µm) (Fleischer et al., 2007). Despite the fact that maximum aspect ratios achieved by micro cutting are considerably smaller in comparison with the other micro fabrication techniques, high level of automation and continuous improvement in the tools precision as well their ability to work with hard metals make it an appropriate technique for the microstructured insert fabrication.

Another popular direct structuring technique for hard metals and alloys is micro electrical discharge machining (µEDM). The heat in the form of pulsed discharge is applied through the thin metal wire (usually brass), allowing to cut through metal by melting and evaporation (Franssila, 2004). Based on this general description, there are a wide range of the µEDM techniques which may be classified by the type of electrode or/and electrodeworkpeace movement (Franssila, 2004; Uhlmann et al., 2005). Size and precision of the pattern replication, to a great extent, depend on the size and shapes of the electrodes and eventually with optimum conditions aspect ratios of 100 may be realized (Franssila, 2004). For machining of the patterns of more complex shapes, electro discharge grinding (WEDG) with the electrodes of micrometeric range (down to 20 μm) is reported to realize the micro structures of 5 μm and roughness of 0.1 μm. However, a maximum aspect ratio of micro structure is less than that of μEDM and only reaches 30 times of the wire diameter (Uhlmann et al., 2005). Surface resolution obtained by laser ablation is similar to that of micro cutting techniques making this method very popular for machining of the wide range of engineering materials (Gower, 2000). Ultra short pulsed lasers with optimized pulse energy and focus size are able to produce microstructures with size from 10 μm, aspect ratio of 10 and roughness of 0.16 μm (Heyl et al., 2001). It is worth mentioning, however, that in terms of surface quality and minimum achievable dimensions, the output from the micro laser ablation is inferior to the LIGA and μEDM.

2.1.3 Rapid heating/cooling process

As a rule of thumb, in conventional injection molding process mould temperature is far below the injection temperature. At such conditions, the frozen layer forms near the cavity wall while the core is significantly hotter and continues cooling down to ejection temperature at the end of the cycle. When the polymer melt is injected to the micrometric cavity, high surface-to-volume ratio and reduced dimensions of microparts promote the instantaneous melt temperature drop and as a result incomplete filling even when high pressures are applied (Su et al., 2004). To guaranty complete filling of the microcavity, the mould should be heated up to the glass transition temperature (T_g) for amorphous thermoplastics and melt temperature (T_{melt}) for the crystalline ones. This definitely requires implementation of a special rapid heating/cooling (variotherm) system. The latter (Fig. 2) allows for a rise of the mould temperature above T_g /T_{melt} during injection with subsequent cool down to the ejection temperature in order to assist successful part release (Piotter et al., 2008).

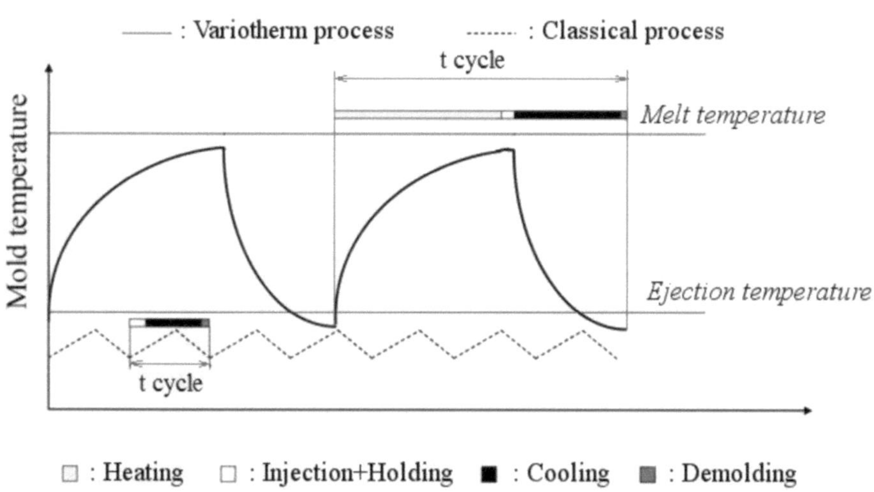

Fig. 2. Comparison of the mould temperature profile in the classical and variotherm processes (Gornik, 2004).

Nowadays, the variotherm process is widely applied for molding of the microparts. When mould temperature is close to processing temperature, even microstructures of the high aspect ratios may be eventually filled. Moreover, isothermal filling induces less residual stresses and surface defects in the micro parts (Chang & Hwang, 2006). How it can be easily noticed from Fig. 2, the temperature gradient for variotherm process is significantly higher than that for the conventional process and may lead to longer cycle times compromising the economic feasibility of the micromolding process (Gornik, 2004).

Considering the variotherm process, two main issues have to be addressed. The first is concerned with the choice of the heating technique capable of rising promptly mold temperature above T_g /T_{melt}. While the second issue is related to the efficient heat removal for the fastest possible cycle times. So far, there are two principal approaches to transport the required power to the mould: external type (infrared radiation and induction heating) and internal/built-in (joule/resistive, high-frequency proximity, water and oil heating). The main limitation to use water as a heating media is that the maximum heating temperature does not exceed 95°C, which is less than the glass transition/melt temperature of the majority of polymers. As a result, water is generally used for cooling in the variotherm process (Su et al., 2004; Chang & Hwang, 2006; Xie & Ziegmann, 2008). Heat retention properties of oil and its ability to be heated up to 140°C makes it a more adequate candidate as heating medium. However, the lower heat transfer coefficient, comparing to water, leads to longer cycle times (Gornik, 2004; Whiteside et al., 2004; Tseng et al., 2005).

Application of the joule/resistive heating in variotherm process has been extensively investigated by many researchers (Su et al., 2004; Xie & Ziegmann, 2008). Eventually, nano features with aspect ratios up to 300 could be realized if molded above the glass transition/melt temperature of polymer. Unfortunately, the cycle times may increase up to several minutes due to the thermal inertia of the mould material (Xie & Ziegmann, 2008). Adequate isolation of the heating elements from the mould base will favor the cycle time reduction and increase economic efficiency of the microinjection molding process (Gornik, 2004). Yao et al. (2006) have implemented a local heating of the mould insert by high-frequency current. Unlike the conventional resistive techniques, mould surface heating rate can reach 40°C/s with an apparent heating power of 93 W/cm^2, while only the local electrical insulation of the mould insert is necessary.

The main advantage of the external heating over the internally built systems is an ability to heat up locally the insert surface allowing for the faster cycle times. In a number of studies infrared heating has been applied for precise localized heating of the microcavity surface Although the local temperature rise with an aid of halogen lamps requires considerably less time in comparison to the resistive heaters, it may lead to uneven distribution of the mould surface temperature and therefore, special attention should be given to more uniform distribution of the heating sources (Gornik, 2004; Chang & Hwang, 2006). During processing, residual oil or resin particles may burn and contaminate the microcavity surface causing surface defects in the micromolded parts. This problem could be eventually solved with periodical cleaning of the mould cavity (Chang & Hwang, 2006).

Direct infrared radiation heating of polymer inside the cavity has been proposed by Saito et al. (2002). As long as CO_2 laser is directed towards the polymer melt through the transparent window in the mould wall, temperature gradient during injection is significantly reduced.

With this technique, the molecular orientation in the surface region of the micropart has been decreased along with significant improvement of the surface replication. Very quick rise of the cavity temperature may be achieved with induction heating. It was reported that heating could be accomplished in several seconds and consequent cooling is also very fast because of minimum amount of heat generated in the mould (Michaeli & Klaiber, 2007a). However, high-temperature cycling may lead to thermal fatigue and eventual shortage of the mould service time (Tseng et al., 2005).

2.1.4 Mould evacuation

Negative effect of the air presence in the mould cavity is a well known phenomenon in the conventional injection molding. The latter is responsible for burn marks on moldings and long-term formation of corrosive residue in the mould which may lead to its permanent damage. While the moulds of conventional size are supplied with special venting channels, this solution cannot be adapted for the micromolding where the part dimensions are frequently comparable in size to the venting grooves. Moreover, high injection speed and complex geometry of micro features may contribute for air entrapment and could eventually lead to incomplete filling, especially in case of the blind hole features with high aspect ratios (Heckele & Schomburg, 2004). In a number of studies, evacuation of air from the cavity prior to injection is referred as an efficient method for improvement of the microparts replication (Despa et al., 1999; Heckele & Schomburg, 2004; Sha et al., 2005; Liou & Chen, 2006; Chang et al., 2007; Sha et al., 2007b). A typical layout of the mould evacuation system is shown on Fig. 3. However, sometimes it is impossible to clearly distinguish the effect of cavity evacuation on the part filling. For example, it could be attributed to the fact that micro cavities are aligned with the parting line of the mould and therefore air can escape easily through the partition (Sha et al., 2005). To assist more efficient air evacuation, the mould platens must be properly adjusted for hermetic sealing of the cavity (Despa et al., 1999).

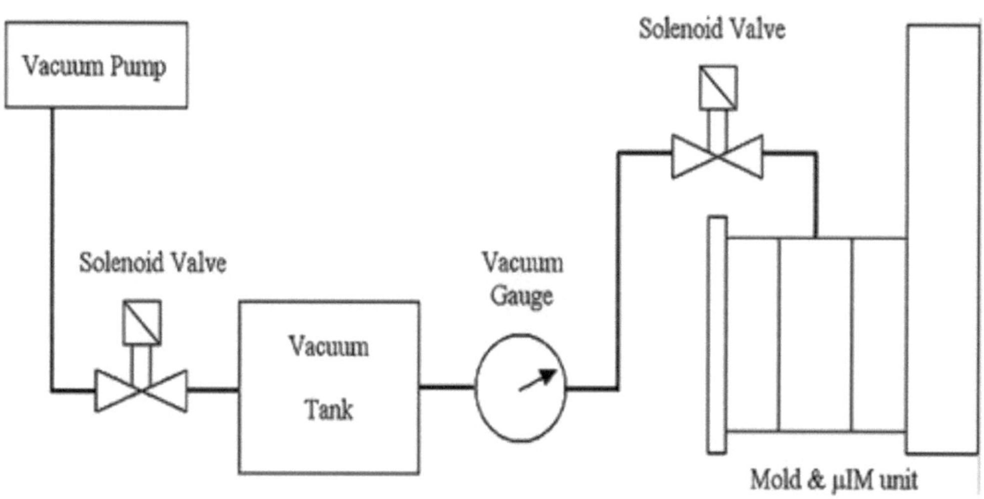

Fig. 3. Layout of the vacuum mould unit (Chang et al., 2007)

It is also worth mentioning that air suction prior to injection may decrease the cavity temperature. The latter may have a negative effect on the mould filling, being especially critical in the case of the thermally sensitive polymers as polyoxymethylene (POM) (Sha et al., 2005; Sha et al., 2007b).

Implementation of the vacuum assisted cavity evacuation might be carefully considered in every case of the microinjection molding taking into account polymer type, geometry and surface finishing of the cavity.

2.1.5 Demolding

Demolding of microparts becomes a critical issue as failure frequently occurs at the onset of ejection. The interactions of polymer shrinkage and the coefficient of friction between the polymer and the molding tool may have detrimental effects on the component failure during ejection (Pouzada et al., 2006). Moreover, the micro features with high aspect ratios have a larger contact surface between the mould and the polymer which results in higher frictional resistance during part release. Smooth surface finishing is highly desirable for decreasing the friction on the polymer/insert interface (Attia et al., 2009). Coatings are frequently used to improve the roughness properties of the mould cavity and achieve superior surface quality of microparts. For example, if a cavity is coated with diamond-like carbon (DLC), fewer forces are required for the PC and ABS microparts ejection (Griffiths et al., 2008). In order to avoid breaking during ejection, the adhesive bond between polymer and stainless steel mould tool should not exceed the tensile strength of the polymer (Navabpour et.al., 2006). In addition to the experimental techniques, the magnitude of the stress on the polymer/metal interface can be accessed via finite elements stress analysis (Grave et al., 2007).

Considering fragility of the microstructured parts, conventional ejection with pins becomes unlikely and may eventually lead to their irreversible damage. In order to overcome those weaknesses, alternative solutions have to be considered. For example, if geometry of micro parts allows for a positive draft angle, the latter could assist the proper demolding of the microparts (Grave et al., 2007; Wu & Liang, 2005). A concept of more even distribution of ejection forces has been applied when ejection with pins was substituted with the micro ejection block used as striper plate to thrust the microparts (Wu & Liang, 2005). Ejection of the high aspect ratio micro structures, molded within the insert of rough surface finishing, may be successfully accomplished with vacuum assisted demolding (Michaeli et al., 2000). Choice of demolding system at micro scale will eventually depend on the cavity geometry, surface finishing and the material to be molded. In addition, a special attention should be given to correct calculation of the ejection forces and proper design of the ejection system, so far as both of those factors could be detrimental for the micropart quality.

2.2 Process monitoring and control

Microinjection molding is inherently a black box process, where the extreme processing conditions and rapid process variations do not lead to the linear correlation of the inputs and outputs. Monitoring of the micro molding process parameters is critical to assess the rheological state of polymer throughout the molding cycle. Hitherto, the majority of sensors currently available at the market are too big to be installed into the micro cavity. To overcome those limitations, an indirect monitoring is frequently applied to gather the information about the mould temperature, heat flux, injection pressure, injection speed and displacements (Yan et al., 2003; Zhao et al., 2006). Frequently pressure sensors are embedded into the plunger of the microinjection molding machine for monitoring of the metering size and pressure evolution. It, however, does not represent the polymer behavior inside the microcavity (Zhao et al., 2006). Cavity pressure monitoring is considered to be a better approach to verify the process variation and control the quality of the molded parts. For example, more relevant information about polymer melt behavior can be acquired by single axis force sensor, embedded into the mould-core extrusion mechanism (Yan et al., 2003). In a number of studies, an integral approach has been adapted towards the monitoring of the pressure and temperature inside the cavity. Pressure monitoring has been performed with piezoelectric force transducer embedded into the ejector pin. Whereas temperature, which has early been mentioned to affect the microcavity filling, may be measured with the aid of the J type thermocouples embedded at depths of several millimeters in the cavity wall. This way, the polymer solidification inside the cavity may be recorded (Whiteside et al., 2004).

When the direct monitoring of the micromolding is concerned, only a few sensors have been referred to be appropriate for the micro applications. The majority of the above-mentioned sensors hardly satisfy the micro criteria with the contact area diameter varying from 1 to 4 millimeters (Luo & Pan, 2007; Ono et al., 2007). However, those sensors were reported to provide reliable data about the temperature and pressure evolution in the cavity and if properly located, could be used to control the micromolding process. Further miniaturization of the sensors is one of the key factors to provide an insight to rheology of the polymer flow in the microcavity. The obtained data could be applied for the better process control as well for the improvement of the rheology models used for numerical simulation of the micro molding process. So far, by the author's knowledge, the market offer of the micro pressure and temperature sensors is very limited. For example, a combined sensor for measuring of the mould cavity pressure und contact temperature in the cavity with the front diameter of 1 mm, was recently launched by KISTLER (Kistler, 2011). Further miniaturization has been accomplished by PRIAMOS, for measurement of the melt temperature at contact with the cavity wall, with the sensor of 0.6 mm of diameter (PRIAMUS®, 2011).

The real time monitoring and control of the micromolding process are highly dependent on the sensors capable of reliably monitor the instant process variations. With further sensors miniaturization and an increase in precision, it will eventually be possible to achieve the higher level of automation and superior economic viability of the micromolding process.

2.3 Effect of the process/tool/polymer interaction on the quality of micromolded components

Quality of the micromolded parts is, to a great extent, determined by complex interaction of the process parameters, polymer and mould tool properties. Considering variability of the microparts design and purpose, their quality criteria are much diversified as well, including filling length, dimensional stability, surface finishing and a range of the mechanical properties of interest (Attia et al., 2009). Currently, two main approaches exist for correlation of the process/polymer/tool factors with the quality output: one factor at time experiment (OFAT) and design of experiment approach (DOE). While the former method highlights the leading trends, the latter addresses the optimization of the process by identifying the interactions among factors. Both methods are extensively explored in recent research work (Zhao et al., 2003; Liou & Chen, 2006; Michaeli et al., 2007b; Sha et al., 2007b; Theilade & Hansen, 2007; Tofteberg & Andreassen, 2008).

Variotherm mould cycling along with the high injection temperature could significantly improve the weld line strength of the micro tensile test samples (Theilade & Hansen, 2007). Moreover, at higher mould temperature viscosity of the polymer melt significantly decreases and therefore requires less injection pressure and speed (Despa et al., 1999). Even in the absence of the variotherm cycling, higher mould temperature was reported to promote better cavity filling for a wide range of the tested polymers (Zhao et al., 2003; Sha et al., 2005; Michaeli et al., 2007b; Sha et al., 2007a; Sha et al., 2007b; Tofteberg & Andreassen, 2008). In the case of over-molded micro needles, an increase of bonding strength has been accomplished by the combination of the high holding pressure, high mould and low melt temperature (Michaeli et al., 2007b). An attempt of the empirical correlation of the part quality with process parameters has been proposed by Tofteberg & Andreassen (2008). In their study, replication of the micro features of Cyclic Olefin Copolymer (COC) and Poly(methyl methacrylate) (PMMA) has been significantly improved with an increase of the mould and melt temperature. It is also worth mentioning that with an increase of the micropart complexity the mould temperature factor was found out to prevail over the other process parameters (Attia et al., 2009).

Premature solidification of polymer may be reduced, to some extent, by injection at high injection speed. The latter is reported to assist in filling of the micro pins, having on the other hand, an adverse effect on the surface finishing (Sha et al., 2007a). The positive effect of the injection speed has also been confirmed by the other researchers. High injection speed was decisive for high-quality replication of the micro walls as well as its interaction with the injection temperature (Theilade & Hansen, 2007). When moulded with cold runners, it is not rare to encounter that volume of the feeding system may contain several times the volume of the micropart. Such discrepancies make precise metering unlikely, increasing the probability of incomplete filling and invalidating the holding pressure effect (Zhao et al., 2003). Nonetheless, some difficulties related to the metering precision and process fluctuations can be attenuated by applying higher holding pressure (Liou & Chen, 2006). Although with the holding pressure increases improvement of filling the micro and sub-micron structures has been reported, it seems uneasy to differentiate single holding pressure influence from its interaction with the mould temperature effect (Liou & Chen, 2006).

In the recent research, discrepancies in rheological behavior of the different grades of plastics have been widely reported. In order to guarantee the proper filling and acceptable surface quality, the easy flow grades should be preferred (Despa et al., 1999; Zhao et al., 2003; Michaeli et al., 2007b; Tofteberg & Andreassen, 2008). Adhesion forces between polymer and mould tool may vary significantly for different polymer grades and should be accounted for successful demolding of the microparts. The latter may be achieved by fine surface finishing of the mould cavity, positive draft angles and by using release agents (Wu & Liang, 2005; Grave et al., 2007). The variety of factors involved into the transformation of the polymer melt within the microcavity makes the interpretation of the cause-effect relationship uneasy task even for neat (unfilled) thermoplastics. Furthermore this relationship may become substantially more complicated for enhanced thermoplastics as their new properties will eventually influence rheological behavior at micro scale.

3. Enhanced thermoplastics

A brief introduction to the current meaning of enhanced thermoplastics will be provided. It will be given some emphasis to enhanced thermoplastics through nanoparticles loading. Then it will be presented research work done on this issue, giving special evidence to the effect of mixing carbon nanotubes (CNT) to a thermoplastic material. The consequences on the rheology and interaction with the molding tools of these materials will also be discussed.

3.1 Definition

Thermoplastics are long chain polymers than can be structurally amorphous or semi-crystalline. These polymers possess long chains, where macromolecules are bonded through the weak van der Waals forces. Their general properties are toughness, resistance to chemical attack and recyclability, i.e. they can be re-processed as many times as needed, till their degradation, due to processing. Despite that the continuum search for better and cheaper materials has been guiding the scientists to the development of new thermoplastics with enhanced properties, such as better resistance to water and UV; better mechanical properties (toughness, stiffness); and enhanced electrical conductivity. These new class of thermoplastics are known as "enhanced thermoplastics".

There are different techniques to manipulate polymeric materials in order to obtain enhanced properties, such as: (a) by modifying their molecular structure, the hard and soft segments. Through this technique, the molecular structure of the polymer is changed by different combinations of: chain flexibility and hard segments, chain entanglement, the orientation of different segments, the hydrogen bonds and their intermolecular interactions. Through the modifications of the hard/soft segments ratios of the polymer, it is possible to obtain different physical, thermal and mechanical properties (Chattopadhyay & Raju 2007); (b) the incorporation of plasticizers, which are substances, usually plastics or elastomers, that are incorporated in a material in order to increase its flexibility, workability or extensibility, and modify the thermal and mechanical properties (Wang et al. 1997; Rahman & Brazel 2006); (c) throughout the incorporation of particles into a polymeric matrix (Tjong 2006). Some of the most commonly used reinforced particles are: $CaCO_3$, glass, carbon fibres, and in the last years, carbon nanotubes have been frequently used due to their extraordinary thermal, electrical and mechanical properties (Wang et al., 1998; Coleman et al., 2004).

3.2 The effect of incorporating carbon nanotubes (CNTs) into thermoplastic materials

The incorporation of carbon nanotubes (CNTs) in thermoplastic materials has been one of the hottest topics in materials science in the last years, since their discovery by Iijima (1991), essentially due to the extraordinary properties of CNTs

CNTs can be classified has single-walled carbon nanotubes (SWCNT), which consist on a single grapheme sheet wrapped into a cylindrical shape and are characterized by a small diameter (0.4 - 3 nm) and lengths up to centimeters; and multi-walled carbon nanotubes (MWCNT) that detain a number of grapheme layers coaxially rolled together to form a cylindrical tube, and their outer diameter ranges from 1.3 - 100 nm and their length can be as long as tens of micrometers (Baughman et al., 2002). It has been reported that CNTs possess an elastic modulus in a range of TPa (Yu et al., 2000) and detain high aspect ratios (>100), due to it small diameter (in the range of nanometers) and long length (as long as 100 micrometers). However, the incorporation of CNTs into a polymeric material is quite difficult due to the Van der Waals forces that tend to clump nanotubes together, leading to a poor dispersion of them into polymeric matrix. For an effective reinforcement there are essentially four important requirements: 1) the CNTs must have high aspect ratio; 2) they must be uniformly dispersed; 3) the alignment of CNTs through a preferable direction, and 4) should be verified interfacial stress transference. The distribution of CNTs into polymeric matrix could be the most important parameter to obtain an enhanced thermoplastic reinforced with CNTs. The CNTs must be individually dispersed and should the coated by polymeric matrix, in order to achieve efficient load transfer to the nanotube network, which is of utmost importance. The aggregation of CNTs is generally accompanied by a decrease in strength and modulus.

In order to improve the dispersion of the CNTs, and to prevent the agglomeration, many techniques has been developed, such as, ultrasonic activation, the addition of surfactants and chemical treatment with surfuric and nitride acids (Esumi et al., 1996; Gong et al., 2000; Deng et al., 2002; Safadi et al., 2002). The present authors applied a chemical treatment, with sulphuric and nitride acids in a proportion of (3:1) to modify the surface of CNTs and the results has shown that CNTs maintain their physical integrity (diameter, length) and that they are no longer entangled as previously to chemical treatment (Fig.4).

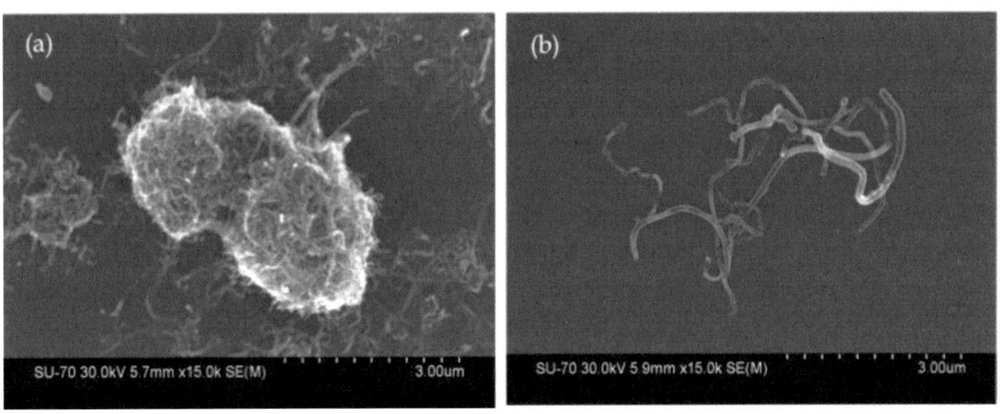

Fig. 4. Scanning electron microscopy of the (a) non-treated CNTs and (b) chemically treated CNTs.

There are essentially three methods to incorporate CNTs into polymeric matrixes: 1) solution mixing or film casting of suspensions of CNTs in dissolved polymers; 2) in situ polymerization of CNTs-polymer monomer mixture; and 3) melt mechanical mixing of CNTs with polymer (Jia et al., 1999; Haggenmueller et al., 2000; Jin et al., 2002). Recently, Fonseca et al. (2011) proposed a simpler technique based upon the general methodology described by method 3, where the CNTs and polymeric matrix are prepared by mechanical ball-milling.

The solution mixing is probably the most applied method to produce polymer nanotube composites. The basis of this method is the mixing of the CNTs and the polymer in a suitable solvent before its evaporation occurs and, thus, forming a composite film. This method has the advantage of the agitation of the CNTs powder in a solvent that could facilitate the de-aggregation and dispersion. The in-situ polymerization has been used because the obtained composite presents a good dispersion of CNTs into polymeric matrixes. This method is particularly important for the preparation of insoluble and thermally instable polymers that cannot be processed by solution or melt processing. The main disadvantages of this method are the time consumed and the complexity involved in the process to obtain the composite material. The melt processing method is commonly used for thermoplastic polymers. The advantages of this technique are the speed and simplicity, and essentially, its compatibility with standard industrial techniques (Andrews et al., 2002; Potschke et al., 2003).

The melt mechanical technique is characterized for the melting of the thermoplastic and then the addition of CNTs by shear mixing. However, the processing conditions should be optimized for each thermoplastic and desired composite, because CNTs can affect melt properties, of the pure thermoplastic, such as viscosity (Potschke et al., 2003). As it was mentioned earlier, Fonseca et al. (2011) has prepared ultra-high molecular weight polyethylene (UHMWPE) reinforced with CNTs through a variation of the melt mechanical processing technique. Previously to melt processing, they used the mechanical ball-milling to mix the raw UHMWPE with the chemically treated CNTs. After mixing the powder composite was processed by compression molding technique. The results have shown a homogenous distribution of the CNTs into the polymeric matrix (before and after melt processing), and an enhancement of the main mechanical properties, such as the elastic modulus and toughness.

The incorporation of CNTs in polymeric matrices has been extensively studied in last years. Different polymeric matrices have been considered, such as PMMA (Jia et al., 1999; Gorga & Cohen, 2004), polystyrene (PS) (Qian et al., 2000; Yang et al., 2005), polypropylene (PP) (Dondero & Gorga, 2006; Bao & Tjong, 2008), polyurethanes (PU) (Koerner et al., 2005; Xiong et al., 2006; Zhang et al., 2011) and polyethylene (PE; high density polyethylene - HDPE, ultrahigh molecular weight polyethylene - UHMWPE) (Ruan et al., 2003; McNally et al., 2005; Kanagaraj et al., 2007; Fonseca et al., 2011).

In 1998, Schadler et al. (1998) studied the effects of the dispersion of 5 wt% of MWCNT in an epoxy resin, and the results have shown an increment of the modulus in tension and in compression of about 20% and 24%, respectively. Jia et al. (1999) prepared PMMA/CNT composites with different weight fractions of CNTs. The composites were prepared by an improved in-situ process and the results reveal an enhancement of the mechanical properties as well as an increase of the heat deflection temperature, as the CNT loading increases. However, for weight percentages higher than 7% the mechanical properties of the PMMA/CNT composite starts to decrease and for 10 wt% the composite becomes brittle. Later, Gorga & Cohen (2004) presented results consistent with the previous one. They studied the mechanical properties of MWCNT/PMMA as a function of nanotube orientation, length, concentration and type. The results reveal a good dispersion of the MWCNT into PMMA matrix for weight percentages ranging from 0.1 - 10 wt%, although, for loadings higher than 5 wt% there was evidence of aggregation. The orientation of MWCNT in PMMA seems to be the better way to improve the toughness of the material. Qian et al. (2000) mixed 1 wt% MWCNT to polystyrene (PS) and the results have shown an increase of about 36 - 42% in elastic modulus and about 25 % in the tensile strength, for the

PS/MWCNT composite as compared with pure PS. Yang et al. (2005) obtained an enhancement in microhardness, of about 40%, for the composites of PS/CNT with CNT loading lower than 1.5 wt%. They also observed a considerably decreasing in wear rate and friction coefficient as CNT loading increases. Dondero & Gorga (2006) studied the mechanical properties and the morphology of the MWCNT/PP composites, prepared by melt mixing, as a function of nanotube orientation and concentration. The results reveal an enhancement in toughness and in modulus of about 32% and 138%, respectively, for 0.25 wt% CNT loading, as compared with pure PP. The effects of loading rate of CNT on thermal and mechanical properties was studies by Bao & Tjong (2008), for PP/CNT composites. The x-ray diffraction (XRD) reveals that the presence of CNT did not influence the crystal structure of PP/CNT composites and differential scanning calorimetry (DSC) shows that the glass transition (T_g) and the activation energy (ΔE) increases with the increase of the amount of CNTs, demonstrating that the mobility od the polymer chains is reduced with the presence of CNTs. The tensile tests reveal an enhancement on Young's modulus from 1570 MPa for pure PP to 2107 MPa for 0.3 wt% of CNTs loading.

The incorporation of CNTs into polyurethanes (PU) has been studied by many groups. Koerner et al. (2005) incorporate different volume fractions of CNTs (from 0.5 to 10 vol.%) into thermoplastic PU. The results reveal an enhancement in electrical conductivity, and in mechanical properties including the modulus and yield stress. Chen et al. (2006) reported an enhancement in elastic modulus, from 4.96 MPa for pure PU to 135.1MPa for the composite with higher weight fraction of CNTs (17.7 wt%). On the other hand, the maximum for tensile strength was occurred for the composite with 9.3 wt% of CNT, decreasing for higher concentrations, which can be ascribed to the increased frequency of localized clusters or aggregations. A study on the thermal properties of the PU/CNT composites was performed by Xiong et al. (2006) and the results reveal an increase on the glass transition temperature of the composite of about 12°C, as compared with pure PU. Results of the tensile testes also reveal an improvement of the mechanical properties of the composites as compared with pure polymer.

The reinforcement of polyethylene (PE, HDPE, UHMWPE) with CNTs is probably the most studied issue in what concerns to the enhancement of thermoplastics, due to its enormous range of applications. McNally et al. (2005) reinforced PE with CNTs with weight fractions ranging from 0.1 to 10 wt% by melt blending. The results have shown that the addition of CNTs did not affect the temperature of melting (T_m) of the PE, however, the temperature of crystallization (T_c) increases for the composite with 10 wt% of CNTs. This indicates that the CNT have a nucleation effect on PE. These composites have shown a decreasing in toughness with increased CNT addition, which can be associated with the nucleation of CNTs into polymeric matrix. On the other hand, Kanagaraj et al. (2007) has shown a good load transfer effect and interface link between CNTs and high density polyethylene (HDPE). The results of tensile tests have shown a linear increase of Young's modulus (with maximum of about 22% for a volume fraction of 0.44 % CNT). This linearity is suggested to be due to a good load transfer effect and interface link between CNT and HDPE. The thermal analysis reveals that the melting point and oxidation temperatures of CNT/HDPE composites are not affected by the addition of CNTs, although the results reveals an increasing on crystallinity of the composites. Ruan et al. (2003) reported an enhancement of toughness in UHMWPE films with the addition of 1 wt% MWCNT. Their results reveal an increase in strain energy density of about 150% for the composites as compared with pure UHMWPE. They also reported an increase of about 140% in ductility and up to 25% in tensile strength. An analysis by nanoindentation and atomic force microscopy (AFM) of UHMWPE/MWCNT composites has been reported by Wei et al. (2006). They have observed a decrease of the friction coefficient with MWCNT content increase. Recently, Fonseca et al. (2011) prepared UHMWPE/CNT composites, with different volume fractions of CNTs, through mechanical ball-milling and processed by compression molding. In this work, for this specific materials and mould geometry, it was optimized the time of mixture, in order to obtain an homogeneous distribution of CNT into polymeric matrix, as well as the processing conditions. The results reveal that mechanical ball-milling is well suited for mix the CNTs with UHMWPE. CNTs were well dispersed and did no loose their physical integrity (as can be shown in Fig. 5). These scanning electron microscopy (SEM) pictures shown different CNT/HDPE powder composites mixed for different time (15, 45 and 60 min). Though the SEM pictures and the tensile tests, it was observed that the optimized time of mixture was for 45 min. To process the mixed composites, it was used the compression molding technique, and the cycle was optimized (Fig. 6 (a)). The SEM analysis of the processed composites has shown that CNTs were well dispersed into the polymeric matrix (Fig. 6(b)). The tensile tests of these composites have shown an enhancement on the mechanical properties of about 20% on the elastic modulus, for 0.2 % vol. of CNTs, and about 80% for higher concentrations (0.4 to 1% vol.). Similar results were reported by other groups (Wang et al., 2005; Mierczynska et al., 2007).

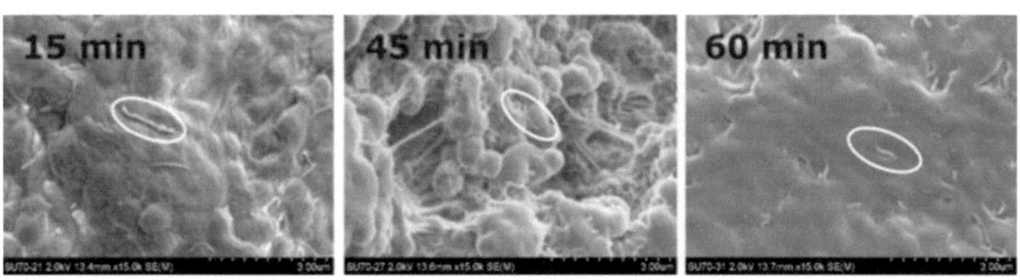

Fig. 5. SEM pictures of the 0.2% vol. MWCNT composite mixed for 15, 45 and 60 min.

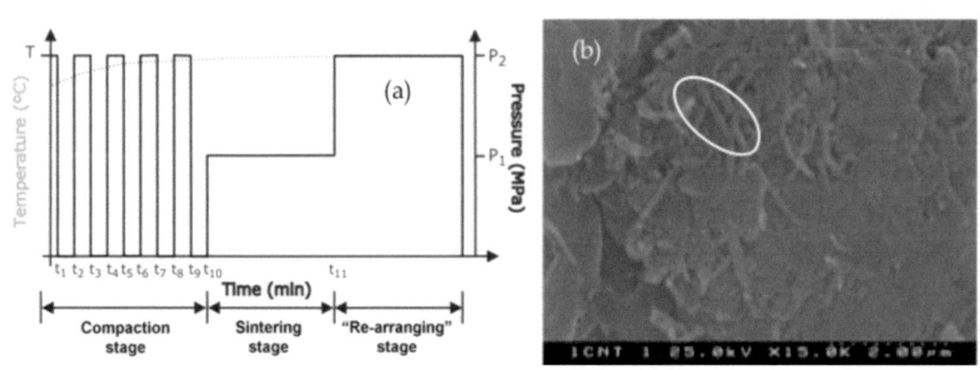

Fig. 6. (a) Optimized compression molding cycle applied for processing the composites; (b) SEM picture of the processed CNT/UHMWPE composite with 1% vol. fraction of CNTs. The round circle gives evidence for the presence of the CNTs.

The enhancement of UHMWPE properties is of utmost importance, essentially for medical applications, as UHMWPE is largely used for that purpose. However, the incorporation of CNTs into UHMWPE, for medical applications, requires special attention in what concerns to toxicity and biocompatibility of these composites. Reis et al. (2010) describe the response of human osteoblasts-like MG63 cells in contact with particles generated from UHMWPE/CNT composites. The results show the absence of significant elevation of the osteolysis inductor IL-6 values, pointed out for possible use of this superior wear-resistant composite for future orthopaedic applications.

3.3 The rheology properties and molding tools

As it was discussed earlier, with the addition of particles to thermoplastics it is possible to modify the physical, thermal and mainly the mechanical properties of the raw thermoplastic. With these enhanced thermoplastics it is of utmost importance to evaluate the relationship of the new materials with the processing parameters, and optimize them, especially when new processing techniques are arising, like microinjection molding. In the bibliography it is possible to read some studies on microinjection molding of neat polymers (Chien, 2006; Sha et al., 2007). Few studies did consider polymer compounds containing fillers, such as glass fibers, glass particles, nanoceramic materials and carbon nanotubes (Huang et al., 2005; Huang, 2006; Hanemann et al., 2009; Abbasi et al., 2011). Huang et al. (2005) study the moldability and wear particles of composites of polymeric matrices reinforced with nanoceramic particles. Their rheology analysis reveals that with increasing nanoparticle loading the shear viscosity increases, meaning that, in microinjection moulding, high pressure is needed to produce high quality parts with micro-features.

Abbasi et al., (2011) prepared PP/CNT and PC/CNT composites and studied the effects of processing conditions on its structure, mechanical properties and electrical conductivity. They concluded that the high deformation values of the microinjection molding only slightly changed the overall crystallinity due to the short cycle time of process. They also observed that the crystals were all oriented in the flow direction. Other interesting result is that the type of processing strongly affects the electrical conductivity of the composites.

4. High performance molding tools

In this section, it will be presented the results of surface modification of the molding blocks. The performance of diamond coated molding tools will be compared with the performance of other typically used coatings and non-coated tools. Different morphologies of the diamond coating will also be presented in order to evaluate their influence on the molding process.

4.1 Surface engineering in molding tools

Although the advances acquired in the past couple of years in the microinjection molding technology process, there are still some problems on the downstream that must be overcome. The micro parts may become statically charged and tend to adhere to surfaces around the molding area, making free fall extraction difficult or even impossible. In the conventional thermoplastic injection molding, the wear of molding tools is known to be one of the main sources of breakdown failures, resulting in production losses. In microinjection of thermoplastic parts, the moulds for components that are miniature complex and require high precision tolerances are not wear free. On the contrary, the molding surface wear can be even much more critical than in conventional molding. The wear out of the molding tools creates demolding problems, compromising the molding finish quality, speeding up the corrosion of the tools and resulting in costly maintenance stops. Polymer abrasion, adhesion and corrosion are the catalysers of these mechanisms. Furthermore, the increasing usage of polymers reinforced with glass fibers, minerals, or even carbon nanotubes, enhance the abrasive power of polymers.

Coatings technologies have been strongly developed in the last decades, as did their application on mould and die tools. Ion implantation and unbalanced magnetron sputtering PVD (Bienk & Mikkelsen, 1997), High Velocity Oxy Fuel (HVOF) and Atmospheric Plasma Spraying (APS) (Gibbons & Hansell, 2008), diamond-like carbon and silicon carbide (Griffiths et al., 2010), nanostructured TiB2 (Martinho et al., 2011), Oxide coating (Alumina) and Nitride coatings (AlN, CrN, NiCr(N), TiN) (Navabpour et al., 2006), among other surface engineering coating have been tested in order to evaluate their performance to avoid molding surface wear and assist the demolding process.

Chemical vapor deposition (CVD) of polycrystalline diamond, in microcrystalline and nanocrystalline morphology, detain a number of extreme properties that point it as a technology suitable for exploitation in numerous industrial applications. It possesses a high mechanical hardness and wear resistance, high thermal conductivity and is very resistant to chemical corrosion. Its properties and their optimization by means of the deposition process, in order to fulfill the application requisite have been investigated by several research teams (Ahmed et al., 2006; Das & Singh, 2007; Gracio et al., 2010). Some successful work has already been performed on the evaluation of diamond coating on molding tools in special for micro-featured tools (Neto, 2008b, 2008c, 2009). As refereed above, in this sub-chapter, the application and evaluation of CVD diamond thin films as a surface engineering technique to improve operation and durability of microinjection mould tools will be highlighted.

4.2 Diamond coatings

Polycrystalline diamond, in microcrystalline or nanocrystalline morphology, detains a number of extreme properties that point it as a technology suitable for exploitation in numerous industrial applications. It detains an extreme mechanical hardness (ca. 90 GPa) and wear resistance, one of the highest bulk modulus (1.2×10^{12} N.m^{-2}), the lowest compressibility (8.3×10^{-13} m^2 N^{-1}), the highest room temperature thermal conductivity (2×10^3 W m^{-1} K^{-1}), a very low thermal expansion coefficient at room temperature (1×10^{-6} K) and is very resistant to chemical corrosion (Das & Singh, 2007; Grácio et al., 2010).

Most of these properties are attractive for the application on cavities and mould tools; nevertheless coating an entire cavity with polycrystalline diamond is presently an utopia. CVD systems are considerably size limited due to the means of activating (thermal, electric discharge, or combustion flame) the gas phase carbon-containing precursor molecules.

A second problem related to the usage of CVD on mould tools is concerned with the fact that diamond cannot be directly coated onto ferrous substrates, the widest used row material to produce mould tools. Carbon, the precursor element of diamond, easily diffuses into the ferrous matrix, leaving behind no matter to start the diamond nucleation process.

To bypass the latter, appropriate interlayers can be used. A suitable interlayer is the one that promotes a diffusion block from and to the substrate material, enhances the adhesion between the diamond coating and the mould, and does not affect the properties of the diamond film or those of the mould tool, as pointed by Neto et al. (2008a).

A third constraint is the typical diamond deposition temperatures. The process temperature could be a limitation, since high temperatures may change the heat treatment induced in the steel material and alter its properties. Nevertheless, in the past years, deposition has been achieved at lower temperatures then the typical 800 to 900 °C (Dong et al., 2002; Petherbridge et al., 2001).

Not all molding tools are made out of steel material. Hybrid moulds or multi-material moulds are currently used for injection molding prototyping or to enhance mould heat extraction effectiveness. Aluminium, copper, silicon, silicon carbide, among others, are also used. The uses of different tool material place new constrains but also open new possibilities. The use of a silicon insert as the molding block, per example, may benefit from the microfabrication technology attained in the electronic industry. As the demand for smaller devices continues to increase, current manufacturing processes will find it more challenging to meet cost, quantity, and dimensional requirements. By merging silicon microfabrication techniques with appropriated surface engineering techniques, nano-scale features may be produced to replicate nano-features devices (Gourgon et al., 2005; Guo, 2007). Additionally, the deposition of diamond coating on silicon material is vastly reported by the scientific community, making it an excellent candidate material to be surface engineered with tuned diamond films.

CVD diamond deposition is possible by means of activating hydrogen and hydrocarbon. The growth of a diamond film starts from distinct nucleation sites. As individual randomly oriented nuclei grow larger, its diameters equal the average distance between the nucleation sites and start to form a continuous film. The subsequent film growth is dominated by competitive growth between differently oriented grains. With increasing film thickness, more and more grains are overgrown and buried by adjacent grains. Only those crystals with the direction of fastest growth perpendicular to the surface will survive. Thermal or plasma energy is the key factor to promote the fluctuations of the density to achieve small aggregates and to promote the thermodynamic environment to lead to the crystal growth. Nevertheless, it has been gradually recognized that the superequilibrium concentration of atomic hydrogen has also an important role on diamond growth. Various activating methods are used, such as DC-plasma, RF-plasma, microwave plasma, electron cyclotron resonance-microwave plasma CVD, and their modifications. More recently, nanocrystalline (NCD) and ultrananocrystalline diamond (UNCD) deposition have been researched. Many synthesis processes are described in the literature. Most of these processes require an extra gas source such as Ar, N_2 or He. The basic idea behind the deposition process is to enhance the diamond secondary nucleation rate during the deposition, thus leading to the formation of the films. (Sharda & Bhattacharyya, 2004; Spears & Dismukes, 1994)

In the following subsections, the conditions for the different diamond films deposited on steel and silicon will be presented, characterized and evaluated as polymer moulding surfaces.

4.3 Performance of diamond coatings on steel substrates

A set of AISI P20 modified steel molding inserts was prepared and pre-coated with a 2 μm thick PVD chromium nitride (CrN) film, in order to block the mutual diffusion between the ferrous substrate and the diamond growth atmosphere. All the steel plates had the molding surface polished with silicon carbide paper till grit #2000. The samples to be diamond coated were ultrasonic abraded with a diamond solution and then cleaned with isopropyl alcohol. Diamond growth was performed in a hot-filament CVD reactor, using time-modulated CVD process.

In a first experiment, four different samples were prepared to evaluate their performance has molding surfaces. Sample AC1 was submitted to 4h30 of diamond deposition and sample AC2 to 9h00. The full deposition conditions used in steel substrates can be seen in Neto et al. (2008b). Sample AC3 had only the CrN film an AC4 was a bare steel substrate. The two last samples were used in order to compare their performance with the diamond coated inserts.

Figure 7 shows the SEM images of inserts AC1, AC2, AC3 and AC4 before injection molding. Both diamond coated samples exhibited diamond crystallites mainly displaying (111) crystal orientation, although (100) oriented crystals were also observed. These growth directions are typical for the processing temperatures employed in this investigation. The average crystal size of these films is 1.74 and 1.25 μm, and the measured average roughness is 0.18 and 0.16 μm, for samples AC1 and AC2, respectively. Inserts AC3 and AC4 displayed an average roughness of 0.10 and 0.11 μm, respectively. Raman spectroscopy was used to assess the diamond Raman quality, on the diamond coated samples, as proposed by Kulisch et al. (1996) and to estimate the residual stresses of the diamond film according to Ralchenko et al. (1995). Calculated quality factor values for the diamond coatings were 56.0 and 58.3%, for samples AC1 and AC2, respectively. Samples AC1 and AC2 presented diamond peak shifts ($\Delta\varphi$) of 11 and 13 cm^{-1}, respectively. Calculated residual stress (σ) values for the diamond coatings were 6.2 and 7.4 GPa, for samples AC1 and AC2, respectively.

After this preliminary analysis using the molding inserts, they were placed in a mould tool specially designed to accommodate the 10 x 10 x 3 mm inserts and mounted in an injection molding machine to perform a cycle of 500 high-density polyethylene (HDPE) sample plates.

The molded samples were analyzed using optical microscopy. Figure 8 displays 150 times magnified images of HDPE objects from the run number 1, 100, 300 and 500. Whatever the type of insert surface, the first injected object presents more heterogeneous surface than the objects injected in cycles 100, 300 or 500. After the first set of injections, the polymeric objects molded with both diamond coating are very alike. The molded pieces by AC3 and AC4, present some surface scratches, but maintain the optical brightness (observed by the naked eye). The samples molded with the diamond coated inserts presented a slightly more tarnished surface than the samples molded with bare steel or with the CrN coating.

This may be due to the slight increased roughness that the diamond coated inserts presented, compared to the non-diamond coated samples, and also due to the crystalline nature of the diamond coatings. It is also worth mentioning that the injected samples 100 by the bare steel insert was very clean, but as the number of injections are increased, the surface of the plastic sample became heterogeneous.

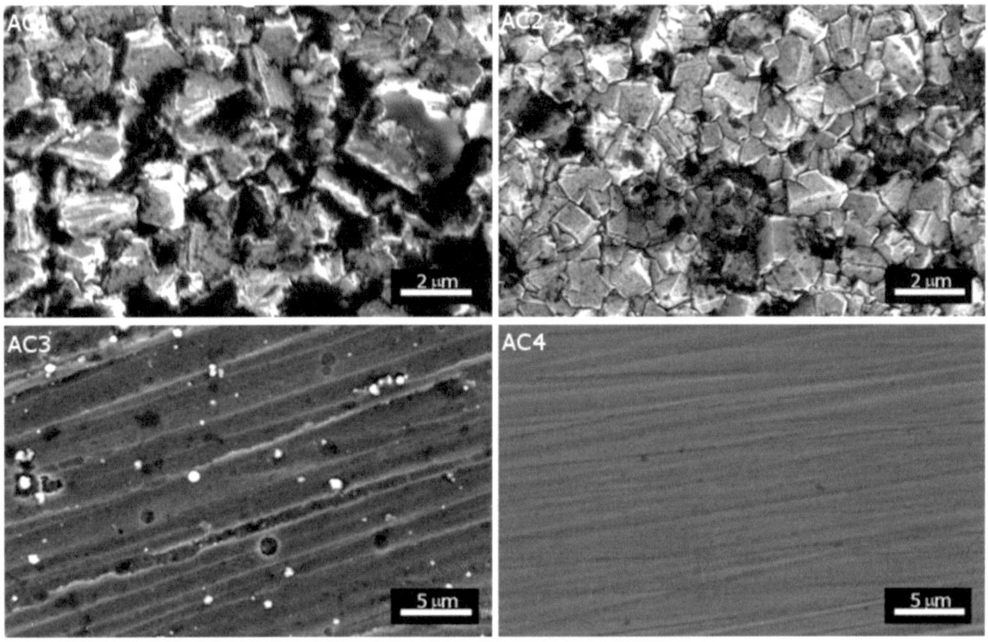

Fig. 7. SEM images of inserts AC1, AC2, AC3 and AC4 before injection molding

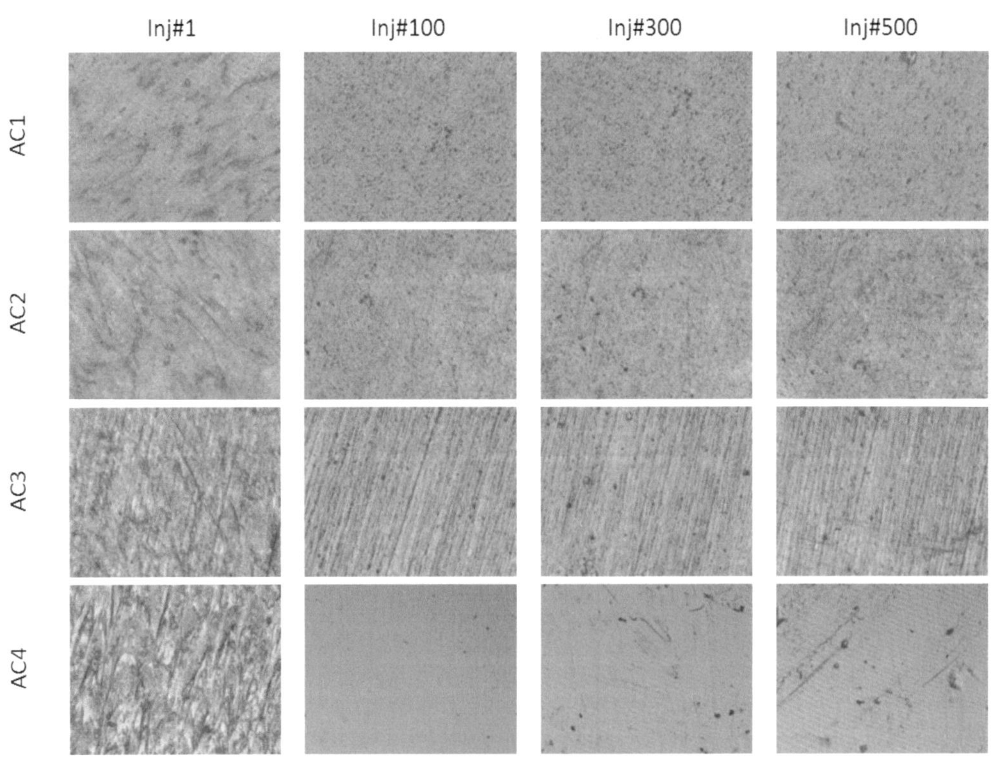

Fig. 8. Optical microscopy images of the HDPE molded surfaces

Since it was found out that the samples molded with the diamond coated inserts presented a slightly more tarnished surface than the samples molded with bare steel or with CrN coating and it is expected that if the average roughness and crystal size of the diamond film is reduced, this problem may be overcome, a new group of samples were prepared. Diamond average roughness and crystal size reduction may be accomplished by the deposition of sub-microcrystalline or nanocrystalline diamond. In order to obtain homogeneous coatings with an average crystal size of about 1 μm or less, different deposition condition where used. The deposition conditions can be seen in Neto et al. (2008c). Three different films were produced. All the as-grown films exhibited sub-micron diamond crystallite size, mainly displaying (111) crystal orientation. The average diamond crystallite sizes of the deposited films were 0.61, 0.71 and 0.83 μm, for sample FD1, FD2 and FD3, respectively. The three coated samples and a bare steel plate (sample F1) was also used in the adapted mould, to serve as a reference sample.

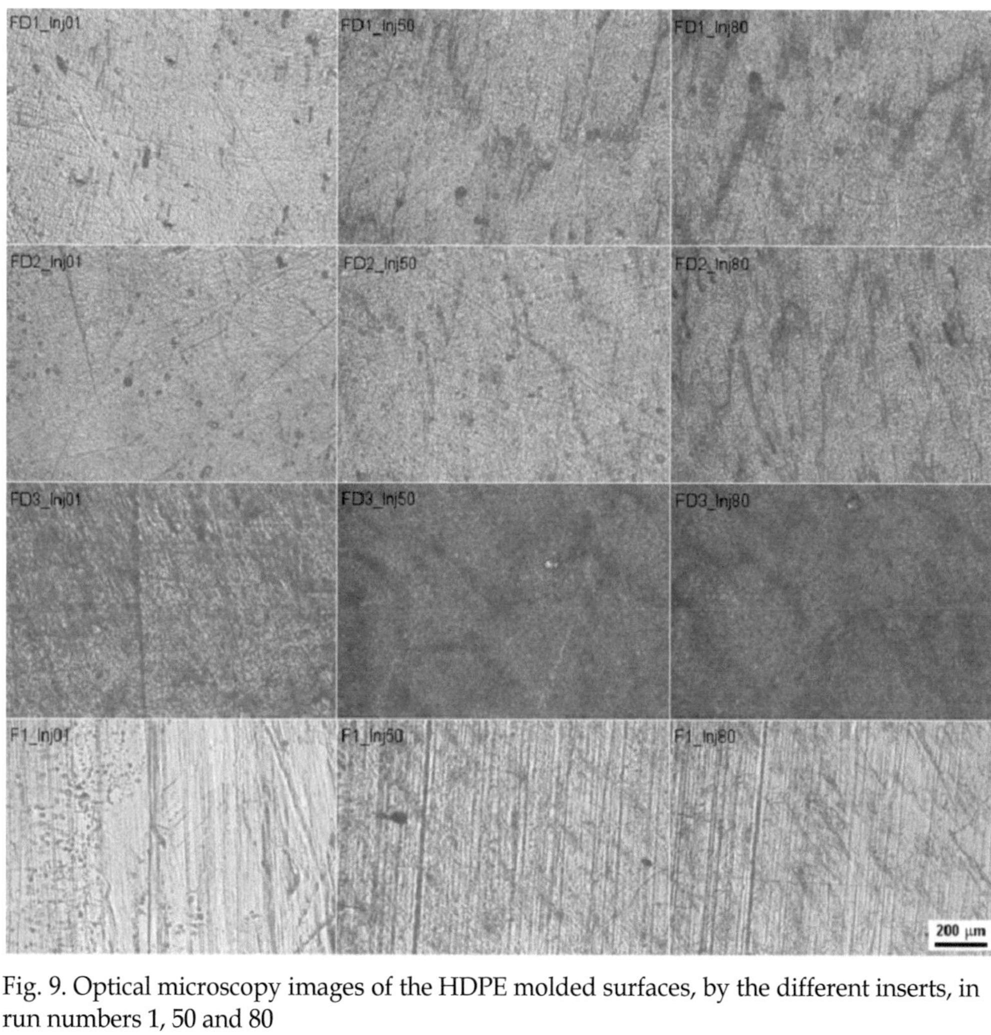

Fig. 9. Optical microscopy images of the HDPE molded surfaces, by the different inserts, in run numbers 1, 50 and 80

Figure 9 displays micrographs of the polymeric molded surface by the different inserts, in run number 1, 50 and 80. Apart from the first set of samples, that present marks of the demolding spray used in the beginning of the processing work, the injected parts are identical. All samples presented a good finishing surface, not showing the tarnished surface that the microcrystalline diamond coated insert originated. From the images, it is evident that the molded objects with the diamond-coated inserts, present more homogeneous surfaces than the ones molded by the insert without coating. It should be noted that insert F1 has the same or better surface finish than the samples that were used to deposit diamond, because their surface was not diamond abraded as the ones pre-treated for diamond coating. The initial steel samples were not surface polished to achieve a mirror surface, in order to obtain optical smooth surfaces and, hence, good quality plastic components/parts. Therefore, the results indicate that polishing time may be saved when using diamond-coated surfaces.

For the evaluation of the degradation of the molding surface, the injected parts by the different systems were analyzed. These parts had simple features which were measured throughout the injection process. Samples molded by diamond coated inserts showed a degradation trend of 0.0001 mm/injection, but the samples molded by samples with only the CrN film or without film presented a degradation trend of 0.0004 mm/injection. This may be due to the polymer aggregation to the cavity.

4.4 Performance of diamond coatings on steel substrates

Silicon (Si) and silicon carbide (SiC) samples were prepared to be diamond coated and then tested to be used as molding blocks. The diamond growth was performed using conditions to achieve both microcrystalline and nanocrystalline diamond morphology. Microcrystalline conditions can be seen in Neto et al. (2008c). Nanocrystalline was achieved by adding Argon gas to the reactor chamber. The full deposition conditions for the latter can be seen in Neto et al. (2011).

Figure 10 presents the SEM images of the as coated diamond films on Si and SiC with the nanocrystalline conditions. Si (a) is a low magnification image showing the rip and the nearby surface of the featured Si substrate. Si (b) is a high magnification image from the top surface and Si (c) from inside a feature. SiC (a) is a 600 times magnified image of one of the SiC samples, and SiC (b) and SiC (c) are also from the top surface from inside one of the rips, respectively. All films are almost fully coalescent and are composed of nanocrystalline diamond particles with a size inferior to 100 nm. The roughness assessment by means of a profilometer lead to an average surface roughness of 0.12 and 0.19 μm for the deposited diamond films on Si and SiC, respectively.

From the coated samples, 100 of HDPE parts were molded. It seems to replicate the molding surface in a proper way, displaying a fine shining surface.

The Si samples coated with microcrystalline diamond presented similar results as the ones presented by the steel coated substrates.

Different coated systems were tested to reproduce high-density polyethylene (HDPE) components, namely microcrystalline, sub-microcrystalline and nanocrystalline diamond films. Each coated systems were tested for a number of injection cycles in order to evaluate the polymeric produced parts and also the degradation of the coated inserts. All coated samples presented good stability at least till 500 runs (the maximum that a single diamond coated insert was subjected to, under laboratory conditions). The HDPE thermoplastic molded objects presented good quality and reproduced well the molding surface. Microcrystalline diamond coated inserts produced slightly tarnished plastic parts. The latter, seems to be considerably dimmed with the use of sub-microcrystalline or nanocrystalline films. The diamond coated featured inserts presented a good performance and a reduced degradation trend comparatively to the non-coated surfaces.

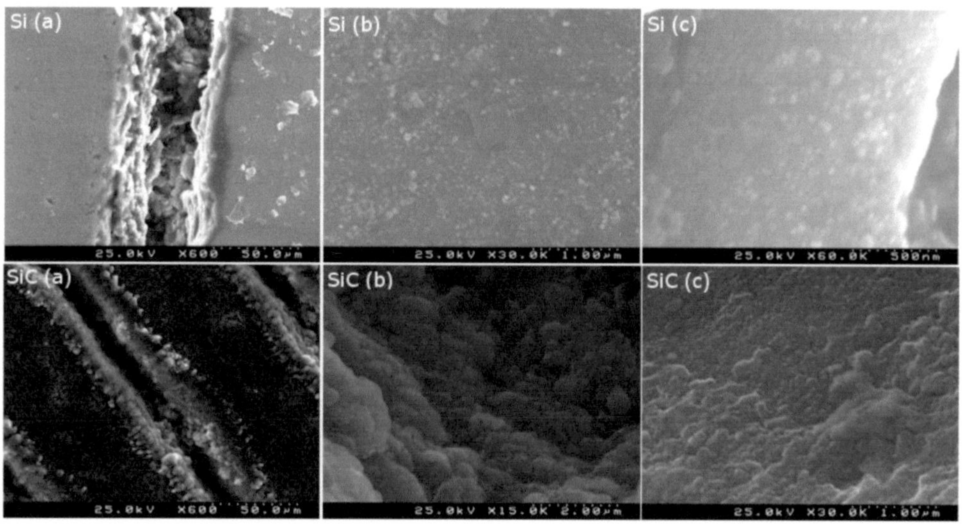

Fig. 10. Nanocrystalline diamond deposited on Si and SiC structured substrates

5. Conclusion

Injection molding enables the large scale production of polymeric components with accuracy. This technology has been progressively used for the production of micro-components in quantity and quality at low cost, which supports the development of micro-electro-mechanical systems. Nevertheless, the dimensional reduction of the components requires a higher control of the dimensional accuracy of these devices. It is also known that in the molding blocks of this type of objects, the wear is amplified due to the fact that the surface roughness is dimensionally very close to the dimensions being controlled. Reinforcing materials such as glass or metallic fibers, or carbon nanotubes in the polymeric matrix enhances the wear capability of the material being injected, compromising the service life time of the tools. Reinforced thermoplastics, the so called nano-composites, where developed in order to test the tools and the process, using carbon nanotubes as enhancers.

A gateway to reduce the deterioration of the molding impressions, increase their durability and reduce the need of corrective intervention on the tool may be by the use of appropriate surface engineering processes. The use of nanocrystalline diamond or other allotropic carbon coatings is considered a potential surface engineering coating for this type of application, since it detains high hardness and high thermal conductivity, being both properties very interesting to apply to the thermoplastic injection molding process. The coating with thin carbon based films is possible by means of chemical vapor deposition techniques. The roughness of these films and the intrinsic characteristics of diamond lead to a reduced friction coefficient, which will be important to minimize the wear of micro-components on the extraction stage.

6. References

Abbasi, S., Derdouri, A., & Carreau, P. J. (2011). "Properties of microinjection molding of polymer multiwalled carbon nanotube conducting composites." *Polymer Engineering & Science* 51(5), 992-1003.

Ahmed, W., Ahmed, E., Maryan, C., Jackson, M., Ogwu, A., Ali, N., Neto, V., & Gracio, J. (2006). Time-modulated CVD process optimized using the taguchi method. *Journal of Materials Engineering and Performance*, 15(2), 236-241.

Andrews, R., Jacques, D., Minot, M., & Rantell, T. (2002). "Fabrication of Carbon Multiwall Nanotube/Polymer Composites by Shear Mixing." *Macromolecular Materials and Engineering* 287(6), 395-403.

Attia, U., Marson, S. & Alcock, J. (2009). "Micro-injection moulding of polymer microfluidic devices." *Microfluidics and Nanofluidics* 7(1), pp. 1-28.

Bao, S. P., & Tjong, S. C. (2008). "Mechanical behaviors of polypropylene/carbon nanotube nanocomposites: The effects of loading rate and temperature." *Materials Science and Engineering A* 485, 508-516.

Baughman, R. H., Zakhidov, A. A., & de Heer, W. A (2002). "Carbon Nanotubes - the Route Toward Applications." *Science* 297(5582), 787-792.

Bienk, E. J., & Mikkelsen, N. J. (1997). Application of advanced surface treatment technologies in the modern plastics moulding industry. *Wear*, 207(1-2), 6-9.

Chang, P., & Hwang, S. (2006). "Experimental investigation of infrared rapid surface heating for injection molding." *Journal of Applied Polymer Science* 102(4), 3704-3713.

Chang, P., Hwang, S., Lee, H. & Huang, D. (2007). "Development of an external-type microinjection molding module for thermoplastic polymer." *Journal of Materials Processing Technology* 184(1-3), 163-172.

Chattopadhyay, D. K., & Raju, K. V. S. N. (2007). "Structural engineering of polyurethane coatings for high performance applications." *Progress in Polymer Science* 32(3), 352-418.

Chen, W., Tao, X., & Liu, Y. (2006). "Carbon nanotube-reinforced polyurethane composite fibers." *Composites Science and Technology* 66(15), 3029-3034.

Chien, R.-D. (2006). "Micromolding of biochip devices designed with microchannels." *Sensors and Actuators A: Physical* 128(2), 238-247.

Coleman, J. N, Cadek, M., Blake, R., Nicolosi, V., Ryan, K. P., Belton, C., Fonseca, A., Nagy, J. B., Gun'ko, Y. K., Blau, W. J. (2004). "High-performance nanotube-reinforced plastics: understanding the mechanism of strength increase." *Advanced Functional Materials*, 14(8), 791-798.

Das, D., & Singh, R. N. (2007). A review of nucleation, growth and low temperature synthesis of diamond thin films. *International Materials Reviews*, 52(1), 29-64.

Davim, J., & Jackson, M. (Eds.). (2009). *Nano and Micromachining*. London: Wiley-Iste.

Deng, J., Ding, X., Zhang, W., Peng, Y., Wang, J., Long, X., Pei, L., Chan, A. S. C. (2002). "Carbon nanotube-polyaniline hybrid materials." *European Polymer Journal* 38(12), 2497-2501.

Despa, M., Kelly, K. & Collier, J. (, 1999). "Injection molding of polymeric LIGA HARMs." Microsystem Technologies Vol.6(No.2): pp. 60-66.

Dondero, W. E. and R. E. Gorga (2006). "Morphological and mechanical properties of carbon nanotube/polymer composites via melt compounding." Journal of Polymer Science Part B: Polymer Physics 44(5): 864-878.

Dong, L., Ma, B., & Dong, G. (2002). Diamond deposition at low temperature by using CH4/H2 gas mixture. *Diamond and Related Materials*, 11(9), 1697-1702.

Esumi, K., Ishigami, M., Nakajima, A., Sawada, K., & Honda, H. (1996). Chemical treatment of carbon nanotubes. *Carbon*, 34, 279-281.

Fleischer, J., Halvadjiysky, G., & Haupt, S. (2007). "The manufacturing of micro molds by conventional and energy-assisted processes." *The International Journal of Advanced Manufacturing Technology* 33(1), 75-85.

Fonseca, A., Kanagaraj, S., Oliveira, M. S. A., & Simões, J. A. O. (2011). "Enhanced UHMWPE reinforced with MWCNT through mechanical ball-milling." *Defects and Diffusion Forum* 312-315, 1238-1243.

Franssila, S., Ed. (2004). *Introduction to Microfabrication*. Chichester, John Wiley & Sons Ltd.

Gibbons, G. J., & Hansell, R. G. (2008). Thermal-sprayed coatings on aluminium for mould tool protection and upgrade. *Journal of Materials Processing Technology*, 204(1-3), 184-191.

Giboz, J., Copponnex, T. & Mélé, P. (2007). "Microinjection molding of thermoplastic polymers: a review." *Journal of Micromechanics and Microengineering* 17(6), 96-109.

Gong, X., Liu, J., Baskaran, S., Voise, R. D., & Young, J. S. (2000). "Surfactant-Assisted Processing of Carbon Nanotube/Polymer Composites." *Chemistry of Materials* 12(4), 1049-1052.

Gorga, R. E., & Cohen, R. E. (2004). "Toughness enhancements in poly(methyl methacrylate) by addition of oriented multiwall carbon nanotubes." *Journal of Polymer Science: Part B: Polymer Physics* 42, 2690-2702.

Gornik, C. (2004). "Injection Moulding of Parts with Microstructured Surfaces for Medical Applications." *Macromolecular Symposia* 217(1), 365-374.

Gourgon, C., Perret, C., Tallal, J., Lazzarino, F., Landis, S., Joubert, O., & Pelzer, R. (2005). Uniformity across 200 mm silicon wafers printed by nanoimprint lithography. *Journal of Physics D: Applied Physics*, 38(1), 70.

Gower, M. (2000). "Industrial applications of laser micromachining." *Optics Express* 7(2), 56-67.

Grácio, J. J., Fan, Q. H., & Madaleno, J. C. (2010). Diamond growth by chemical vapour deposition. *Journal of Physics D: Applied Physics* 43(37), 374017.

Grave, A., Eriksson, T. & Hansen, H. (2007). Demouldability of Microstructures in Polymer Moulding. *International Conference on Multi-Material Micro-Manufacture*, Borovets, Bulgaria, October 3-5, 2007.

Griffiths, C., Dimov, S., Brousseau, E., Chouquet, C., Gavillet, J. & Bigot, S. (2008). Micro-Injection moulding: surface treatment effects on part demoulding. *International Conference on Multi-Material Micro Manufacture*, Cardiff, UK, September 09-11, 2008.

Griffiths, C., Dimov, S., Brousseau, E., Chouquet, C., Gavillet, J., & Bigot, S. (2010). Investigation of surface treatment effects in micro-injection-moulding. *The International Journal of Advanced Manufacturing Technology* 47(1), 99-110.

Guo, L. J. (2007). Nanoimprint Lithography: Methods and Material Requirements. *Advanced Materials* 19(4), 495-513.

Haggenmueller, R., Gommans, H. H., Rinzler, A. G., Fischer, J. E., & Winey, K. I. (2000). "Aligned single-wall carbon nanotubes in composites by melt processing methods." *Chemical Physics Letters* 330, 219-225.

Hanemann, T., Hau√üelt, J. R., & Ritzhaupt-Kleissl, E. (2009). "Compounding, micro injection moulding and characterisation of polycarbonate-nanosized alumina-composites for application in microoptics." *Microsystem Technologies* 15(3), 421-427.

Heckele, M., & Schomburg, W. (2004). "Review on micro molding of thermoplastic polymers." *Journal of Micromechanics and Microengineering* 14(3), 1-14.

Heyl, P., Olschewski, T. & Wijnaendts, R. (2001). "Manufacturing of 3D structures for micro-tools using laser ablation." *Microelectronic Engineering* 57-58, 775-780.

Hormes, J., Gottert, J., Lian, K., Desta, Y. & Jian, L. (2003). "Materials for LiGA and LiGA-based microsystems." *Nuclear Instruments and Methods in Physics Research Section B: Beam Interactions with Materials and Atoms* 199, 332-341.

Huang, C. K., Chen, S. W., & Yang, C. T. (2005). "Accuracy and mechanical properties of multiparts produced in one mold in microinjection molding." *Polymer Engineering & Science* 45(11), 1471-1478.

Huang, C. K. (2006). "Filling and wear behaviors of micro-molded parts made with nanomaterials." *European Polymer Journal* 42(9), 2174-2184.

Iijima, S. (1991). "Helical microtubules of graphitic carbon." *Nature* 354, 56-58.

Jia, Z., Wang, Z., Xu, C., Liang, J., Wei, B., Wu, D., & Zhu, S. (1999). "Study on poly(methyl methacrylate)/carbon nanotube composites." *Materials Science and Engineering A* 271, 395-400.

Jin, Z., Pramoda, K. P., Goh, S. H., & Xu, G. (2002). "Poly(vinylidene fluoride)-assisted melt-blending of multi-walled carbon nanotube/poly(methyl methacrylate) composites." *Materials Research Bulletin* 37, 271-278.

Kanagaraj, S., Varanda, F. R., Zhil'tsova, T. V., Oliveira, M. S. A., & Simões, J. A. O. (2007). "Mechanical properties of high density polyethylene/carbon nanotube composites." *Composites Science and Technology* 67(15-16), 3071-3077.

Kim, D., Lee, H., Lee, B., Yang, S., Kwon, T. & Lee, S. (2006). "Replications and analysis of microlens array fabricated by a modified LIGA process." *Polymer Engineering & Science* 46(4), 416-425.

Kistler. (2011). "p-T-Sensor for mold cavity pressure and temperature with front ø1 mm, cable length 0,4 m." Retrieved 30.08. 2011, from http://www.kistler.com/ca_en-us/13_Productfinder/App.6188AA0,4/p-T-Sensor-for-mold-cavity-pressure-and-temperature-with-front-1-mm-cable-length-0-4-m.html.

Koerner, H., Liu, W., Alexander, M., Mirau, P., Dowty, H., & Vaia, R. A. (2005). "Deformation-morphology correlations in electrically conductive carbon nanotube--thermoplastic polyurethane nanocomposites." *Polymer* 46(12), 4405-4420.

Kulisch, W., Ackermann, L., & Sobisch, B. (1996). On the Mechanisms of Bias Enhanced Nucleation of Diamond. *Physica Status Solidi (A)* 154(1), 155-174.

Liou, A., & Chen, R. (2006). "Injection molding of polymer micro- and sub-micron structures with high-aspect ratios." *The International Journal of Advanced Manufacturing Technology* 28(11), 1097-1103.

Luo, R., & Pan, Y. (2007). "Rapid Manufacturing of Intelligent Mold With Embedded Microsensors." *IEEE/ASME Transactions on Mechatronics* 12(2), 190-197.

Martinho, R. P., Silva, F. J. G., Alexandre, R. J. D., & Baptista, A. P. M. (2011). TiB_2 Nanostructured Coating for GFRP Injection Moulds. *Journal of Nanoscience and Nanotechnology*, 11(6), 6.

McNally, T., P^tschke, P., Halley, P., Murphy, M., Martin, D., Bell, S. E. J., Brennan, G. P., Bein, D., Lemoine, P., & Quinn, J. P. (2005). "Polyethylene multiwalled carbon nanotube composites." *Polymer* 46(19), 8222-8232.

Meyer, P., Schulz, J.Hahn, L. & Saile, V. (2008). "Why you will use the deep X-ray LIG(A) technology to produce MEMS?" *Microsystem Technologies* 14(9), 1491-1497.

Michaeli, W., Rogalla, A., & Ziegmann, C. (2000). "Processing technologies for the injection moulding of hybrid microstructures." *Macromolecular Materials and Engineering* 279(1), 42-45.

Michaeli, W., & Klaiber, F. (2007a). Investigations in Injection Moulding of Micro Structures and Microstructured Surfaces. *Conference on Multi-Material Micro Manufacture* Borovets, Bulgary, October 3-5, 2007.

Michaeli, W., Opfermann, D., & Kamps, T. (2007b). "Advances in micro assembly injection moulding for use in medical systems." *The International Journal of Advanced Manufacturing Technology* 33(1), 206-211.

Mierczynska, A., Mayne-L'Hermite, M., Boiteux, G., & Jeszka, J. K. (2007). "Electrical and mechanical properties of carbon nanotube/ultrahigh-molecular-weight polyethylene composites prepared by a filler prelocalization method." *Journal of Applied Polymer Science* 105(1), 158-168.

Munnik, F., Benninger, F., Mikhailov, S., Bertsch, A., Renaud, P., Lorenz, H. & Gmür, M. (2003). "High aspect ratio, 3D structuring of photoresist materials by ion beam LIGA." *Microelectronic Engineering* 67-68, 96-103.

Navabpour, P., Teer, D. G., Hitt, D. J., & Gilbert, M. (2006). Evaluation of non-stick properties of magnetron-sputtered coatings for moulds used for the processing of polymers. *Surface and Coatings Technology* 201(6), 3802-3809.

Neto, V. F., Shokuhfar, T., Oliveira, M. S. A., Grácio, J., & Ali, N. (2008a). Polycrystalline diamond coatings on steel substrates. *International Journal of Nanomanufacturing* 2(1/2), 17.

Neto, V. F., Vaz, R., Ali, N., Oliveira, M. S. A., & Grácio, J. (2008b). Diamond coatings on 3D structured steel. *Diamond and Related Materials* 17(7-10), 1424-1428.

Neto, V., Vaz, R., Ali, N., Oliveira, M., & Grácio, J. (2008c). Performance of sub-micron diamond films coated on mould inserts for plastic injection moulding. *Journal of Materials Science*, 43(10), 3392-3399.

Neto, V. F., Vaz, R., Oliveira, M. S. A., & Grácio, J. (2009). CVD diamond-coated steel inserts for thermoplastic mould tools - Characterization and preliminary performance evaluation. *Journal of Materials Processing Technology* 209(2), 1085-1091.

Neto, V. F., Oliveira, M. S. A., & Grácio, J. (2011). Performance of nanocrystalline diamond coated micromolding tools. *Materials Science & Technology 2011 - Surface Protection for Enhanced Materials Performance: Science and Technology*, Columbus, Ohio, USA.

Ono, Y., Whiteside, B., Brown, E., Kobayashi, M., Cheng, C., Jen, C. & Coates, P. (2007). "Real-time process monitoring of micromoulding using integrated ultrasonic sensors." *Transactions of the Institute of Measurement and Control* 29(5), 383-401.

Petherbridge, J. R., May, P. W., Pearce, S. R. J., Rosser, K. N., & Ashfold, M. N. R. (2001). *Low temperature diamond growth using CO_2/CH_4 plasmas: Molecular beam mass spectrometry and computer simulation investigations* (Vol. 89): AIP.

Piotter, V., Bauer, W., Hanemann, T., Heckele, M. & Müller, C. (2008). "Replication technologies for HARM devices: status and perspectives." *Microsystem Technologies* 14(9), 1599-1605.

Potschke, P., Bhattacharyya, A. R., Janke, A., & Goering, H. (2003). "Melt mixing of polycarbonate/multi-wall carbon nanotube composites." *Composite Interfaces* 10(4-5), 389-404.

Pouzada, A., Ferreira, E. & Pontes, A. (2006). "Friction properties of moulding thermoplastics." *Polymer Testing* 25(8), 1017-1023.

PRIAMUS®. (2011). "Cavity temperature sensors." 30.08.2011, from http://www.priamus.com/english/pdf/Datenblaetter_pdf_e/4011B-e.pdf.

Rahman, M., & Brazel, C. S. (2006). "Ionic liquids: New generation stable plasticizers for poly(vinyl chloride)." *Polymer Degradation and Stability* 91(12), 3371-3382.

Ralchenko, V. G., Smolin, A. A., Pereverzev, V. G., Obraztsova, E. D., Korotoushenko, K. G., Konov, V. I., Lakhotkin, Y. V., & Loubnin, E. N. (1995). Diamond deposition on steel with CVD tungsten intermediate layer. *Diamond and Related Materials*, 4(5-6), 754-758.

Reis, J., Kanagaraj, S., Fonseca, A., Mathew, M. T., Capela-Silva, F., Potes, J., Pereira, A., Oliveira, M. S. A., & Simões, J. A. (2010). "In vitro studies of multiwalled carbon nanotube/ultrahigh molecular weight polyethylene nanocomposites with osteoblast-like MG63 cells." *Brazilian Journal of Medical and Biological Research* 43, 476-482.

Rötting, O., Röpke, W., Becker, H. & Gärtner, C. (2002). "Polymer microfabrication technologies." *Microsystem Technologies* 8(1), pp. 32-36.

Ruan, S. L., Gao, P., Yang, X. G., & Yu, T. X. (2003). "Toughening high performance ultrahigh molecular weight polyethylene using multiwalled carbon nanotubes." *Polymer* 44(19), 5643-5654.

Safadi, B., Andrews, R., & Grulke, E. A. (2002). "Multiwalled carbon nanotube polymer composites: Synthesis and characterization of thin films." *Journal of Applied Polymer Science* 84(14), 2660-2669.

Saito, T., Satoh, I. & Kurosaki, Y. (2002). "A new concept of active temperature control for an injection molding process using infrared radiation heating." Po*lymer Engineering & Science* 42(12), 2418-2429.

Schadler, L. S., Giannaris, S. C., & Ajayan, P. M. (1998). "Load transfer in carbon nanotube epoxy composites." *Applied Physics Letters* 73(26), 3842-3844.

Sha, B., Dimov, S., Pham, D. & Griffiths,C. (2005). Study of Factors Affecting Aspect Ratios Achievable in Micro-injection Moulding. *International Conference on Multi-Material Micro Manufacture* Karlsruhe, Germany, 29 June-1July, 2005.

Sha, B., Dimov S., Griffiths C. & Packianather, M. (2007a). "Investigation of micro-injection moulding: Factors affecting the replication quality." *Journal of Materials Processing Technology* Vol.183(No.2-3): pp. 284-296.

Sha, B., Dimov, S., Griffiths, C. & Packianather, M. (2007b). "Micro-injection moulding: Factors affecting the achievable aspect ratios." *The International Journal of Advanced Manufacturing Technology* 33(10), 147-156.

Sharda, T., & Bhattacharyya, S. (2004). Diamond Nanocrystals. In H. S. Nalwa (Ed.), *Encyclopedia of Nanoscience and Nanotechnology* (Vol. 2, pp. 337-370): American Scientific Publishers.

Spears, K. E., & Dismukes, J. P. (Eds.). (1994). *Synthetic diamond: emerging CVD science and technology*. New York: Wiley Inter-Science.

Su, Y., Shah, J. & Lin, L. (2004). "Implementation and analysis of polymeric microstructure replication by micro injection molding." *Journal of Micromechanics and Microengineering* 14(3), 415-422.

Theilade, U., & Hansen, H. (2007). "Surface microstructure replication in injection molding." *The International Journal of Advanced Manufacturing Technology* 33(1), 157-166.

Tjong, S. C. (2006). "Structural and mechanical properties of polymer nanocomposites." *Materials Science and Engineering* R 53, 73-197.

Tofteberg, T., & Andreassen, E. (2008). Injection moulding of microfeatured parts. *Polymer Processing Society 24th Annual Meeting* Salerno, Italy, June 15-19, 2008.

Tseng, S., Chen, Y., Kuo, C. & Shew, B. (2005). "A study of integration of LIGA and M-EDM technology on the microinjection molding of ink-jet printers' nozzle plates." *Microsystem Technologies* 12(1), 116-119.

Uhlmann, E., Piltz, S. & Doll, U. (2005). "Machining of micro/miniature dies and moulds by electrical discharge machining - Recent development." *Journal of Materials Processing Technology* 167(2-3), 488-493.

Wang, A., Lin, R., Polineni, V. K., Essner, A., Stark, C., & Dumbleton, J. H. (1998). "Carbon fiber reinforced polyether ether ketone composite as a bearing surface for total hip replacement." *Tribology International* 31(11), 661-667.

Wang, C.-C., Zhang, G., H. Shah, N., Infeld, M. H., Waseem Malick, A., & McGinity, J. W. (1997). "Influence of plasticizers on the mechanical properties of pellets containing EudragitÆ RS 30 D." *International Journal of Pharmaceutics* 152(2), 153-163.

Wang, Y., Cheng, R., Liang, L., & Wang, Y. (2005). "Study on the preparation and characterization of ultra-high molecular weight polyethylene-carbon nanotubes composite fiber." *Composites Science and Technology* 65(5), 793-797.

Wei, Z., Zhao, Y.-P., Ruan, S. L., Gao, P., & Yu, T. X. (2006). "A study of the tribological behavior of carbon-nanotube-reinforced ultrahigh molecular weight polyethylene composites." *Surface and Interface Analysis* 38(4), 883-886.

Whiteside, B., Martyn, M., Coates, P., Allan, P., Hornsby, P. & Greeenway, G. (2003). "Micromoulding: process characteristics and product properties." *Plastics, rubber and composites* 32(6), 231-239.

Whiteside, B., Martyn, M., Coates, P., Allan, P., Hornsby, P. & Greeenway, G. (2004). "Micromoulding: process measurements, product morphology and properties." *Plastics, Rubber and Composites* 33(1), 11–17.

Wu, C., & Liang, W. (2005). "Effects of geometry and injection-molding parameters on weld-line strength." *Polymer Engineering & Science* 45(7), 1021-1030.

Xie, L., & Ziegmann, G. (2008). "A visual mold with variotherm system for weld line study in micro injection molding." *Microsystem Technologies* 14(6), 809-814.

Xiong, J., Zheng, Z., Qin, X., Li, M., Li, H., & Wang, X. (2006). "The thermal and mechanical properties of a polyurethane/multi-walled carbon nanotube composite." *Carbon* 44(13), 2701-2707.

Yan, C., Nakao, M., Go, T., Matsumoto, K. & Hatamura, Y. (2003). "Injection molding for microstructures controlling mold-core extrusion and cavity heat-flux." *Microsystem Technologies* 9(3), 188-191.

Yang, R., Jiang, J., Meng, W. J. & Wang, W. (2006). "Numerical simulation and fabrication of microscale, multilevel, tapered mold inserts using UV-Lithographie, Galvanoformung, Abformung (LIGA) technology." *Microsystem Technologies* 12(6), 545-553.

Yang, Z., Dong, B., Huang, Y., Liu, L., Yan, F.-Y., & Lo, H.-L. (2005). "Enhanced wear resistance and micro-hardness of polystyrene nanocomposites by carbon nanotubes." *Materials Chemistry and Physics* 94, 109-113.

Yao, D., Kimerling, T & Kim, B. (2006). "High-frequency proximity heating for injection molding applications." *Polymer Engineering & Science* 46(7), 938-945.

Yu, M.-F., Files, B. S., Arepalli, S., & Ruoff, R. S. (2000). "Tensile Loading of Ropes of Single Wall Carbon Nanotubes and their Mechanical Properties." *Physical Review Letters* 84(24), 5552-5555.

Zhang, L., Yilmaz, E. D., Schjodt-Thomsen, J., Rauhe, J. C., & Pyrz, R. (2011). "MWNT reinforced polyurethane foam: Processing, characterization and modelling of mechanical properties." *Composites Science and Technology* 71(6), 877-884.

Zhao, J., Mayes, R., Chen, G., Xie, H. & Chan, P. (2003). "Effects of process parameters on the micro molding process." *Polymer Engineering & Science* 43(9), 1542-1554.

Zhao, J., Lu, X. , Chen, G., Liu, S., Juay, Y. & Yong, M. (2006). "Micromold filling behavior studies of polymer materials." *Materials Research Innovations* 10, 394-397.

12

Investigation of the Physical Characteristics of Polypropylene Meltblown Nonwovens Under Varying Production Parameters

Deniz Duran
Ege University
Turkey

1. Introduction

Nonwovens are a unique class of textile materials, formed by bonding the fibers by various techniques. The demand through the nonwovens is increasing on the world day by day and nonwovens are getting integrated to more application areas rapidly. Today, nonwovens have achieved an excellent position among the products of daily use like hygiene, medicine, household and more. (Ebeling et al., 2006)

Meltblown nonwovens are a relatively new class of thermoplastic based nonwoven materials which integrate to new application areas to replace conventional textile materials. Due to their unique micro structure, low porosity, absorbency, light weight and high surface area, microfiber meltblown nonwovens are promising materials of the future. Meltblowing is a one step process which enables to produce microfiber nonwovens directly from thermoplastic polymers with the aid of high velocity air to attenuate the melt filaments. In this process high velocity air blows a molten thermoplastic resin from an extruder die tip onto a conveyor or take up screen to form a fine fiberous and self-bonding web. (Bang One Lee et al., 2010) It has become an important industrial technique in nonwovens because of its ability to produce fabrics of microfiber structure suitable for filtration media, thermal insulators, battery seperators, oil absorbents, wipes, apparel, medical applications and many lamination applications. (Zhang et al., 2002 , Duran&Perincek, 2010; Dutton, 2008)

Meltblowing is a unique process since it is used almost exclusively to produce microfibers rather than fibers the size of conventional textile fibers. Meltblown microfibers generally have diameters in the range of 2 to 4 μm, although they may be as small as 0.1 μm and as large as 10 to 15 μm. Differences between meltblown nonwoven fabrics and other nonwoven fabrics, such as degree of softness, cover or opacity, and porosity can generally be traced to differences in filament size. (Bang One Lee et al., 2010)

Meltblowing technology is used for producing light fiber webs directly from polymers. This process process allows the production of ultrafine filament nonwovens under very economical conditions. In the basic melt blowing process, a thermoplastic fibre forming polymer is melted in an extruder, pumped through die holes and then the melt enters high-speed, hot air streams. Streams of hot air exiting from the left and right sides of the die nosepiece rapidly attenuate the extruded polymer streams to form extremely fine filaments. The filaments then are blown by high-velocity air onto a collector screen, thus forming a fine-filtered, self-bonded nonwoven web. After that, they are bonded and wound to form roll goods for further processing. Web structure begins to develop when fiber entanglement first occurs, but network structure becomes fixed only when fibers contact the collector and their motion ceases. (Duran&Perincek, 2010; Randall et al, 2004; Rupp, 2008)

Because the meltblowing process employs high velocity air to impinge upon the polymer as it exits the orifices, it elongates the polymer strands from 500 μm diameter to diameters as small as 0.1 μm. Extreme entanglemet of fibres, characterizing meltblown fibrous webs, produces coherent fiber webs. Unique micrometer characteristics of meltblown structures produce a high surface area per unit weight and fine porosity. Therefore meltblown technology can be used to produce efficient filter materials that are able to filter particles as small as 0.5 μm. (Farer et al., 2003)

The melt blown process generally consists of five major components, which are the extruder, metering pump, die assembly, web formation, and winding. The extruder, which consists of a heated barrel with a rotating screw inside is one of the important elements in all polymer processing. In the extruder, the polymer pellets or granules are heated and melted until appropriate temperature and viscosity are reached. The extruder is divided in to three different zones. (Dahiya et al., 2004) The feed zone, is the zone where the polymer pellets are preheated and pushed to the next zone. The transition zone has a decreasing depth channel in order to compress and homogenize the melted polymer. The melting of the pellets in the extruder is due to the heat and friction of the viscous flow and the mechanical action between the screw and the walls of the barrel. The molten polymer is then fed to the metering pump to ensure uniform polymer feed to the die assembly. (Duran&Perincek,2010; Dutton, 2008; Dahiya et al., 2004) The metering zone is the last zone in the extruder whose main purpose is to generate maximum pressure in order to pump the molten polymer in the forward direction. (Dahiya et al., 2004) The metering pump is a device for uniform melt delivery to the die assembly. It ensures consistent flow of clean polymer mix under process variations in viscosity, pressure, and temperature. The metering pump also provides polymer metering and the required process pressure. At this point the breaker plate controls the pressure generated with a screen pack placed near to the screw discharge. The breaker plate also filters out any impurities such as dirt, foreign particle metal particles and melted polymer lumps. (Dahiya et al., 2004)

The die assembly is the most important element of the melt blown process. It has three distinct components: polymer-feed distribution, die nosepiece, and air manifolds. The feed distribution is usually designed in such a way that the polymer distribution is less dependent on the shear properties of the polymer. This feature allows the melt blowing of widely different polymeric materials with one distribution system. The feed distribution balances both the flow and the residence time across the width of the die. From the feed distribution channel the polymer melt goes directly to the die nosepiece. The web uniformity hinges largely on the design and fabrication of the nosepiece. Therefore, the die nosepiece in the melt blowing process requires very tight tolerances, which have made their fabrication very costly. The die nosepiece is a wide, hollow, and tapered piece of metal having several hundred orifices or holes across the width. The polymer melt is extruded from these holes to form filament strands which are subsequently attenuated by hot air to form fine fibers.

As soon as the molten polymer is extruded from the die holes, high velocity hot air streams blown from the top and bottom sides of the die nosepiece, attenuate the polymer streams to form microfibers. As the hot air stream containing the microfibers progresses toward the collector screen, it draws in a large amount of surrounding air that cools and solidifies the fibers. The solidified fibers subsequently get laid randomly onto the collecting screen, forming a self-bonded nonwoven web. (Duran&Perincek,2010; Dutton, 2008; Dahiya et al., 2004) Usually, a vacuum is applied to the inside of the collector screen to withdraw the hot air and enhance the fiber laying process. The combination of fiber entanglement and fiber-to-fiber bonding generally produce enough web cohesion so that the web can be readily used without further bonding. However, additional bonding and finishing processes may further be applied to these melt-blown webs. (Dahiya et al., 2004)

The properties of the meltblown webs are affected by various production parameters including air temperature, polymer/die temperature, die to collector distance (DCD), collector speed, polymer throughput, air throughput, die hole size and air angle. All of these affect the final properties of the nonwoven web. Both polymer throughput and air flow rate control the final fiber diameter, fiber entanglement, basis weight and the attenuating zone. Polymer/die and air temperatures combined with air flow rate affect the uniformity, shot formation which can be described as large globules of nonfibrous polymer larger in diameter than fibers in webs, rope and fly formation, fabric appearance and touch. (Dahiya et al., 2004) Because meltblowing uses an attenuating air stream to draw and orient the fibres, the distance between the polymer orifice and the collection surface includes the fiber characteristics and resulting fabric properties. This distance is commonly referred to as the die-to-collector distance (DCD). The period of time that fibres spend in flight before being collected influences fiber orientation, strength and surface properties. A varying DCD permits the production of structures with varying properties from stiff and brittle at close distance to soft and bulky at large distance. (Farer et al., 2003) Die hole size along with die set back affects the fiber size. Air angle controls the nature of air flow, i.e. as the air angle approaches 90° it results in a high degree of fiber separation or turbulence that leads to random fiber distribution. At an angle of 30°, roped or parallel fibers deposited as loosely coiled bundles of fibers are generated. This structure is undesirable. At angles greater than 30°, attenuation as well as breakage of fibers occurs. (Dahiya et al., 2004)

The melt blowing process is amenable to a wide range of polymers in terms of viscosities and blends. The type of polymer or resin used defines the elasticity, softness, wetability, dyeability, chemical resistance and other related properties of formed webs. Some polymers, which can be used for the formation of melt-blown webs are listed as polypropylene (PP), polyethylene (PE), polyethylene terephthalate (PET), polybutylene terephthalate (PBT), polyamide (PA) and polylacticacid (PLA). Polypropylene (PP) is the most widely used polymer for meltblowing process, because it has a low viscosity, a low melting point and is easy to draw into fibers. A lot of efforts have been made in the last 30 years on the process study, new resin and product development and process improvements. (Zhang et al., 2002; Dahiya et al., 2004)

The main characteristics and properties of melt-blown nonwovens can be summarised as follows:

- Random fiber orientation.
- Lower to moderate web strength
- Generally high opacity and a high cover factor
- Low porosity
- Good absorbency
- Fiber diameter in the range of 0.5 - 30 μm, typically 2-7 μm
- Basis weight in the range of 8-350 g/m^2, typically 20-200 g/m^2
- High surface area due to microfibers
- Good insulation and filtration characteristics
- Fibers with smooth surface texture
- Generally fibres with circular in cross-section but it can also be variable in some cases variable, ranging from circular to a flat configuration and other variations
- Variable fiber length over a broad range depending from a few millimeters to several hundred centimeters (Dahiya et al., 2004; Duran&Perincek, 2010; Dutton, 2008)

Meltblowing is a unique system since the process generates a fine fiber not available to the other nonwoven processes. Because the micro-denier fiber (less than 0.1 denier per filament) is not really available as a nonwoven fibrous raw material, the meltblown process, which can produce such a fiber, opens new vistas of products and applications. (Dahiya et al., 2004) Microfiber nonwovens produced via meltblowing technique are suitable to be used as thermal insulators, filtration media, oil absorbents, battery separators, medical area, wipes, apparel, laminates and many other applications. (Duran&Perincek, 2010, Dutton, 2008) Meltblown nonwovens find extensive use especially in absorbent cloths and wipes; oil absorption; and filtration for liquids, gas and air. Also, very important end uses of meltblowns include sanitary applications such as hygiene and incontinence products for babies, adults and feminine hygiene. The application areas of meltblown nonwovens can be summarised as follows: (Dahiya et al., 2004)

- Filtration media: surgical face mask filter media, liquid and gaseous filtration, cartridge filters, clean room filters and others
- Medical fabrics: disposable gown and drape market and sterilization wrap segment
- Sanitary products: feminine sanitary napkin, disposable adult incontinence absorbent products
- Oil adsorbents: sorbents to pick up oil from the surface of water, such as encountered in an accidental oil spill and for mats in machine shops and in industrial plants
- Apparel: thermal insulation, disposable industrial apparel and substrate for synthetic leather
- Hot melt adhesives
- Electronic Specialities: liner fabrics in computer floppy disks, battery separators and as insulation in capacitors
- Manufacture of tents and elastomeric nonwoven fabrics which have the same appearance as continuous filament spunbonded products

Beside many advantages they provide, meltblowns have limited strength characteristics and poor abrasion resistance. (Farer et al., 2003) Therefore, they often are combined with other nonwovens. Spunbonds and other types of nonwovens also can be covered and refined with meltblowns. Today it is possible to implement production lines not only with spunbonds, but also with a meltblown component, to produce different types of nonwovens with an even more of a textile feeling. Approximately 40 percent of meltblowns are manufactured using a stand-alone process. The remaining meltblowns are combined with spunbonds or laminated to form other materials. Combinations with spunbonded fabrics are primarily used to make nonwoven materials with barrier properties. Another variation is the combination of meltblown with cellulose or an absorbent powder to produce a soft, strong, but still absorbent material that can retain absorbed liquid while still keeping its strength. The outcome is a combination of spunbonds (S) and meltblowns (M), resulting in SMS, or even SSMMMSS or other combinations, depending on the final product. So many different products can be manufactured and implemented from SMS to SSMMMSS. Some examples of their applications are hygiene, baby and adult diapers, medical products, protective masks for medical use, general use as a barrier layers, paper composites, work safety, protective clothing, breathing masks and combination with other nonwovens. (Rupp, 2008)

2. Material and method

In this chapter, the results of a study about polypropylene microfibre meltblown nonwovens were covered. Aspects related to the production phase, properties of the materials produced and application areas were discussed. In the content of this study, effect of some production parameters namely die air pressure, collector drum speed, collector vacuum and extruder pressure on the physical properties of the web structure such as thickness, basis weight, air permeability, tensile properties, surface friction and fiber diameter were investigated.

As raw material, PP with 1100 melt flow rate (MFR), 0.75 g/cm^3 density and 335 F melting point was used. The main production settings namely; extruder temperatures, die temperature, air temperature (the air fed to the spinerette to spin the fibers), die hole diameter and die-to-collector distance were given in Table 1.

Production Setting	Value
Extruder zone 1 temperature (°F)	270
Extruder zone 2 temperature (°F)	320
Extruder zone 3 temperature (°F)	370
Die temperature (°F)	336
Air temperature (°F)	350
Die hole diameter (inches)	0.09
Die-to collector distance (cm)	50

Table 1. Main production settings and their values used in experiments

The sample codes and production parameters of the PP melt blown nonwovens investigated in this study can be seen in Table 2.

The produced samples were characterized for thickness, basis weight, air permeability, fiber diameter and tensile properties.

Fabric No	Collector drum speed (ft/min)	Collector vacuum (%)	Die air pressure (psi)	Extruder pressure (psi)
1	10	15	6	530
2	10	30	6	523
3	10	60	6	539
4	20	15	6	514
5	20	30	6	525
6	20	60	6	524
7	30	15	6	525
8	30	30	6	526
9	30	60	6	532
10	10	15	7	515
11	10	30	7	531
12	10	60	7	492
13	20	15	7	515
14	20	30	7	517
15	20	60	7	512
16	30	15	7	521
17	30	30	7	516
18	30	60	7	516
19	10	15	8	550
20	10	30	8	540
21	10	60	8	542
22	20	15	8	535
23	20	30	8	532
24	20	60	8	541
25	30	15	8	535
26	30	30	8	544
27	30	60	8	525

Table 2. Sample codes and production parameters of PP nonwovens

Thickness was tested by using SDL Thickness Gauge according to TS 7128 EN ISO 5084 standard, with 20 cm^2 measurement area under 200 g weight. Air permeability test was performed by using FX 3300 Air Permeability test device according to TS 391 EN ISO 9237 standard, with 20 cm^2 measurement area and 100 Pa air pressure. Fiber diameter was measured by using Leica DM EP light microscope with 400 zoom. Tensile properties were tested according to TS EN ISO 13934-1 standard by using Zwick Z010 Universal tensile strength with 200 mm measurement distance and 100 mm/min measurement speed.

The results were evaluated statistically by using SPSS software. To evaluate the differences between the subgroups in the collector drum speed, the collector vacuum and the die air pressure Student-Newman-Kleus test were performed. For the investigation of the effect of the extruder pressure 2-Tailed Pearson Correlation tests were performed.

3. Results and discussions

In this study, influence of some crucial production parameters namely die air pressure, extruder pressure, collector drum speed, and collector vacuum on the thickness, basis weight, air permeability, fiber diameter and tensile properties of polypropylene meltblown nonwoven webs were investigated.

3.1 Thickness

The results obtained from the thickness measurements were presented in Figure 1. The statistical analysis have shown that thickness of polypropylene meltblown nonwovens were effected mostly by the drum speed and the collector vacuum. An increase in the drum speed and an increase in the vacuum caused a decrease in the thickness. In other words, thicker surfaces were obtained with lower collector drum speeds and lover vacuum values. The highest values were obtained with 10 ft/min drum speed and %15 vacuum.

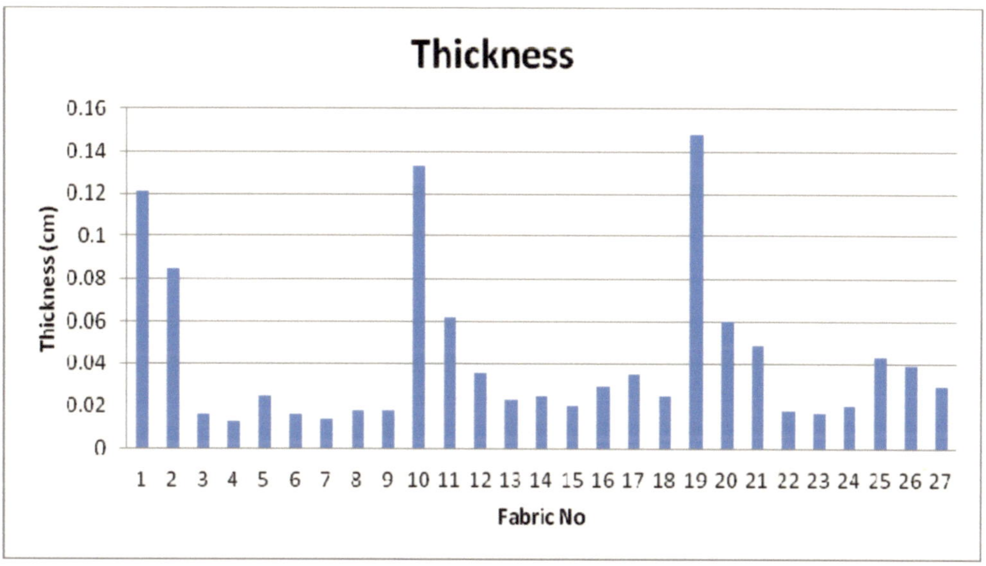

Fig. 1. Thickness values of polypropylene meltblown nonwovens

Statistical analyses also proved that the drum speed and collector vacuum on the thickness of meltblown nonwovens were statistically significant. The results of the tests between subgroups were shown in Table 3 and Table 4. As it can be seen in Table 3 and Figure 1, as a general trend the thickness increased with decreasing collector drum speed. Even though the thicknesses of the samples produced with 30 ft/min which were 0,0278 cm on average are slightly higher than the ones produced with 20 ft/min which were 0,0197 cm, the main trend did not change. The highest results were obtained with 10 ft/min as 0,0789 cm and the lowest results were obtained with 20 ft/min as 0,0197 cm.

As it can be seen in Table 4 and Figure 1, the effect of collector vacuum on the thickness was found to be statistically significant and the thickness increased with decreasing collector vacuum. The highest thickness results were obtained with 15% as 0,0602 cm and the lowest results were obtained with 60 ft/min as 0,0254 cm.

Collector DrumSpeed (ft/min)	N	Subset		
		1	2	3
20	45	,0197		
30	45		,0278	
10	45			,0789
Sig.		1,000	1,000	1,000
Alpha = 0,05				

Table 3. Student-Newman-Keuls test results related with the effect of collector drum speed on to the thickness

Collector Vacuum (%)	N	Subset		
		1	2	3
60	45	,0254		
30	45		,0407	
15	45			,0602
Sig.		1,000	1,000	1,000
Alpha = 0,05				

Table 4. Student-Newman-Kleus test results related with the effect of collector vacuum on to the thickness

The effect of air pressure on the thickness was also found to be statistically significant, when the difference between 6 psi and 8 psi were considered. Student-Newman-Keuls test results showed that an increase in the die air pressure caused the increase in the thickness. The results of the tests between subgroups were presented in Table 5. As it can be seen in Table 5, the thickness value increased from 0.362 cm to 0.431 cm by the increase of the die air pressure from 6 psi to 7 psi. Similarly, when the pressure increased to 8 psi, the average thickness increased to 0,047 cm.

Die Air Pressure (psi)	N	Subset	
		1	2
6	45	,0362	
7	45	,0431	,0431
8	45		,0470
Sig.		,061	,287
Alpha = 0,05			

Table 5. Student-Newman-Keuls test results related with the effect of die air pressure on to the thickness

The results obtained from the 2-Tailed Pearson Correlation test were given in Table 6. The results showed that there was a positive correlation of 19,7% between the die air pressure and the thickness, which means that increase in the die air pressure caused an increase in the thickness in 19,7%.

Correlations		Extruder Pressure	Thickness
Extruder pressure (psi)	Pearson Correlation	1	,197*
	Sig. (2-tailed)		,022
	N	135	135
Thickness (cm)	Pearson Correlation	,197*	1
	Sig. (2-tailed)	,022	
	N	135	135
*. Correlation is significant at the 0.05 level (2-tailed).			

Table 6. 2-Tailed Pearson Correlation test results related with the effect of extruder pressure on to the thickness

3.2 Basis weight

The results obtained from the basis weight measurements were given in Figure 2. The basis weight of the polypropylene meltblown nonwovens were effected mostly by the die air pressure, collector drum speed and collector vacuum.

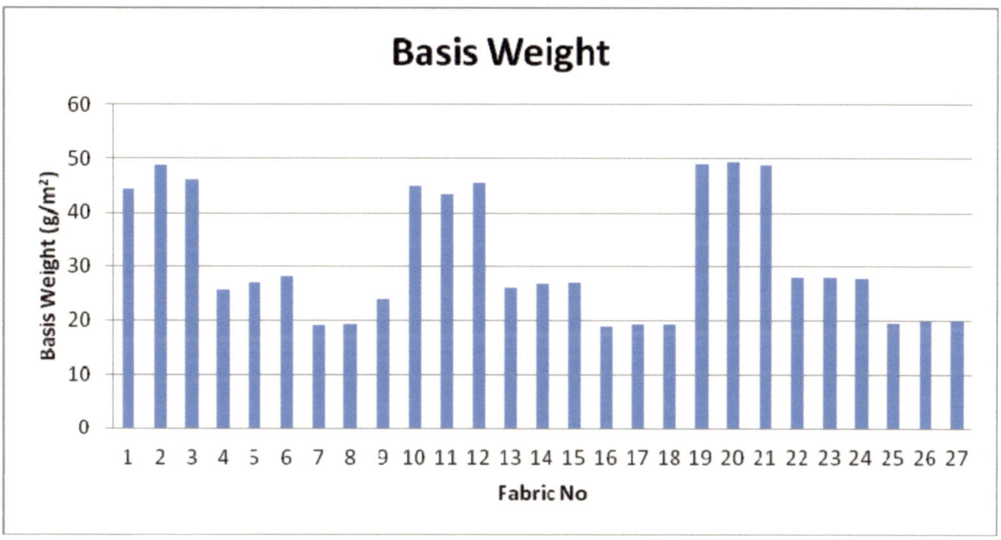

Fig. 2. Basis weight values of polypropylene meltblown nonwovens

Collector Drum Speed (ft/min)	N	Subset		
		1	2	3
30	45	137,033		
20	45		190,824	
10	45			239,769
Sig.		1,000	1,000	1,000
Alpha = 0,05				

Table 7. Student-Newman-Keuls test results related with the effect of the collector drum speed on to the basis weight

Table 7 and Figure 2 showed that the basis weight increased gradually with decreasing collector drum speed. The highest values were obtained with 10 ft/min collector drum speed as 239,769 g/m^2 on average and the lowest values were obtained with 30 ft/min collector drum speed as 137,033 g/m^2 on average.

As it can be seen in Figure 2 and Table 8, basis weight increased also with increasing die air pressure. The highest values were obtained with 8 psi as 322,711 g/m^2, whereas the lowest values were obtained with 6 psi pressure as 31,411 g/m^2.

Die Air Pressure (psi)	N	Subset		
		1	2	3
6	45	31,411		
7	45		213,504	
8	45			322,711
Sig.		1,000	1,000	1,000
Alpha = 0,05				

Table 8. Student-Newman-Keuls test results related with the effect of the die air pressure on to the basis weight

Student-Newman-Keuls tests have also shown that the collector vacuum had a significant effect on basis weight and also that when the vacuum increased the weight also increased. So the highest values were obtained with 60% vacuum as 220,720 g/m and the lowest values were obtained with 15% vacuum as 171,902 g/m as it can be seen in Table 9 and Figure 2.

Collector Vacuum (%)	N	Subset		
		1	2	3
15	45	171,902		
30	45		175,004	
60	45			220,720
Sig.		1,000	1,000	1,000
Alpha = 0,05				

Table 9. Student-Newman-Kleus test results related with the effect of the collector vacuum on to the basis weight

Correlations		Extruder Pressure	Weight
Extruder Pressure (psi)	Pearson Correlation	1	,151
	Sig. (2-tailed)		,080
	N	135	135
Weight (g/m)	Pearson Correlation	,151	1
	Sig. (2-tailed)	,080	
	N	135	135

Table 10. 2-Tailed Pearson Correlation test results related with the effect of extruder pressure on to the basis weight

The results of the 2-Tailed Pearson Correlation test applied to see the relationship between the extruder pressure and the basis weight were given in Table 10. The results showed that there was not any significant correlation, which means that the effect of extruder pressure on the basis weight was not statistically significant.

3.3 Air permeability

Air permeability is an important property for meltblown nonwovens, that effect their performance in many applications especially in filtration. The air permeability property of the meltblown nonwovens are mostly effected by the die air pressure, the collector drum speed, the collector vacuum and extruder pressure. The results obtained from air permeability results were presented in Figure 3.

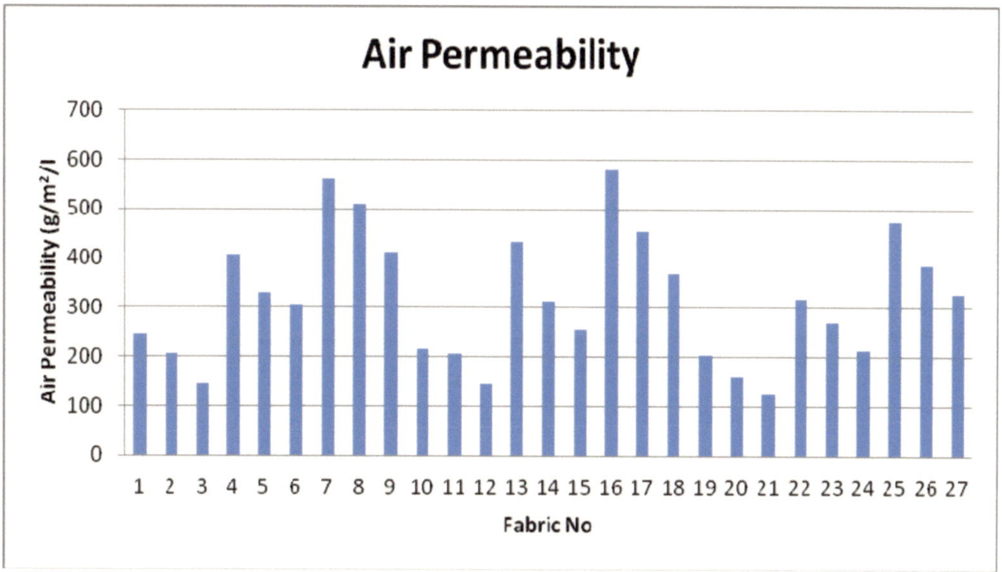

Fig. 3. Air permeability values of polypropylene meltblown nonwovens

As it can be seen in Table 11, the die air pressure had a significant effect on the air permeability of the polypropylene meltblown nonwovens. Table 11 and Figure 3 showed that the air permeability increased with increasing die air pressure. It can be also said that the air permeability was effected by the weight of the sample and it increased by decreasing weight. The highest results were obtained with 6 psi as 347,065 g/m²/l, whereas the lowest values were obtained with 8 psi as 275,444 g/m²/l.

Die Air Pressure	N	Subset		
		1	2	3
8	45	275,444		
7	45		330,789	
6	45			347,056
Sig.		1,000	1,000	1,000

Alpha = 0,05

Table 11. Student-Newman-Kleus test results related with the effect of the die air pressure on to the air permeability

Figure 3 and Table 12 show that, the air permeability increased with increasing collector drum speed. This result was thought to be related to the decrease in thickness and basis weight. The highest results were obtained with 30 ft/min as 452,267 g/m²/l, whereas the lowest values were obtained with 10 ft/min as 184,122 g/m²/l.

Collector Drum Speed	N	Subset		
		1	2	3
10	45	184,122		
20	45		316,900	
30	45			452,267
Sig.		1,000	1,000	1,000
c. Alpha =0,05				

Table 12. Student-Newman-Kleus test results related with the effect of the collector drum speed on to the air permeability

Regarding the effect of the collector vacuum on the air permeability, it can be said that the air permeability increased with decreasing vacuum. This can be due to the increase in the basis weight with increasing vacuum. The highest results were obtained with 15% as 382,156 g/m²/l, whereas the lowest values were obtained with 60% as 255,689 g/m²/l.

The results of the statistical subgroup analysis about the effect of the collector vacuum on the air permeability were presented in Table 13.

Collector Vacuum	N	Subset		
		1	2	3
60	45	255,689		
30	45		315,444	
15	45			382,156
Sig.		1,000	1,000	1,000
Alpha = 0,05				

Table 13. Student-Newman-Kleus test results related with the effect of the collector drum speed on to the air permeability

Correlations		Extruder Pressure	Air Permeability
Extruder_pressure	Pearson Correlation	1	-,174*
	Sig. (2-tailed)		,044
	N	135	135
Air_permeability	Pearson Correlation	-,174*	1
	Sig. (2-tailed)	,044	
	N	135	135
*. Correlation is significant at the 0.05 level (2-tailed).			

Table 14. 2-Tailed Pearson Correlation test results related with the effect of the extruder pressure on to the air permeability

Table 14 shows the correlation test results regarding the effect of the extruder pressure on the air permeability. As it can be seen in the table, the correlation was found to be statistically significant, evet though it was not very strong considering the value. It can be commented that there was a negative correlation of 17,4% between the extruder pressure and the air permeability. The air permeability decreased with the increasing extruder pressure.

3.4 Fibre diameter

Fibre diameter of meltblown nonwovens plays a critical role in some physical properties of the meltblown nonwovens, as it determines the surface area, which is a very important parameter for such applications as filtration and cleaning. Better filtration and cleaning performances are achieved with smaller fibre diameter, due to the increased surface are. Fibre diameter was effected by the collector vacuum and the extruder pressure. Figure 4 showed the fibre diameter properties of polypropylene meltblown nonwovens. In this study meltblown nonwovens with fibre diameter of 5-9 μm were achieved.

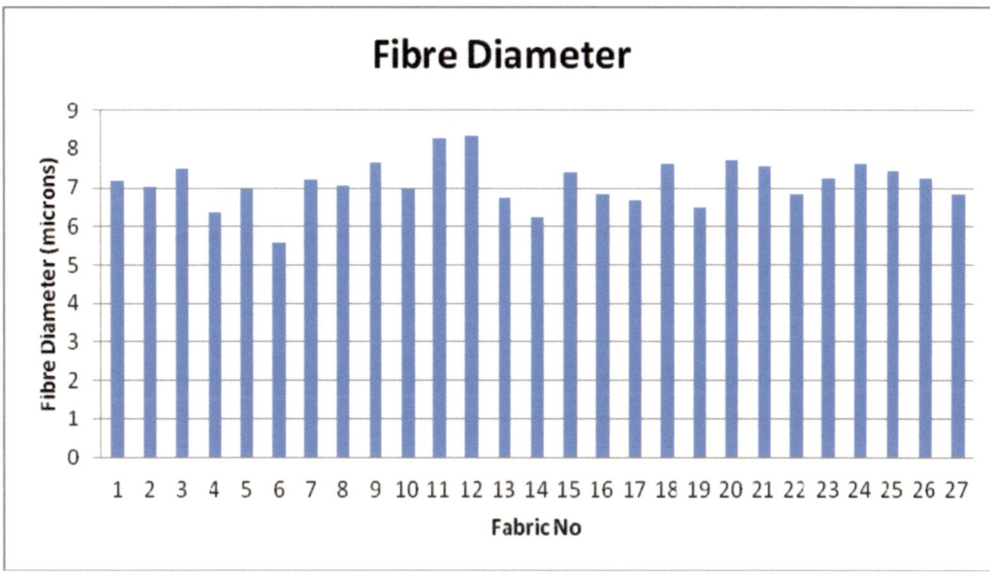

Fig. 4. Fibre diameter values of polypropylene meltblown nonwovens

As it can be seen in Table 15, results of the statistical analysis have shown that the die air pressure did not have a significant effect on the fibre diameter. It is also possible that the effect was not clear since the die air pressure values were very close to each other.

Die Air Pressure	N	Subset
		1
6	45	6,944
8	45	7,224
7	45	7,235
Sig.		,055
Alpha = 0,05.		

Table 15. Student-Newman-Kleus test results related with the effect of the die air pressure on to the air fibre diameter

The effect of the collector vacuum on the fibre diameter was found to be statistically significant. As it can be seen in Table 16, the fibre diameter slightly increased from 6,89 μm to 7,15 μm, when the collector vacuum increased from 15% to 30%. The reason of this increase might be the increasing pressure on the fibres. It can also be seen in Figure 4 and Table 15 that the fibre diameter did not change significantly with an increase in the collector vacuum from 30% to 60%. The highest results were obtained with 60% as 7,345 μm, whereas the lowest values were obtained with 15% as 6,900 μm.

Collector Vacuum	N	Subset	
		1	2
15	45	6,900	
30	45		7,159
60	45		7,345
Sig.		1,000	,140
Alpha = 0,05			

Table 16. Student-Newman-Keuls test results related with the effect of the collector vacuum on to the fibre diameter

Table 17 shows the correlation between the extruder pressure and the fibre diameter. As it can be seen in the table, the correlation was not found to be statistically significant, so the fibre diameter of the meltblown nonwovens investigated in this study were not influenced by the extruder pressure.

Correlations		Extruder Pressure	Fibre Diameter
Extruder Pressure	Pearson Correlation	1	,025
	Sig. (2-tailed)		,769
	N	135	135
Fibre Diameter	Pearson Correlation	,025	1
	Sig. (2-tailed)	,769	
	N	135	135
**. Correlation is significant at the 0.01 level (2-tailed).			

Table 17. 2-Tailed Pearson Correlation test results related with the effect of the extruder pressure on to the fibre diameter

3.5 Tensile properties

Breaking load and % elongation were investigated to evaluate the tensile properties of the meltblown nonwovens. The results of the measurements have shown that the breaking load and the elongation were significantly affected by the collector drum speed, collector vacuum, die air pressure and extruder pressure in production direction, where a the extruder pressure did not appear to be a significant factor for the tensile properties in the width direction.

Figure 5 shows the breaking load values of the polypropylene meltblown nonwovens in production and width directions. It can be seen in Figure 5 that the breaking load results in production direction were slightly higher than the results in the width direction. This was because orientation of the fibre were more towards the production direction around the collector drum and therefore the strength was more enhanced in this direction.

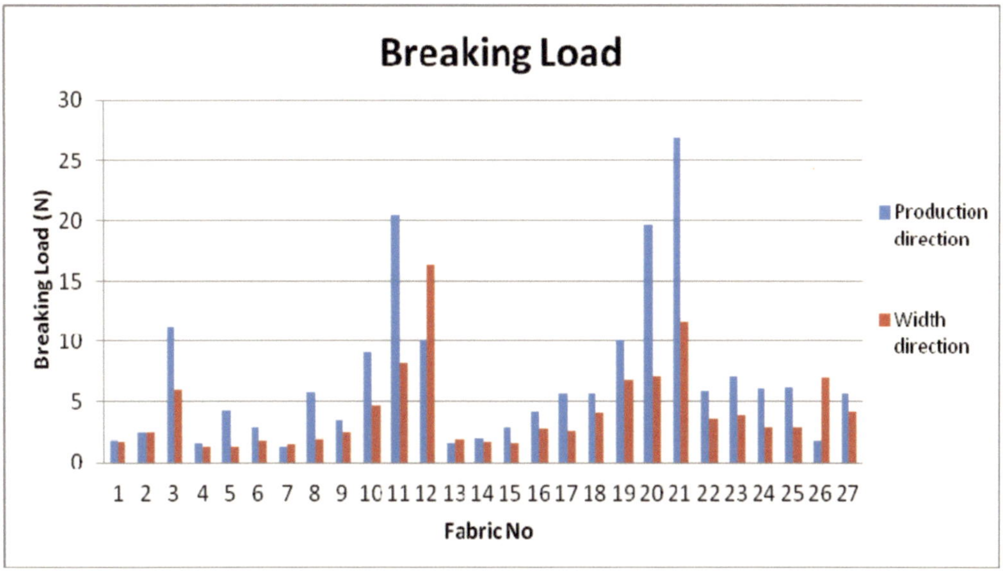

Fig. 5. Breaking load values of polypropylene meltblown nonwovens

Figure 6 shows the elongation values in the production and the width directions. As it can be seen in Figure 6, the elongation results in production direction were slightly lower than the results in the width direction. In general, the elongation decreased with increasing breaking load.

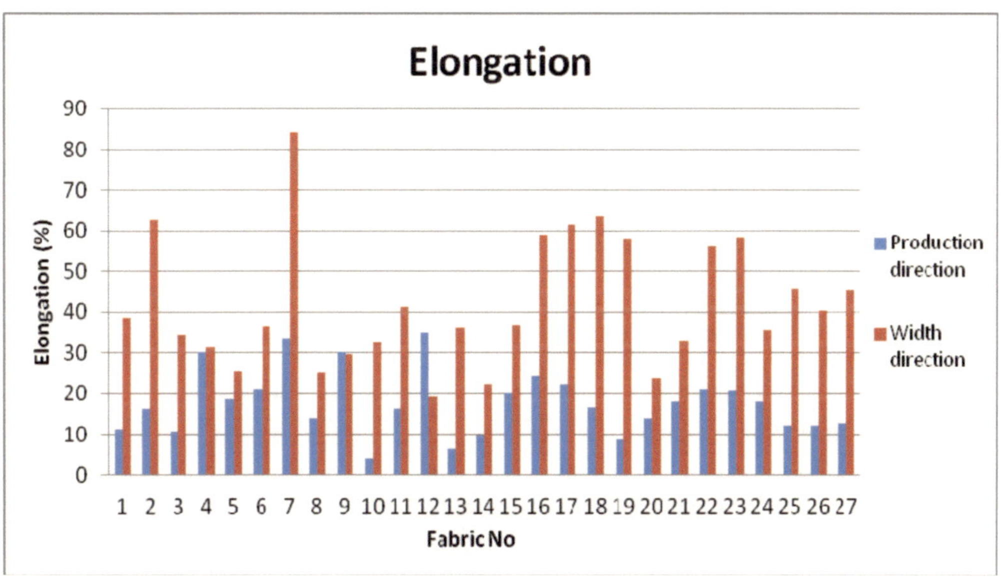

Fig. 6. Elongation values of polypropylene meltblown nonwovens

It can be seen in Figure 5 and Table 18 that the breaking load increased with increasing die air pressure both in the production and the width directions. This was caused by the increase in basis weight and thickness. The highest values were obtained with 8 psi as 9,904 N and the lowest values were obtained with 6 psi pressure as 3,859 N in the production direction, whereas highest values were obtained with 8 psi as 5,555 N and the lowest values were obtained with 6 psi pressure as 2,252 N in the width direction.

Die Air Pressure	N	Subset (production direction)			Subset (width direction)		
		1	2	3	1	2	3
6	45	3,859			2,252		
7	45		6,832			4,870	
8	45			9,904			5,555
Sig.		1,000	1,000	1,000	1,000	1,000	1,000
Alpha = 0,05							

Table 18. Student-Newman-Keuls test results related with the effect of the die air pressure on to the breaking load in the production and width directions

Table 19 shows the effect of the die air pressure on elongation both in production and width directions. It can be seen in Table 19 and Figure 6 that elongation decreased with increasing die air pressure in the production direction, due to incresing breaking load. The difference between 6 psi and other die air pressures were found to cause a statistically significant difference in the elongation whereas the difference between 7 psi and 8 psi were not found to be statistically significant. In the width direction this trend was not valid; increasing pressure caused an increase in the elongation in this direction. The highest values were obtained with 6 psi as 20,640 % in the production direction and as 44,040 % in the width direction, whereas the lowest values were obtained with 8 psi as 15,310 % in the production direction and as 40,993 % in the width direction.

Die Air Pressure	N	Subset (production direction)		Die Air Pressure	N	Subset (width direction)	
		1	2			1	2
8	45	15,310		6	45	40,993	
7	45	17,137		7	45	41,380	
6	45		20,640	8	45		44,040
Sig.		,090	1,000	Sig.		,742	1,000
Alpha = 0,05							

Table 19. Student-Newman-Keuls test results related with the effect of the die air pressure on to the elongation in the production direction and width directions.

Table 20 shows the effect of collector drum speed on the breaking load in production and width directions. For both of the directions as a general trend the breaking load decreased with increasing collector drum speed, due to the decrease in the basis weight. It can be seen both in the production and the width directions that when the collector drum speed increased from 10 ft/min to 20 ft/min the breaking load decreased sharply and slightly increased again when the collector drum speed increased to 30 ft/min. The reason of this slight increase was thought to be the increase in fibre strength due to the drawing with higher speed. The highest values were obtained with 10 ft/min as 12,408 N in the production direction and as 7,211 N in the width direction, whereas the lowest values were obtained with 20 ft/min as 3,787 N in the production direction and as 2,212 N in the width direction.

Collector Drum Speed	N	Subset (production direction)			Subset (width direction)		
		1	2	3	1	2	3
20	45	3,787			2,212		
30	45		4,400			3,254	
10	45			12,408			7,211
Sig.		1,000	1,000	1,000	1,000	1,000	1,000
Alpha = 0,05							

Table 20. Student-Newman-Kleus test results related with the effect of the collector drum speed on to the breaking load in the production and width directions

The effect of the collector drum speed on the elongation can be seen in Figure 6 and Table 21. The increase in collector drum speed caused an increase in the elongation both in the production and width directions, due to the decreasing basis weight and breaking load. In the production direction significantly lower elongation results were obtained with 10 ft/min compared to the other drum speeds. The difference between the results obtained with 20 ft/min and 30 ft/min were not found to be statistically significant. In the width direction significantly higher elongation results were obtained with 30 ft/min compared to the other drum speeds, whereas the difference between the results obtained with 10 ft/min and 20 ft/min were not found to be statistically significant. The highest values were obtained with 30 ft/min as 19,780% in the production direction and as 50,539% in the width direction, whereas the lowest values were obtained with 10 ft/min as 14,952% in the production direction and as 37,675% in the width direction.

Collector Drum Speed	N	Subset (production direction)		Collector Drum Speed	N	Subset (width direction)	
		1	2			1	2
10	45	14,952		20	45	37,675	
20	45		18,355	10	45	38,199	
30	45		19,780	30	45		50,539
Sig.		1,000	,185	Sig.		,656	1,000

Alpha = 0,05

Table 21. Student-Newman-Kleus test results related with the effect of the collector drum speed on to the elongation in the production direction and width directions.

The effect of the collector vacuum on the breaking load was shown in Table 22 and Figure 5. The breaking load increased with increasing vacuum in both directions, due to stronger bounding of the fibres in the web as a result of the increasing air pressure applied to the web by the vacuum. The highest values were obtained with 60% as 8,309 N in the production direction and as 5,654 N in the width direction, whereas the lowest values were obtained with 15% as 4,613 N in the production direction and as 2,994 N in the width direction.

Collector Vacuum	N	Subset (production direction)			Subset (width direction)		
		1	2	3	1	2	3
15	45	4,613			2,994		
30	45		7,673			4,029	
60	45			8,309			5,654
Sig.		1,000	1,000	1,000	1,000	1,000	1,000
Alpha = 0,05							

Table 22. Student-Newman-Kleus test results related with the effect of the collector vacuum on to the breaking load in the production direction.

Table 23 shows the effect of collector vacuum on the elongation in the production and the width directions. As it can be seen in Figure 6 and Table 23, the elongation increased with increasing collector vacuum in both directions, even though there has been a slight decrease in 30% in the width direction. The highest values were obtained with 60% vacuum as 20,253% elongation in the production direction and as 49,139% elongation in the width direction, whereas the lowest values were obtained with 15% vacuum as 15,978% elongation in the production direction and as 37,181% elongation in the width direction.

Collector Vacuum	N	Subset (production direction)		Collector Vacuum	N	Subset (width direction)		
		1	2			1	2	3
15	45	15,978		20	45	37,181		
30	45	16,856		10	45		40,094	
60	45		20,253	30	45			49,139
Sig.		,412	1,000	Sig.		1,000	1,000	1,000

Alpha = 0,05

Table 23. Student-Newman-Kleus test results related with the effect of the collector vacuum on to the elongation in the production direction and width directions

The correlation test results regarding the effect of the extruder pressure on the breaking load in the production and the width directions were presented in Table 24. As it can be seen in the table, there was a significant positive correlation of 33,3% between the extruder pressure and the breaking load in the production direction, whereas the correlation in the width direction was not found to be statistically significant. In other words, in the production direction the breaking load increased with increasing extruder pressure, but it did not have a significant effect in the width direction. Since the fibres were oriented more towards the production direction, the increase in the extruder pressure caused an increase in the fiber strength and it was reflected to the strength of the web in the production direction, but not in the width direction.

	Correlations		Extruder Pressure	Breaking Load
Production Direction	Extruder Pressure	Pearson Correlation	1	,333**
		Sig. (2-tailed)		,000
		N	135	135
	Breaking Load	Pearson Correlation	,333**	1
		Sig. (2-tailed)	,000	
		N	135	135
Width Direction	Extruder Pressure	Pearson Correlation	1	-,002
		Sig. (2-tailed)		,979
		N	135	135
	Breaking Load	Pearson Correlation	-,002	1
		Sig. (2-tailed)	,979	
		N	135	135

**. Correlation is significant at the 0.01 level (2-tailed).

Table 24. 2-Tailed Pearson Correlation test results related with the effect of the extruder pressure on to the breaking load in the production and width directions

Table 25 shows the correlations between the extruder pressure and elongation of the meltblown webs in the production and the width directions. As it can be seen in the table, there was a significant negative correlation of 30,5% between these two factors in the production direction, which means that the elongation decreased with increasing extruder pressure in this direction. The correlations between the extruder pressure and elongation were not found to be statistically significant in the width direction.

	Correlations		Extruder Pressure	Breaking Load
Production Direction	Extruder Pressure	Pearson Correlation	1	-,305**
		Sig. (2-tailed)		,000
		N	135	135
	Elongation	Pearson Correlation	-,305**	1
		Sig. (2-tailed)	,000	
		N	135	135
Width Direction	Extruder Pressure	Pearson Correlation	1	,124
		Sig. (2-tailed)		,153
		N	135	135
	Elongation	Pearson Correlation	,124	1
		Sig. (2-tailed)	,153	
		N	135	135

**. Correlation is significant at the 0.01 level (2-tailed).

Table 25. 2-Tailed Pearson Correlation test results related with the effect of the extruder pressure on to the elongation in the production direction

4. Conclusion

The MB technique for making nonwoven products has been forecast in recent years as one of the fastest-growing in the nonwovens industry. With the current expansion and interest, a strong and bright future is forecasted for this technology. The scope and utility of this technology will increase and meltblowing will become a major technique in nonwoven technology. The application of speciality polymer structures will no doubt offer new nonwoven materials unobtainable by other competitive technologies. (Dahiya et al., 2004)

In this chapter the results of a study regarding the investigations of the effect of die air pressure, extruder pressure, collector drum speed, and collector vacuum on the physical properties, namely thickness, basis weight, air permeability, fiber diameter and tensile properties of polypropylene meltblown nonwoven webs were presented.

The results have shown that thickness of polypropylene meltblown nonwovens were effected mostly by the drum speed and the collector vacuum. An increase in the drum speed and an increase in the vacuum caused a decrease in the thickness. Thicker surfaces were obtained with lower collector drum speeds and lover vacuum values.

The basis weight of the polypropylene meltblown nonwovens were mostly influenced by the die air pressure, collector drum speed and collector vacuum. The basis weight increased gradually with decreasing collector drum speed and increased with increasing die air pressure. The collector vacuum had a significant effect on basis weight; when the vacuum increased the basis weight also increased. The effect of extruder pressure on the basis weight was not statistically significant.

Air permeability is an important property for meltblown nonwovens, that effect their performance in many applications especially in filtration. The air permeability property of the meltblown nonwovens were influenced by the die air pressure, the collector drum speed, the collector vacuum and extruder pressure. The air permeability increased with increasing die air pressure, collector drum speed and decreasing vacuum. The air permeability decreased with the increasing extruder pressure.

Fibre diameter of meltblown nonwovens is a very important parameter for such applications as filtration and cleaning. Fibre diameter was effected by the collector vacuum and the extruder pressure. The die air pressure did not have a significant effect on the fibre diameter. The fibre diameter slightly increased, when the collector vacuum increased from 15% to 30%, but it did not change significantly with an increase in the collector vacuum from 30% to 60%. The fibre diameter of the meltblown nonwovens investigated in this study were not affected by the extruder pressure.

Breaking load and elongation were significantly influenced by the collector drum speed, collector vacuum, die air pressure and extruder pressure in production direction. The extruder pressure did not appear to be a significant factor for the tensile properties in the width direction. The breaking load results in production direction were slightly higher and therefore the elongation results were slightly lower than the results in the width direction. This is because orientation of the fibres were more towards the production direction around the collector drum and therefore the strength were more enhanced in this direction. The breaking load increased with increasing die air pressure both in the production and the width directions, due to the increase in basis weight and thickness. The elongation decreased with increasing die air pressure in the production direction, due to incresing breaking load. In the width direction this trend was not valid; increasing pressure caused an increase in the elongation in this direction. For both of the directions as a general trend the breaking load decreased with increasing collector drum speed, due to the decrease in the basis weight. The increase in collector drum speed caused an increase in the elongation both in the production and width directions, due to the decreasing basis weight and breaking load. The breaking load increased with increasing vacuum in both directions, due to stronger bounding of the fibres in the web as a result of the increasing air pressure applied to the web by the vacuum. The elongation increased with increasing collector vacuum in both directions. In the production direction the breaking load increased with increasing extruder pressure, but it didn't have a significant effect in the width direction. The elongation decreased with increasing extruder pressure in the production direction, whereas the correlations between the extruder pressure and elongation were not found to be statistically significant in the width direction.

Polypropylene meltblown nonwovens can be used in various application areas such as surgical face masks filter media, liquid and gaseous filtration, cartridge filters, clean room filters, hot melt adhesives, cleaning wipes and others. The properties of such materials should be investigated deeply for different applications.

5. References

[1] Ebeling, H.; Fink, H. P.; Luo, M.& Geus, H. G. (2006). Cellulose Meltblown Nonwovens Using The Lyocell-Process, In: *Lenzinger Berichte 86,* 21.11.2011, Available from <http://www.lenzing.com/fileadmin/template/pdf/konzern/lenzinger_berichte/ausgabe_86_2006/LB_2006_Ebeling_16_ev.pdf>

[2] Duran, D.; Perincek, S. (2010). The Effect of Various Production Parameters on the Physical Properties of Polypropylene Meltblown Nonwovens, *Industria Textila,* Vol.61, No.3, 2010, pp. 117-123, ISSN: 12225347

[3] One Lee, B.; Anko, J.& Won Han, S.(2010). Characteristics of PP/PET Bicomponent Meltblown Nonwovens as Sound Absorbing Material, *Advanced Materials Research,* Vols. 123-125, August 2010, pp. 935-938, DOI: 10.4028/www.scientific.net/AMR.123-125.935

[4] Zhang, D.; Sun, C.; Beard, J.; Brown, H.; Carson, I.; Hwo, C., (2002). Innovative Polytrimethylene terephthalate (PTT) Polymers for Technical Nonwovens, *Journal of Industrial Textiles,* Vol. 31, No.3, January 2002, pp. 159-178, DOI: 10.1106/152808302025393

[5] Dutton, K. C. (2008). Overview and Analysis of the Meltblown Process and Parameters, *In: JTATM Journal of Textile and Apparel* Technology and Management, Vol. 6, No. 1, Fall 2008, 21.11.2011, Available from
< http://ojs.cnr.ncsu.edu/index.php/JTATM/article/viewFile/342/275>

[6] Bresee, R. R., Qureshi, U. A. (2004). Influence of Processing Conditions On Melt Blown Web Structure: Part 1 - DCD, In: *INJ Spring 2004,* pp. 49-55, 21.11.2011, Available from: http://www.inda.org/subscrip/inj04_1/p49-55-bresee.pdf

[7] Rupp, J. (2008), Spunbond and Meltblown Nonwovens, In: *Nonwovens and Technical Textiles, Textile World,* 21.11.2011, Available from:
http://www.textileworld.com/Articles/2008/May_2008/Nonwovens/Spunbond_x_Meltblown_Nonwovens.html

[8] Farer, R.; Seyam, A.M.; Ghosh, T.K.; Batra, S.K.; Grant, E.& Lee, G. (2003). Forming Shaped/Molded Structures by Integrating Meltblowing and Robotic Technologies,

Textile Research Journal, Vol. 73, No.1, January 2003, pp. 15-21, DOI: 10.1177/004051750307300103

[9] Dahiya A.; Kamath, M. G. & Hegde, R. R. (2004). Meltblown Technology, In: *The Official Website of the University of Tennessee*, 21.11.2011, Available from: http://web.utk.edu/~mse/Textiles/Melt%20Blown%20Technology.htm

13

Thermoplastic Extrusion in Food Processing

Caroline Joy Steel, Maria Gabriela Vernaza Leoro, Marcio Schmiele, Reinaldo Eduardo Ferreira and Yoon Kil Chang
University of Campinas
Brazil

1. Introduction

Extrusion cooking was first introduced in food and feed processing in the late 1950s. Since then, the systems involved have grown in popularity, efficiency and flexibility. Extrusion cooking technology is most used for cereal and protein processing in the food industry and is closely related to the pet food and feed sectors. In the last decade, the development of extruders has evolved to yield sophisticated products, new flavour generation, encapsulation and sterilisation.

Thermoplastic extrusion is considered a HTST (High-Temperature, Short-Time) process in the food industry, and it permits, with little or no modification of the basic equipments and appropriate process control, the production of a great variety of food and feed products (Camire et al., 1990; Chang et al., 2001; El-Dash, 1981). This technique has been widely used with raw materials such as corn, wheat, rice and, especially in recent years, with soy (Chang et al., 2001; Kadan & Pepperman, 2002).

Depending on the raw materials and of the characteristics desired for the final product, extruders operate with low, medium or high shear; however, thermoplastic extruders are used for high shear. As examples, pasta and processed meat products are produced with low shear (cold extrusion); meat analogues and some pet foods are produced with medium shear; and expanded snack products, breakfast cereals and textured vegetable proteins are produced with high shear (thermoplastic extrusion).

For Fellows (2000), the two main factors that influence the characteristics of extruded products are: raw material characteristics and operational conditions of the extruder. As main characteristics of the raw material, the following can be highlighted: type of material, moisture content, physical state, chemical composition (quantity and type of starch, proteins, fats and sugars) and pH of the material. The operational parameters that can be pointed out as important are: temperature, pressure, die diameter and shear force, with the latter being influenced by the internal design of the extruder and by its length; as well as screw geometry and rotation speed.

Guy (2001) and Stanley (1986) relate the following advantages to the thermoplastic extrusion process: versatility, low costs, high production yields, good quality products and no effluents.

2. Equipments

The use of thermoplastic extrusion in food processing is facilitated by the dynamism of extruders, which can be divided into two types: single-screw and twin-screw extruders (Riaz, 2000).

Extruders are composed of five main parts: (i) the pre-conditioning system; (ii) the feeding system; (iii) the screw or worm; (iv) the barrel; (v) the die and the cutting mechanism (El-Dash, 1981), which can be seen in Figure 1. Also, they can vary with respect to screw, barrel and die configuration. The selection of each of these items will depend on the raw material used and the final product desired (Riaz, 2000).

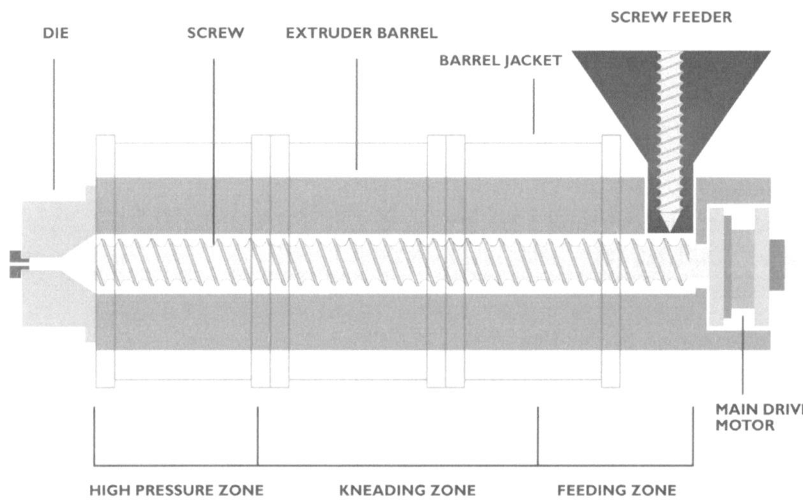

Fig. 1. Schematic representation of an extruder including its main parts and zones

In the extrusion process, the dry or pre-conditioned material (generally between 15 and 30% moisture content) is fed to the extruder through a screw feeder, reaching the feeding zone. The screw in this zone presents greater depth and pitch of the worm flight, and has as main function the transportation and homogenizing of the raw material. The material is conducted from the feeding zone to the compression zone. In the compression zone, there is a reduction in screw depth and pitch, with a consequent increase in shear rate, temperature (110 - 180°C) and pressure (20 – 30 atm). In this zone, the conversion from a solid material to a fluid melt starts to occur. In the subsequent high pressure zone, the screw has its depth and pitch reduced even more, resulting in higher shear and maximum heat generation. Thus, the extruded mass reaches maximum temperature and pressure and a reduction in viscosity immediately before exiting the extruder (Fellows, 2000; Riaz, 2000). The material, under high pressure, is expelled through the die and, in contact with ambient pressure, expands to its final format and cools rapidly through water flash-off (Fellow, 2000). In material that is not previously conditioned, water is added, in liquid or vapour form, during the process (El-Dash, 1981). The product that leaves the extruder is generally submitted to a drying process, reaching values close to 3% moisture content, as is the case of extruded snacks (Riaz, 2000).

2.1 Pre-conditioning

Pre-conditioning with steam or water has always been an important part of the extrusion process. Recent research has shown that efficient throughput of the extruder is almost doubled if the starting material is pre-conditioned with steam or water (Guy, 2001). There are many applications of extruded cooked food products where pre-conditioning plays a key role in the overall extrusion process. These products include direct expanded and flaked breakfast cereals, pre-cooked pasta, textured vegetable proteins, meat analogues, extruded bread crumbs, and third-generation snacks.

Pre-conditioning is not applied to all extrusion processes. In general, this step is applied when moisture contents around 20 to 30% and long residence times of the material are used. Pre-conditioning favours uniform particle hydration, reduces retention times within the extruder and increases throughput, increasing the life of the equipment, due to a reduction in the wearing of barrel and screw components, also reducing the costs of energy involved in the process (Huber & Rokey, 1990). Depending on screw configuration, the residence time of the material inside the extruder can vary from 5 seconds to more than 2 minutes, with the average residence time of the material in the pre-conditioner being 3 minutes. Pre-conditioning occurs with the addition of hot water (80-90°C) or steam, through spray nozzles, with the use of steam reducing energy consumption of the equipment up to 60% during the process. The most commonly used pre-conditioners have 2 axis of different diameters and rotation speeds, guaranteeing a residence time between 2 to 4 minutes and a production capacity between 300 and 18,000 $kg.h^{-1}$. When it is necessary to add melted fats or oils during pre-conditioning, it is best to do it at the end of the equipment, because if addition is done at the beginning, a coating may be formed over the particles, making water penetration more difficult. The main aim of pre-conditioning is to uniformly hydrate the raw material in order to eliminate any dry core (Strahm, 2000).

2.2 Feeding system

Most raw materials used in food extrusion are solid. The feeding system is normally composed of a holding bin where the material is loaded and the discharge of the material can occur through a vertical feeding screw, a horizontal feeding screw, a horizontal vibrating trough system, a disk feeder or a volumetric belt feeder. It is necessary to guarantee a constant and non-interrupted feeding of the raw materials into the extruder for an efficient and uniform functioning of the extrusion process (El-Dash, 1981). When liquids are added, they can be dosed using a rotameter, orifice and Venturi meters, positive displacement meter, magnetic flow meter or metering pumps (Chessari & Sellahewa, 2000).

2.3 Screw

The screw of the extruder is certainly its most important component, not only to determine cooking degree, gelatinization and dextrinization of starch and protein denaturation, but also to ensure final product quality. Screws can be mono-piece (composed of a unique piece) or multi-piece (composed of various elements) (El-Dash, 1981). Screw elements can vary in number and shapes, each segment is designed for a specific purpose. Some elements only convey raw or pre-conditioned material into the extruder barrel, while other segments compress and degas the feedstock. Others must promote kneading, backflow and shear.

Some kneading screws have interrupted flights to improve dispersive mixing, increase backflow, or increase mechanical energy dissipation into the extruder (Huber, 2001). Main characteristics of screw design include: (i) screw length; (ii) screw diameter; (iii) screw channel depth; (iv) screw channel width; (v) axial flight land width; (vi) clearance between screw and barrel; (vii) screw helix angle; (viii) leading flank angle; (ix) trailing flank angle; (x) screw pitch; (xi) direction of drag flow; (xii) direction of pressure flow and (xiii) direction of leakage flow (Figure 2) (El-Dash, 1981). The screw pitch (t) is the distance between corresponding points on adjacent thread profiles, and the number of parallel screw channels or leads (n) is defined as the number of screw pitches in the axis distance that the helix advances in one turn.

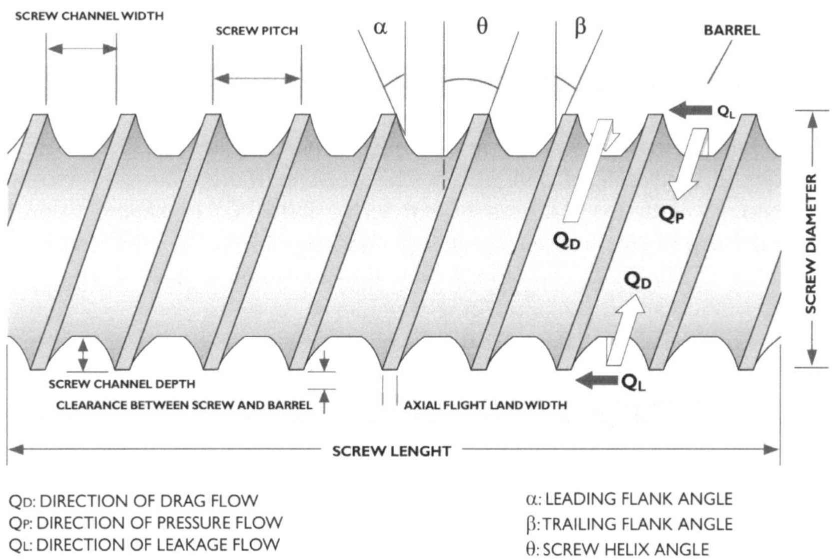

Fig. 2. Main characteristics of screw design

2.4 Barrel or sleeves

The barrel is divided into feeding, kneading and high pressure zones (Figure 1).

The sleeves surrounding the screw can be solid, but they are often jacketed to permit circulating of steam or superheated oil for heating or water or air for cooling, thus enabling the precise adjustment of the temperature in the various zones of the extruder. And most sleeves are equipped with pressure and temperature sensing and temperature control mechanisms as well (El-Dash, 1981).

In twin-screw extruders, the sleeves are usually smooth but can be constructed with longitudinal or helical grooves (Huber, 2000). In single-screw extruders, the sleeves are usually fluted on the inside, with either straight or spiral grooves. Parallel grooves are often cut or more often cast into the barrel. Spiral grooves provide high forward flow, while straight grooves hinder it. The latter thus result in a lower flow rate, but more mechanical shear. The clearance between the screw and its sleeve is usually kept to a minimum to reduce leakage flow (El-Dash, 1981).

2.5 Die

The die presents two main functions: give shape to the final product and promote resistance to material flow within the extruder permitting an increase in internal pressure. The die can present various designs and number of orifices (El-Dash, 1981). Dies may be designed to be highly restrictive, giving increased barrel fill, residence time and energy input. Die design and its effects on functional properties and quality of a final product are many times overlooked.

2.6 Cutting mechanism

The cutting mechanism must permit obtaining final products with uniform size. Product size is determined by the rotation speed of the cutting blades. This mechanism can be horizontal or vertical (El-Dash, 1981).

2.7 Types of extruders

Two types of extruders are used for food production: (i) single-screw extruders and (ii) twin-screw extruders. Single-screw extruders are the most common extruders used in the food industry. Twin-screw extruders are used for high-moisture extrusion, products that include higher quantities of components such as fibres, fats, etc. and more sophisticated products.

2.7.1 Single-screw extruders

Single-screw extruders are the most common extruders applied in the food industry. The classification of single-screw extruders can be defined based on process or equipment parameters such as: conditioning moisture content (dry or wet), solid or segmented screw, desired degree of shear and heat source. From a practical point of view, the main classification used considers the degree of shear and the heat source (Riaz, 2000).

Regarding screw configuration, there are screws made up of only one piece or screws of multiple pieces. Single element screws may present different configurations: (i) screw with constant depth and flight - straight -; (ii) screw with constant flight and variable depth - tapered - (conical from the feeding extremity to the die extremity); (iii) screw with a reduction in depth just after feeding, becoming constant at the end - tapered-straight - and (iv) screw with flight openings - interrupted flight - to increase shear force due to the increase in leakage flow and turbulence of the material (El-Dash, 1981).

Screws of multiple elements can be built up to desired configuration due to the great number of possible formats, varying screw flight and depth. Usually this type of screw is divided into five sections, where the first section presents wide flight and great depth with the objective of homogenizing and conveying the material. In the second section, also known as the intermediate section, there is a reduction in parallel screw flight (or adjacent screw flight) and depth, resulting in even greater mixing of the material and beginning of shear, while the material is transported to the next section. The third section is responsible for an increase in shear and pressure, promoting structural changes in the material, which passes to the viscoelastic state. The increase in shear force in the third section can be reached with interruptions in screw flight favouring material turbulence. In the fourth section, due to the small clearance available for the material, there is high shear and an increase in the temperature of the molten mass, resulting in cooking of the product. In the last section, due to an even greater reduction in screw flight and depth, shear and heat generation promote final cooking of the product (El-Dash, 1981).

Extrusion conditions when using extruders that have single screws of multiple elements can be controlled varying the number of sections, screw configuration in each section and through the inclusion of shearlocks (El-Dash, 1981).

Single-screw extruders can be classified in four different types based on the degree of shear, as follows:

Cold forming extruders - operate with moderate conditioning moisture contents (30 - 40%), low shear and smooth internal barrel surface, deep flight and low screw speed. These are not used for thermoplastic extrusion. They are used to form compact products such as pasta, cookies, pastry doughs, processed meats and certain candies (Riaz, 2000).

High-pressure forming extruders - operate with low shear, grooved barrel and compression screw. They are used to produce pre-gelatinized flours and pellets (for post expansion by hot air or frying) (Riaz, 2000). The latter are considered 3rd generation products.

Low-shear cooking extruders - operate with moderate degree of shear, high compression screw and grooved barrel (straight or helicoidal) to favour mixing. Usually involve external heating (steam jacket or electric resistance) to improve cooking with the objective of pasteurization, enzymatic inactivation, protein denaturation and/or starch gelatinization (Riaz, 2000).

Collet extruders - operate with high shear, grooved barrel and screw with an increase in compression (through multiple shallow flights). Commonly used to produce expanded snacks from corn grits. Conditioning moisture content must be low (12 - 14%) and temperature high (150 - 175°C), resulting in partial dextrinization and gelatinization of starch. Due to the high pressure formed within the extruder, when exiting the die there is immediate expansion of the material (Riaz, 2000).

2.7.2 Twin-screw extruders

Twin-screw extruders are composed of two axis that rotate inside a single barrel; usually the internal surface of the barrel of twin-screw extruders is smooth. Depending on the position of the screws and their direction of rotation, four different types of configurations are possible: (i) co-rotating intermeshing screws; (ii) co-rotating non-intermeshing screws; (iii) counter-rotating intermeshing screws; and (iv) counter-rotating non-intermeshing screws. Conical intermeshing extruders also exist. Although intermeshing screws result in greater residence time of the material in the extruder, non-intermeshing screws cause greater degrees of shear, especially if they rotate in opposite directions. However, this type of extruder is little used in the food industry, even though they present more efficient displacement properties (El-Dash, 1981). The intermeshing configuration is more effective, as the two screws function as a positive pump, increasing the drag flow and reducing the slipping of material in the extruder. Non-intermeshing screws provide higher shear than intermeshing screws because of the open channel between them.

When the material enters the barrel, the ingredients are thoroughly mixed before further processing in the other zones of the extruder. In this initial step, the screw is designed with a large screw channel depth to provide enough space between the root of the screw and the barrel for sufficient mixing to take place, and often, the screws are reverse-threaded to permit intensive mixing and longer residence times before delivery. In the next zone, the diameter of the root increases rapidly while the channel depth becomes shallower in order to provide material cooking, thus increasing the pressure applied to the product, and the starchy content of food is gelatinized and the proteinaceous material denatured (El-Dash, 1981).

When needed, after the cooking zone, the material is forced to the depressurizing zone were the screw root diameter is much smaller, while the channel depth is much deeper than in the previous zone. To promote pressure reduction, a depressurizing valve can be opened to atmospheric pressure or to a vacuum pump. In the forming zone, the diameter of the screw root increases, reducing channel depth and resulting in an increase in the pressure applied. This pressure must be high enough to permit the extrusion of the product in the appropriate form through the die (El-Dash, 1981).

3. Raw materials and changes in major components

3.1 Raw materials

The most used raw materials in the extrusion process are starch and protein based materials. The structure of the extruded products may be formed from starch or protein polymers. Most products, such as breakfast cereals, snacks and biscuits are formed from starch, while protein is used to produce products that have meat-like characteristics and that are used either as full or partial replacements for meat in ready meals, dried foods and many pet food products (Guy, 2001).

In general, the chemical or physicochemical changes in biopolymers that can occur during extrusion cooking include: binding, cleavage, loss of native conformation, fragment recombination and thermal degradation. The composition of raw materials can be altered by physical losses including leakage of oil and evaporation of water and volatile compounds at the die. Since most chemical reactions occur in the high-pressure zone of the barrel, thermally labile compounds such as flavours and vitamins may be injected immediately before the die to minimize exposure to heat and shear (Riaz, 2000).

The structure of an extruded product is created by forming a fluid melt from a polymer and blowing bubbles of water vapour into the fluid to form a foam. The bubbles rapidly expand as the superheated water is released very quickly at atmospheric pressure. In the extruded structure, the fluid melt of the polymers forms the cell walls of the gas bubbles. After gas expansion, the rapid drop of temperature caused by water evaporation and the rapid rise in viscosity due to moisture loss, solidifies the cell structure. The rapid increase in viscosity is followed by the formation of a glassy state. Starch polymers are very good at this function and also expand well. Structure forming polymers must have a minimum molecular weight sufficient to give enough fluid viscosity to prevent or control shrinkages of an extrudate after it reaches its maximum expansion (Guy, 2001).

3.2 Starch

The major difference between extrusion processing and conventional food processing is that in the former starch gelatinization occurs at much lower moisture contents (12-22%).

Starch is contained in a large variety of plant crops, such as cereals (50-80% db starch), legumes (25-50% db), and tubers (60-95% db) (Colonna et al., 1998). The three major cereals in order of world production are wheat, rice and maize; moreover, other important crops are barley, rye, oats and sorghum. All these cereals are available as grains, which are milled to form flours rich in starch once first separated the endosperm from the hull or pericarp (Guy, 2001). Starch is present in endosperm cells, is insoluble in cold water and its main nutritional property is to supply energy (4 kcal/g) (Caldwell et al., 2000; Cheftel, 1986). Each cereal has a different composition of its flour which basically depends on the level of the non-starch components such as protein and fibres. For example, maize and rice flours are generally richer in starch than wheat flour due to the lower protein and fibre contents. Oat flours are high in both oil and fibre presenting the lowest starch content (Guy, 2001). Native starch is found in the form of discrete particles or granules, with defined sizes and shapes depending on the botanical source. The starch granule consists of two different glucose polymers: amylose and amylopectin, responsible for its physicochemical and functional properties (Bornet, 1993; Caldwell et al., 2000). The highly branched structure of amylopectin is more prone to shear, but both amylose and amylopectin molecules may decrease in weight (Collona et al., 1998). Amylose is a basically linear polymer, with linear α-1-4 glucosidic bonds, polymerization degree of 600 to 6000 glucose units and molecular weight of 10^5-10^6 Da. Amylopectin is a branched polymer, with linear α-1-4 glucosidic bonds and, at branching points, α-1-6 glucosidic bonds, in a proportion of 5-6%; it consist of approximately 10^6 glucose units, with a molecular weight of 10^8 Da. Starch contains different proportions of amylose and amylopectin, depending on its botanical origin. Starch from cereals has an amylose content that varies from 15 to 28% (Bornet, 1993).

Thermoplastic extrusion, depending on process conditions and raw material composition, causes swelling and rupture of the starch granule, completely or partially destroying the organized granule structure, reducing viscosity and releasing amylose and amylopectin (Camire et al., 1990; El-Dash et al., 1983). During thermoplastic extrusion, amylose and amylopectin are partially hydrolysed to maltodextrins, due to the high temperatures and shear inside the extruder (Cheftel, 1986; El-Dash et al., 1983).

An important consequence of starch degradation is the reduction in expansion. Highly expanded products may crumble easily due to thin cell walls, while dense products are often hard (Riaz, 2000).

Larger amylopectin molecules in corn flour had the greatest molecular weight reduction. High molecular weight (>10^7) starches disappeared during extrusion, and there was a general increase in starch molecules of 10^5 - 10^7 (Guy, 2001).

The physical nature of the cereal flour may affect the final extruded product. Some cereals present soft and floury endosperm. In this material, the starch granules and protein layer are only loosely bound together and the endosperm is broken down easily on milling to provide a mixture of separated starch and protein bodies. In the extruder, soft flour will create less mechanical energy between its particles and require less mechanical energy to process through the same screw configuration, a longer time being necessary before melt formation and less time for the transformation of the melt in the shearing section (high-pressure zone). On the other hand, in certain cereals, such as hard wheat, durum wheat, vitreous flint maize and some varieties of barley there is a strong bonding between the starch granules and the protein layers forming a hard particle of flour that requires more energy to break down and will generate more heat in the extruder. Therefore, if high expansion is required in a low moisture product, finely milled forms of harder endosperm types will give excellent results. If the product requires low to medium expansion, some of the hard material may be replaced by soft flour; and for low expansion in a dense product such as breading crumb, soft flour may be used (Guy, 2001).

Inside the extruder, starch goes through several stages. First, the initial moisture content is very important to define the desired product type. Once inside the extruder, and at relatively high temperatures, the starch granules melt and become soft, besides changing their structure that is compressed to a flattened form (Guy, 2001). The application of heat, the action of shear on the starch granule and water content destroy the organized molecular structure, also resulting in molecular hydrolysis of the material (Mercier et al., 1998). The starch polymers are then dispersed and degraded to form a continuous fluid melt. The fluid polymer continuum retains water vapour bubbles and stretches during extrudate expansion until the rupture of cell structure. The starch polymer cell walls recoil and stiffen as they cool to stabilize the extrudate structure. Finally, the starch polymer becomes glassy as moisture is removed, forming a hard brittle texture (Guy, 2001). The final expanded product presents air cells that are formed due to superheated water vapour pressure. When the temperature of the extrudate is reduced below its glass transition temperature (Tg), it solidifies and maintains its expanded form (Riaz, 2000).

According to Colonna et al. (1998), maximum expansion degree is closely related to starch content. Maximum expansion is obtained with pure starches (an increase of 500% in product diameter), followed by whole grains (400%) and with lower expansions for seeds or germ (150-200%); the starch content of these products is 100, 65-78, 40-50 and 0-10, respectively. The minimum starch content for expansion is 60-70% (Riaz, 2000).

3.3 Proteins

Proteins are biopolymers with a great number of chemical groups when compared to polysaccharides and are therefore more reactive (Mitchel & Areas, 1992) and undergo many changes during the extrusion process, with the most important being denaturation (Camire, 2000). Proteins are formed from chains of amino acids and have a wide range of physical sizes and forms in native raw materials. Proteins in general are classified, with respect to their solubility, in albumins, globulins, prolamines and glutelins with solubility in water, saline solution, alcohol solution and acid or alkaline solutions, respectively (Pereda et al., 2005).

During extrusion, disulfide bonds are broken and may re-form. Electrostatic and hydrophobic interactions favour the formation of insoluble aggregates. The creation of new peptide bonds during extrusion is controversial. High molecular weight proteins can dissociate into smaller subunits (Guy, 2001).

Enzymes, also proteins, lose their activity after being submitted to the extrusion process due to high temperatures and shear. Also, proteins lose their solubility in water and saline solution due to the temperature and specific mechanical energy to which the product is submitted (Camire, 2000).

One of the main applications of extrusion in high protein content foods is protein texturization. Texturization processes by extrusion can be used to obtain products that imitate the texture, taste, and appearance of meat or seafood with high nutritional value (Cheftel et al., 1992).

The use of raw materials with high protein contents in extrusion began around the 1970s, with the use of soy for the production of texturized soy products and meat analogues (Ledward & Mitchell, 1988; Mitchell & Areas, 1992).

In extrusion, the proteins that have been found to form a continuous structure are globular proteins from oilseeds such as soybeans, sunflower seeds, common beans, peas and cottonseed and from cereals, especially wheat gluten proteins (Riaz, 2000; Strahm, 2006).

The extrusion process, physically, converts protein bodies into a homogeneous matrix, while chemically, the process recombines storage proteins in some way into structured fibres (Stanley, 1998). Low moisture (up to 35%) extrusion of vegetable protein can be used to elaborate products to partially or totally substitute meat. Usually, these products are expanded and need to be re-hydrated before consumption. On the other hand, high moisture (>50%) extrusion results in products that do not need to be re-hydrated and can be consumed directly. In general, dry extrusion is applied when the aim is to produce meat extenders and wet extrusion is used for meat analogues (Noguchi, 1998). In dry extrusion, when the conditioned material passes through the die at a high temperature, the water in the material is changed into superheated steam, which expands the extrudate immediately. Water also makes the extrudate very soft, by reducing its viscosity drastically, so the material just after the die is not self-supporting. Therefore, cooling at the die is essential to increase the viscosity of the hot melt and reduce its fluidity so that the necessary pressure and temperature before the die can be maintained. When cooling is done appropriately, the correct amount of extrudate elasticity and fluidity can be obtained to allow a continuous "rope" structure without explosive puffing and the destruction of product integrity (Noguchi, 1998). The mechanism for structure creation with proteins is similar to that with starch in the sense that proteins must be dispersed from their native bodies into a free flowing continuous mass. Texturization occurs between the molecules as they flow in the streamlines to form laminar cross-linked products. Evaporation of water in the mass creates gas bubbles that form alveolar structures held in place by cross-linking in the protein layers (Guy, 2001).

Denaturation during the extrusion process of proteins results in reduction of protein solubility, favours digestibility and inactivates antinutritional factors (such as antitrypsin factor, lectins, etc.). Also, the extrusion of soy protein reduces the bitter taste and the undesirable volatile compounds related to this protein (Areas, 1992; Kitabatake & Doi, 1992).

During extrusion, protein structures are disrupted and altered under high shear, pressure, and temperature (Harper, 1984). In the extrusion of proteins, disulfide bonds are cleaved and undergo reorganization and polymerization. Disulfide bonds, non-specific hydrophobic and electrostatic interactions are the main bonds and interactions responsible for protein texturization by extrusion (Areas, 1992). Protein solubility decreases and cross-linking reactions occur and possibly, some covalent bonds form at high temperatures (Areas, 1992). Thermal plasticization of the protein mix at high moisture contents (60%) is possible at relatively high extrusion temperatures (>150°C). At moisture levels lower than 60%, plasticization requires higher temperatures. Apart from this, hydrophobic and electrostatic interactions favour the formation of insoluble aggregates, like the fibrous structure of meat analogues, for example (Li & Lee, 1996; Tilley et al., 2000; Li-Chan, 2004; Sluimer, 2005).

Protein reactions, including both non-covalent and disulfide bonds, form upon cooling. Protein–protein interactions may be enhanced by decreased temperature and by macromolecular alignment. Crystalline aggregation leads to parallel fibre formation of varying length and thickness. A wide range of interaction energy is possible for protein cross-linking with protein and other molecules due to the diversity of amino acids. Therefore, hydrophobic, cation–mediated electrostatic interactions, and covalent bonds also contribute to the stabilization of the three dimensional network formed during extrusion (Areas, 1992).

Also, during the extrusion process high temperatures are normally used, and these favour the Maillard reaction. Reducing sugars can be produced during the process and they can react with the free amine groups of lysine or other amino acids (Camire, 2000).

3.4 Lipids

Fats and oils can be described as lipids. Lipids have a powerful influence in extrusion cooking processes by acting as lubricants, because they reduce the friction between particles in the mix and between the screw and barrel surfaces and the fluid melt (Guy, 2001). In the extruder, fats and oils become liquid at temperatures > 40°C, being mixed with the other materials, and are rapidly dispersed as fine oil droplets.

The presence of lipids in quantities lower than 3% does not affect expansion properties, however, in amounts above 5%, reduction in expansion rate is considerable (Harper, 1994). Collona et al. (1998) suggest that the increase in lipid content can be corrected through the reduction in conditioning moisture content, so as not to affect the expansion index of second generation products (directly expanded snacks).

The type of starch and lipid present in the raw material influences the formation of the amylose-lipid complex, with free fatty acids and monoglycerides being more favourable to the formation of this complex than triglycerides (Mitchel & Areas, 1992; Harper, 1994; Camire, 2000).

Moreover, in wet protein extrusion, the presence of lipids does not support protein fibre formation since the lubricating effect of lipids decreases the shear effects and particle alignment (Akdogan, 1999).

3.5 Fibres

The term "fibres" covers a great variety of substances with different physical, chemical and physiological properties. Dietary fibre consists of fractions of vegetable cells, polysaccharides, lignin, and associated substances, which are resistant to hydrolysis by enzymes present in the digestive system of humans; however, some types of fibres may be fermented by bacteria in the colon. As many physiological effects of fibres seem to be related to their solubility in water, they are frequently classified as "soluble" and "insoluble" (Roberfroid, 1993; Stark & Madar, 1994).

Soluble fibres form a gel network or a viscous network, in determined physicochemical conditions, and thus bond water increasing viscosity, retarding gastric transit, reducing glucose, lipid and sterol absorption rates (Gorinstein et al., 2001). Soluble fibres are also seen as fermentable substrates, as they can modify the pH and the microflora of the colon, leading to a reduction or modification of mutagenic agents (Thebaudin et al., 1997). Insoluble fibres increase faecal volume, thus diluting its contents, which reduces the interaction between the intestinal mucosa and any carcinogenic component present. Apart from this, insoluble fibres reduce intestinal transit time, avoiding that mutagenic agents in the faeces interact with the intestinal epithelium (Thebaudin et al., 1997).

Research has shown that cooking fibres by extrusion can produce changes in their structural characteristics and physicochemical properties, with the main effect being a redistribution of insoluble fibre to soluble fibre (Camire et al., 1990; Guillon et al., 1992; Larrea et al., 2005). This effect would be the result of the rupture of covalent and non-covalent bonds between carbohydrates and proteins associated to the fibre, resulting in smaller molecular fragments, that would be more soluble (Fornal et al., 1987; Wang et al., 1993).

Various researchers have reported a reduction in expansion index (EI) when dietary fibre is added to the formulation (Hsieh et al., 1989; Ilo et al., 1999; Vernaza et al., 2009). The reduction in the expansion index due to fibre addition can be explained through different mechanisms: (i) fibrous materials found in the formulation of extruded products include materials composed of hemicellulose, cellulose and lignin. In normal extrusion conditions, these materials tend to remain firm and stable during processing, without size reduction. The physical presence of fibres in air cell walls reduces the expansion potential of the starchy film (Guy, 2001); larger particles, such as bran, tend to rupture air cell walls of the extruded product, causing a reduction in expansion index (Riaz, 2000); (ii) according to Colonna et al. (1998), maximum degree of expansion is closely related to starch content, with maximum expansion being obtained for pure starches. As bran contains high fibre content, it reduces the starch content of the formulations; (iii) non-starch polysaccharides, such as fibres, may bind water more strongly than proteins and starch during extrusion. This water binding capacity inhibits water loss at the die, that is, at the exit of the extruder, reducing expansion (Camire & King, 1991); (iv) the starch present cannot be totally gelatinized in the presence of fibre and is thus not capable of supporting expansion (Camire & King, 1991); and (v) the porous structure of the extrudate depends of the plasticity of the mass before the die, for which starch is mainly responsible. Porosity, defined by the existence of fine pores and a tender structure, is influenced by alterations in the plasticity of the mass, affected by the composition of the mix. Formulations can be enriched by plasticizing substances or by non-plasticizing substances that retard expansion by diluting starch, as is the case of fibres (Colonna et al., 1998).

3.6 Moisture and temperature

In the extrusion process of expanded products with low moisture, the expansion of the final product is inversely related to the moisture of the raw material and directly related to the increase in extrusion temperature; however, the effect of moisture is more significant (Harper, 1994).

Water acts as a plasticizer for the starchy material that displaces itself within the extruder, reducing viscosity and mechanical energy, producing higher density products and inhibiting bubble growth. Studies carried out with corn grits demonstrated that expansion is inversely proportional to the moisture content of the material being extruded (Chinnaswamy, 1993; Colonna et al., 1998). With higher moisture, starch gelatinization is reduced and bubble growth is retarded, resulting in denser and less crunchy final products (Ding et al., 2005).

Inside the extruder, the product that contains molten starch in its composition, when leaving the extruder, has part of its water rapidly evaporated. This water loss is of 3 to 5% and contributes to cooling the product. Subsequent cooling occurs more slowly due to the low thermal conductivity of the extrudate. Also, when emerging from the die, the extrudate undergoes an abrupt pressure fall that also contributes to its expansion (Colonna et al., 1998). The expanded final product presents air cells that are formed due to superheated water vapour pressure. As the temperature of the extrudate is reduced below its glass transition temperature (Tg), it solidifies and maintains its expanded form (Riaz, 2000).

In high moisture extrudates, expansion occurs when the product exits the die, but the structure collapses before the necessary cooling, resulting in a dense and hard product (Harper, 1994).

Another important parameter for extrudate expansion is process temperature. Products do not expand if the temperature does not reach 100°C. Expansion increases with the increase in temperature when moisture content of the material is close to 20%, due to lower viscosity, permitting a more rapid expansion of the molten mass, or due to an increase in water vapour pressure. At low extrusion temperatures, expansion is reduced because starch is not completely molten. Radial expansion degree is proportional to temperature up to a certain value, decreasing at much higher temperatures. The reduction of expansion at very high temperatures is attributed to an increase in dextrinization, weakening starch structure (Colonna et al., 1998).

At high temperatures, the gel is more elastic, forming a matrix with small uniform cells, while at low moisture, the gel formed is not very elastic and the extruded material has large, not very uniform cells. It is expected that an increase in temperature should reduce viscosity of the molten material, favouring bubble growth and producing low density extrudates, with finer cells and greater crunchiness (Ding et al., 2005).

In high moisture extrusion, the properties of the protein extrudates are strongly influenced by extruder conditions (Thiebaud et al., 1996; Noguchi, 1998). Hayashi et al. (1991) reported that extruder barrel temperature was the most important parameter for the texturization of dehulled soybean. Melt temperature is a critical factor in protein cross-linking reactions. Increasing temperature from 140 to 180°C results in a proportional decrease in disulfide linkages formed in extruded soy protein isolates (Areas, 1992). Temperatures lower than 90°C hinder expansion and layer formation (Cheftel et al., 1992). At a given temperature, higher moisture contents result in softer and less texturized extrudates due to reduced protein–protein interactions and lower viscosity. At relatively lower moisture contents, higher barrel temperatures (140 to 180°C) result in better textures. At higher moisture levels, temperature needs to be decreased as moisture flash-off may cause considerable water loss if a cooling die is not used (Thiebaud et al., 1996).

High moisture levels combined with elevated temperatures yield extrudates that are very soft and not self-supporting after the die. However, a specially designed die which provides cooling at this section will increase the viscosity of the hot extrudate before exiting, contributing to the correct elasticity and fluidity required for texturization (Noguchi, 1998). The temperature at which solidification occurs is related to the plasticization temperature.

Low moisture (15 to 30%) extrusion tends to result in processes with greater generation of mechanical energy and products with lower density, while high moisture (50 to 70%) extrusion results in products with higher density, and is normally used in pellet production (Guy, 2001).

4. Influence on product quality

Extrusion-cooking is a process widely used in the food industry to manufacture snacks, crackers and expanded cereals. The degree of expansion in extruded products is an important characteristic which relates to the texture and sensory properties of extrudates (Lue et al., 1990). Extrusion-cooking has an important influence on product quality, emphasizing features like expansion, texture, shelf-life, colour and flavour.

For extrusion cooking, changes in ingredients such as sugar, salt and fibre, or processing parameters like screw design or speed and temperature, can affect extrusion system variables and product characteristics such as texture, structure, expansion and sensory attributes (Mendonça et al., 2000).

Products obtained with high temperatures and short extrusion process times normally present a porous, open structure, what confers to them a "crunchy" texture (Barrett, 2003).

Expansion occurs in both radial and axial directions, at different degrees, depending on the viscoelastic properties of the melt. Vaporization of moisture and cooling of the extrudate serve to bring the product from a molten to a rubbery state; and further drying is usually used to produce the brittle, fracturable texture typical of these products (Barrett, 2003).

Colour in extruded products is influenced by temperature, raw material composition, residence time, pressure and shear force (Guy, 2001; Mercier et al., 2001). During the extrusion process, several reactions happen and they in general affect the colour of the products. Among the most important, the most common are non-enzymatic browning (Maillard and caramelization) and pigment degradation. The process conditions normally used in thermoplastic extrusion (high temperature and low moisture content) are known to favour the reaction among reducing sugars and amino acids, that results in the formation of coloured compounds and the reduction of the amino acid lysine. If the browning is too intense, colours and flavours can be produced. Besides, changes in colour during the extrusion process can be an indicator to evaluate the intensity of the process in terms of chemical and nutritional changes (Ilo et al., 1999).

5. Influence on nutritional quality

The effects of extrusion cooking on nutritional quality are ambiguous. Benefits include destruction of antinutritional factors, gelatinization of starch, increased soluble dietary fibre and reduction of lipid oxidation. On the other hand, Maillard reactions between protein and sugars reduce the nutritional value of the protein, depending on the raw material types, their composition and process conditions. Besides, heat-labile vitamins may be lost to varying extents (Singh et al., 2007).

Starch digestibility is largely dependent on complete gelatinization. High starch digestibility is essential for specialized nutritional foods such as infant and weaning foods. Creation of resistant starch by extrusion may have value in reduced calorie products (Guy, 2001; Riaz, 2000).

The nutritional value of vegetable proteins is generally enhanced by mild extrusion cooking conditions due to the increase in digestibility (Asp and Björck 1989; Arêas, 1992), probably a result of protein denaturation and the inactivation of enzyme inhibitors present in raw materials, by the exposure of new active sites for enzyme attack (Colonna et al., 1989).

Processing nutritional food products at moisture levels below 20% has been proven to be uneconomical and nutritionally undesirable. Low-moisture extrusion results in production of certain undesirable dextrins as a result of increased shear energy inputs. Losses of vitamins and reduced amino acid availability are greatly accelerated as extrusion moistures are decreased. For this reason, vitamins and heat-sensitive nutrients are usually added post extrusion when processing at low moisture conditions (Huber, 2001).

Mild extrusion conditions (high moisture content, low residence time, low temperature) improve nutritional quality, while high extrusion temperatures (higher than 200°C), low moisture contents (lower than 15%) and/or improper formulation (e.g. presence of high-reactive sugars) can affect nutritional quality adversely. Also, to obtain a nutritionally balanced extruded product, careful control of process parameters is essential (Singh et al., 2007).

A benefit derived from extrusion-cooking is the partial or total destruction of potentially antinutritional factors, especially protease inhibitors, haemagglutinins, tannins and phytates, which limit utilization of nutrients in legume seeds. However, chemical alteration produced by thermal treatment could also result in decreased nutrient assimilation, including lower apparent absorption of certain minerals (Alonso et al., 2001).

Vitamin losses in extruded foods vary according to the type of food, moisture content, processing temperature and retention time. Generally, losses are minimal in cold extrusion. The HTST conditions in extrusion cooking, the short residence time of the extrudate and the rapid cooling as the product emerges from the die, cause relatively small losses of vitamins and essential amino acids (Fellows, 2000).

Extrusion cooking was reported by Saalia & Phillips (2011) as an efficient process to destroy or inactivate aflatoxins, if special conditions (high shear, high temperature, and adequate pH) are used.

Zhu et al. (1996) emphasized that by extrusion cooking, soybeans can be converted into high quality food ingredients. The short residence time and high temperature in an extruder reduce the damage to nutritional properties, but still adequately inactivate the enzymes responsible for the development of the undesirable off-flavour.

Minerals are heat stable and unlikely to become lost in the steam flash-off at the die. Extrusion can improve the absorption of minerals by reducing other factors that inhibit absorption, like phytates and condensed tannins. In addition, extrusion cooking usually increases the amount of iron available for absorption. For foods fortified with minerals prior to extrusion, some problems can be verified, like the formation of iron complexes with phenolic compounds that are dark in colour and detract from the appearance of foods; added calcium hydroxide can contribute to decrease expansion and increase lightness in colour of some products (Singh et al., 2007).

6. Influence on microbiological quality

One of the most important consumer requirements is the microbiological safety of food products. Most conventional extruded products such as snack foods and breakfast cereals are safe to eat because the raw materials are subjected to high temperatures (higher than 130°C) and the water activity of the product is low because the product is dried to a moisture content of less than five per cent. Although it is well known that most vegetative organisms, yeast and moulds are destroyed under typical extrusion conditions, the operating conditions under which spores are inactivated are not well understood (Guy, 2001).

The reduction of antinutritional factors, the increase of product microbiological safety and much better consumer acceptability are also related to extrusion cooking (Sumathy et al., 2007).

The extrusion process, as it is carried out at high temperatures, even with short residence times, eliminates a high amount of microorganisms (Baik et al., 2004). The extrusion process also allows obtaining lower water activity values in the final product, with values between 0.1 and 0.4. Therefore, it is possible to extend the shelf life of products (Fellows, 2000).

Fraiha et al. (2011) emphasize studies with heat-resistant microorganisms showing that shear stress may be involved in microbial load reduction during the extrusion process, predicting that mechanical forces might cause cell rupture. These authors studied a pre-treatment of *Bacillus stearothermophilus* spores in a 99% CO_2 modified atmosphere and checked that it did not affect cell viability during food extrusion. For them, heat was not the sole phenomena to explain cell death during extrusion, a mechanical damage of cells might be involved.

Okelo et al. (2006), studying the optimization of extrusion conditions for elimination of mesophilic bacteria during thermal processing of animal feed recipes, pointed out that, in general, thermal processing is designed to eliminate mesophilic organisms and not thermophilic organisms such as *Bacillus stearothermophilus*. It was predicted that most pathogenic organisms in feed would be inactivated by extrusion cooking through selecting extruder conditions within the experimental variable ranges that maximized spore destruction of thermophilic bacteria. This reduction would also likely include members of mesophilic *Bacillus cereus* group.

Mycotoxins are a risk to human health mainly via the intake of contaminated foods of plant origin, such as corn and wheat, which are consumed worldwide. The risk of exposure is therefore high for humans, and industrial processing methods that are effective in reducing mycotoxin contents in processed samples have received increased attention. Among them, extrusion cooking may be one of the most effective ways to reduce mycotoxin levels in processed products, especially if glucose or other additives such as ammonia or sodium bisulphite are included as ingredients. This is especially important since extruded products are highly popular in the food and feed market (Castells et al., 2005).

7. Products

7.1 Second and third generation snacks

The evolution of snacks occurred rapidly and can be divided into three generations. In the first, the raw material, such as whole grains, is processed through the combination of moisture, cooking temperature and drying. Only second and third generation snacks are produced by thermoplastic extrusion, and a flow diagram is shown in Figure 3.

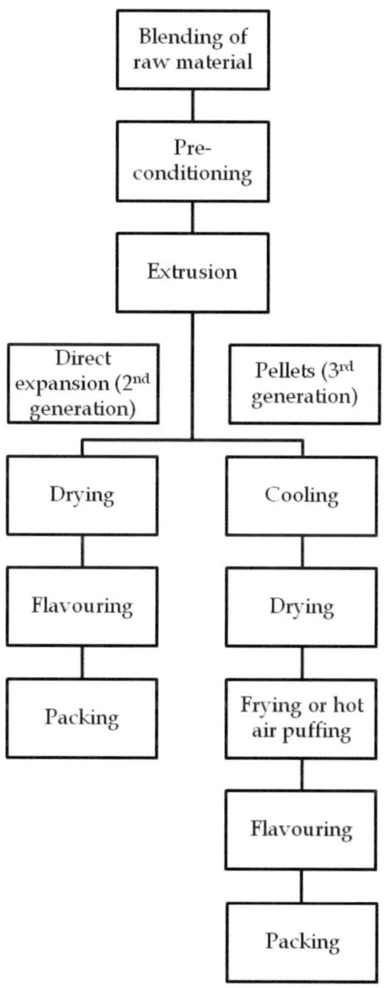

Fig. 3. Flow diagram for the production of second and third generation snacks

Second generation snacks or expanded snacks, where most extruded snacks are classified, are usually low in bulk density and are often marketed as high-fibre, low-calorie, high-protein and nutritional products (Lazou & Krokida, 2011). Different raw materials used to produce these kinds of snacks (i.e. flours and/or cereals and tubers starches and proteins) are processed in a extruder resulting in a continuous mass, that is cut into pieces of uniform size, being afterwards dried, flavoured and stored (Booth, 1990).

Third generation snacks or pellets are normally called "half products". These snacks are produced almost the same way as second generation snacks, however, when the product exits the extruder, it has the form of the die, that is, it is not expanded, being dried in this form. The expansion of the product occurs afterwards through frying, heating by hot air or in a microwave oven (Riaz, 2000; Carvalho et al., 2009). This kind of product presents a low moisture content (between 7 and 10%), high density and stability to be stored for a long time without microbial damage (Carvalho et al., 2009).

7.2 Breakfast cereals

Breakfast cereals (BCs) have been defined as "processed grains for human consumption". The breakfast cereal industry, in the United States, appeared at the beginning of the 20th century and has grown rapidly in the last few years, making BCs important economically viable products. Basic processes for the production of BCs include flaking, oven and gun puffing, baking, shredding and direct expansion (extrusion cooking). These processes convert raw and dense grains (7.7 kg/100 cm^3) into friable, crunchy or chewable products, adequate for human consumption, with apparent density in the range of 0.6 to 1.6 kg/100 cm^3 (Fast, 2000). Thermoplastic extrusion presents various advantages over the conventional processes used for BC production, such as shorter processing times and lower costs; less physical space necessary; greater flexibility for the production of a variety of products, simply changing the die, process conditions, initial formulation and final enrobing of the product, and particle size of the raw material (Huang, 1998; Bailey et al., 1991; Riaz, 2000).

The most commonly used cereals in extruded BC formulations are rice, wheat, oats and corn (Fast, 2000; Riaz, 2000). In the formulation of extruded BCs, a mixture of these cereals can be used, in the form of flours, grits or whole grain flours, and they can also be mixed with other ingredients such as starches, sugar, salt, malt extract or other liquid sweeteners, heat stable vitamins and minerals, flavourings, colorants and water, to vary appearance, texture, taste, aroma and other product characteristics (Riaz, 2000).

To produce BCs by extrusion cooking two different processes can be used: (i) pellet or shred production for manufacturing cereals such as cornflakes or shredded cereals with additional laminating and toasting processes after extrusion cooking, and (ii) directly expanded cereals production. Extruded flakes can be obtained by adding a blend of grains (flour, starches, etc.) and liquids (water, sweetener, colorants, etc.) directly to the extruder. The blend is forced through the barrel, where heat is applied to cook the dough, and then the product passes through a cooling section to prevent expansion, obtaining pellets after the die. Then the pellets pass through an equipment to be flaked and toasted. The process to obtain shredded cereals is similar to that of the flake process, with the exception that pellet sizes are not as critical, because they will be shredded without additional drying and tempering. Finally, the manufacturing process for extruded, puffed or expanded BCs follows the same steps, differing only at the end of the extruder, when the dough passes through a die that is designed to allow expansion after leaving the extruder because the moisture in the formula (whether natural or added) is released from a zone of elevated temperature and pressure to ambient conditions. The die holes control the shape of the finished cereal pieces, once they are cut with a rotating knife on the outer face of the die. The extruded products can be sugar-coated or coloured and flavoured to produce a variety of products for various tastes (Fast, 2000; Eastman et al., 2001). In expanded extruded products two important characteristics are expansion and texture. Extruded products are characterized for their expansion and usually maximum expansion is desired for expanded extruded snack products. For expanded extruded breakfast cereals a different structure is desired. It is necessary to obtain products with higher apparent density, lower porosity and thicker cell walls, as these products will be immersed in an aqueous medium, such as milk, and must maintain their texture during the longest time possible, absorbing the lowest quantity of moisture possible (Collona et al., 1998).

BCs can be considered protein sources (even though cereal proteins are incomplete due to limiting essential amino acids such as lysine), as they are often formulated with various different types of cereals and consumed with milk. When produced with whole grain flours, they can be considered sources of fatty acids and fibres. Breakfast cereals are sources of vitamins and minerals, as grains contain significant amounts of B-group vitamins, tocopherols, and minerals such as iron, zinc and copper, apart from being normally consumed with milk or yoghurt, considered important sources of calcium (Jones, 2001).

Nowadays, the development of new products in the BC segment aims at the production of breakfast cereals with a functional appeal. Recent research reports the use of antioxidants such as tocopherol and lycopene (Paradiso et al., 2008; Dehghan-Shoar et al., 2010) and fibres, such as β-glucans, gums and oat, wheat and passion fruit brans (Holguín-Acuña et al., 2008; Vernaza et al., 2010; Yao et al., 2011; Ryan et al., 2011).

7.3 Meat analogues

To human diets, especially those of vegetarians, the ingestion of high protein content products has been incorporated, including, for example, meat extenders and meat analogues, obtained through the extrusion process of vegetable proteins, resulting in a product with appearance and texture similar to the fibrilar structure of meat (Strahm, 2006; Macdonald et al., 2009). Meat extenders are obtained by thermoplastic extrusion at low moisture contents (20-35%) and meat analogues are obtained by thermoplastic extrusion at high moisture contents (50-70%). The raw materials commonly used to produce meat extenders are defatted soy flour and soy protein concentrate (SPC), whereas for the production of meat analogues, soy protein concentrate (SPC) and soy protein isolate (SPI) are used.

Amongst the main vegetable proteins used to produce meat analogues are proteins from legumes such as soybeans, common beans and peas, and from cereals, especially wheat proteins responsible for the formation of the gluten network (Riaz, 2000; Strahm, 2006).

The use of raw materials with high protein contents in extrusion began around the 1970s, with the use of soy for the production of texturized soy protein and meat analogues. Although researchers agree that, during extrusion of high protein content raw materials, denaturation, melting and alignment during mass flow occur, there is still a need to understand the physicochemical and rheological changes involved, once the phenomena that lead to the formation of the fibrilar structure from extruded vegetable proteins are not completely elucidated at a molecular level (Ledward & Mitchell, 1988; Mitchell & Areas, 1992).

Denaturation during the extrusion process of proteins results in a reduction of protein solubility, favours digestibility and inactivates antinutritional factors. Also, the extrusion of soy protein eliminates and/or reduces the bitter taste and the undesirable aroma and volatiles related to this protein (Areas, 1992; Kitabatake & Doi, 1992).

8. Future trends

In the last three decades, the development of extruders has advanced greatly. However, technical innovations will be continuously needed for the evolution of new generation extruders and complementary equipments targeting greater efficiency and higher productivity, increasing throughput, easing process control, enabling the production of numerous sophisticated snacks and improving final product quality. Also, it will always be vital to attend consumer requirements, which nowadays are closely related to nutrition and health foods that promote well-being and a positive life style.

Extruders permit the production of many foods of nutritional importance. The ability of extruders to blend diverse ingredients in novel foods can be exploited in the development of functional foods. Traditional snacks or breakfast cereals can be enhanced by the addition of extra fibres or whole grain flour as ingredients during extrusion, transformed into palatable cereal-based products that also promote beneficial physiological effects. Functional ingredients such as soy and botanicals (fruit, vegetables, cereals, etc.) that present high amounts of bioactive compounds can be used in the extrusion process to develop novel products with phytochemicals and other healthful food components. Improved chemical and immunoassay methods will undoubtedly facilitate research in this area. The extrusion process may have value in the formation of resistant starch and modified starch to promote reduced calories in food products. Extruders may also be interesting tools to obtain micro-encapsulated materials that have the objective of protecting sensitive additives (such as flavours, etc.), increasing their shelf-life and controlling the release of food ingredients at the right place and time. Carbohydrate matrices, such as hydrophobically-modified starches in the glassy state, have good barrier properties and extrusion is a convenient process enabling the encapsulation of flavours in such matrices.

In the future science and technology of the extrusion field, scientists and engineers should focus on the relationship between composition changes and product quality, evaluating and enhancing nutritional, sensory and functional properties of extruded foods.

9. References

Akdogan, H. (1999). High moisture food extrusion. *International Journal of Food Science and Technology*, Vol.34, No.3, (June 1999), pp.195-207, ISSN 1365-2621.

Alonso, R.; Rubio, L.; Muzquiz, M. & Marzo, F. (2001). The effect of extrusion cooking on mineral bioavailability in pea and kidney bean seed meals. *Animal Feed Science and Technology*, Vol.94, No.1-2, (November 2001), pp.1-13, ISSN 0377-8401.

Areas, J. A. (1992). Extrusion of food proteins. *Critical Reviews in Food Science and Nutrition*, Vol.32, No.4, pp.365-392, ISSN 1040-8398.

Asp, N. G. & Björck, I. (1989). Nutritional properties of extruded foods. In: *Extrusion cooking*, C. Mercier; P. Linko & J. M. Harper, (Eds.), pp.399-434, American Association of Cereal Chemists, ISBN 978-091-3250-67-8, Saint Paul, United States of America.

Baik, B.; Powers, J. & Nguyen, L. (2004). Extrusion of regular and waxy barley flours for production of expanded cereals. *Cereal Chemistry*, Vol.81, No.1, (January/February 2004), pp.94-99, ISSN 0009-0352.

Bailey, L. N.; Huack, B. W.; Sevatson, E. S.; Singer, R. E. (1991). Systems for manufacture of ready-to-eat breakfast cereals using twin-screw extrusion. *Cereal Foods World*, Vol.36, No.10, pp.863-869, ISSN 0146-6283.

Barrett, A. (2003). Characterization of macrostructures in extruded products. In: *Characterization of cereal and flours: properties, analysis and applications*, G. Kaletunç & K. Breslauer, (Eds.), pp.369-386, CRC Press, ISBN 978-0-8247-0734-7, Boca Raton, United States of America.

Booth, R. (1990). *Snack food*, Van Nostrand Reinhold, ISBN 978-812-3905-06-8, New York, United States of America.

Bornet, F. (1993). Technological treatments of cereals. Repercussions on the physiological properties of starch. *Carbohydrates Polymers*, Vol.21, No.2-3, pp.195-203, ISSN 0144-8617.

Caldwell, E. F.; Fast, R. B.; Ievolella, J.; Lauhoff, C.; Levine, H.; Miller, R. C.; Slade, L.; Strahm, B.S. Whalen, P. J. (2000). Unit operation and equipment. I. Blending and cooking. In: *Breakfast cereals and how they are made*, (2nd ed.) , R. B. Fast & E. F. Caldwell, (Eds.), pp.165-216, American Association of Cereal Chemists, ISBN 978-189-1127-15-1, Saint Paul, United States of America.

Camire, M. E.; Camire, A. & Krumhar, K. (1990). Chemical and nutritional changes in foods during extrusion. *Critical Reviews in Food Science and Nutrition*, Vol.19, No.1, pp.35-57, ISSN 1040-8398.

Camire, M. E. (2000). Chemical and nutritional changes in food during extrusion, In: *Extruders in food applications*, M. N. Riaz, (Ed.), pp.127-147, CRC Press, ISBN 978-156-6767-79-2, Boca Raton, United States of America.

Camire, M. E. & King, C. C. (1991). Protein and fiber supplementation: effects on extrudate cornmeal snack quality. *Journal of Food Science*, Vol.56, No.3, (May 1991), pp.760-763, ISSN 1750-3841.

Carvalho, A.; Vasconcelos, M.; Silva, P. & Aschieri, J. (2009). Produção de snacks de terceira geração por extrusão de misturas de farinhas de pupunha e mandioca. *Brazilian Journal of Food and Technology*, Vol.12, No.4, (October/December 2009), pp.277-284, ISSN 1516-7275.

Castells, M.; Marín, S.; Sanchis, V. & Ramos, A. (2005). Fate of mycotoxins in cereal during extrusion cooking: a review. *Food Additives and Contaminants*, Vol.22, No.2, (February 2005), pp.150-157, ISSN 1944-0049.

Chang, Y. K.; Hashimoto, J. M.; Moura-Alcioli, M. & Martínez-Bustos, F. (2001). Twin-screw extrusion of cassava starch and isolated soybean protein blends. *Molecular nutrition and food research*, Vol.45, No.4, (August 2001), pp.234-240, ISSN 1613-4133.

Cheftel, J. C.; Kitagawa, M. & Queguiner, C. (1992). New protein texturization processes by extrusion cooking at high moisture levels. *Food Reviews International*, Vol.8, No.2, pp.235-275, ISSN 1525-6103.

Cheftel, J. C. (1986). Nutritional effects of extrusion-cooking. *Food Chemistry*, Vol.20, No.4, pp.263-283, ISSN 0308-8146.

Chessari, C. J. & Sellahewa, J. N. (2001). Effective control processing. In: *Extrusion cooking: technologies and application*, R. Gay, (Ed.), pp.83-107, CRC Press, ISBN 978-084-9312-07-6, Boca Raton, United States of America.

Chinnaswamy, R. (1993). Basis of cereal starch expansion. *Carbohydrate Polymers*, Vol.21, No.2-3, pp.157-167, ISSN 0144-8617.

Colonna, P.; Tayeb, J. & Mercier, C. (1998). Extrusion cooking of starch and starchy products. In: *Extrusion cooking*, C. Mercier; P. Linko & J. M. Harper, (Eds.), pp.247-319, American Association of Cereal Chemists, ISBN 978-091-3250-67-8, Saint Paul, United States of America.

Dehghan-Shoar, Z.; Hardacre, A. K. & Brennan, C. S. (2010). The physico-chemical characteristics of extruded snacks enriched with tomato lycopene. *Food Chemistry*, Vol.123, No.4, (December 2010), pp.1117-1122, ISSN 0308-8146.

Ding, Q; Ainsworth, P.; Plunkett, A.; Tucker, G. & Marson, H. (2005). The effect of extrusion conditions on the physicochemical properties and sensory characteristics of rice-base expanded snacks. *Journal of Food Engineering*, Vol.66, No.3, (February 2005), pp.283-289, ISSN 0260-8774.

Eastman, J.; Orthofer, F. & Solorio, S. Using extrusion to create breakfast cereal products. *Cereal Foods World*, Vol.46, No.10, (October 2001), pp.468-471, ISSN 0146-6283.

El-Dash, A. A. (1981). Application and control of thermoplastic extrusion of cereals for food and industrial uses. In: *Cereals: a renewable resource, theory and practice*, Y. Pomeranz & L. Munich, (Eds.), pp.165-216, American Association of Cereal Chemists, ISBN 978-091-3250-22-8, Saint Paul, United States of America.

El-Dash, A. A.; Gonzales, R. & Ciol, M. (1983). Response surface methodology in the control of thermoplastic extrusion of starch. *Journal of Food Engineering*, Vol.2, No.2, pp.129-152, ISSN 0260-8774.

Fast, R. B. (2000). Manufacturing technology of ready-to-eat cereals. In: *Breakfast cereals and how they are made*, (2nd ed.), R. B. Fast & E. F. Caldwell, (Eds.), pp.15-86, American Association of Cereal Chemists, ISBN 978-091-3250-70-9, Saint Paul, United State of America.

Fellows, P. (2000). *Food processing technology: principles and practice*, (2nd ed.), CRC Press, ISBN 978-084-9308-87-1 Boca Raton, United States of America.

Fornal, L.; Soral-Smietana, M. & Szpenelowski, J. (1987). Chemical characteristics and physicochemical properties of the extruded mixtures of cereal starches. *Starch/Stärke*, Vol.39, No.2, pp.75-78, ISSN 0038-9056.

Fraiha, M.; Ferraz, A. & Biagi, J. (2011). Pre-treatment of thermoduric spores in CO_2 modified atmosphere and their survivability during food extrusion. *Ciência e Tecnologia de Alimentos*, Vol.31, No.1, (January/March 2011), pp.167-171, ISSN 0101-2061.

Gorinstein, S.; Zachwieja, Z.; Folta, M.; Barton, H.; Piotrowicz, J.; Zemser, M.; Weisz, M.; Trakhtenberg, S. & Martin-Belloso, O. (2001). Comparative contents of dietary fiber, total phenolics, and minerals in persimmons and apples. *Journal of Agriculture and Food Chemistry*, Vol.49, No.2, (February 2001), pp.952-957, ISSN 0021-8561.

Guillon, F.; Barry, J. L. & Thibault, J. F. (1992). Effect of autoclaving sugar-beet fibre on its physico-chemical properties and its *in vitro* degradation by human faecal bacteria.

Journal of the Science of Food and Agriculture, Vol.60, No.1, (September 1992), pp.69-79, ISSN 1097-0010.

Guy, R. (2001). *Extrusion cooking: technologies and applications*, Woodhead Publishing, ISBN 978-185-5735-59-0, Cambridge, United Kingdom.

Harper, J. M. (1994). Extrusion processing of starch. In: *Developments in carbohydrate chemistry*, (2nd ed.), R. J. Alexander & H. F. Zobel, (Eds.), pp.37-64, American Association of Cereal Chemists, Saint Paul, United States of America.

Hayashi, N.; Abe, H.; Hayakawa, I. & Fujio, Y. (1991). Texturization of dehulled whole soybean with a twin screw extruder and texture evaluation. In: *Food processing by ultra high pressure twin-screw extrusion*, A. Hayakawa, (Ed.), pp.133–146, CRC Press, ISBN 978-087-7628-21-7, Boca Raton, United States of America.

Holguín-Acuña, A. L.; Carvajal-Millán, E.; Santana-Rodríguez, V.; Rascón-Chu A.; Márquez-Escalante, J.; León-Renova, N. E. P. & Gastelum-Franco, G. (2008). Maize bran/oat flour extruded breakfast cereal: A novel source of complex polysaccharides and an antioxidant. *Food Chemistry*, Vol.111, No.3, (December 2008), pp.654-657, ISSN 0308-8146.

Hsieh, F.; Mulvaney, S. S; Huff, H. E.; Lue, S. & Brent, J. (1989). Effect of dietary fiber and screw speed on some extrusion processing and products variables. *Lebensmittel Wissenschaft und Technologie*, Vol.22, pp.204-207, ISSN 0023-6438.

Huang, W. N. (1998). Comparing cornflake manufacturing processes. *Cereal Foods World*, Vol.43, No.8. pp.641-643, ISSN 0146-6283.

Huber, G. R. (2000). Twin-screw extruders. In: *Extruders in food applications*, M. N. Riaz, (Ed.), pp.81-114, CRC Press, ISBN 978-156-6767-79-8, Boca Raton, United States of America.

Huber, G. R. & Rokey, G. J. (1990). Extruded snacks. In: *Snack food*, R. G. Booth, (Ed.), pp.107-138, Van Nostrand Reinhold, ISBN 978-044-2237-45-5, New York, United States of America.

Huber, G. R. (2001). Snack foods from cooking extruders. In: *Snack foods processing*, E. W. Lusas & R. W. Rooney, (Eds.), pp.315-368, CRC Press, ISBN 978-156-6769-32-9, Boca Raton, United States of America.

Ilo, S. & Berghofer, E. (1999). Kinetics of colour changes during extrusion cooking of maize grits. *Journal of Food Engineering*, Vol.39, No.1, (January 1999), pp.73-80, ISSN 0260-8774.

Ilo, S.; Liu, Y. & Berghofer, E. (1999). Extrusion cooking of rice flour and amaranth blends. *Lebensmittel Wissenschaft und Technologie*, Vol.32, No.2, (March 1999), pp.79-88, ISSN 0023-6438.

Jones, J. M. (2001). The benefits of eating breakfast cereals. *Cereal Foods World*, Vol.46, No.10, (October 2001), pp.461-467, ISSN 0146-6283.

Kadan , R. & Pepperman, A. (2002). Physicochemical properties of starch in extruded rice flours. *Cereal Chemistry*, Vol.79, No.4, (August 2002), pp.476-480, ISSN 0009-0352.

Kitabatake, N. & Doi, E. (1992). Denaturation and texturization of food protein by extrusion cooking. In: *Food extrusion: science and technology*, J. L. Kokini; C-T. Ho & M. V. Karwe, (Eds.), Marcel Dekker, pp.361-371, ISBN 978-082-4785-42-0, New York, United States of America.

Larrea, M. A; Chang, Y. K. & Bustos, F. M. (2005). Effect of some operational extrusion parameters on the constituents of orange pulp. *Food Chemistry,* Vol.89, No.2, (February 2005), pp.301-308, ISSN 0308-8146.

Lazou, A. & Krokida, M. (2011). Thermal characterization of corn-lentil extruded snacks. *Food Chemistry,* Vol.127, No.4, (August 2011), pp.1625-1633, ISSN 0308-8146.

Ledward, D. A. & Mitchell, J. R. (1988). Protein extrusion - More questions as answers? In: *Food structure: its creation and evaluation,* J. M. V. Blanshard & J. R. Mitchell, (Eds.), pp.219-229, Butterworth-Heinemann , ISBN 978-040-8029-50-6, London, United Kingdom.

Li-Chan, E. C. Y. (2004). Properties of proteins in food systems: an introduction. In: *Proteins in food processing,* R. Y. Yada, (Ed.), pp.2-26, Woodhead Publishing Limited, ISBN 978-084-9325-36-6, Cambridge, London.

Li, M. & Lee, T-C. (1996). Effect of extrusion temperature on solubility and molecular weight distribution of wheat flour protein. *Journal of Agricultural and Food Chemistry,* Vol.44, No.3, (March 1996), pp.763-768, ISSN 0021-8561.

Linko, P.; Hakulin, S. & Linko, Y-Y. (1983). Extrusion cooking of barley starch for the production of glucose syrup and ethanol. *Journal of Cereal Science,* Vol.1, No.4, (October 1983), pp.275-289, ISSN 0733-5210.

Lue, S.; Hsieh, F.; Peng, I. & Huff, H. (1990). Expansion of corn extrudates containing dietary fiber: a microstructure study. *Lebensmittel Wissenschaft und Technologie,* Vol.23, No.2, pp.165-173, ISSN 0023-6438.

MacDonald, R. S.; Pryzbyszewshi, J. & Hsieh, F. H. (2009). Soy protein isolate extruded with high moisture retains high nutritional quality. *Journal of Agricultural and Food Chemistry,* Vol.57, No.9, (May 2009), pp.3550-3555, ISSN 0021-8561.

Mendonça, S.; Grossmann, M. & Verhé, R. (2000). Corn bran as a fibre source in expanded snacks. *Lebensmittel Wissenschaft und Technologie,* Vol.33, No.1, (February 2000), pp.2-8, ISSN 0023-6438.

Mercier, C.; Linko, P. & Harper, J. M. (1998). *Extrusion cooking,* (2nd ed.), American Association of Cereal Chemists, ISBN 978-091-3250-67-8, Saint Paul, United States of America.

Mitchel, J. R. & Areas, J. A. G. (1992). Structural changes in biopolymers during extrusion. In: *Food extrusion: science and technology,* J. L. Kokini; C-T. Ho & M. V. Karwe, (Eds.), pp.345-360, Marcel Dekker, ISBN 978-082-4785-42-0, New York, United States of America.

Noguchi, A. (1998). Extrusion cooking of high-moisture protein foods. In: *Extrusion cooking,* C. Mercier; P. Linko & J. M. Harper, (Eds.), pp.343-370, American Association of Cereal Chemists, ISBN 978-091-3250-67-8, Saint Paul, United States of America.

Okelo, P.; Wagner, D.; Carr, L.; Wheaton, F.; Douglass, L & Joseph, S. (2006). Optimization of extrusion conditions for elimination of mesophilic bacteria during thermal processing of animal feed mash. *Animal Feed Science and Technology,* Vol.129, No.1-2, (August 2006), pp.116-137, ISSN 0377-8401.

Paradiso, V. M.; Summo, C.; Trani, A. & Caponio, F. (2008). An effort to improve the shelf life of breakfast cereals using natural mixed tocopherols. *Journal of Cereal Science,* Vol.47, No.2, (March 2008), pp.322-330, ISSN 0733-5210.

Pereda, J. A. O.; Rodríguez, A. I. C.; Álvarez, L. F; Sanz, M. L. G.; Minguillón, G. D. G. F.; Perales, L. H. & Cortecero, M. D. S. (2005). *Tecnologia de alimentos: componentes dos alimentos e processos,* Vol.1, Artmed, ISBN 9788536304366, Porto Alegre, Brazil.

Riaz, M. N. (2000). Introduction to extruders and their principles. In: *Extruders in food applications,* M. N. Riaz, (Ed.), pp.1-23, CRC Press, ISBN 978-156-6767-79-8, Boca Raton, United States of America.

Roberfroid, M. (1993). Dietary fiber, inulin and oligofructose: a review comparing their physiological effects. *Critical Reviews in Food Science and Nutrition,* Vol.33, No.2, pp.103-148, ISSN 1040-8398.

Ryan, L.; Thondre, P. S. & Henry, C. J. K. (2011). Oat-based breakfast cereals are a rich source of polyphenols and high in antioxidant potential. *Journal of Food Composition and Analysis,* Vol.24, No.7, (November 2011), pp.929-934, ISSN 0889-1575.

Saalia, F. & Phillips, R. (2011). Degradation of aflatoxins by extrusion cooking: effects on nutritional quality of extrudates. *LWT-Food Science and Technology,* Vol.44, No.6, (July 2011), pp.1496-1501, ISSN 0023-6438.

Singh, S.; Gamlath, S. & Wakeling, L. (2007). Nutritional aspects of food extrusion: a review. *International Journal of Food Science and Technology,* Vol.42, No.8, (August 2007), pp.916-929, ISSN 0950-5423.

Sluimer, P. (2005). *Principles of breadmaking: funtionality of raw material and process steps,* American Association of Cereal Chemists, ISBN 978-1891127458, Saint Paul, United States of America.

Stanley, D.W. (1998). Protein reactions during extrusion processing. In: *Extrusion cooking,* C. Mercier, P. Linko & L. M. Harper, (Eds.), pp.321–341, American Association of Cereal Chemists, ISBN 978-091-3250-67-8, Saint Paul, United States of America.

Stanley, D. W. (1886). Chemical and structural determinants of texture of fabricated foods. *Food Technology,* Vol.40, No.3, (March 1986), pp.65-68, ISSN 0015-6639.

Stark, A. & Madar, Z. (1994). Dietary fiber. In: *Functional foods: designer foods, pharmafoods, nutraceuticals,* I. Goldberg, (Ed.), pp.183-201, Springer, ISBN 978-083-4216-88-4, New York, United States of America.

Strahm, B. S. (2006). Meat alternatives. In: *Soy applications in food,* M. N. Riaz, (Ed.), pp.135-154, CRC Press, ISBN 978-084-9329-81-4, Boca Raton, United States of America.

Strahm, B. S. (2000). Preconditiong. In: *Extruders in food applications,* M. N. Riaz, (Ed.), pp.115-126, CRC Press, ISBN 978-156-6767-79-2, Boca Raton, United States of America.

Sumathy, A.; Ushakumari, S. & Malleshi, N. (2007). Physico-chemical characteristics, nutritional quality and shelf-life of pearl millet based extrusion cooked supplementary foods. *International Journal of Food Sciences and Nutrition,* Vol.58, No.5, pp.350-362, ISSN 0963-7486.

Tadmor, Z. & Gogos, C. G. (1979). *Principles of polymer processing,* Wiley, ISBN 978-047-1843-20-2, New York, United States of America.

Thebaudin, J. Y.; Lefebvre, A. C.; Harrington, M. & Bourgeois, C. M. (1997). Dietary fibres: nutritional and technological interest. *Trends in Food Science and Technology,* Vol.8, No.2, (February 1997), pp.41-48, ISSN 0924-2244.

Thiebaud, M., Dumay, E. & Cheftel, J. C. (1996). Influence of process variables on the characteristics of a high moisture fish soy protein mix textured by extrusion cooking. *Lebensmittel Wissenschaft und Technologie*, Vol.29, No.7, pp.526–535, ISSN 0023-6438.

Tilley, K. A.; Benjamin, R. E.; Bagorogoza, K. E.; Okot-Kotber, B. M.; Prakash, O. & Kwen, H. (2001). Tyrosine cross-links: molecular basis of gluten structure and function. *Journal of Agricultural Food Chemistry*, Vol.49, No.5, (May 2001), pp.2627-2632, ISSN 0021-8561.

Vernaza, M. G.; Chang, Y. K. & Steel, C. J. (2009). Efeito do teor de farelo de maracujá e da umidade e temperatura de extrusão no desenvolvimento de cereal matinal funcional orgânico. *Brazilian Journal of Food Technology*, Vol.12, No.2, (April/June 2009), pp.145-154, ISSN 1516-7275.

Vernaza, M. G.; Pedrosa, M. T.; Chang, Y. K. & Steel, C. J. (2010). Evaluation of the in vitro glycemic index of a fiber-rich extruded breakfast cereal produced with organic passion fruit fiber and corn flour. *Ciência e Tecnologia de Alimentos*, Vol.30, No.4, (October/December 2010), pp.964-968, ISSN 0101-2061.

Wang, W. M.; Klopfenstein, C. F. & Ponte, J. G. (1993). Effects of twin-screw extrusion on the physical properties of dietary fiber and other components of whole wheat and wheat bran and on the baking quality of wheat bran. *Cereal Chemistry*, Vol.70, No.6, pp.707-711, ISSN 0009-0352.

Wong, D. W. S. (1995). *Food enzymes: structure and mechanism*. Springer, ISBN 978-041-2056-91-8, New York, United States of America.

Yao, N.; White, P. & Alavi, S. (2011). Impact of beta-glucan and other oat flour components on physico-chemical and sensory properties of extruded oat cereals. *International Journal of Food Science and Technology*, Vol.46, No.3, (March 2011), pp.651-660, ISSN 0950-5423.

Zhu, S.; Riaz, M. & Lusas, E. (1996). Effect of different extrusion temperatures and moisture content on lipoxygenase inactivation and protein solubility in soybeans. *Journal of Agricultural and Food Chemistry*, Vol.44, No.10, (October 1996), pp.3315-3318, ISSN 0021-8561.

14

Lightweight Plastic Materials

Marek Kozlowski
Wroclaw University of Technology
Poland

1. Introduction

Lightweight constructions are increasingly used in automotive, aerospace and construction sectors, because using the low density materials allows reducing the structural weight of products. That may result in substantial fuel savings and a lower carbon footprint in transportation and facilitates manipulation of details in the house construction applications. Moreover, the low material density leads to conservation of natural resources, since less material is required for manufacturing consumer goods.

In polymer engineering the lightweight solutions include:

- selection of polymers with a density lower than their counterparts of comparable properties,
- using composites filled with natural fibers instead of glass fibers,
- composite sandwich panels with cellular/honeycomb structure,
- hollow components manufactured by the gas or water assisted injection molding,
- polymer foaming.

Cellular and hollow structure polymeric materials offer additional advantages resulting from their thermal insulating properties, thus allowing additionally energy savings.

Preference given to low density materials belong to the factors deciding on a success of polypropylene (PP) in the automotive sector. Being lighter than other plastics for 15-20% PP allows substantial fuel savings – it is assumed that a weight reduction in a car body of 100 kg brings about 0.3-0.5 litres of fuel savings per 100 km.

2. Biocomposites

Polymer composites constitute a broad group of materials, composed of the macromolecular matrix and various fillers. Currently the filler market for plastic composites is dominated by calcium carbonate (40%) and glass fiber (31%) and some other inorganic fillers such as talc, mica and clay. Although the conventional fillers offer property changes in the composites, their high density is not beneficial to fuel savings in automotive applications. Polymer composites with cellulose fillers are growing rapidly, mainly in the construction and automotive industry. The main advantage of such composites is lower density in comparison to that of glass fiber reinforced plastics. In Fig. 1 the density of polypropylene and PP filled with wood flour (WF), without or with of a compatibilizer (PP/WF/comp) has been compared to that of PP composites with glass fibers (GF). One can observe an increase in density with the filler content for all composites, however markedly higher density is that of PP/GF materials.

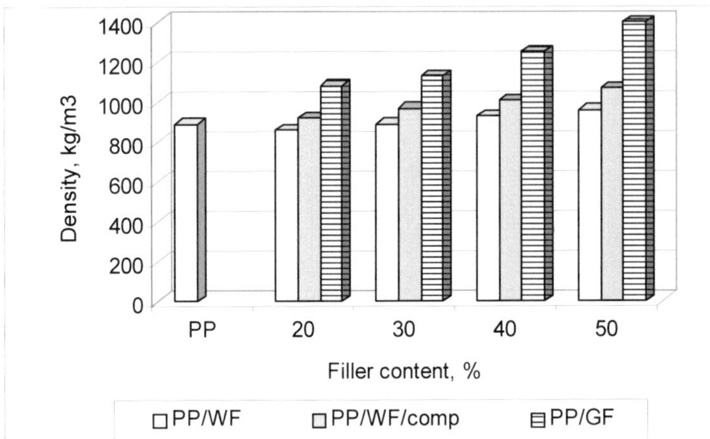

Fig. 1. Density of polypropylene and PP composites with wood flour

Low density of polymer composites filled with natural fibers was received due to a specific hollow structure of the fibers (Fig. 2), which is totally different from a bulky structure of glass fibers. Cellulose fibers do not exist in nature as separate items, but they form bundles. Each bundle contains 10-60 fibers diameter of 10-17 microns, linked together with pectines.

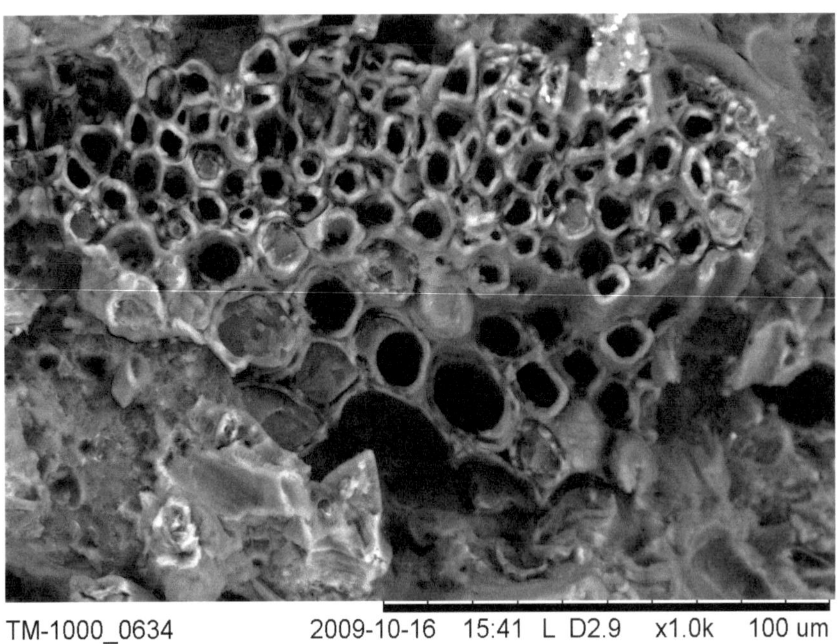

Fig. 2. Flax fibers bundle cross-section

Natural fillers are biodegradable and derive from renewable resources, which is higly advantageous for sustainable development (Ellison et al., 2004; Bledzki et al., 2008; Oksman Niska & Sain, 2008; Klesov, 2007; Kozlowski & Kozlowska, 2005). Apart from the ecological reasons, natural fillers offer possibility of reinforcing thermoplastic matrices (Figs. 3 and 4) and provide weight savings in comparison to glass fibers (Table 1).

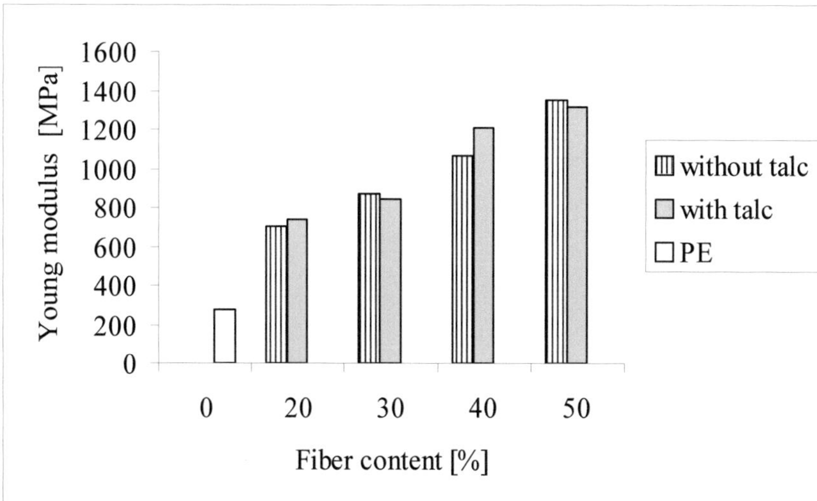

Fig. 3. Young modulus of LDPE and LDPE composites with hemp fibers

Increase in the stiffness of thermoplastic composites filled with cellulose fibers depends on a matrix polymer nature - it is particularly significant to composites based on the low density polyethylene (LDPE). In case of LDPE filled with 30wt.% of hemp fibers Young modulus increased for 300%, whereas at 50% loading it is for 5-times higher (Fig. 3).

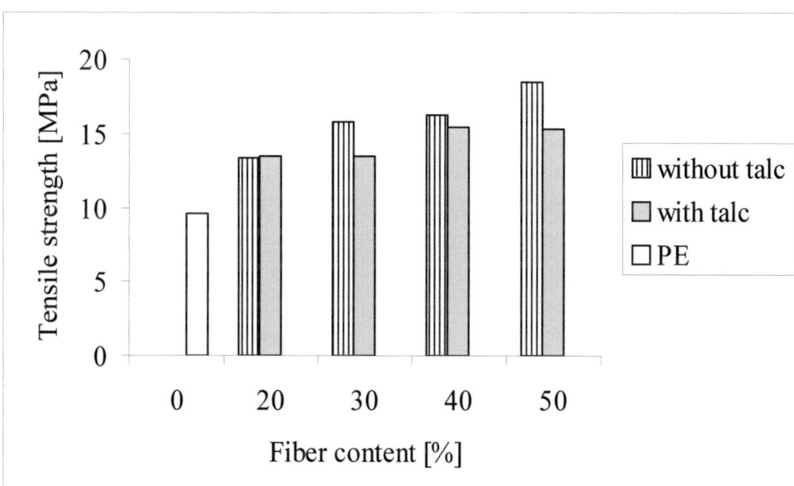

Fig. 4. Tensile strength of LDPE and LDPE composites with hemp fibers

Tensile strength of LDPE composites filled with natural fibers also enhances as a function of the filler content - at 30wt.% an increase for 60% was observed in comparison to neat LDPE, while at 50wt.% of hemp fibers the improvement reached 100% (Fig. 4).

Looking at the data presented in Table 1 and considering that the glass-fibre reinforced polypropylene composites are widely used in automotive components (front/end carriers, door panel supports, dashboards, consoles, seat backs, headliners, package trays, under-body shields etc.) one should expect significant environmental profits from replacing glass fibers with natural fibers (NF).

	PP	**Natural fibers**	**Glass fibers**
Price [€ / kg]	1.4	2.2-3.0	1.5-2.2
Density [g/cm^3]	0.92	1.2-1.5	2.5-2.8
Energy [MJ/kg]	101.1	3.4	48.3
CO_2 [kg/kg]	3.11	0.64	2.04
SO_x [g/kg]	22.2	1.2	8.8
NO_x [g/kg]	2.9	0.95	2.9
BOD [mg/kg]	38.37	0.265	1.75
PM [g/kg]	4.37	0.2	1.03
COD [g/kg]	1.14	3.23	0.02

Table 1. Comparison of data related to materials used for manufacturing modern composites

The fibers used most frequently for reinforcement of polymer composites are the bast fibers grown in the climate of Europe (flax and hemp) and sub-tropical fibers (kenaf, jute and sisal) imported mainly from Asia. Composites filled with wood/natural fibers are called biocomposites or wood polymer composites (WPC), sometimes also a term "artificial wood" is applied. Their properties combine higher stiffness, hardness and better dimensional stability in comparison to plastics and lower density in comparison to mineral fillers. Currently biocomposites are being used by Fiat, Ford, Opel, Daimler Chrysler, Saturn, BMW, Audi, Peugeot, Renault, Mercedes Benz and Volvo.

Different polymers can be used as a matrix for biocomposites, and the loading of cellulose fillers usually vary within a range of 10 to 70%. Although thermosets were used first as a matrix for wood composites, currently thermoplastics use to be applied for manufacturing of biocomposites. Most frequently used matrices are polypropylene, polyethylene, polyvinylchloride and polystyrene. Depending on a chemical structure of the matrix the interaction at the polymer-filler interface is of different strength. This is crucial for the mechanical properties and melt viscosity of composites. High adhesion is expected if both components have polar groups. Unfortunately, this is not a case of the most popular biocomposites, which are composed of hydrophobic polyolefines and hydrofilic cellulose fibers (-OH groups). Therefore, extensive research has been performed to enhance interfacial adhesion and improve dispersion of cellulosic fibers in polyolefines. Frequently used solution is addition of compatibilizers composed of blocks interacting on a physical or chemical way with each component of the composite. Good results are reported on using maleated polypropylene (PP-g-MAH) for compatibilization of PP-based composites with natural fillers (Fig. 1). Another approach might be hydrophobisation of cellulose fibers (esterification). Although polymer composites with natural fillers have been commercialized, their industrial applications are in some sectors limited because of their low impact strength and high density compared to natural wood and polyolefines.

The original processing technology of biocomposites was nonwoven technology, which is a normal production precursor to compression moulding. Further developments extended the processing technology to extrusion or injection molding of composites reinforced with short cellulose fibers. What should be considered at processing of thermoplastic composites is the melt viscosity, which increases after addition of fillers. Melt flow rate (MFR) of LDPE and composites with hemp fibers has been presented in Fig. 5. Flowability of composite filled with 30wt.% of fibers is for 2.5 times lower than that of neat LDPE, droping down at 50wt.% of hemp for 6 times in comparison to MFR of polyethylene. For that reason either higher pressure or higher processing temperature has to be used in order to shape high quality products.

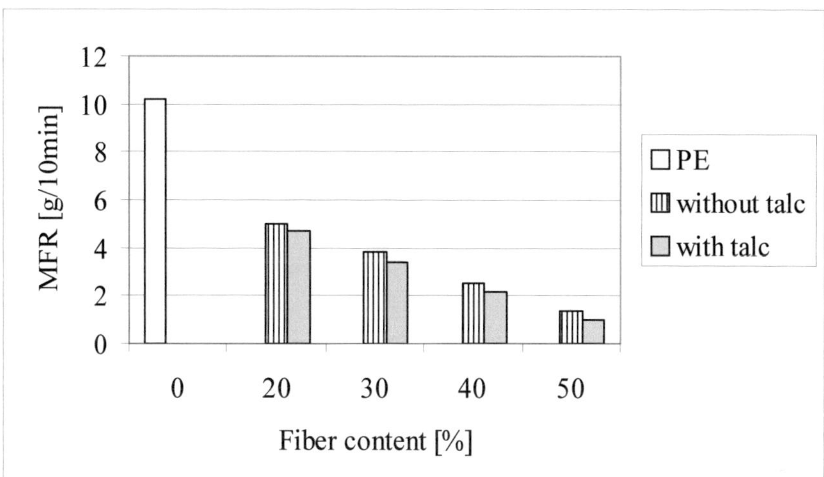

Fig. 5. Melt flow rate of LDPE and LDPE composites with hemp fibers

Particular type of biocomposite is the particle board, which is a composite material made from small pieces of wood or other lignocellulosic material (branches, stem, saw dust, straw or bagasses) that are mechanically pressed into sheets being bonded with a resin. Similar sandwich structure have composites which external skin layers are made of a bulky plastic, whereas a core is lightweight (density of 40-70 kg/m3), made either as a foam or honeycomb structure. Such sandwich materials (Panelplus panels) while used in the manufacture of truck bodies weigh 60% less than the equivalent plywood panels (Institute of Materials, Minerals and Mining, 2004).

Hollow, lightweight plastic components may be manufactured also by the gas or water assisted injection molding. In this technology the mould is only partially filled with a polymer melt, afterwards a gas or water is injected to pack the mould completely. After mold cooling and solidification of a polymer the gas or water is evacuated from the cavity and the hollow part is ejected (Fig. 6).

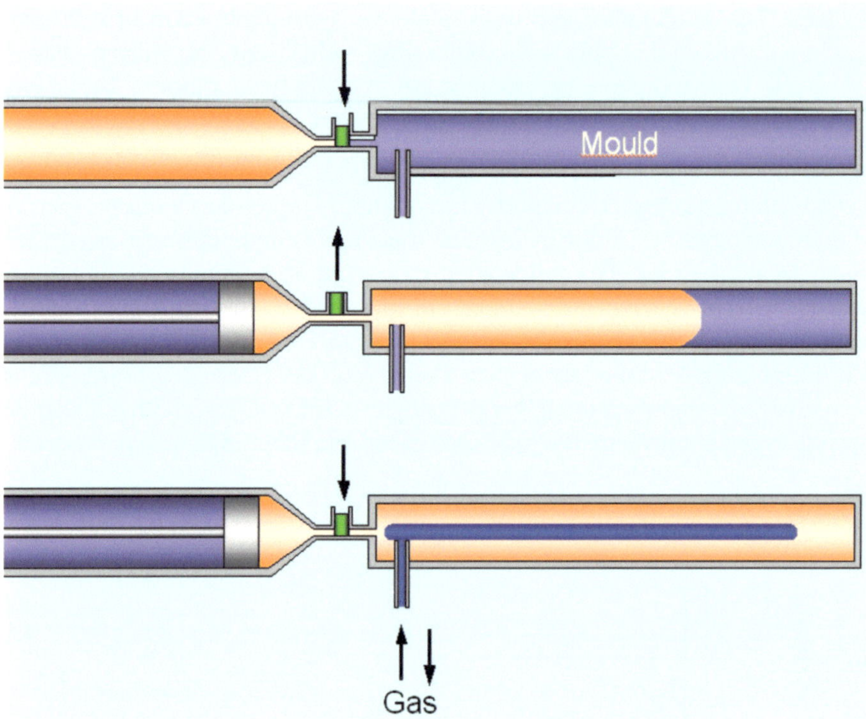

Fig. 6. Principle of gas-assisted injection molding (acc. to www.cinpres.net)

3. Polymer foaming

In parallel with other technologies a fast development in foaming technology of plastic parts is observed, driven primarily by the transportation sector demands. The main reason is that porous components allow to reduce the amount of raw materials and fuel consumption at the same time. Another characteristic of foams is thermal insulation, therefore an important application field is insulation of buildings and industrial constructions. A typical example is polystyrene foam, which is used in the construction industry for over 50 years. Polyurethane and polyethylene foams have been used for insulations of pipelines, air ducts, containers, solar collectors etc. (Fig. 7). Plastic foams are used extensively for thermal insulation of refrigerators and freezers.

Fig. 7. Foam insulation made of synthetic rubber ARMAFLEX HT (www.azflex.pl)

Cellular structure of foams allows also sound and vibration damping, which has been used in sound insulating panels, upholstery in furniture, car seats, protective pads etc.

Cellular plastics may be manufactured either by the periodic or continuous technology, using chemical or physical foaming agents. The conventional foams have cells of large size (0.1-1 mm) and broad size distribution (Fig. 8), therefore their mechanical properties are inferior to that of bulky polymers. The cell density of conventional foams is in a range of 10^4 - 10^6 cells/cm^3.

The microfoams contain much higher number of cells (>10^9/cm^3), which size is markedly smaller (ca. 10 μm). Such materials exhibit better performance than the conventional foams, with higher mechanical properties and better thermal insulating characteristics.

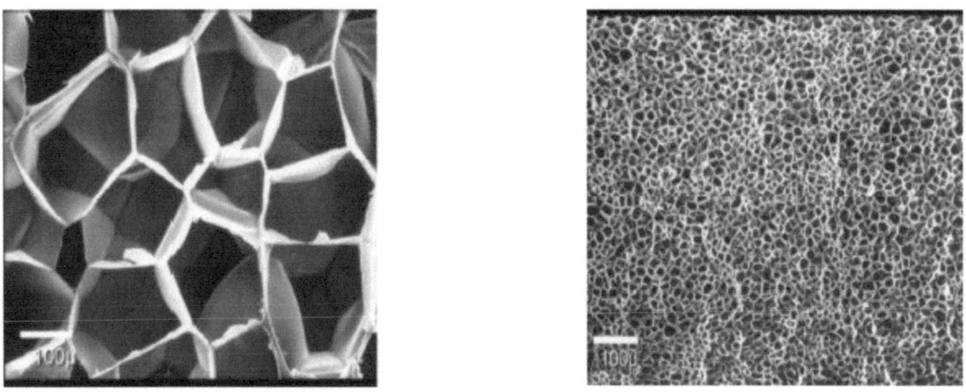

Fig. 8. Polystyrene foams - conventional (left) and microfoam (www.trexel.com)

3.1 Methods of foaming

Foaming techniques can be divided into several groups. Similar to the plastics processing technology, they can be divided on continuous (extrusion foaming) and periodic (injection molding or press foaming) processes. Periodic technology requires a long time and excluding manufacturing of expanded polystyrene (EPS), they are rarely applied.

Other method of rather seldom use is manufacturing of polymer composites filled with easily soluble compounds, like salt or sugar. After the filler is eluted with an appropriate solvent, the empty holes form cells of the resulted foam.

Cellular structure can be also formed by sintering of polymer powders at high temperature. Soft surfaces of neighboring spheres stick each other, whereas the free volumes between them create foam cells.

Akzo Nobel offers a foaming method based on mixing of a matrix polymer with thermoplastic spheres filled with volatile hydrocarbons (Expancel). At heating the polymer becomes soft, while the hydrocarbon evaporates, expanding the material. Initial sphere diameter is 12 μm, which after expansion increases to 40 μm. The material of spheres should be compatible with the matrix polymer, whereas a hydrocarbon is selected depending on a required decomposition temperature. The spheres of Expancel are added to a polymer in an amount of 2 - 8% and such mixture is processed by extrusion or injection molding technology. Decrease in a density for 30% was reported after addition of 3% microspheres, however the cells were of diverse size (Fig. 9).

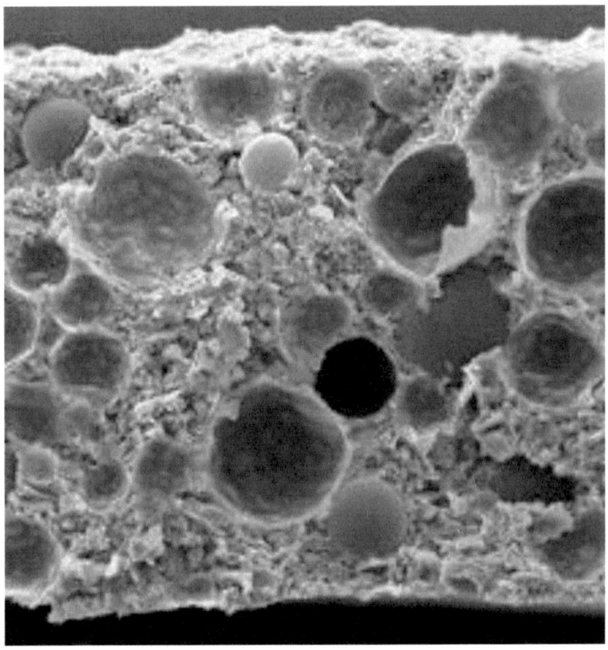

Fig. 9. Structure of foam manufactured with Expancel technology (www.akzonobel.com)

Undoubtedly the principal polymer foaming technology is that involving a gas delivered to a polymer by means of the chemical (CFA) or physical foaming agent (PFA). Low density foams (2 - 500 kg/m^3) are manufactured with physical blowing agents, whereas chemical blowing agents produce foams density of 500 - 750 kg/m^3.

3.2 Foaming agents

Physical blowing agents comprise of gases and low boiling hydrocarbons or their halogenated derivatives. Initially used blowing agents (pentane, butane, chlorofluoro hydrocarbons) are withdrawn because of ecological reasons (the Montreal Protocol Agreement) and fire hazard and replaced by noble gases (argone, nitrogen, carbon dioxide). Interesting properties have also hydrofluoro olefines, which have been used for manufacturing of polyurethane and EPS foams (Rosato, 2010).

Unfortunately several safe blowing agents exhibit either too low solubility in polymers or too high heat coefficient, which deteriorates thermo-insulating properties of foams. Thermal insulation expressed with the heat transfer coefficient λ depend on the cell size and density (Schellenberg & Wallis, 2010), but also on a nature of gas in the cells. Nitrogen and oxygen have comparable λ values (at 0°C respectively 22,7 and 23,2 mW/m K), however that of carbon dioxide equals to 13,7 mW/m K. Thus, the foams filled with CO_2 exhibit much better thermo-insulating properties than others.

Foaming with gases results mostly in foams of large cell size, however using supercritical liquids bring about manufacturing of microfoams (Cooper, 2000). At critical conditions (temperature and pressure) the density of a liquid and a gas equals. Above the critical temperature condensation of a gas is impossible, independing on a pressure applied. From that reason carbon dioxide is most appropriate for transportation, storage and dosing conditions, since its critical temperature is +31.1, whereas that for nitrogen is -146,9°C and argone -122,3°C.

Chemical blowing agents decompose within a specific temperature range, emiting a stechiometric amount of gases (usually nitrogen or carbon dioxide). Chemical blowing agents are classified as egzo- or endothermic, depending on the effect of a decomposition process. Due to a vigorous character of the decompositin reaction, egzothermic CBAs produce large size cells (>100 μm) of a non-uniform size distribution and cause a high overall expansion of the material (Fig. 10).

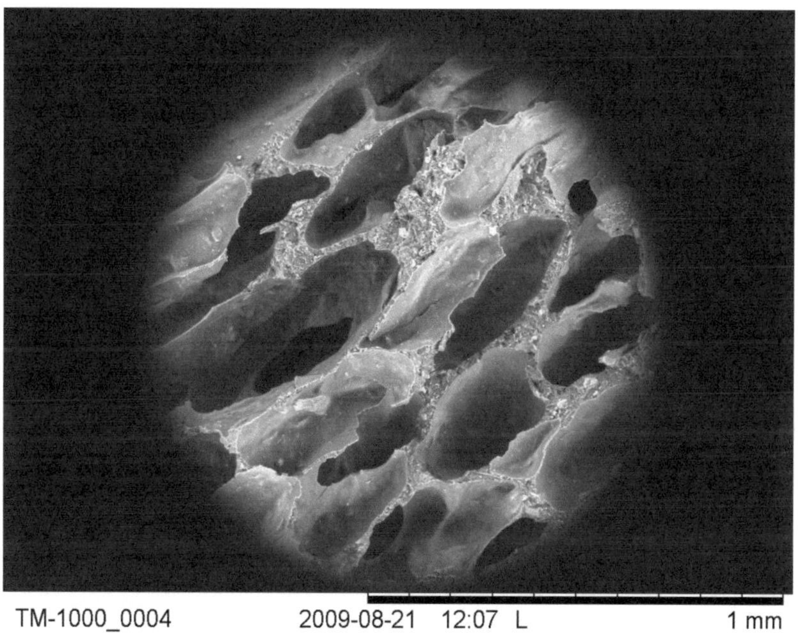

Fig. 10. Stucture of polypropylene foam produced with egzothermic CBA

The most popular egzothermic chemical blowing agent is azodikarbonamide (ADC) H_2N-NH-(CO)-N=N-(CO)-NH_2. ADC decomposes at 200-220°C with emission of gases in the amount of 220 cm^3/g. The mixture of gases comprises of nitrogen (65%), carbon monoxide (24%), carbon dioxide (5%) and ammonia (5%).

Endothermic chemical blowing agents need heat to continue decomposition, therefore it is easier to control the process just by changing its temperature. For that reason one can produce with endothermic CBA foams of lower cell size. Most popular endothermic blowing agent is a mixture of sodium hydrogen carbonate and citric acid. It decomposes to carbon dioxide and water in a two-stage reaction: first at 130-140°C, second at 180-200°C (Fig. 11).

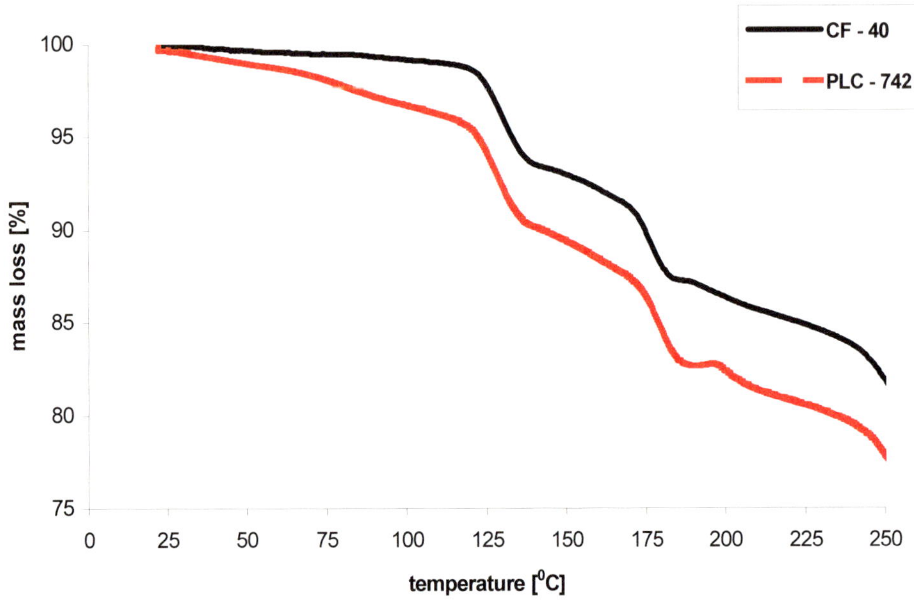

Fig. 11. TG analysis of endothermic chemical blowing agents

Decomposition takes place according to a reaction:

$$C_6H_8O_7 + 3\,NaHCO_3 = \left(C_6H_5Na_3O_7\right)\cdot 2\,H_2O + 3\,CO_2 + H_2O \tag{1}$$

Total amount of gases emitted in the above reaction equals to 120 cm^3/g.

Solubility of gases in polymers is a crucial factor for foaming. If it is high, the saturation time of a polymer with a gas is shorter and lower pressure level is required to keep the gas in the melt. Solubility of carbon dioxide in several polymers is higher than that of nitrogen, therefore it generates more cell nuclei, which is essential for the foam structure. The size of a foaming gas particle is also important for the foaming technology. Since CO_2 particle is small, it diffuses fastly through a polymer, which means that the cell growth rate should be high. However, from the other side, carbon dioxide may escape more easily to atmosphere, thus causing a collapse of the foamed material.

3.3 Technology of polymer foaming

Foaming process consists of four stages:

1. dissolving of a gas in a polymer under high pressure (if polymer is in a solid state) or at high temperature and high pressure (e.g. in a molten polymer);
2. cell nucleation due to a sudden change in the thermodynamic state of a material resulting from its decompression or temperature change;
3. cell growth – their size and density depend on the blowing agent content, process parameters and properties of the polymer;
4. morphology fixation by polymer solidification, e.g. cooling below the glass temperature or crosslinking.

Cellular polymers may be manufactured by saturation of a solid polymer with gas in a high pressure vessel at elevated temperature. First trials concerned foaming of polystyrene with carbon dioxide. It was evidenced that the equilibrium amount of CO_2 adsorbed by PS at 80°C under 240 bar pressure equals to 11.8%. The foam density varied within 0.05-0.85 g/cm^3 depending on the applied temperature, pressure and pressure drop rate. The cell size amounted to 1-70 µm (Arora et al., 1998).

The foam morphology is related to a structure of polymers - the most important are branchings in the polymer chain and crystallinity (Huang et al., 2008; Rachtanpun et al., 2004; Su & Huang, 2010; Li et al., 2007). It is well established that foaming is easier with amorphous polymers like PS than with crystalline ones like polyolefines. Amorphous polymers usually have higher melt strength and are more viscous, therefore cell growth is more difficult, but they hold gas pores better. Crystalline resins are less viscous but difficult to foam due to their chain entanglement and crystals formation at cooling, which disturbs the cell growth process. Foaming of semicrystalline polymers is more complicated than the amorphous polymer foaming, because the gas dissolves exclusively in amorphous regions. That causes a non uniform cell nucleation and irregular foam morphology.

Temperature range of the efficient foaming is limited from above by the polymer degradation temperature and from a bottom by the polymer melt viscosity, which allows for a cell growth. Because it takes place only in the amorphous regions, fast increase in a viscosity of the semi-crystalline polymers upon cooling makes the available temperature range small in comparison to that of amorphous polymers.

Technology of a direct polymer saturation with a gas is useful rather for niche products due to a long time required for saturation because diffusivity of a gas in a solid polymer is low. Nevertheless, it may be used a. o. for scaffold manufacturing. Mooney et al. (1996) have shown that after treating copolymer of D,L lactic acid and glycolic acid with carbon dioxide under pressure of 5.5 MPa for 72 hours followed by fast decompression one observes cells in a polymeric material. Their size equals to ca. 100 µm, while the cell density depends on the process parameters and crystallinity of a polymer, reaching 93%.

Cell nucleation in a polymer starts spontaneously after a sudden change of a thermodynamic state of the system (homogeneous nucleation) or may be induced with addition of small amount of a filler (heterogeneous nucleation) (Lee, 2000; Lee et al., 2006).

Technology of foaming by means of extrusion or injection molding is more widely used, because a gas diffusion in molten polymers is faster and it is facilitated by mixing. These technologies were applied at late nineties at Massachussetts Institute of Technology (MIT) after the successful research on manufacturing PS microfoams by a solid state saturation with supercritical CO_2. Since then the foaming injection molding received an industrial maturity, providing remarkable material savings. Chen et al. (2006) presented an example of polypropylene foaming, showing that the material savings for thin wall (0.5 mm) items equals to 4-9% and for the thick wall (15 mm) samples reached 50%. Unfortunately, in parallel a deterioration of mechanical properties was reported. Similar findings were presented by Bielinski (2004), who tested foaming of polypropylene and polyvinyl chloride using chemical blowing agents. He has found that depending on the CBA content (0,5-2 wt.%) and injection molding parameters the cell size varied in a range of 10-350 µm.

The material savings and lower amount of waste cause that the chemical blowing agents are widely used. In Fig. 12 a yogurt cup made of PS foamed with CBA has been presented. The mass of a cup was decreased for 15-20% in comparison to the non-foamed item. Even if a foam structure is not uniform, the economical and ecological advantages are obvious.

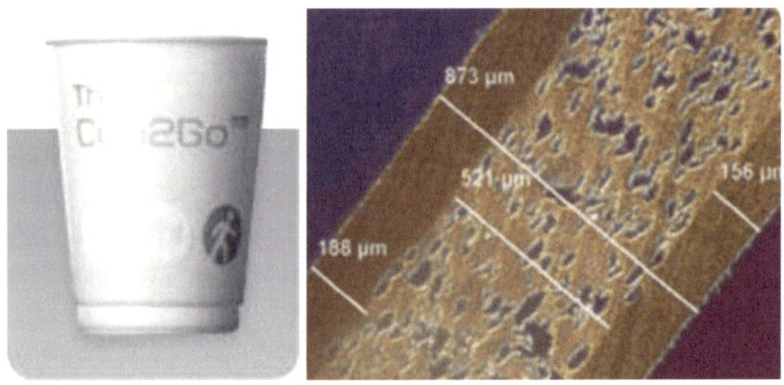

Fig. 12. Foamed polystyrene cup and its wall structure (acc. to www.adeka-palmarole.com)

Technology of microcellular foam extrusion has been extensively studied a.o. by Park and co-workers (Park, 2000; Lee & Park, 2006). After injection of a gas into the polymer melt its diffusion is intensified by mixing, which results in a complete gas dissolution. The equilibrium reached in the extruder is lost after the melt exits the die (Fig. 13). Sudden pressure decrease causes also decrease in a gas solubility, which has to evolve from a polymer in a form of microcells. These sites form nuclei, of which the larger cells grow as more gas appears in the system as the polymer melt-gas solution decompression proceeds. That process is continued until the new equilibrium state is reached or the polymer solidifies.

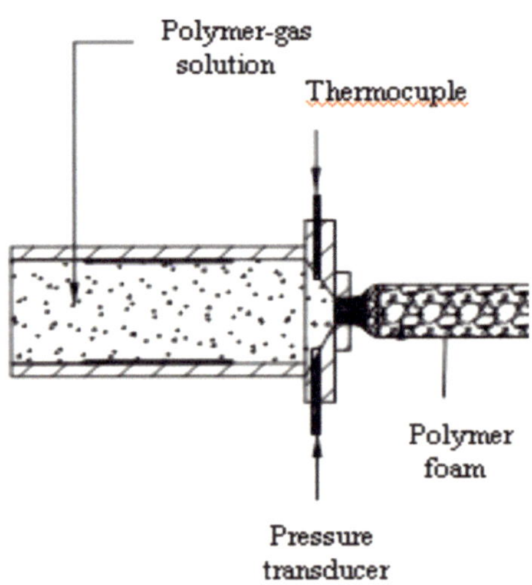

Fig. 13. Polymer foaming extrusion principle

Since foaming involves several processes occuring in a fluid state, therefore knowledge of the viscoelastic characteristics of a molten polymer is very important. Because the dominating stresses are related with shearing and elongational forces, therefore a knowledge of a dependence of the polymer melt viscosity on temperature and on the shear rate is essential. In Figs. 14 and 15 the basic characteristics for three different LDPE grades have been presented.

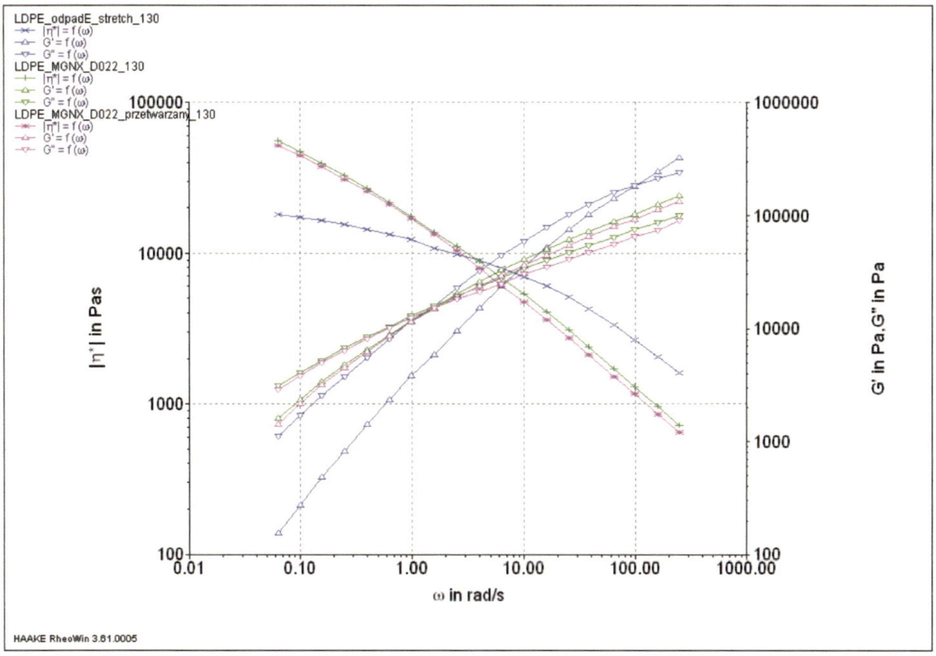

Fig. 14. Viscoelastic characteristics of different LDPE grades (130°C)

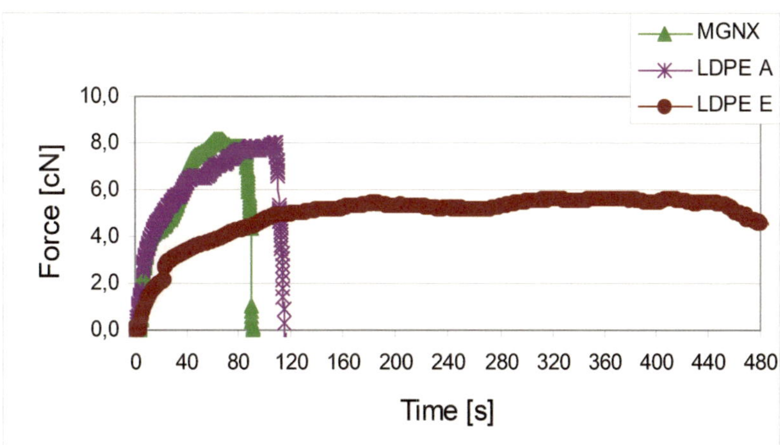

Fig. 15. Melt strength and elongation for different LDPE grades (130°C)

The examples show that different grades of the same polymer differ in pseudoplasticity, therefore they should exhibit diverse melt reaction at low and high deformation rates. That means a different mixing efficiency of the melt with a gas (torque level and gas diffusion rate) and easy or more difficult cell grow (melt viscosity) and resistance to rupture (melt strength). One can expect that significant differences in viscoelastic properties should have an impact on morphology of cellular plastics.

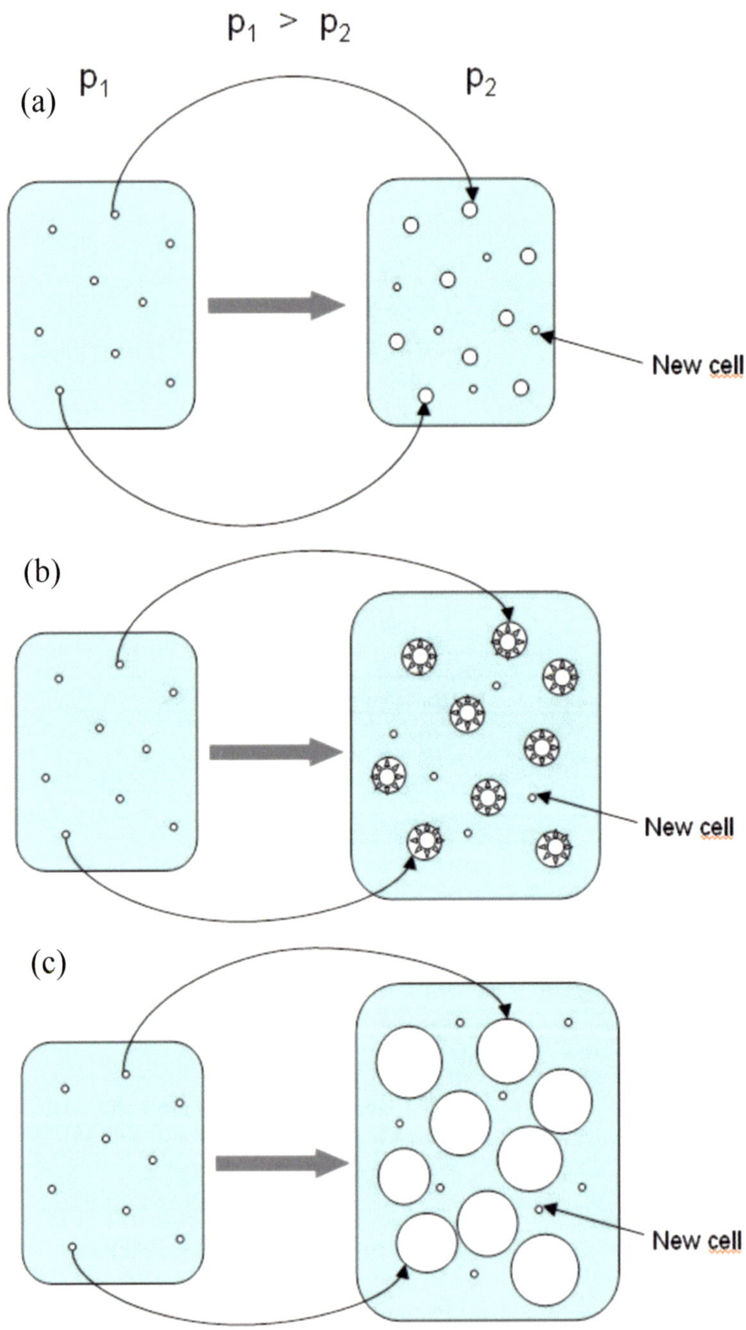

Fig. 16. Schematic representation of extrusion foaming stages

The most important stages of the foaming process (e.g. cell nucleation and growth) have been presented schematically in Fig. 16.

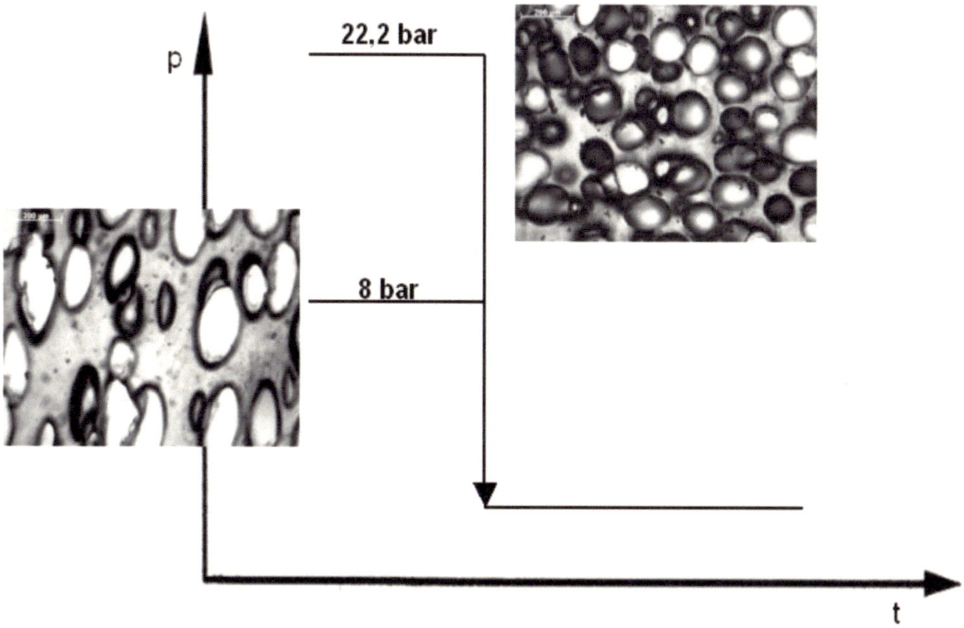

Fig. 17. Cell size and amount in LDPE foamed at different melt pressure

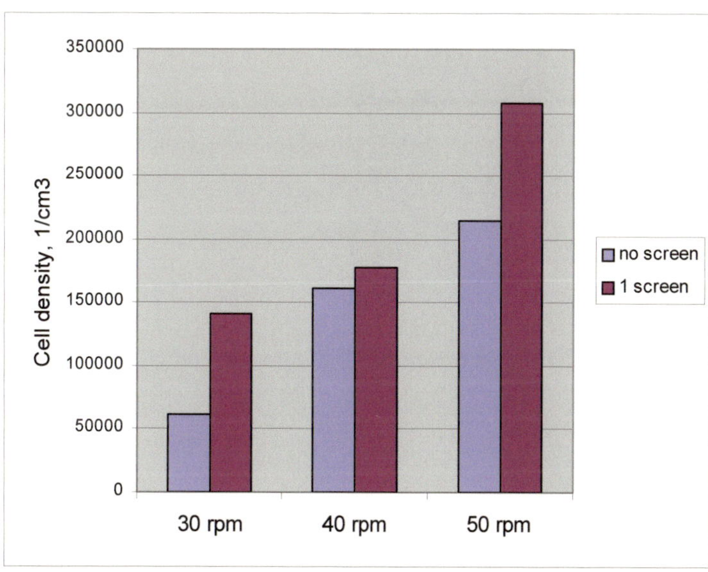

Fig. 18. Cell density in LDPE foam at different processing parameters

Critical factor for foaming is the cell nucleation rate, which is related to a change in the thermodynamic state of a system and phase separation. The gas dissolved formerly in a molten polymer is evolving simultaneously at several sites of the material (a). Since the nucleation rate is much higher than the diffusion rate, the cell nuclei arise first, and only after some time they start growing due to a diffusion of next gas particles which appear as the gas solubility in a polymer melt falls due to the pressure and temperature decrease in a material after it exits the extrusion die.

Number of cells nucleated in the polymer depend on the pressure difference in the melt and atmospheric pressure. High difference developed by a change in the processing parameters or equipment configuration facilitates generation of higher cell density and their smaller size (Figs. 17 and 18).

Cell growth process (b) and (c) depends on several parameters, a.o. on the melt viscosity, melt strength and dynamics of cooling. For lower viscous melts the cells grow faster and are larger than these in a more viscous system (Fig. 19).

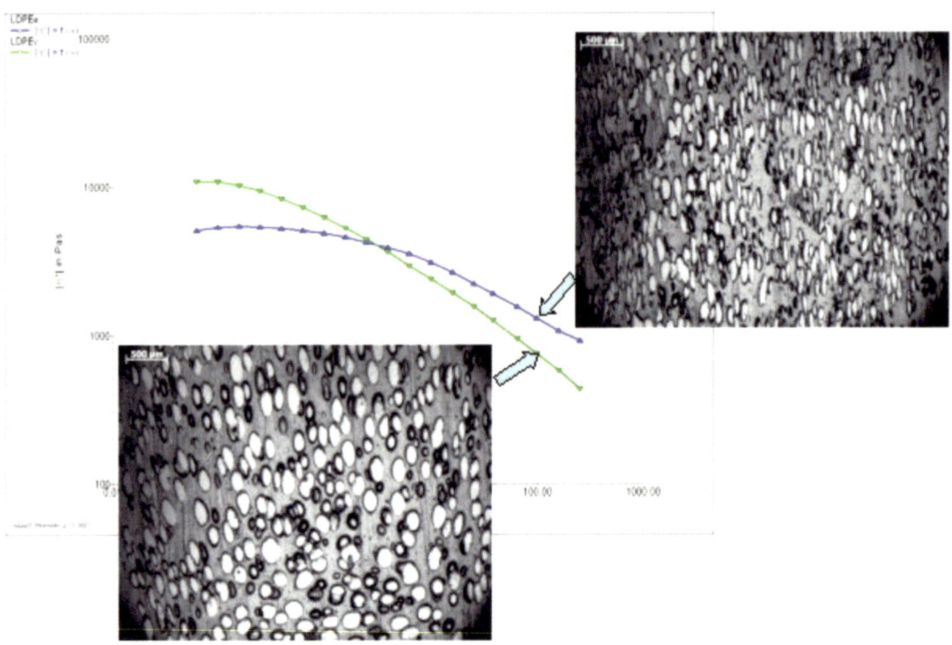

Fig. 19. Cell size and population for different LDPE grades (130°C)

One should consider that the gas dissolved in a polymer causes its plasticization, therefore the melt viscosity decreases markedly, thus modifying the foaming progress. In Fig. 20 a drop in the torque measured at kneading of the polymer melt after addition of an endothermic chemical blowing agent has been presented.

Total amount of cells and their size are related to the polymer melt strength. In case it is high the neighbouring cells may grow individually, however if it is low, the cell walls may disrupt due to an internal gas pressure and coalescence of cells occurs (Fig. 21). Another issue is a gas escape from the polymer melt to atmosphere and a resulting surface warpage.

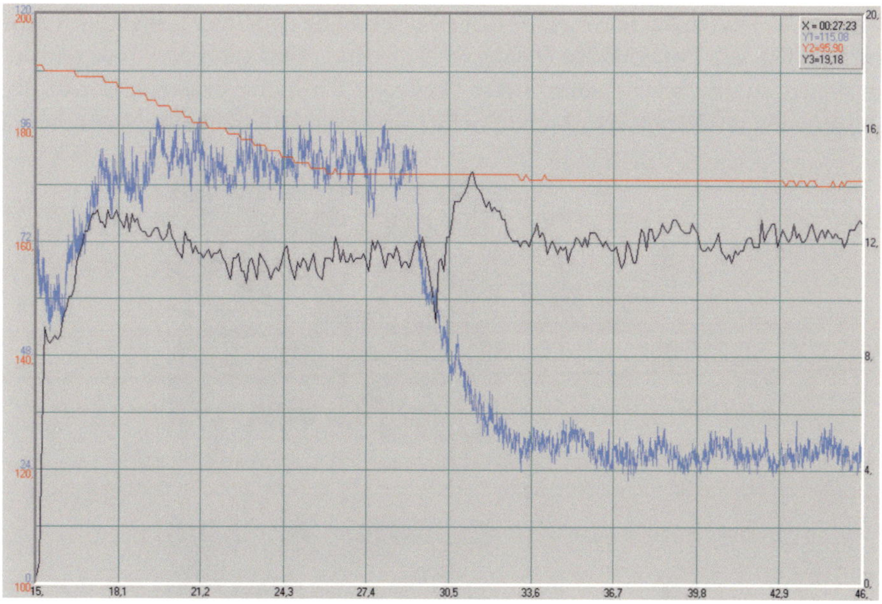

Fig. 20. Mixing torque and temperature change after addition of endothermic CBA to polymer

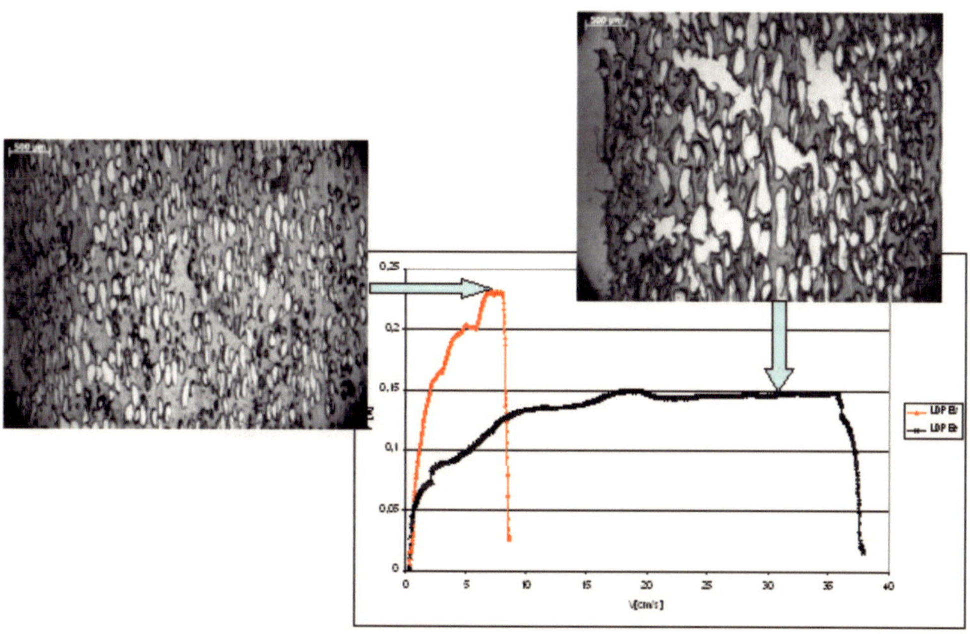

Fig. 21. Foam morphology generated in LDPE of different melt strength (130°C) - foam density 0.576 g/cm³ (left) and 0.447 g/cm³ (right)

The process of cell growth outlined in Fig. 16 stops as soon as the polymer glass temperature reaches the lower foaming limit. It may proceed fastly, because T_g increases with decreasing concentration of a gas in the polymer. At that stage a cooling dynamics is essential for the foam morphology (Fig. 22). Provided the cooling is fast in the cooled calibrator, cells remain at the size generated shortly after exiting a die. However, if cooling proceeds slowly, the cells continue growing until the material solidifies by heat exchange with the ambient atmosphere.

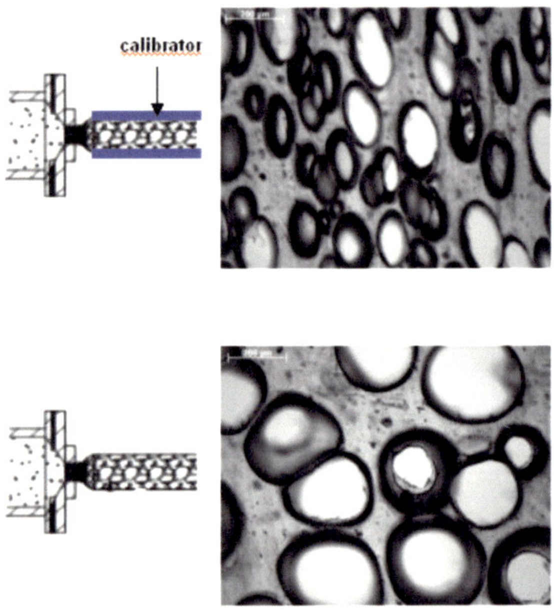

Fig. 22. LDPE foam morphology after cooling in calibrator and cooled freely (bottom)

Research on extrusion microfoaming of polystyrene with supercritical carbon dioxide has shown high influence of the extrusion die temperature and the dynamics of cooling on the resulting foam density and structure. The foaming level amounted to 15-25% (Sauceau et al., 2007).

Extruded microfoams have small cell size and exhibit low cell size distribution (Fig. 23).

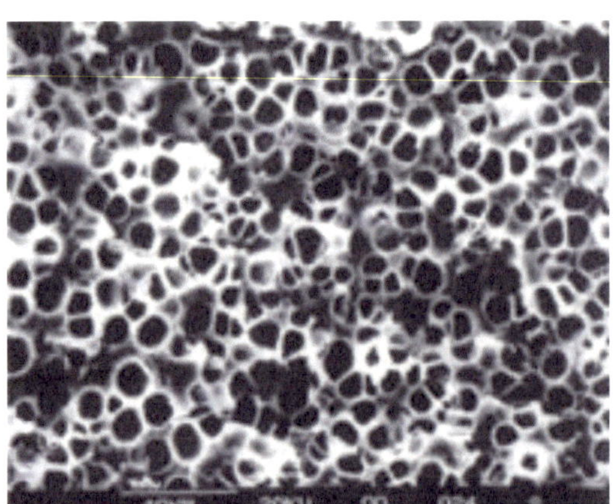

Fig. 23. Structure of microfoam manufactured by extrusion foaming (Park, 2000)

Cell size depend on the pressure within the cells, the melt strength and interfacial tension. The higher is the pressure, the smaller is the melt strength and interfacial tension, the larger cells are generated. As the cell grow and they wall thickness decreases, the coalescence probability of neighbouring cells increases. Behravesh at al. (1998) have found that a coalescence is facilitated with a high shear stress during processing, however its probability decreases with lower melt temperature. Therefore cooling of the gas-polymer solution in a heat exchanger or within a die is advantageous.

Polymer foaming may be performed with single screw extruders of high L/D ratio equiped with mixing elements at a last section of the screw, or with the twin screw or tandem extruders. In any case very important is a precise dosing of a gas, since its surplus causes large cells formation.

In a tandem system (Fig. 24) the first extruder serves for polymer melting and mixing it with the injected gas. In the second extruder further homogenisation of the temperature and gas distribution within a polymer melt should be performed.

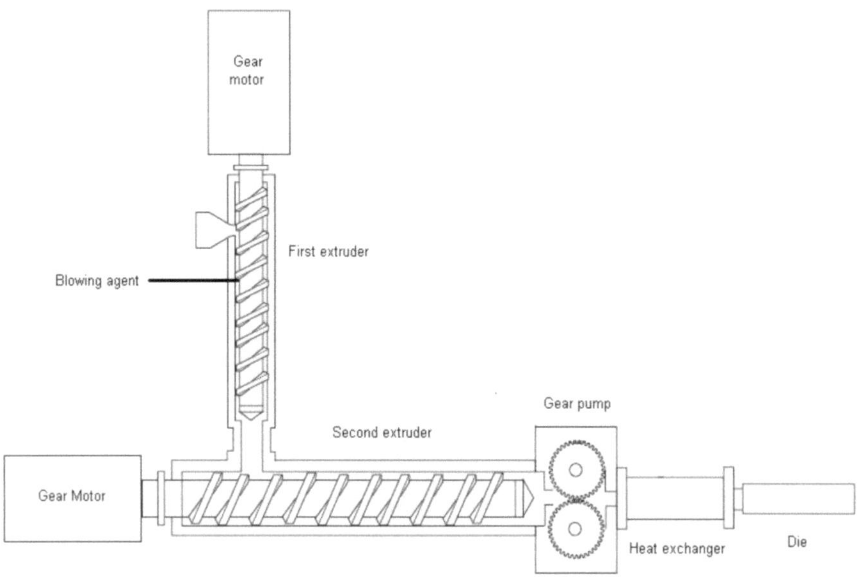

Fig. 24. Tandem extruder system for foaming (acc. to Lee & Park, 2006)

Proper design of every detail of the extrusion foaming set, the processing parameters and composition of the foamed material are crucial factors for the final morphology of a foam. The die geometry is of high importance, because the cell nucleation takes place there. Extensive discussion of the die role for generation of a high cell density has been presented by Xu et al. (2003) at the example of PS foaming with carbon dioxide.

Pressure in a die depends on the temperature and shear forces in the molten polymer. Provided the die temperature is high, the melt viscosity is low, which causes a low pressure drop in the melt after it exits the die and a low number of nuclei. However, at the low die temperature the melt pressure becomes high and the cell density is also high.

Shear stress level in the melt also influences pressure in the die. High pressure is beneficial to the cell density, however its too high level may cause the melt instabilities and cells rupture.

The nuclei number in a foamed polymer melt may be increased by means of fillers addition (Antunes et al., 2009; Khorasani et al., 2010). Foaming of polypropylene with carbon dioxide and addition of talc has shown that the cell density depends on both the filler and foaming agent content, however the nucleating effect of talc was observed only at low talc loading. Similarly, the positive effect of talc on the cell nucleation process and total cell density has been noted at foaming of PP with isopentane. It has been concluded that the nucleation mechanism depends on a size of the gas particles used for foaming. Recently also nanofillers have been reported as efficient cell nucleants for PP and HDPE foams (Khorasani et al., 2010).

Interesting properties exhibit cellular plastics manufactured from polymer blends. Basing on a knowledge on polymer melt rheology and blends morphology one can generate bi-modal distribution of cells located in different polymer domains. That idea has been presented for PPE/SAN blend foaming with carbon dioxide (Ruckdaeschel et al., 2007).

Another modification of foaming technology presents mixing of polymer with CBA, partial crosslinking of the polymer and than decomposition of the blowing agent with evolution of a gas. Cell structure can be limited by means of the crosslinking level, thus microfoams can be manufactured with such technology (Rodriguez-Perez et al., 2008). Microcellular polyethylene is used a.o. for pipe insulations, gaskets and in a healthcare as the wound dressing or sensitive skin protection.

4. Foaming devices

Extensive research on polymer foaming which was performed at MIT, Massachusetts (USA) received commercialisation of obtained results. For that purpose Trexel Inc. company was established, which has developed MuCell® technology of microfoams manufacturing by means of a supercritical gas [9]. Its main advantage is low consumption of polymer, high mechanical properties (including the impact strength), good thermoinsulating properties, as well as high surface quality of manufactured details. Such characteristics have been received due to the low cell size (5 - 50 μm), their regular shape and high cell density (10^8 cells/cm^3).

Applying MuCell® technology for injection molding requires mounting the gas injector in a proper place of the cylinder (Fig. 25), selecting a screw geometry enabling thorough mixing of a gas with molten polymer and adjusting the gas injection with the overall injection cycle time. Decompression of the gas-melt solution should take place just in a mould. MuCell® allows for a shorter cycle time (15-35%), lower density of details (6-12%), reduction of internal stresses and high dimension stability.

While applying MuCell® technology for extrusion foaming one should consider also the gas injector, proper screw design and well selected die geometry. Proper matching of all parameters should result in a high pressure in the melt, which is crucial for efficient foaming. Foaming extrusion by MuCell® allows a.o. manufacturing HDPE pipes with cell size of 15 μm and material density of 0,54 g/cm^3, EPDM gaskets for automotive applications etc.

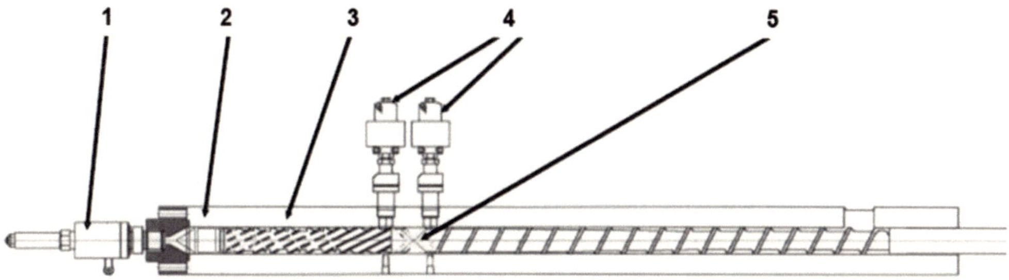

Fig. 25. Gas injection system in MuCell® technology (acc. to www.trexel.com)

1. die closure
2. control ring
3. screw mixing section
4. gas injector
5. valves

Another technology of cellular profiles extrusion offers Sulzer Chemtech (Switzerland). Optifoam™ technology anticipates a gas injector between the screw and extruder head (www.sulzerchemtech.com). Gas (CO_2 or N_2) is injected into the gas melt through a fluid injection nozzle made of sintered metal. The mixture of a gas and a polymer is next thoroughly homogenized in the static mixer section (Fig. 26).

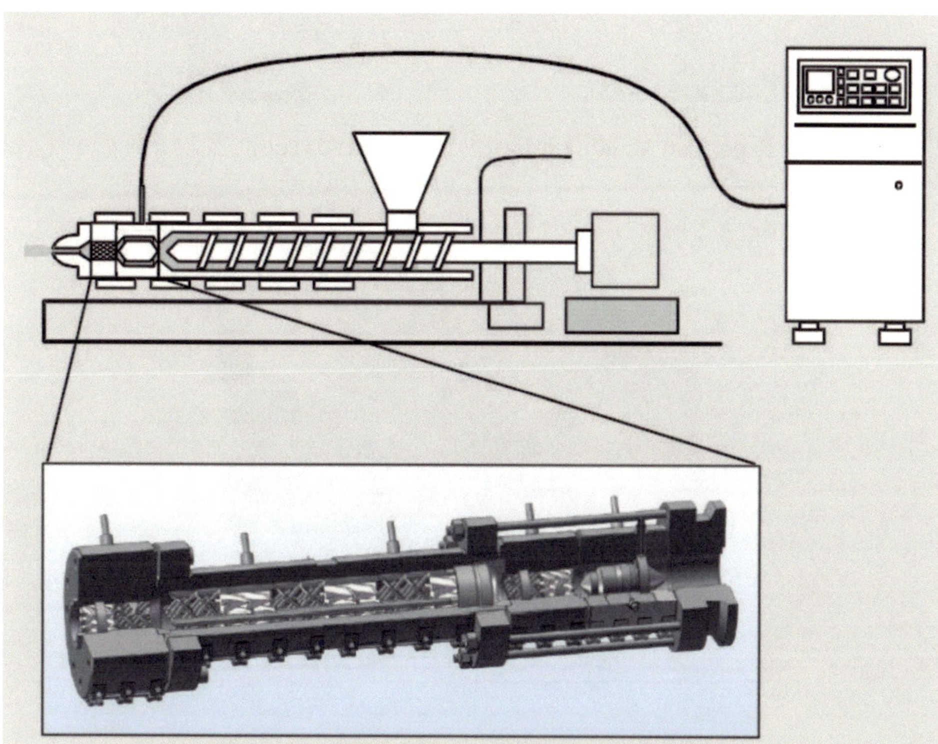

Fig. 26. Extrusion foaming set for Optifoam™ technology (www.sulzerchemtech.com)

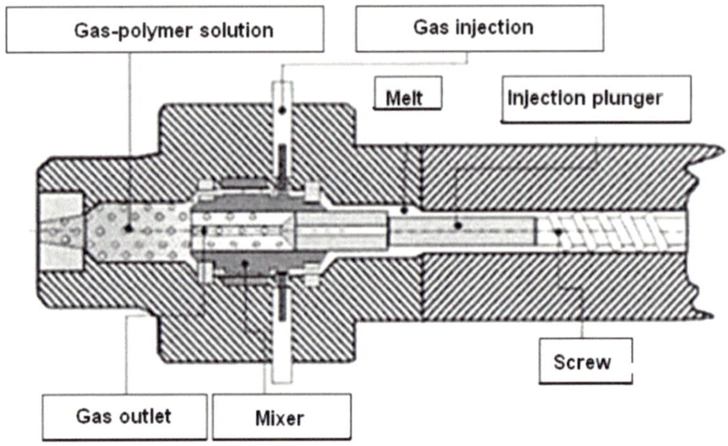

Fig. 27. Mixing unit of the ErgoCell® foaming injection molding (Sauthof, 2003)

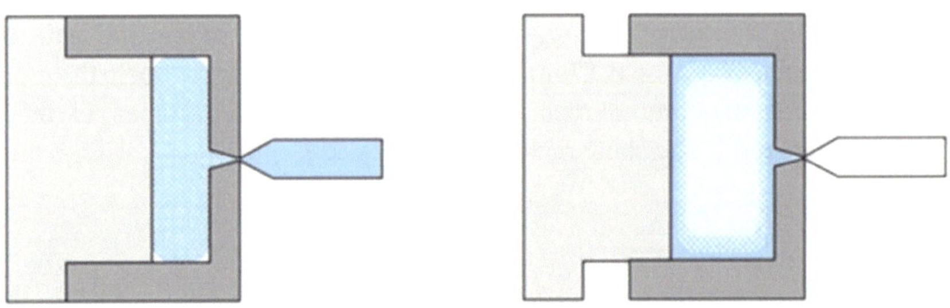

Fig. 28. Core Back Expansion Molding principle (www.mazda.com)

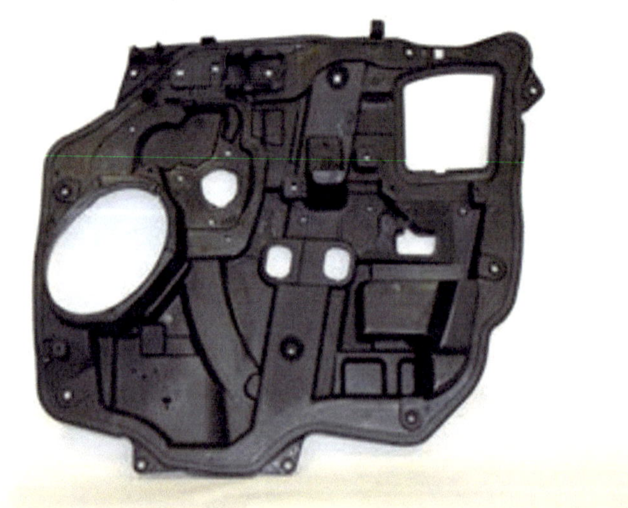

Fig. 29. Mazda2 door panel by Core Back Expansion Molding ((www.mazda.com)

ErgoCell® foaming technology developed by Demag Ergotech GmbH (Mapleston, 2002) assumes two stage process, with injection of a gas into mixing device located between the stationary screw and a melt accumulator equiped with an injection plunger (Fig. 27). ErgoCell® allows for a shorter cycle time, lower weight details (6-25%), reduction of internal stresses and surface defects.

Car manufacturer Mazda also applied solution of the supercritical gas (CO_2 or N_2) in the polymer melt. According to the Core Back Expansion Molding technology the material is injected into a mold and once the foamed polymer has filled up the mould, its volume is increased by moving the back of the mould (Fig. 28). Weight of the door panel for Mazda2 (Fig. 29) made with such technology is lower for 20%, while its stiffness is for 16% higher.

5. Foaming of biocomposites

Biocomposites used in the construction and automotive sector are frequently called „artificial wood" because their many properties and appearance are like wood (Matuana et al., 1998; Migneault et al., 2008; Bledzki et al., 2008). Unfortunately, the density of biocomposites, even if markedly lower than that of glass fiber reinforced composites, is still twice as high as the natural wood density. That drawback can be reduced by foaming of biocomposites that are lighter and feel more like real wood (Rodrigue et al., 2006; Guo et al., 2004, Bledzki & Faruk, 2006, Kozlowski et al., 2010). The earliest known foamed and wood-filled thermoplastics were based on polystyrene (PS) - this amorphous polymer is a perfect bubble catcher. Wood flour itself has been proved as an efficient nucleating filler in polyethylene foamed with azodicarbonamide (Rodrigue et al. 2006). As far as length of natural fibers is concerned, short fibers (75-125 µm) are favorable for foaming, since they do not disturb the cell growth process, like do the long fibers (4-25 mm).

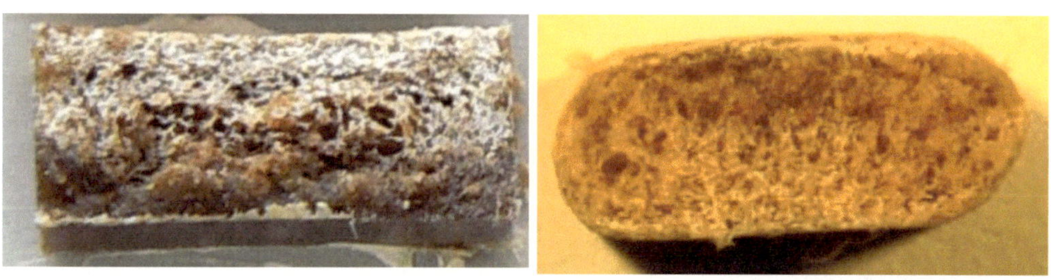

Fig. 30. Cross section of foamed PP filled with 30 wt.% of wood flour - injection molded (left) and extruded profile (right)

Selection of the polymer matrix is very important for properties of biocomposites. Because cellulose fibers are polar, the hydrophobic matrices (like polyolefines) need addition of adhesion promoters in order to facilitate regular fiber distribution and efficient stress transfer across the composite during deformation in a molten state and during/after solidification.

Cellular biocomposites can be manufactured both by the extrusion or injection molding technology, however the extrusion foaming provides better results (Fig. 30), as it allows for a more precise process control.

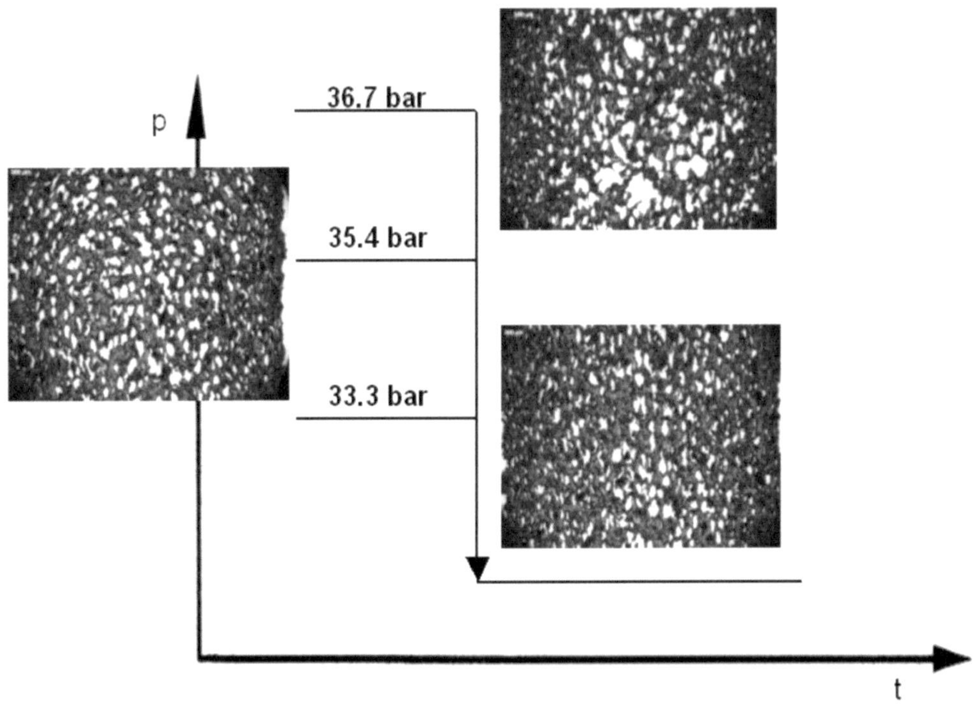

Fig. 31. Dependence of foam morphology on melt pressure for LDPE/wood flour composite

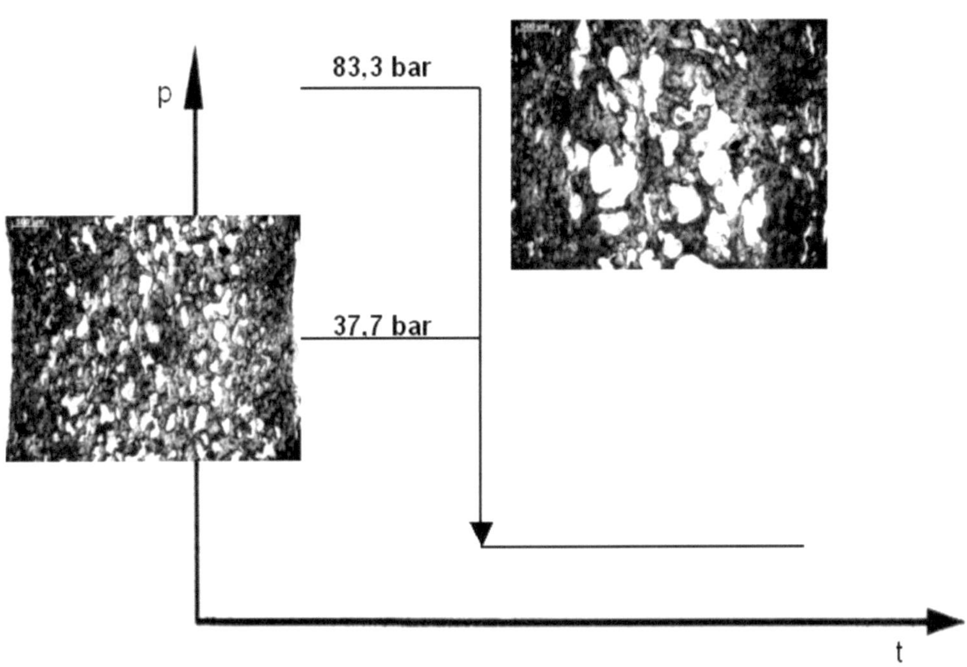

Fig. 32. Dependence of foam morphology on melt pressure for PP/wood flour composite

The study on foaming of biocomposites confirmed the findings formulated for non-filled polymers. High drop in the melt pressure between that in a die and ambient is favourable for manufacturing foams of fine, regular cells (Fig. 31), however if it is too high, that causes a cell damage and foam collapse (Fig. 32).

In general foaming is more difficult due to the high melt viscosity of biocomposites and low melt strength, however the results of research reported in recent years for foaming of wood composites with the chemical or physical blowing agents are promising. The technology of extrusion foaming seems to be fully controled and the profiles manufactured looks like wood outside (Fig. 33) and in a cross-section (Fig. 34). The possibility of foaming composites filled with cellulose fibers make them ideal candidates for the low weight and thermal insulating engineering materials in all transport modes (Fig. 35).

Fig. 33. Extrusion foaming of PP filled with 30 wt.% of wood flour

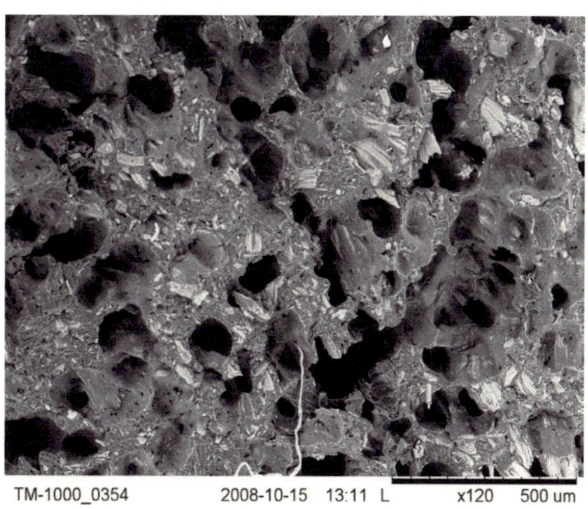

Fig. 34. Cross-section of PP/woof flour composite

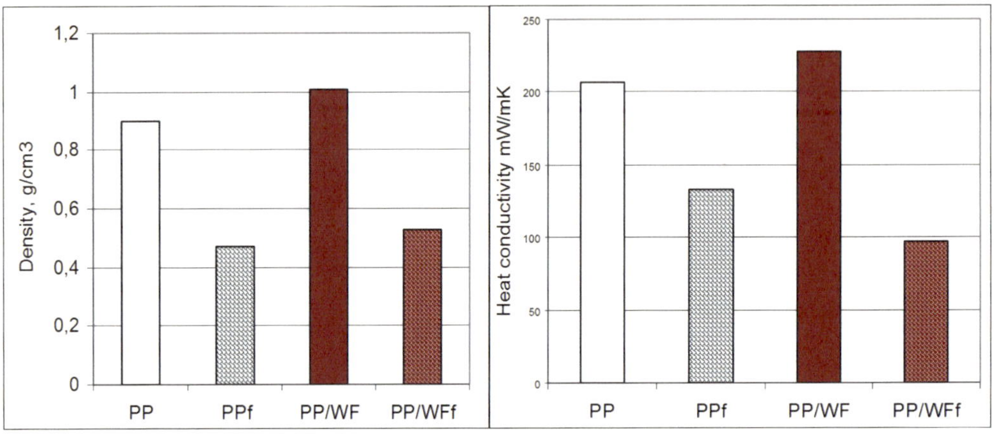

Fig. 35. Properties of PP, PP biocomposite and their foams

6. Acknowledgements

This work has been sponsored by the Polish structural grants POIG.01.03.01-00-018/08-00 under the Innovative Economy scheme. The Author wish to thank Dr. A. Kozlowska and Dr. S. Frackowiak for their valuable help in selected measurements.

7. References

Antunes, M.; Velasco, J.I.; Realinho, V. & Solorzano E. (2009). Study of the cellular structure heterogeneity and anisotropy of polypropylene and polypropylene nanocomposite foams. *Polymer Engineering and Science*. Vol.49, No.12, pp. 2400-2413, ISSN 1548-2634

Arora, K. A.; Lesser, A. J. & McCarthy, T. J. (1998). Preparation and Characterization of Microcellular Polystyrene Foams Processed in Supercritical Carbon Dioxide. *Macromolecules*, Vol.31, No.14, pp. 4614-4620, ISSN 0024-9297

Behravesh, A. H.; Park, C. B. & Venter R. D. (1998). Approach to the Production of Low-Density, Microcellular Foams In Extrusion. *Proceedings of ANTEC'98*, pp. 1958-1967, ISBN 1566766699, Atlanta, USA, April 26-30, 1998

Bledzki A. K., Faruk O., (2006). Microcellular injection molded wood fiber-PP composites: Part II - Effect of wood fiber length and content on cell morphology and physico-mechanical properties. *Journal of Cellular Plastics*, Vol.42, No.1, pp. 77-88, ISSN 1530-7999

Bledzki, A. K.; Faruk O. & Mamun A. A. (2008). Influence of compounding processes and fibre length on the mechanical properties of abaca fibre-polypropylene composites. *Polimery*, Vol.53, No.2, pp. 120-125, ISSN 0032-2725

Bieliński, M. (2004). *Techniki porowania tworzyw termoplastycznych*, Wydawnictwo Uczelniane ATR, ISBN 83-89334-86-0, Bydgoszcz, Poland

Chen, X.; Feng J. J. & Bertelo C. A. (2006). Plasticization effect on bubble growth during polymer foaming. *Polymer Engineering and Science*. Vol.46, No.1, pp. 97-107, ISSN 1548-2634

Cooper, A. I. (2000). Synthesis and processing of polymers using supercritical carbon dioxide. Journal of Materials Chemistry, Vol.10, No.2, pp. 207-234, ISSN 0959-9428

Ellison, G. C.; McNaught, R. & Eddleston E. P. (2004). *The use of natural fibres in nonwoven structures for applications as automotive component substrates,* R&D Report NF0309, Ministry of Agriculture Fisheries and Food, London, United Kingdom

Guo, G.; Rizvi, G. M.; Park, C. B. & Lin, W. S. (2004). Critical Processing Temperature in Manufacture of Fine-Celled Plastic/Wood-fiber Composite Foams. *Journal of Applied Polymer Science.* Vol.91, No.1, pp. 621-629, ISSN 0021-8995

http://www.adeka-palmarole.com

http://www.akzonobel.com/expancel/application_areas/benefits/porous_structure/index.aspx

http://www.azflex.pl/izolacje/producent/armacell.php

http://www.cinpres.net/index.php/cinpres-technologies/igm

http://www.mazda.com/mazdaspirit/env/engine/mold.html

http://www.specialchem4polymers.com/resources/articles

http://www.sulzerchemtech.com/en/portaldata/11/Resources//brochures/mrt/Optifoam_e.pdf

http://www.trexel.com/injection-molding-solutions/index.php

Huang, H.-X.; Wang, J.-K. & Sun, X.-H. (2008). Improving of Cell Structure of Microcellular Foams Based on Polypropylene/High-density Polyethylene Blends. *Journal of Cellular Plastics*, Vol.44, No.1, pp. 69-85, ISSN 1530-7999

Institute of Materials, Minerals and Mining. (2004). Lightweight Platic Panels Manufactured from Recycled Plastics - Panels for Commercial Vehicles. *Materials World*, Vol. 12, No. 6, pp. 25-27, ISSN: 0967-8638

Khorasani, M. M.; Ghaffarian, S. R.; Babaie, A. & Mohammadi N. (2010). Foaming Behavior and Cellular Structure of Microcellular HDPE Nanocomposites Prepared by a High Temperature Process. *Journal of Cellular Plastics*, Vol.46, No.2, pp. 173-190, ISSN 1530-7999

Klesov, A. A. (2007). *Wood-plastic composites*, Wiley, ISBN 978-0470148914, Hoboken, USA

Kozlowski M. & Kozlowska, A. (2005). Biocomposites from waste plastics, *Proceedings of Global Plastics Environmental Conference 2005*, Atlanta, USA, February 23-25, 2005

Kozłowski, M.; Kozłowska, A. & Frąckowiak, S. (2010). Materiały polimerowe o strukturze komórkowej. *Polimery.* Vol.55, No.10, pp. 726-739, ISSN 0032-2725

Lee, J. W. S. & Park C. B. (2006). Use of Nitrogen as a Blowing Agent for the Production of Fine-Celled High-Density Polyethylene Foams. *Macromolecular Materials and Engineering.* Vol.291, No.10, pp. 1233-1244, ISSN 1438-7492

Lee S.T. (2000). Foam Nucleation in Gas-Dispersed Polymeric Systems, In: *Foam Extrusion. Principles and Practice,* Lee, S.T. (Ed.), CRC Press, Boca Raton, USA

Lee, S. T.; Park, C. B. & Ramesh, N.S. (Eds.). (2006). *Polymeric Foams: Science and Technology,* CRC Press, Boca Raton, USA

Li, G.; Wang, J.; Park, C.B. & Simha R. (2007). Measurement of Gas Solubility in Linear/Branched PP Melts. *Journal of Polymer Science Part B-Polymer Physics.* Vol.45, No. 17, pp. 2497-2508, ISSN 0887-6266

Mapleston, P. (2002). Modern Plastics Worldwide, November, p.31, ISSN 0026-8275

Matuana, L. M.; Park, C. B. & Balatinecz J. J. (1998). Cell Morphology and Property Relationships of Microcellular Foamed PVC/Wood-Fiber Composites. *Polymer Engineering and Science.* Vol.38, No.11, pp. 1862-1872, ISSN 1548-2634

Migneault, S.; Koubaa, A.; Erchiqui, F.; Chaala, A.; Englund, K.; Krause, C. & Wolcott M. (2008). Effect of fiber length on processing and properties of extruded wood-fiber/HDPE composites. *Journal of Applied Polymer Science*. Vol.110, No.2, pp. 1085-1092, ISSN 0021-8995

Mooney, D.J.; Baldwin, D.F.; Suh, N.P.; Vacanti, J.P. & Langer, R. (1996). Novel approach to fabricate porous sponges of poly(D,L-lactic-co-glycolyc aicd) without the use of organic solvents. *Biomaterials*. Vol.17, No.14, pp. 1417-1422, ISSN 0142–9612

Oksman Niska, K. & Sain, M. (Eds.). (2008). *Wood-polymers Composites*, CRC Press, ISBN 978-1-4200-7611-0, Boca Raton

Park, C. B. (2000). Continuous Production of High-Density and Low-Density Microcellular Plastics, In: *Extrusion* w *Foam Extrusion. Principles and Practice*, Lee S.-T. (Ed.), CRC Press, Boca Raton, USA

Rachtanpun, P.; Selke, S. E. M. & Matuana L. M. (2004). Effect of the high-density polyethylene melt index on the microcellular foaming of high-density polyethylene/polypropylene blends. *Journal of Applied Polymer Science*, Vol.93, No.1, pp. 364-371, ISSN 0021-8995

Rodrigue, D.; Souici, S. &, Twite-Kabamba, E. (2006). Effect of wood powder on polymer foam nucleation. *Journal of Vinyl and Additive Technology*. Vol.12, No.1, pp. 19-24, ISSN 1548-0585

Rodriguez-Perez M. A., Almanza O., Ruiz-Herrero J. L., de Saja J. A. (2008). The effect of processing on the structure and properties of crosslinked closed cell polyethylene foams. *Cellular Polymers*. Vol.27, No.3, pp. 179-200, ISSN 0262-4893

Rosato, D. (April 2010). Progress in Sustainable Foam Blowing Agent Technology. Available from: SpecialChem - Apr. 19,2010

Ruckdaeschel, H.; Altstaedt, V. & Mueller A.I I. (2007). Foaming of polymer blends - chance and challenge. *Cellular Polymers*. Vol.26, No.6, pp. 367-380, ISSN 0262-4893

Sauceau, M.; Nikitine, C.; Rodier, E. & Fages J. (2007). Effect of Supercritical Carbon Dioxide on Polystyrene Extrusion. *Journal of Supercritical Fluids*, Vol.43, No.2, pp. 367-373, ISSN 08968446

Sauthof, R. (2003). Physical foaming with Ergocell. Proceedings of Blowing Agents and Foaming Processes, Munich 19-20 May, 2003, p. 91, paper 10

Schellenberg, J. & Wallis, M. (2010). Dependence of Thermal Properties of Expandable Polystyrene Particle Foam on Cell Size and Density. *Journal of Cellular Plastics*, Vol.46, No.3, pp. 209-222, ISSN 1530-7999

Su, F.-H. & Huang, H.-X. (2010). Rheology and melt strength of long chain branching polypropylene prepared by reactive extrusion with various peroxides. *Polymer Engineering and Science*. Vol.50, No.2, pp. 342-351, ISSN 1548-2634

Xu, X.; Park, C.B.; Xu, D. & Pop-Iliev, R. (2003). Effects of Die Geometry on Cell Nucleation of PS Foams Blown with CO2. *Polymer Engineering and Science*. Vol.43, No.7, pp. 1378-1390, ISSN 1548-2634

15

The Performance Envelope of Spinal Implants Utilizing Thermoplastic Materials

Daniel J. Cook, Matthew S. Yeager,
Shamik Chakraborty, Donald M. Whiting and Boyle C. Cheng
Drexel University – Allegheny General Hospital
USA

1. Introduction

Back pain is the second most common reason for physician visits in the United States, and affects up to 84 percent of patients at some point in their lives (Deyo & Tsui-Wu 1987). For many patients, neck and back pain is often due to injury or spinal instability through trauma, disease or degeneration. This can cause impingement of neural structures resulting in pain, numbness, or weakness. Also, this can be a consequence of changes in the vertebral bodies or degenerative changes in the intervertebral cartilaginous discs. Treatment of such disorders may require surgery if the pain or neurologic symptoms prove to be intractable to conservative treatments such as physical therapy or pain medications.

The spine is a load-bearing structure made of ligaments, tendons, and bone that allows for a functional range of motion while protecting the spinal cord. The vertebral bodies are joined by two bilateral facet joints in the posterior aspect of the spinal column and are separated at each level by a cartilaginous intervertebral disc. This three-joint complex, illustrated in Figure 1, comprises the smallest mobile segment of the spine, commonly referred to as the functional spinal unit (FSU). The posterior joints are diarthrodial joints, which include articular cartilage, synovial membrane, and a joint capsule (Yong-Hing et al. 1980).

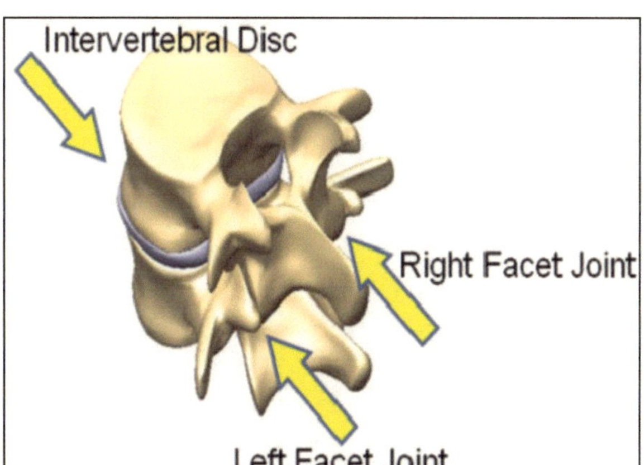

Fig. 1. Illustration of Three-Joint Complex

Spinal stability refers to the ability of the spine segment to limit excessive displacement that can possibly result in incapacitating deformity or neurologic symptoms due to impingement on the spinal cord and nerve roots (White & Panjabi 1990). Factors that can contribute to instability may affect the bones, disc, joints, or ligaments. These include trauma, tumors, infections, inflammatory diseases, connective tissue disorders, congenital disorders and degenerative disorders (Yong-Hing & Kirkaldy-Willis 1983). It is important to note that the surgical procedures performed in the interest of relieving neurologic symptoms may themselves contribute to instability.

Degenerative changes may affect the vertebrae and the intervertebral disc as well. The viscoelastic properties of the cartilaginous disc allow it to act as a shock absorber for the spine. As the disc degenerates, the ability for it to handle such stresses also changes (Panagiotacopulos et al. 1987). As peripheral fibers of the disc are stressed, rupture may occur resulting in herniation of disc material outwards to compress nerve roots or the spinal cord, potentially causing neurological symptoms including pain and weakness (Benzel 2004).

Surgical treatment of the spine often involves widening of the spinal canal and decompression of the spinal cord, the removal of osteophytes or disc material impinging on neurologic structures, and careful consideration of alignment of the spinal segment (Benzel 2004). These procedures may lead to destabilization of the FSU. Vertebral fusion helps provide stabilization of the spine, and can prevent further deformation and additional neurologic symptoms (Gibson & Waddell 2005). If spinal fusion is performed, the intervertebral disc is removed and the endplates of the vertebral bodies are bridged by a bone graft, often with an interbody device that allows for preservation of disc height and appropriate sagittal balance. Bony union connects the two vertebrae together into a continuous structure providing the needed stability; this structure along with any new bone is often referred to as a "fusion mass." Clinically, if the fusion is successful, it is referred to as an arthrodesis. Immobilization of the vertebral bodies has been shown to significantly enhance the success of fusion (Bridwell et al. 1993). Currently, internal spinal instrumentation is thought to be the best therapy to ensure a good clinical outcome in this regard.

2. Spinal instrumentation

Spinal instrumentation is designed to stabilize the spine and prevent excessive motion at the affected segment as the bones fuse together. It provides immediate stability of the affected segment of the spine and obviates the need for external fixation devices such as rigid collars, braces, and halo traction. This allows for a greater quality of life for patients immediately following surgery. These internal fixation devices such as screws, plates, and rods are affixed to the vertebral bodies and combine to form an instrumentation construct. This construct takes the load-bearing responsibility of the affected spinal segment until fusion has occurred. The instrumentation construct is therefore a temporary load-bearing adjunct to fusion (Benzel 2004). When arthrodesis is achieved, the fusion mass will become the principal bearer of load on the FSU, and the instrumentation becomes obsolete. Therefore, a principle of spine surgery is that spinal instrumentation is to maintain spinal stability until fusion occurs.

Modern spinal fixation devices have been made of biocompatible metallic alloys such as stainless steel and titanium. While these have become standard due to their strength and fatigue resistance, they possess radiographic challenges and biomechanical limitations. First, and foremost, metallic materials are radiologically problematic. Artifacts created by metals in computerized tomography (CT) and magnetic resonance (MR) imaging, as well as obfuscation of bone by metals in planar x-rays, inhibit visualization of new bone growth thus, making it difficult for physicians to assess progression of arthrodesis. Second, implants that are stiffer than natural bone are subject to so-called stress shielding. When two bones are fused together, the state of compressive forces between them determines the extent of modeling or remodeling. According to Wolff's law, bone apposition within a fusion graft is governed in part by the state of compressive stresses within it. Because of this it has been proposed that stiff metallic implants may shield the graft from the stress required for fusion, delaying or preventing the process. This can induce iatrogenic effects such as device-related osteopenia, intervertebral device protrusion into a neighboring vertebral body, and fracture or instability (Lippman et al. 2004). Implants developed out of less stiff materials may prevent this stress shielding and foster better clinical outcomes compared to traditional devices.

3. Thermoplastic spine implants

Thermoplastic polyetheretherketone (PEEK) polymers, a subset of the polyaryletherketone or PAEK class of polyaromatic polymers, were initially considered for use in spinal applications to overcome the aforementioned limitations due to their inherent material properties, such as biocompatibility, radiolucency, and the ability to tailor their elastic moduli with fiber reinforcement such as carbon filament. Neat or unfilled PEEK has an elastic modulus of approximately 3.6 GPa, however this can be tailored to closely match cortical bone (18 GPa) or even titanium alloy (110 GPa) (Kurtz & Devine 2007). These properties, along with validation of their strength, wear, creep and fatigue resistance have pushed PEEK implants to be the most viable and attractive alternatives to metallic spinal implants to date. (Brown et al. 1990; Brantigan & Steffee 1993; Kurtz & Devine 2007). Figure 3 is an x-ray taken from a patient following implantation of a PEEK cage along with a pedicle screw and titanium rod system. Notice the radiolucency of the PEEK cage; only the radiopaque markers inside the implant are visible.

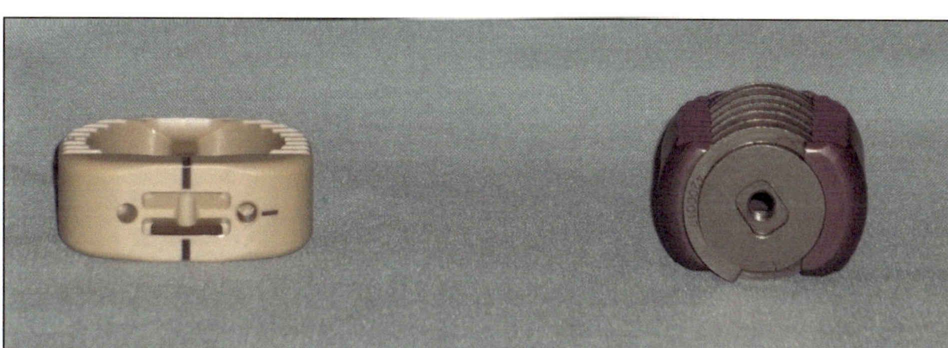

Fig. 2. PEEK Lumbar Interbody Cage (Left) alongside Titanium Lumbar Interbody Cage (Right)

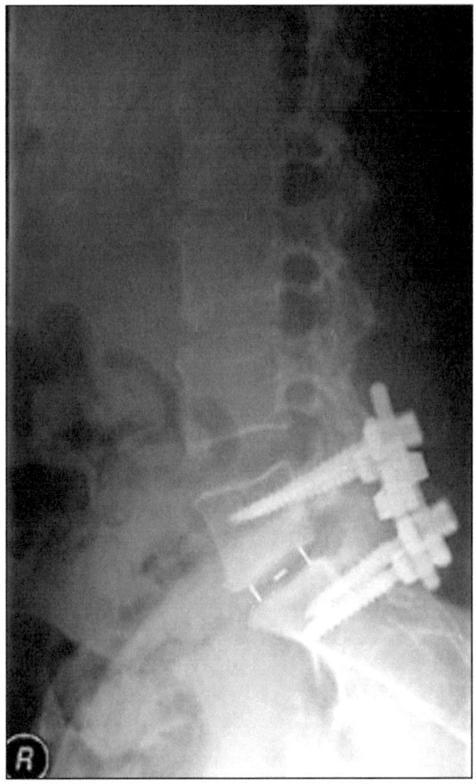

Fig. 3. Planar X-ray of Patient Immediately after Surgery with PEEK Cage and Titanium Rod Construct Implanted in the Lumbar Spine

Although PEEK implants offer improvements over titanium and titanium alloys, they are still subject to such complications as migration, subsidence, extrusion, wear debris and late foreign body reaction due to the fact that they are permanently implanted (Alexander et al. 2002; Smith et al. 2010). Furthermore, additional surgeries may be required to remove the constructs. These limitations have led clinicians to investigate bioresorbable polymers for use in surgical applications. Alpha-polyesters, mainly polylactic acid (PLA) and polyglycolic acid (PGA), are the primary bioresorbable materials used in clinical applications, and over the past decade such compounds have been used extensively in craniomaxillofacial, orthopedic and trauma applications for bone reconstruction and fixation.

Resorbable sutures and films have been found to be extremely useful in many surgical procedures. PLA sheaths have been used to protect neural elements from fibrotic adhesions and promote iliac crest harvest site reconstruction, while sutures and suture anchors have been extensively employed to mitigate foreign body effects and the need for further surgeries (Wang et al. 2002; Welch et al. 2002; Klopp et al. 2004).

Resorbable rod and screw systems have been well documented in the treatment of extremity fracture, however little experience has been gained with these devices in spinal surgery (Tormala et al. 1991; Majola et al. 1992; Springer et al. 1998). Bezer et al. recently employed resorbable self-reinforced PLA rods for use in posterior lumbar fixation in a rabbit model

(Bezer et al. 2005). They were found to be comparable to titanium, however no studies to date have examined the the ability of these devices to withstand higher loads in larger animal models or the human spine.

Resorbable plate and screw systems (RPSSs) have been used as far back as the 1970's, shortly after the introduction of their clinical potential, to fix mandibular fractures (Kulkarni et al. 1966; Kulkarni et al. 1971). More recently, RPSSs have been used to successfully treat distal radial fractures with no statistical difference compared to a titanium control (van Manen et al. 2008). RPSSs have also been shown to successfully treat rib fractures in trauma patients (Mayberry et al. 2003). Once adapted for use in spinal applications, RPSS's have been investigated for supplemental anterior fixation to cervical interbody fusion constructs. Although prospective randomized studies to date are limited, many retrospective analyses have concluded resorbable anterior cervical plating as a reasonable alternative to metal (Vaccaro et al. 2002; Franco et al. 2007; Tomasino et al. 2009). RPSSs have also been considered for supplemental lumbar fixation (DiAngelo et al. 2002), however in-vivo loading conditions of the lumbar spine may prove too intense for this application.

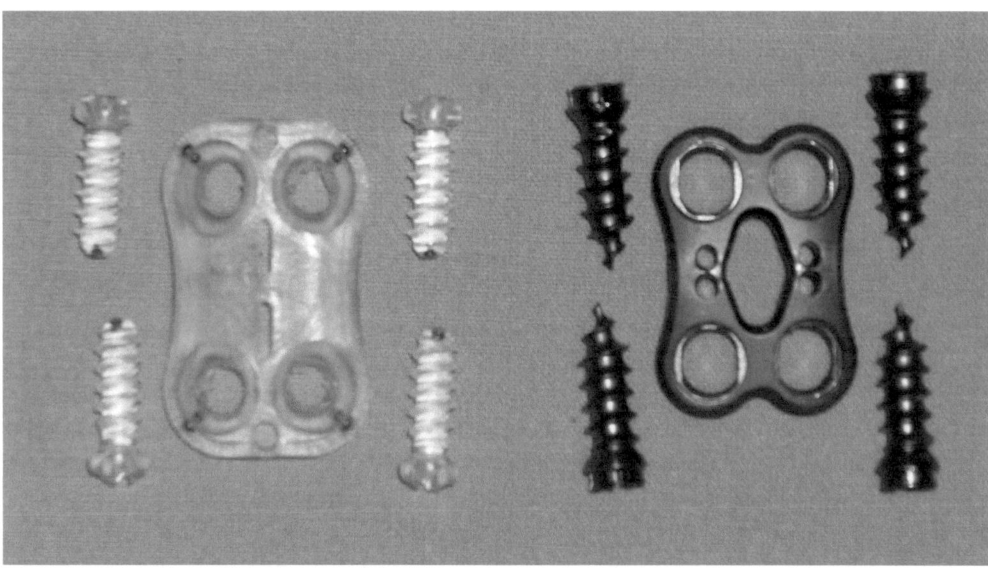

Fig. 4. Resorbable Plate and Screw System (Left) beside Nonresorbable titanium system (Right)

The most pervasive usage of alpha-polyesters within the field of spinal implants has been as resorbable interbody cages which serve as temporary adjuncts to fusion. Resorption of the material allows for gradual load transfer from the implant to the surrounding bone, eventually eliminating effects of stress shielding. Resorption also mitigates risks associated with explantation, corrosion, wear debris, migration or late foreign body reaction associated with non-resorbable devices. Furthermore, as with other thermoplastic materials, alpha-polyesters are radiolucent and do not interfere with the clinicians ability to assess the fusion mass during the course of post-surgical follow up.

From a design perspective the greatest advantage thermoplastics possess over other materials for the construction of spinal implants is the ability to tailor material properties to suit the requirements of the given application. The choice of polymer and specific manufacturing parameters allow for a broad range of material and bioreactive properties. PGA polymers degrade more rapidly than do PLA polymers while copolymers of the two compounds degrade at intermediate rates. Due to the relatively long time period required for successful spinal fusion (6-12 months), PLA has emerged as the material of choice for resorbable spinal implants. The ratio of L- and D-lactides comprising the PLA polymer influences the mechanical and biochemical properties of the material. Poly (L-lactic acid) (PLLA) is semi-crystalline and is characterized by high strength and long resorption time while the racemic poly (D,L- lactic acid) (PDLLA) is amorphous and characterized by lower strength and shorter resorption time. Amorphous polymers are generally preferred for bioresorbable implants because crystalline portions of the polymer degrade much more slowly and may remain in the host much longer, potentially leading to long term complications. As a result most materials used in spinal implants consist of a mixture of L-lactide and racemic D,L-lactide. The most common among these is 70:30 poly(L-lactice-co-D,L-lactice) (PLDLLA) which consists of 70% molar ratio of L-lactide and a 30% molar ratio of D,L-lactide (Smit et al. 2008). This material has been chosen for interbody fusion devices because of its long resorption time (18-36 months), relatively high elastic modulus (3.15 GPa) and high compressive strength (100 MPa) (Alexander et al. 2002; Toth et al. 2002).

While the conceptual advantages outlined above have to some extent successfully been reduced to practice, the materials described are not without their limitations. While significant clinical success in terms of fusion and appropriately timed resorption has been achieved in animals studies (Toth et al. 2002; Thomas et al. 2008), subsequent research in humans has indicated poor clinical outcomes (Smith et al. 2010; Jiya et al. 2011). Particularly, in a prospective cohort study including 81 patients who underwent Transforaminal Lumbar Interbody Fusion (TLIF) surgery with either a resorbable PDLLA cage or a similar non-resorbable carbon fiber cage, Smith et al. found significantly higher incidence of non-union and cage migration in PDLLA cages compared to the carbon fiber cages. Four of the eight patients that exhibited cage migration experienced the adverse event within six months of implantation. Cages that were explanted exhibited moderate plastic deformation, a finding consistent with insufficient material strength discussed previously (Smith et al. 2010). In a randomized prospective human study, Jiya et al. compared PDLLA lumbar interbody fusion cages to similar non-resorbable PEEK implants and found no significant improvement in clinical outcomes upon 2-year follow-up when compared to preoperative values in the PDLLA group. Conversely, significant improvements in all clinical parameters were found in the PEEK group indicating the potential inferiority of the PDLLA implant (Jiya et al. 2011).

Other studies have found anterior cervical plate systems to be inadequate or potentially problematic in providing fixation in both animal and clinical studies. Lyons et al. recently investigated a PDLLA cervical RPSS in an ovine model. Results showed a fusion rate of 25% after 3 months, and plate or screw fracture in 50% of specimens (Lyons et al. 2011). Bindal et al. clinically evaluated a PLDLLA RPSS, supplemental to anterior cervical discectomy and fusion. With a mean follow up time of 15 months, implants were found intact, however psuedoarthrosis and kyphosis was found to be significantly greater in resorbable plates compared to titanium (Bindal et al. 2007).

Through investigations of the time-dependent material properties of 70:30 PDLLA, Smit et al. have shown that the strength of the material decreases with lower loading rates, higher temperature and higher humidity. The authors conducted long-term, static failure experiments at various loads and found that PDLLA cages loaded to approximately 50% of their static compressive strength failed within one day. The authors attribute this time dependent behavior of the polymer to its structure, stating that the material maintains to some extent its behavior as a fluid, resulting in plastic flow under sustained loads (Smit et al. 2008; Smit et al. 2010). These conclusions correlate well to some of the clinical failure experience in PDLLA cages described in the literature. Attempts have been made recently to address this limitation of PLA by combining the polymer with other materials such as beta tricalcium phosphate (βTCP), a commonly used ceramic bone substitute. Debusscher et al. investigated resorbable composite anterior cervical fusion cages consisting of 40% PLLA and 60% βTCP implanted into 20 patients. The authors reported positive clinical and radiologic results in terms of improved pain scores and CT-based fusion evaluation at a mean follow-up of 27 months. No incidences of cage migration or material complications were reported, and an average resorption of 64% was reported at latest follow-up (Debusscher et al. 2009).

4. Conclusion

Thermoplastics have demonstrated a track record of success in medical device applications. The prevalent use of PEEK is evidence of widespread acceptance of thermoplastic materials in spinal fixation procedures. The ability to tailor the properties of the device suitable for each design application continues to present thermoplastics as attractive design materials. Furthermore, the existence of biodegradable thermoplastic polymers has provided an opportunity to develop resorbable medical implants. These materials have exhibited widespread success in applications in which they are subjected to loads of relatively low magnitude and/or short duration, such as in sutures, sheaths and in extremity fixation. However, in the case of biodegradable spine implants, it is important to understand the balance between the benefits of biodegradation and potential strength limitations. In spite of investigations into the use of these materials in a wide array of spinal fixation devices, they have yet to achieve wide clinical acceptance in the field, ostensibly due to their poor performance under the sustained loads borne by the spine.

5. References

Alexander, J. T., C. L. Branch, Jr., et al. (2002). "Applications of a resorbable interbody spacer in posterior lumbar interbody fusion." J Neurosurg 97(4 Suppl): 468-72.

Benzel, E. (2004). Spine Surgery: Techniques, Complication Avoidance, and Management. Philadelphia, Churchill Livingstone.

Bezer, M., Y. Yildirim, et al. (2005). "Absorbable self-reinforced polylactide (SR-PLLA) rods vs rigid rods (K-wire) in spinal fusion: an experimental study in rabbits." Eur Spine J 14(3): 227-33.

Bindal, R. K., S. Ghosh, et al. (2007). "Resorbable anterior cervical plates for single-level degenerative disc disease." Neurosurgery 61(5 Suppl 2): 305-9; discussion 309-10.

Brantigan, J. W. and A. D. Steffee (1993). "A carbon fiber implant to aid interbody lumbar fusion. Two-year clinical results in the first 26 patients." Spine (Phila Pa 1976) 18(14): 2106-7.

Bridwell, K. H., T. A. Sedgewick, et al. (1993). "The role of fusion and instrumentation in the treatment of degenerative spondylolisthesis with spinal stenosis." J Spinal Disord 6(6): 461-72.

Brown, S. A., R. S. Hastings, et al. (1990). "Characterization of short-fibre reinforced thermoplastics for fracture fixation devices." Biomaterials 11(8): 541-7.

Debusscher, F., S. Aunoble, et al. (2009). "Anterior cervical fusion with a bio-resorbable composite cage (beta TCP-PLLA): clinical and radiological results from a prospective study on 20 patients." Eur Spine J 18(9): 1314-20.

Deyo, R. A. and Y. J. Tsui-Wu (1987). "Descriptive epidemiology of low-back pain and its related medical care in the United States." Spine (Phila Pa 1976) 12(3): 264-8.

DiAngelo, D. J., J. L. Scifert, et al. (2002). "Bioabsorbable anterior lumbar plate fixation in conjunction with cage-assisted anterior interbody fusion." J Neurosurg 97(4 Suppl): 447-55.

Franco, A., P. Nina, et al. (2007). "Use of resorbable implants for symptomatic cervical spondylosis: experience on 16 consecutive patients." J Neurosurg Sci 51(4): 169-75.

Gibson, J. N. and G. Waddell (2005). "Surgery for degenerative lumbar spondylosis." Cochrane Database Syst Rev(4): CD001352.

Jiya, T. U., T. Smit, et al. (2011). "Posterior lumbar interbody fusion using non resorbable poly-ether-ether-ketone versus resorbable poly-L-lactide-co-D,L-lactide fusion devices. Clinical outcome at a minimum of 2-year follow-up." Eur Spine J 20(4): 618-22.

Klopp, L. S., W. C. Welch, et al. (2004). "Use of polylactide resorbable film as a barrier to postoperative peridural adhesion in an ovine dorsal laminectomy model." Neurosurg Focus 16(3): E2.

Kulkarni, R. K., E. G. Moore, et al. (1971). "Biodegradable poly(lactic acid) polymers." J Biomed Mater Res 5(3): 169-81.

Kulkarni, R. K., K. C. Pani, et al. (1966). "Polylactic acid for surgical implants." Arch Surg 93(5): 839-43.

Kurtz, S. M. and J. N. Devine (2007). "PEEK biomaterials in trauma, orthopedic, and spinal implants." Biomaterials 28(32): 4845-69.

Lippman, C. R., M. Hajjar, et al. (2004). "Cervical spine fusion with bioabsorbable cages." Neurosurg Focus 16(3): E4.

Lyons, A. S., B. P. Sherman, et al. (2011). "Failure of resorbable plates and screws in an ovine model of anterior cervical discectomy and fusion." Spine J 11(9): 876-83.

Majola, A., S. Vainionpaa, et al. (1992). "Intramedullary fixation of cortical bone osteotomies with self-reinforced polylactic rods in rabbits." Int Orthop 16(1): 101-8.

Mayberry, J. C., J. T. Terhes, et al. (2003). "Absorbable plates for rib fracture repair: preliminary experience." J Trauma 55(5): 835-9.

Panagiotacopulos, N. D., M. H. Pope, et al. (1987). "Water content in human intervertebral discs. Part II. Viscoelastic behavior." Spine (Phila Pa 1976) 12(9): 918-24.

Smit, T. H., T. A. Engels, et al. (2010). "Time-dependent failure in load-bearing polymers: a potential hazard in structural applications of polylactides." J Mater Sci Mater Med 21(3): 871-8.

Smit, T. H., T. A. Engels, et al. (2008). "Time-dependent mechanical strength of 70/30 Poly(L, DL-lactide): shedding light on the premature failure of degradable spinal cages." Spine (Phila Pa 1976) 33(1): 14-8.

Smith, A. J., M. Arginteanu, et al. (2010). "Increased incidence of cage migration and nonunion in instrumented transforaminal lumbar interbody fusion with bioabsorbable cages." J Neurosurg Spine 13(3): 388-93.

Springer, M. A., E. A. van Binsbergen, et al. (1998). "[Resorbable rods and screws for fixation of ankle fractures. A randomized clinical prospective study]." Unfallchirurg 101(5): 377-81.

Thomas, K. A., J. M. Toth, et al. (2008). "Bioresorbable polylactide interbody implants in an ovine anterior cervical discectomy and fusion model: three-year results." Spine (Phila Pa 1976) 33(7): 734-42.

Tomasino, A., H. Gebhard, et al. (2009). "Bioabsorbable instrumentation for single-level cervical degenerative disc disease: a radiological and clinical outcome study." J Neurosurg Spine 11(5): 529-37.

Tormala, P., J. Vasenius, et al. (1991). "Ultra-high-strength absorbable self-reinforced polyglycolide (SR-PGA) composite rods for internal fixation of bone fractures: in vitro and in vivo study." J Biomed Mater Res 25(1): 1-22.

Toth, J. M., B. T. Estes, et al. (2002). "Evaluation of 70/30 poly (L-lactide-co-D,L-lactide) for use as a resorbable interbody fusion cage." J Neurosurg 97(4 Suppl): 423-32.

Vaccaro, A. R., J. A. Carrino, et al. (2002). "Use of a bioabsorbable anterior cervical plate in the treatment of cervical degenerative and traumatic disc disruption." J Neurosurg 97(4 Suppl): 473-80.

van Manen, C. J., M. L. Dekker, et al. (2008). "Bio-resorbable versus metal implants in wrist fractures: a randomised trial." Arch Orthop Trauma Surg 128(12): 1413-7.

Wang, M. Y., A. D. Levi, et al. (2002). "Polylactic acid mesh reconstruction of the anterior iliac crest after bone harvesting reduces early postoperative pain after anterior cervical fusion surgery." Neurosurgery 51(2): 413-6; discussion 416.

Welch, W. C., K. A. Thomas, et al. (2002). "Use of polylactide resorbable film as an adhesion barrier." J Neurosurg 97(4 Suppl): 413-22.

White, A. and M. Panjabi (1990). Clinical Biomechanics of the Spine. Philadelphia, Lippincott.

Yong-Hing, K. and W. H. Kirkaldy-Willis (1983). "The pathophysiology of degenerative disease of the lumbar spine." Orthop Clin North Am 14(3): 491-504.

Yong-Hing, K., J. Reilly, et al. (1980). "Prevention of nerve root adhesions after laminectomy." Spine (Phila Pa 1976) 5(1): 59-64.

16

Application of Thermoplastics in Protection of Natural Fibres

Fauziah Ahmad, Farshid Bateni,
Mohammadreza Samadi Tavana and Ahmad Shukri Yahaya
Universiti Sains Malaysia
Malaysia

1. Introduction

Nowadays thermoplastics are widely used in industries; many products used different specifications of the thermoplastics that are adapted to their requirement. The importance of consideration to the environmental protection leads the mankind to thinking about producing environment-friendly material, and also controlling the waste material by recycling or reducing them. In the meantime the natural materials such as cellulose based material are one of the major resources that can be used to replace with many manufactured materials. However, cellulosed based material such as natural fibres, because of their degradability, need protection from any circumferential agents. The protection may require a special condition in order to utilise in soil, due to water absorption, soil organisms, and minerals. Natural fibres are amenable to modifications as they bear hydroxyl groups from cellulose and lignin. In addition coating the fibres with any chemical materials reduce their water absorptions and protect them from any bacteria and fungi attack. The hydroxyl groups may be involved in the hydrogen bonding within the cellulose molecules. This weakness of the natural material and good characteristics of natural fibres is the basis of bio-composites invention.

Fibres	Cellulose (%)	Hemicelluloses (%)	Lignin (%)	Tensile Strength (%)	Elongation (%)	Toughness (MPa)
OPEFB*	65	-	19	248	14	2,000
Coir	32-43	0.15-0.25	40-45	140	25.0	3,200
Banana	63-64	19	5	540	3.0	816
Sisal	66-72	12	10-14	580	4.3	1,250
Pineapple	81.5	-	12.7	640	2.4	970

*OPEFB : Oil Palm Empty Fruit Bunch

Table 1. Chemical and mechanical properties of some important natural fibres

2. Natural fibres

Many different kinds of natural cellulosic fibre such as cotton, hemp, sisal, coconut fibres and oil palm fibres are used in different composite products. Properties of the natural fibres depend mostly on the nature of the plant, locality in which it is grown, age of the plant, and the extraction method that is used (Sreekala et al., 1997). Coir is a hard and tough multi cellular fibre with a central portion called "lacuna.", On the other hand, banana fibre is weak and cylindrical in shape. Sisal is an important leaf fibre and is strong. Pineapple leaf fibre is soft and has high cellulose content. Many studies have been done on the natural fibre based composite products (Maldas & Kokta, 1990; Pavithran et al., 1987; Shah & Lakkad, 1981; Sreekala et al., 1997). Table 1 summarised the chemical and mechanical properties of some natural fibres.

2.1 Properties of oil palm fibres

Oil palm is one of the most economical and very high-potential perennial oil crops. It belongs to the species of Elaeis guineensis under the family Palmacea, and originated in the tropical forests of West Africa. Major industrial cultivation is in Southeast Asian countries such as Malaysia and Indonesia. Large-scale cultivation has come up in Latin America. In India, oil palm cultivation is coming up on a large-scale basis with a view to attaining self sufficiency in oil production.

Oil palm fibre is non-hazardous biodegradable material, extracted from oil palm's empty fruit bunch (EFB). Oil palm fibre is an important lignocellulosic raw material. OPEFB fibre and oil palm mesocarp fibre are two types of fibrous materials left in the palm-oil mill. The mesocarp fibres are left as a waste material after the oil extraction. These fibres must be cleaned of oily and dirty materials. The only current uses of this highly cellulosic material are as boiler fuel and in the preparation of potassium fertilizers. When left on the plantation floor, these waste materials create great environmental problems. Therefore, economic utilization of these fibres will be beneficial (Sreekala et al., 1997).

Chemical constituents (%)	
Cellulose	65
Hemi cellulose	-
Lignin	19
Ash content	2

Table 2. Chemical constituents of oil palm empty fruit bunch fibre

Physical properties of oil palm fibre	
Diameter (mm)	0.15-0.50
Density (g/mm^3)	0.7-1.55
Linear density (denier)*	2150
Tensile strength (MPa)	100-400
Young's modulus (MPa)	1000-9000
Elongation at break (%)	14
Microfibrillar angle (°)	46

* 1 denier= 1/9000 g/m

Table 3. Physical and mechanical properties of oil palm empty fruit bunch fibre

OPEFB fibre is obtained after the subtraction of oil seeds from fruit bunch for oil extraction. OPEFB fibre is extracted by the retting process of the EFB. Average yield of OPEFB fibre is about 400 g per bunch. Previous studies report the mechanical properties of OPEFB fibres. Table 2 and Table 3 show the summary of oil palm fibre properties (Jacob et al., 2004; Sreekala et al., 2001; Sreekala et al., 1997)

3. Thermoplastic coat

Thermoplastics as a coating can lead to improving the natural fibre performance in two ways: 1. the thermoplastics cover the fibres and keep the fibres from any fungi or bacteria attacks by decreasing the water absorption and contact of the fibres to the soil and any organism inside it, 2. The physical performance of the fibre such as tensile strength and elongation can be affected by modification and coating with any kind of thermoplastics. Therefore, a method was developed to coat the fibres with the thermoplastics. The solvent was used to prepare soluble thermoplastic since the natural fibres cannot reside in high temperature. Different density of the thermoplastic solution was used to evaluate the coated fibres to reach the best strength and resistance. Two types of the fibre were used as a reinforcement of composites such as soil, first the discrete fibres where it needs to be coated one by one, and second is the sheet fibres that were made by compaction of bulk fibres. The fibre was coated by acrylonitrile butadiene styrene (ABS) solution and the characterisation test results for both single and sheet fibres are described in the following sections.

3.1 Acrylonitrile butadiene styrene

ABS is an important engineering copolymer widely used in industry due to superior mechanical properties, chemical resistance, ease of processing and recyclability (Yang et al., 2004). ABS is a common thermoplastic used to make polymeric wood composites, has good physical properties in comparison with other commodity plastics and is cheap in comparison with other engineering plastics (Huang & Mo, 2002).

ABS is derived from acrylonitrile, butadiene, and styrene. The chemical structure of the ABS is shown in Figure 1. Acrylonitrile is a synthetic monomer produced from propylene and ammonia. Butadiene is a petroleum hydrocarbon obtained from butane. Styrene monomers, derived from coal, are commercially obtained from benzene and ethylene from coal. The advantage of ABS is that this material combines the strength and rigidity of the acrylonitrile and styrene polymers with the toughness of the polybutadiene rubber. The most amazing mechanical properties of ABS are resistance and toughness.

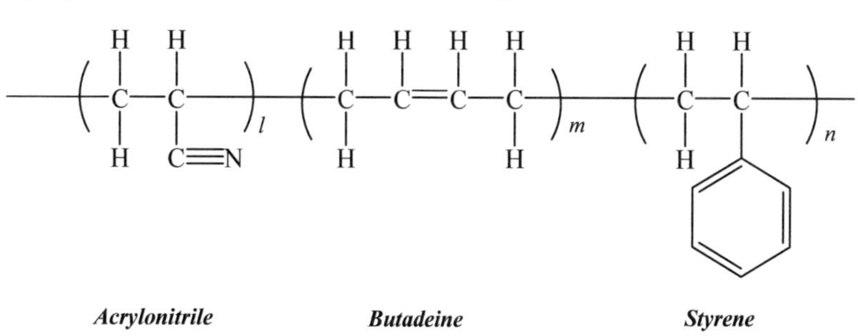

Fig. 1. Chemical structure of ABS

The chemical resistance for ABS is relatively good and it is not affected by water, non organic salts, acids and basic. The material will dissolve in aldehyde, ketone, ester and some chlorinated hydrocarbons. The properties of moulded ABS are shown in Table 4 based on MatWeb (2009) material specification data sheet.

Property	Test Method	Value
Tensile Strength	ASTM D638	44.8 MPa
Flexural Modulus	ASTM D638	2.59 GPa
Tensile Elongation	ASTM D638	15 %
Flexural Yield Strength	ASTM D790	69 MPa
Flexural Modulus	ASTM D790	2.59 GPa

Table 4. Physical properties of moulded ABS

3.2 Characterization of coated fibres

An ABS solution was prepared by adding ABS pieces to methyl ethyl ketone (MEK) solvent. Fibres were chopped into 30 mm length for water absorption tests and 100 mm length for tensile strength tests. The average aspect ratio for 100 mm length fibre was found to be equal to 250. The chopped OPEFB fibres were incubated in the 15% ABS solution to be coated. Coated fibres were dried over a mesh at room temperature.

3.2.1 FTIR

The chemical reactions during fibre coating were characterised using IR spectroscopy. IR spectra of the uncoated and coated OPEFB fibres are given in Figure 2 (Bateni et al., 2011). Series (a) is the IR spectra for ABS, Series (b) shows the IR spectra of OPEFB fibre and series (c) and (d) show the coated OPEFB fibre infrared spectra, for coating incubation times of 6 hour and 24 hour, respectively. ABS coating imparts physical and chemical modifications to the fibre. A band shown in the 3300-3600 cm^{-1} regions in coated and uncoated OPEFB fibre corresponds to O-H stretching of the cellulose and lignin. The intensity of the 1636 cm^{-1} band increased and the 3400 cm^{-1} band was shifted to 3420 cm^{-1}, corresponding to C=O stretching and O-H stretching vibrations after coating of the fibre, respectively. Strong peaks are observed in the IR spectrum of coated fibres at 2239 and 2929 cm^{-1} when compared with the uncoated fibre. Peak detected at 1455 cm^{-1} may correspond to the characteristic peaks of ABS plastic which are the aliphatic C-H stretching (Sreekala et al., 2000).

The presence of a peak at 2929 cm^{-1} may be due to C-H stretching. The peaks at 1039 and 2929 cm^{-1} for coated fibres were increased and shifted, corresponding to C-O stretching and C-H stretching vibrations. The change C=C peak frequency increased with coating. Two peaks were observed at approximately 2347 cm^{-1} due to C≡N stretching. The shifting of the 2347 cm^{-1} band to 2239 cm^{-1} indicates the change in C≡N stretching of the OPEFB fibre after coating by ABS. Those peaks which changed over the time show the increment the presence rate of ABS in coated fibres. The presence of the peaks over the time increment shows that the chemical reactions of ABS and fibres have increases. This increase led to a more physical stability of ABS coating over fibres and thus fibres were more resistant.

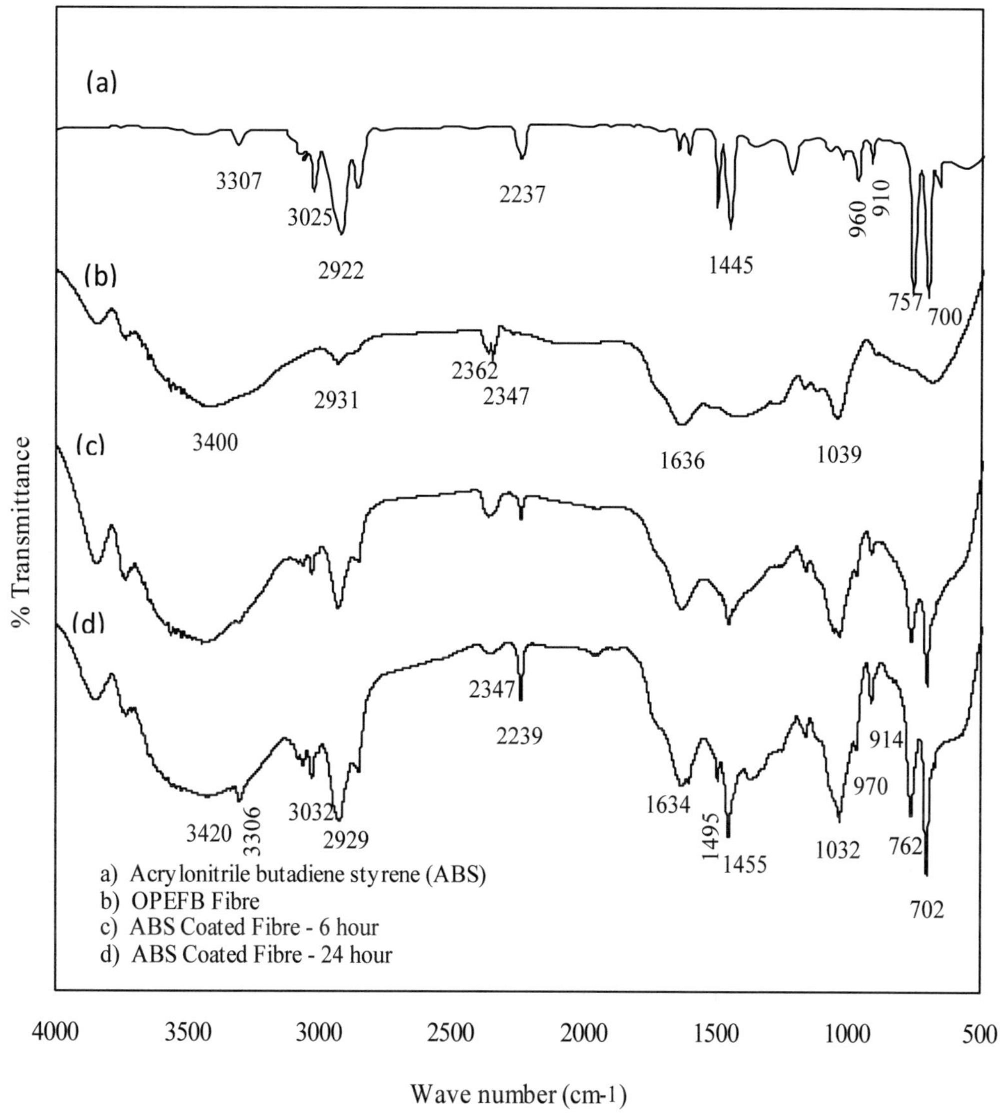

Fig. 2. FTIR of ABS coated and uncoated fibres

3.2.2 Fibre surface topology

The porous surface morphology was useful for better mechanical interlock of fibre with the ABS coating. The SEM micrograph of fibres and the coated fibres clearly shows the surface structure of an uncoated OPEFB fibre and the quality of thermoplastic coat (Figure 3). The micrographs show porous and grooves on the surface of the fibre. The uniform cover and fully coating of the fibre surface is an important factor in protecting the fibres while surface structure of a coated fibre is shown in Figure 3(b). The layer of ABS worked as a surface to protect the fibres from water, degradation and physical damages. The entire fibre appears to be covered and the surface exhibits a smoother surface than the uncoated fibre.

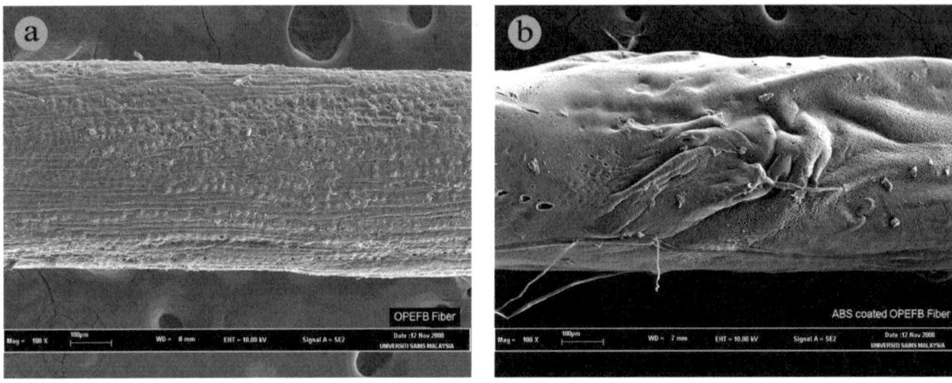

Fig. 3. SEM micrographs of uncoated (left) and coated (right) OPEFB fibre

Figure 4(a) presents the cross section of an uncoated fibre, which exhibits a lacuna-like portion in the middle in comparison with Figure 4(b), the SEM micrograph of the cross-section of a coated fibre. The thickness of the coated layer can be seen in the figure, the structure of portions is indicating the penetration of the ABS into the fibre structure. The ABS filled some of the lacuna like portion in the fibre.

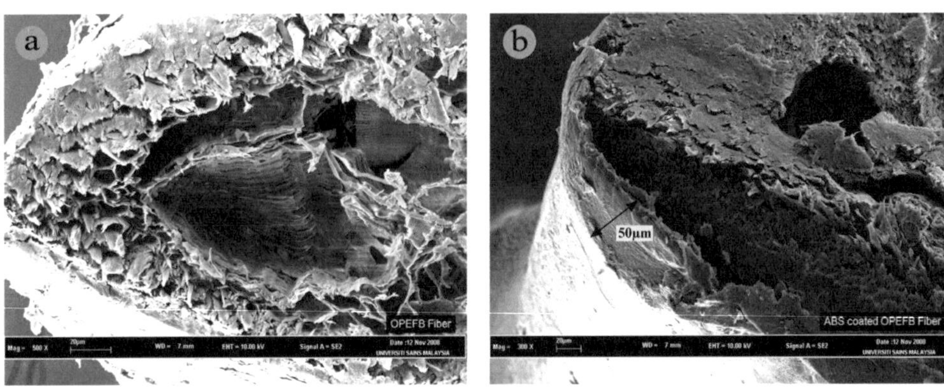

Fig. 4. Cross section of uncoated (left) and coated (right) OPEFB fibre

The uncoated fibre surface was found to be rough and had protruding portions and groove-like structures on its surface (Figure 5(a)). The surface of the coated fibre has an uneven structure, as shown in Figure 5(b). This feature of surface depends on the application of the fibres where it can be positive or negative due to less friction existence within the fibres and composites mass. Otherwise, the ABS coating may increase the diameter and the section area of fibres which can affect the contact surface area between fibres and soil particles. The surface area of the fibres is the most effective parameter in increasing the shear strength of some fibre reinforced composites.

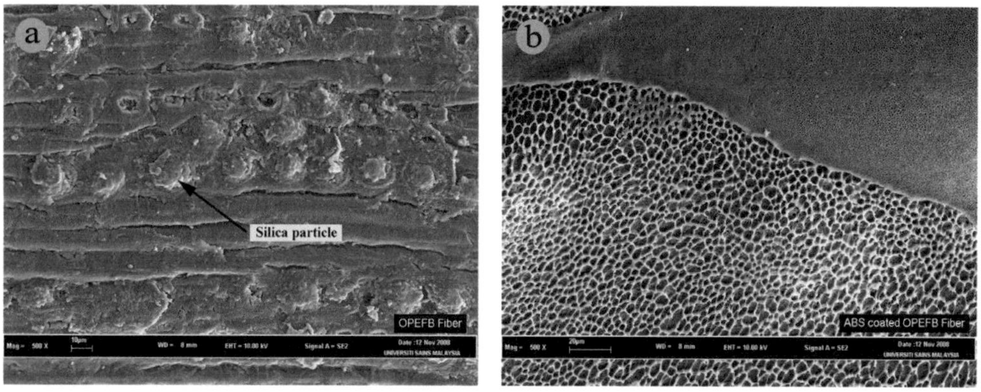

Fig. 5. Surface structure of uncoated (left) and coated (right) OPEFB fibre

3.2.3 Tensile strength of fibre

The tensile strength test result of the coated OPEFB fibre showed an increase in tensile strength of the fibre in breaking point. The elongation of the fibre in tensile test was increased from 15% to near 20% in coated fibre. The main improvement in coated fibres occurs in Young's modulus. Table 5 shows the tensile properties of coated and uncoated OPEFB fibre.

Type of the fibres	Tensile Strength (MPa)	Elongation at Break (%)	Young's modulus (MPa)
OPEFB fibre	283	15.4	5500
Coated OPEFB fibre	306	19.1	6600

Table 5. Summary of the tensile test result on coated and uncoated OPEFB fibre

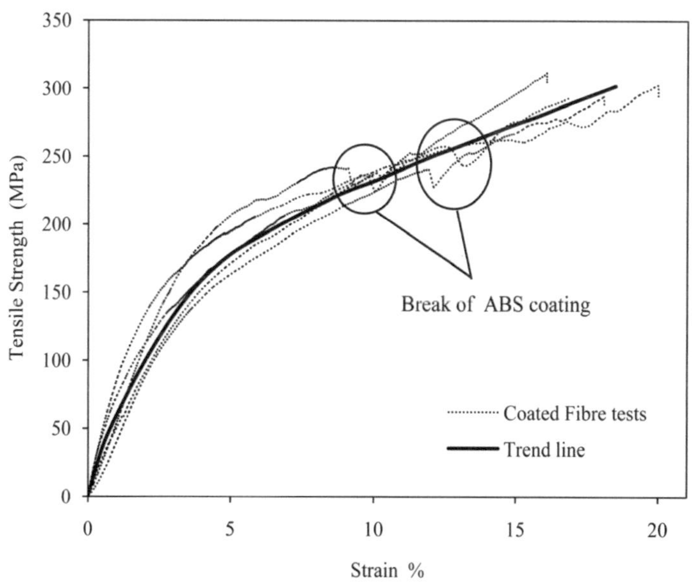

Fig. 6. Stress-strain curve of coated OPEFB fibres on tensile test

The tensile test result showed that the ABS coat was broken before failure of the OPEFB fibre, the gaps in the stress strain curve within the strain of 10% to 15% in Figure 6 describe the weakness of the coated fibres to handle the force. Figure 7 shows the photographs of the coated fibre before and after the tensile test. The split of the ABS coating was shown clearly at different strain of OPEFB fibre and ABS thermoplastic. From the figure the gap in the stress strain curve represented the failure of the ABS coat before the OPEFB fibres. The strain of the ABS thermoplastic (Table 3) also proves this result.

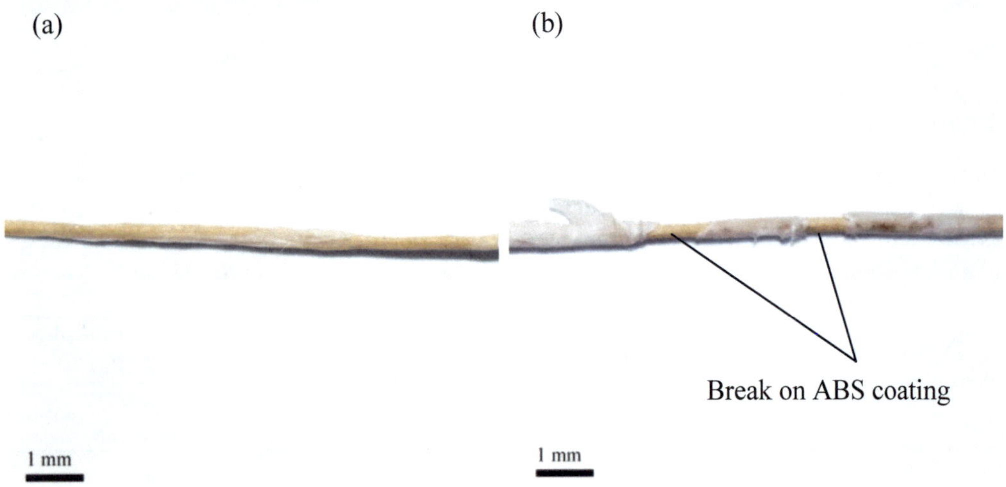

Fig. 7. Coated OPEFB fibre before and after tensile test, a) coated OPEFB fibre, b) Break of the coating after tensile test

3.2.4 Water absorption of fibres

The water absorption of the coated and uncoated OPEFB fibre was presented in Figure 8 as a percentage of dry weight. Figure 8 shows the absorption behaviour of coated and uncoated OPEFB fibre in distilled water at 30°C and 70°C respectively. The results show that the water absorption of the coated fibre was lower than that of the uncoated fibre. As the temperatures increased, the water sorption was generally decreased.

The decrease in sorption value for coated fibre had the same range of treated fibres with different methods reported by Sreekala & Thomas, (2003). Different fibre surface modifications such as mercerization, latex coating, gamma irradiation, silane treatment, isocyanate treatment, acetylation and peroxide treatment were used in their study. It is recommended that the modification techniques were also used before the coating process. The decrease of the water sorption capacity of the fibre reduces the biodegradability of the OPEFB fibre, and also increase in tensile capacity of the fibres.

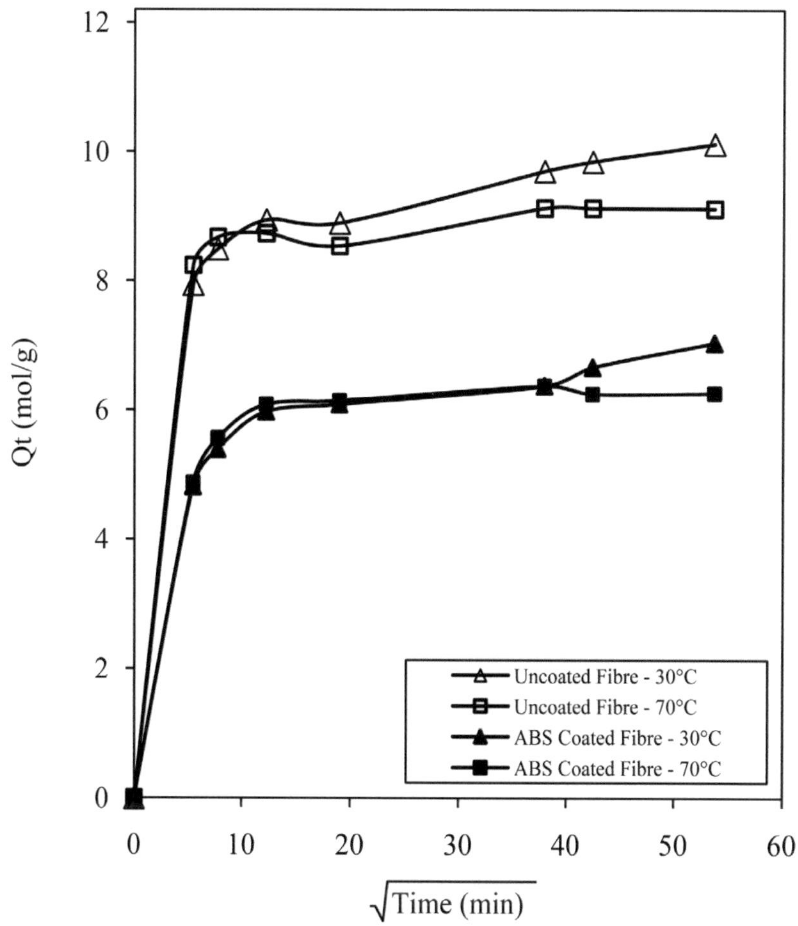

Fig. 8. Sorption curves of uncoated and ABS coated OPEFB fibres

3.3 Characterization of coated fibre sheets

For the first phase of experiments, the ABS solution was prepared in three different percentages of 5, 10 and 15 per cents. The 20% ABS solution was excluded from the experiments since some gel-forming behaviour was observed. The specimens were soaked for one minute in 5%, 10% and 15% of ANS solution and the tensile test conducted for specimens according to ASTM D4595-86 (2001) specifications. The 15% solution was picked for the rest of the experiments because it gives the optimum results. For next stage, the specimens were soaked in 15% ABS solution and repeated the tensile tests to study the effect of soaking duration on tensile strength.

3.3.1 Fibre sheets

OPEFB sheets are commercially available is Malaysia. These sheets are manufactured through a compaction process in which the fibres orient randomly (Figure 9). Sheets are produced only in single direction that is the machine direction, so there is no warp or weft direction, the detailed sheet and cross section are presented in Figure 9 and Figure 10.

The size of merchandised sheets was 3000 mm in length, 1000 mm in width and 10 mm in thickness. The coated or uncoated sheets have a potential to be used as a kind of geotextile for soil reinforcement was named Geo-Mat.

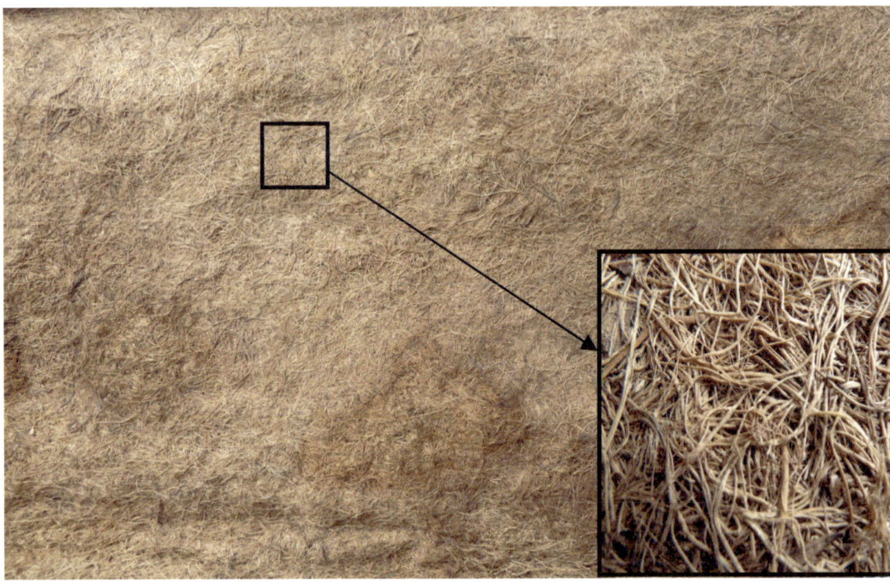

Fig. 9. OPEFB sheet

Fig. 10. Longitudinal cross section of OPEFB sheet

3.3.2 Effects of ABS percentage on OPEFB sheets

The weight variations of sheets are determined before coating and after drying process. The results, which are shown in Figure 11, showed that by increasing the density of ABS solution, larger amounts of ABS were oriented on the fibres.

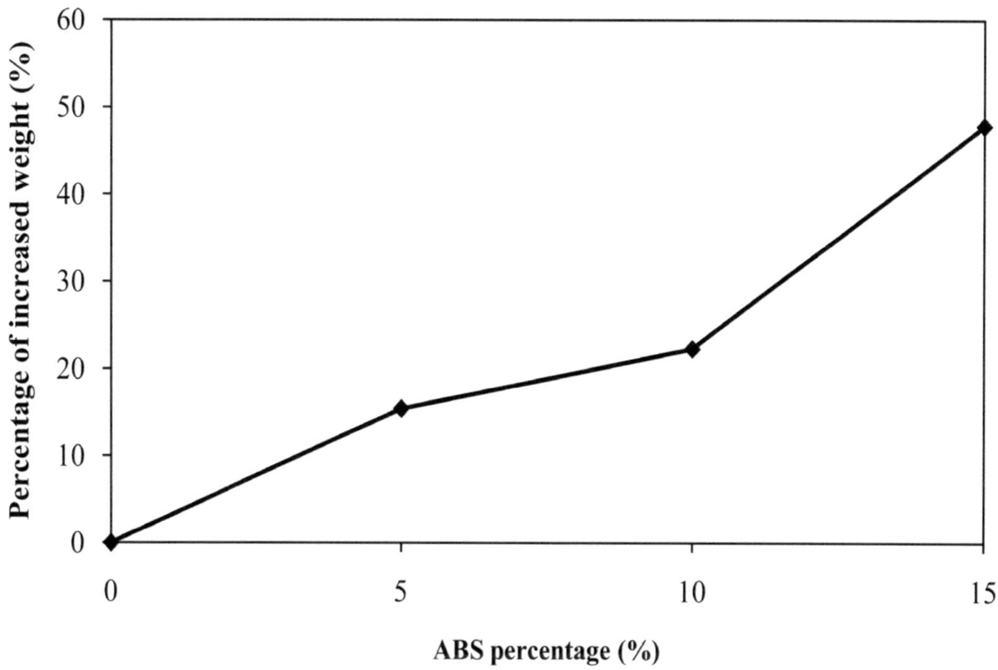

Fig. 11. Weight changes of ABS coated OPEFB sheets

The ABS thermoplastic brings ductility and toughness for the sheets. When the amount of oriented ABS increases, the toughness and ductility of sheets increases too. The influences of such ductility and toughness could become clearer by conducting the tensile test.

3.3.3 Effect of ABS percentage on tensile strength

By comparing the results of untreated sheets and 5%, 10% and 15% ABS coated sheets, it could be realized that the coating resulted to a slight increase in average tensile strength and remarkable decrease of elongation percentages. The effect of those variations is more obvious in tensile modulus, since it was doubled for the coated OPEFB sheets. Enhancement of 15% ABS coating showed better improvements than the 5% and 10% ABS coating; the average tensile strength of sheets reached to approximately 12 kN/m. The obtained average tensile strength is relatively 6 times higher than the average tensile strength of untreated OPEFB sheets. The resulted average tensile modulus of these sheets was around 350.3 kN/m. As it was expected, the ABS improved the tensile properties of OPEFB sheets very significantly. The ductility and toughness of 15% ABS coated sheets were more sensible than the others (Table 6). The ABS covered approximately all of the fibres properly and filled the void areas among the fibres.

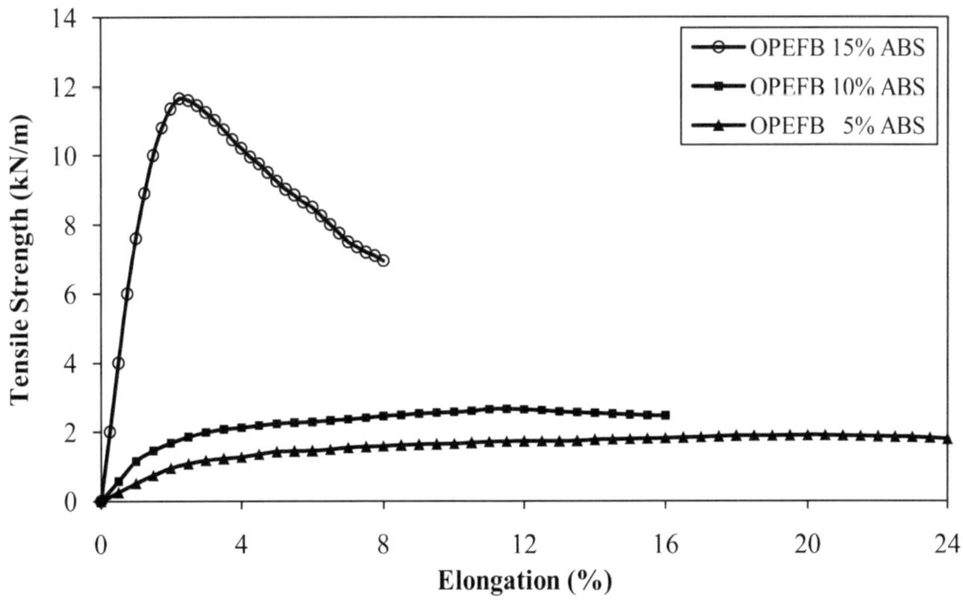

Fig. 12. Comparison of tensile test results for 5%, 10% and 15% ABS coated OPEFB sheets

Since the 15% ABS solution covered the fibres and sheets more properly, it is expectable to experience higher tensile strength and also better durability for them. It is worth to mention that for all specimens, the complete rupture did not occur at the peak force. Although the failure area became clear, but the two parts of sheets were still joined together. The Figure 13 shows the rupture area of the failed specimens of ABS coated sheets.

Specimen	Breaking Force (kN)	Tensile Strength (kN/m)	Elongation (%)	Tensile Modulus (kN/m)
Uncoated	0.359	1.79	37.08	4.80
5 %	0.392	1.93	20.60	9.37
10 %	0.542	2.71	12.27	22.00
15 %	2.41	12.05	3.44	350.30

Table 6. Tensile properties of ABS coated OPEFB sheets

3.3.4 Tensile strength comparison

Among the previous works there is only one work which is comparable. Subaida et al. (2008) conducted experimental investigation on tensile strength of woven coir geotextiles. They reported that the tensile strength of the mesh mattings lies in range of 10 and 20 kN/m. The tensile test was carried out for three types of nonwoven geotextiles with commercial names of MTS 300, MTS 350 and MTS 400. MTS series geotextiles are a technical fabric mechanically bonded nonwoven needle punched made from 100% UV stabilized polyester. The average tensile test results of these fabrics are presented and compared by Figure 14. For all of these fabrics the peak tensile strengths are achieved at relatively large values of elongation which are practically useless. In soil structures large displacements are equal to failure of structures; basically, soil reinforcing materials must be able to reach to the peak tensile strength within the minimum displacement. Between all the materials mentioned in Figure 14, the 15% ABS coated OPEFB sheets are the most suitable choice based on the highest tensile strength along with the small elongation at the peak point.

Fig. 13. Rupture area for 5% ABS coated OPEFB sheet

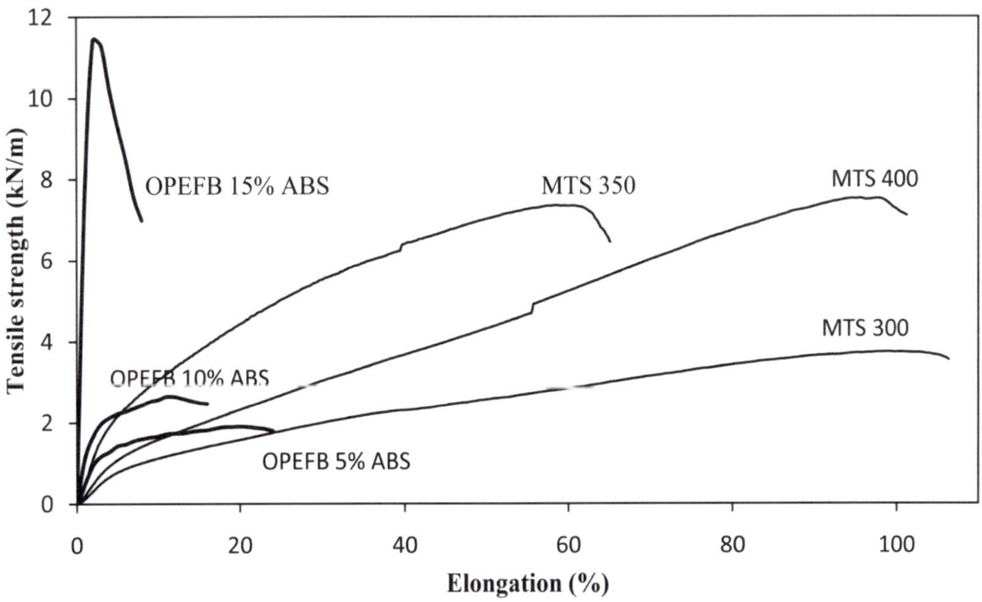

Fig. 14. Tensile test result of MTS fabrics and coated OPEFB sheets

3.3.5 Biodegradation of fibres

The degradable properties of both coated and uncoated OPEFB fibre were monitored through aging in two different soils and in contact with moisture and fungus for about 3 months. Figure 15 shows the effect of fungus in the degradation of the fibre, the black part of the fibre was affected by the fungus. Most part of the uncoated fibres was influenced by fungus and the fungus spread over the surface of the fibres. In comparison the coated fibre was less affected by fungus and it only decayed at the end parts of fibres. Otherwise, the colour of the fibres was shown as the water sorption in the uncoated fibres. Water is the important factor for the growth of the fungus that increased the biodegradation of the fibres.

Fig. 15. Fungus biodegradation of the fibres after 3 month a) Sheet of OPEFB fibre, b) Sheet of coated OPEFB fibre

The fibres were also placed in to the silty sand and peat soil to estimate the weight loss of the fibres in soil. The discrete fibre and fibre sheet were weighed before test. The fibres also were coated with ABS solution for 24 hours and all the specimens were placed on soils for about 3 months.

Figure 16 shows the coated and uncoated OPEFB fibre after aging in the silty sand soil. The fibre sheets decayed after 3 months and it is shown that the uncoated fibre had the separate structure due to the biodegradation of the fibre.The coating was protected the fibres from biodegradation; the shape of the coating fibres was kept.

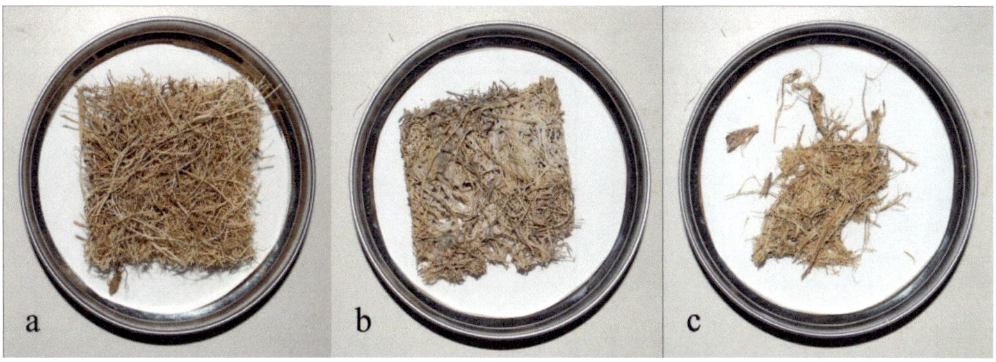

Fig. 16. a) before decay b) decayed ABS coated c) decayed uncoated fibre sheet in the silty sand soil after 3 months

The same results are shown in Figure 17 for the fibres in organic clay soil. The uncoated sheet fibre in organic clay separated from each other and did not have the textile structure. The uncoated fibre lost its weight due to biodegradation in the soil.

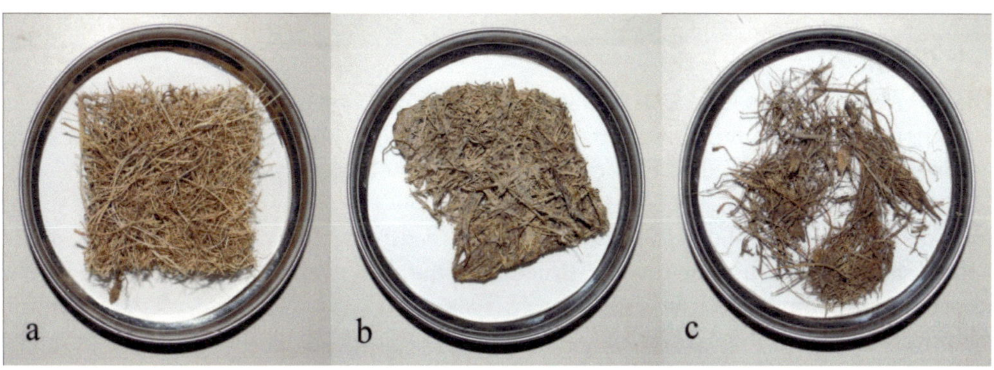

Fig. 17. a) before decay b) decayed ABS coated c) decayed uncoated fibre sheet in the organic clay soil after 3 months

The results of these losses are graphically plotted in Figure 18. Loss of weight of the discrete fibres was higher than the fibre sheet due to their larger contact surface with the soils and environment factors.

The result shows that in all condition coating decreases the biodegradation of the fibres both in discrete fibre and fibre sheet. The decay of the fibres in three conditions had approximately the same result. The weight loss result indicate the influence of coating on protecting OPEFB fibres from biodegradation, around 50% decrease on weight loss were estimated from tests after three months.

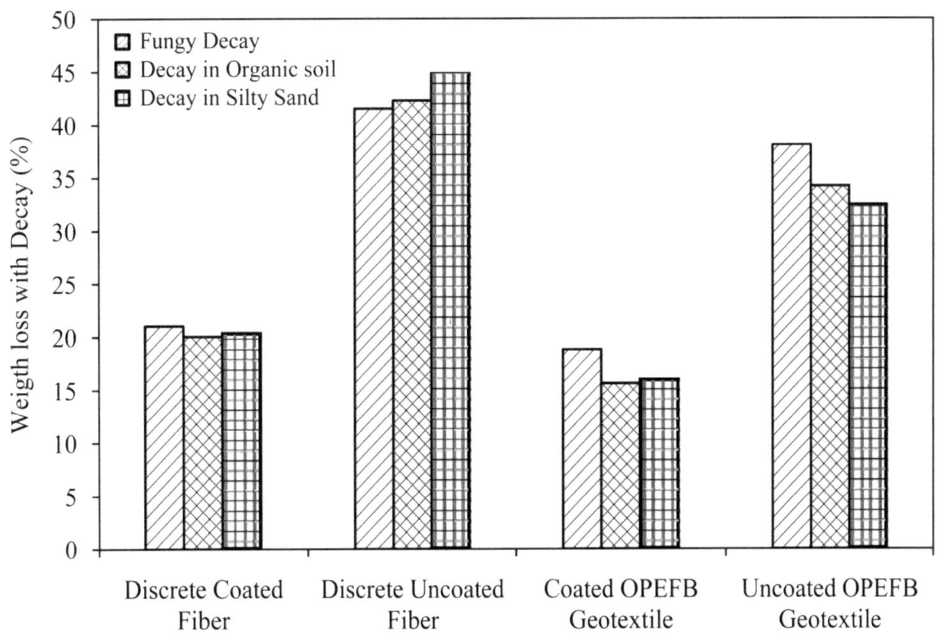

Fig. 18. Biodegradation of the fibres inside Silty sand, Organic soil and be in contact with fungus after 3 months

4. Conclusion

The thermoplastic was used as a treatment of the OPEFB fibres. The ABS coated fibres had an acceptable effect on the protection of fibre as the same as other treatment techniques. The water absorption of the coated fibre decreased due to protection capacity of the coated layer. The ABS coated fibre was found to be more durable compared to uncoated due to the condition of fibre. Morphological studies revealed that the coating modifies and protects the fibre surface entirely and the covered structure of the ABS over fibres can be seen from the respective scanning electron micrographs, also FTIR studies shown the chemical modifications within the ABS thermoplastic and fibres. From the tensile test it was found the Young's modulus of the coated fibre shown improvement due to ABS coating. However, the tensile strength of the fibre indicated less increase in comparison to untreated fibre.

The previous studies describe that inclusion of OPEFB fibres can significantly increase the peak shear strength of silty sand soil (Ahmad et al., 2010). The fibre content increment leads to increasing the shear strength and consequently stabilized the reinforced soil. Coated OPEFB fibres increased the shear strength of silty sand compared to uncoated fibres. Coated fibres shown higher interface friction between fibre and soil particles by increasing the surface area.

5. Acknowledgment

Also to all the collaborative partners and the working group committee especially to all the academicians, technical officers from other organization and technicians who have work closely in this project.

6. References

Ahmad, F.; Bateni, F. & Azmi, M. (2010). Performance evaluation of silty sand reinforced with fibres. *Geotextiles and Geomembranes*, Vol.28, No.1, pp. 93-99, ISSN: 0266-1144.

ASTM D4595-86 (2001) Standard Test Method for Tensile Properties of Geotextiles by the Wide-Width Strip Method. *ASTM International*, West Conshohocken, PA, www.astm.org.

Bateni, F.; Ahmad, F., Yahya, A. S. & Azmi, M. (2011). Performance of oil palm empty fruit bunch fibres coated with acrylonitrile butadiene styrene. *Construction and Building Materials*, Vol.25, No.4, pp. 1824-1829, ISSN: 0950-0618.

Huang, C.-Y. & Mo, W.-W. (2002). The effect of attached fragments on dense layer of electroless Ni/P deposition on the electromagnetic interference shielding effectiveness of carbon fibre/acrylonitrile-butadiene-styrene composites. *Surface and Coatings Technology*, Vol.154, No.1, pp. 55-62, ISSN: 0257-8972.

Jacob, M.; Thomas, S. & Varughese, K. T. (2004). Mechanical properties of sisal/oil palm hybrid fiber reinforced natural rubber composites. *Composites Science and Technology*, Vol.64, No.7-8, pp. 955-965, ISSN 0266-3538.

Maldas, D. & Kokta, B. V. (1990). Effect of Extreme Conditions on the Mechanical-Properties of Wood Fiber Polystyrene Composites .2. Sawdust as a Reinforcing Filler. *Polymer-Plastics Technology and Engineering*, Vol.29, No.1-2, pp. 119-165, ISSN 0360-2559.

MatWeb (Jun 2009). PTS Trilac® ABS-HS4000 General Purpose ABS Resin, In: *Matweb*, 8.6.2009, Available from: http://www.matweb.com/search/DataSheet.aspx ?MatGUID=423e95f8367f42cbad42b991f6744516

Pavithran, C.; Mukherjee, P. S., Brahmakumar, M. & Damodaran, A. D. (1987). Impact Properties of Natural Fiber Composites. *Journal of Materials Science Letters*, Vol.6, No.8, pp. 882-884, ISSN 0261-8028.

Shah, A. N. & Lakkad, S. C. (1981). Mechanical-Properties of Jute-Reinforced Plastics. *Fibre Science & Technology*, Vol.15, No.1, pp. 41-46, ISSN 0015-0568.

Sreekala, M. S.; George, J., Kumaran, M. G. & Thomas, S. (2001). Water-sorption kinetics in oil palm fibers. *Journal of Polymer Science Part B-Polymer Physics*, Vol.39, No.11, pp. 1215-1223, ISSN 0887-6266.

Sreekala, M. S.; Kumaran, M. G., Joseph, S., Jacob, M. & Thomas, S. (2000). Oil palm fibre reinforced phenol formaldehyde composites: Influence of fibre surface modifications on the mechanical performance. *Applied Composite Materials*, Vol.7, No.5-6, pp. 295-329, ISSN 0929-189X.

Sreekala, M. S.; Kumaran, M. G. & Thomas, S. (1997). Oil palm fibers: Morphology, chemical composition, surface modification, and mechanical properties. *Journal of Applied Polymer Science*, Vol.66, No.5, pp. 821-835, ISSN 0021-8995.

Sreekala, M. S. & Thomas, S. (2003). Effect of fibre surface modification on water-sorption characteristics of oil palm fibres. *Composites Science and Technology*, Vol.63, No.6, pp. 861-869, ISSN 0266-3538

Subaida, E. A.; Chandrakaran, S. & Sankar, N. (2008). Experimental investigations on tensile and pullout behaviour of woven coir geotextiles. *Geotextiles and Geomembranes*, Vol.26, No.5, pp. 384-392, ISSN: 0266-1144.

Yang, S.; Rafael Castilleja, J., Barrera, E. V. & Lozano, K. (2004). Thermal analysis of an acrylonitrile-butadiene-styrene/SWNT composite. *Polymer Degradation and Stability*, Vol.83, No.3, pp. 383-388, ISSN: 0141-3910.

17

Characterization of Thermoplastic Elastomers by Means of Temperature Scanning Stress Relaxation Measurements

Vennemann Norbert
University of Applied Sciences Osnabrück
Germany

1. Introduction

Thermoplastic elastomers (TPE) and, in particular thermoplastic vulcanizates (TPV), are a new class of materials, combining the properties of conventional elastomers (rubber) and the processibility of thermoplastics. Compared with conventional rubber elastomers, these materials can be more easily processed and more easily recycled. TPE are often used to replace conventional thermoset rubber, but those are also used for a great variety of new applications and products, particularly in hard/soft combinations with other thermoplastics. Due to the advantages over conventional thermoset rubber, the commercial uses for thermoplastic elastomers are growing rapidly (Schäfer, 2001; Bittmann, 2004).

Beside many advantages, some disadvantages do exist, also. In comparison to chemically crosslinked elastomers (e.g., EPDM or natural rubber), TPE materials have stronger limitations with respect to upper service temperature, which is caused by softening or melting of the hard phase. Furthermore, TPE exhibit higher creep and stress relaxation, than thermoset rubber, even at ambient temperatures (Holden et al., 2004). Thus, new demands on polymer testing arise from the assessment of thermoplastic elastomers (TPE) with respect to their rubber elastic use properties and stress relaxation behaviour, particularly at elevated temperatures.

Due to the complex molecular structure and phase morphology of TPE, traditional test methods normally used for characterization of elastomers give only limited information about the unique properties of TPE. For this reason, temperature scanning stress relaxation (TSSR) test method has been developed, recently (Vennemann et al., 2001, 2003). In this work, the basic principle of the TSSR test method as well as the theoretical background will be described. Furthermore, an overview of numerous results obtained from selected TPE materials, will be presented to demonstrate the versatility of the TSSR method. In addition, further development of the method will be presented, in particular for rapid determination of crosslink density of TPV.

2. Theoretical background

2.1 Stress relaxation and determination of relaxation spectra

Polymers exposed to constant strain exhibit the well known phenomenon of stress relaxation, i.e. a more or less strong decrease of stress as a function of time. The microscopic mechanisms, leading to the macroscopic recognizable decrease of stress, may result from physical and/or chemical processes. In contrast to thermoset rubber, where the thermal - mechanical behaviour is dominated by chemical reactions resulting in cleavage of polymer chains and network junctions, in case of thermoplastic elastomers physically induced stress relaxation processes are most important with respect to usage properties.

For the simple Maxwell - model, as represented in Figure 1, the relaxation time constant τ is defined as the period of time, after the stress has dropped down to the value of σ_0/e. Here, σ_0 indicates the initial stress at time zero when the strain has been applied to the sample. Unfortunately, the real behaviour of materials is more complicated and cannot be described by the simple Maxwell - model. According to the well known theory of linear viscoelasticity the entire relaxation process can be described by means of the generalized Maxwell - model, which consists of an infinite number of individual spring - dashpot - elements. Under iso-thermal conditions (T = const.) the relaxation modulus E_{iso} is a function of time t and given by Eq. (1), for this model (Ferry, 1980).

$$E_{iso}(t) = E_{\infty} + \int_{-\infty}^{\infty} H'(\tau) \cdot e^{\frac{-t}{\tau}} \, d\ln\tau \tag{1}$$

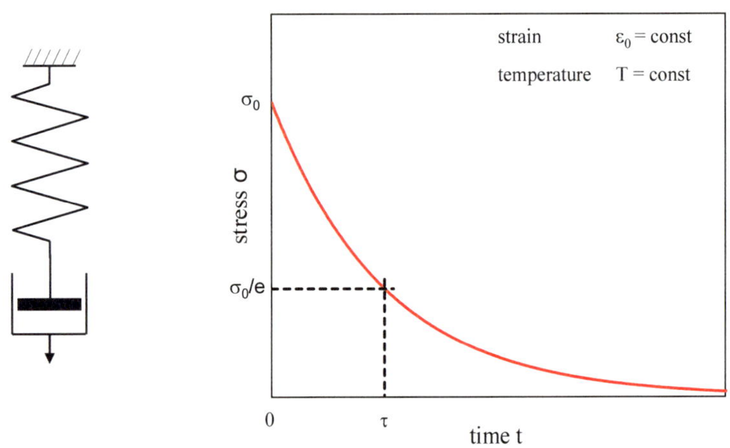

Fig. 1. Maxwell - Model and stress as a function of time after a constant strain ε_0 has been applied at t = 0.

The relaxation modulus is also directly related to the experimentally observable stress $\sigma(t)$ and can be easily calculated by dividing the stress by the applied strain ε_0. In this equation the relaxation spectrum $H'(\tau)$ is a steady function describing the probability of the relaxation time constants τ of the model which may be associated with the population of relaxation mechanisms of the system. The constant E_{∞} is added in Eq. (1) to allow the system to approach an equilibrium state higher than zero, as observed normally for viscoelastic solids.

According to Alfrey's rule (Alfrey & Doty, 1945) the value of the relaxation spectrum $H'(\tau)$ at point τ = t is obtained in first approximation by differentiating $E_{iso}(t)$ with respect to ln*t*, by Eq.(2).

$$H'(\tau) = -\left(\frac{dE_{iso}}{d\ln t}\right)_{t=\tau} = -t \cdot \left(\frac{dE_{iso}}{dt}\right)_{t=\tau} \tag{2}$$

All relaxation mechanisms, and thus the relaxation time constants, are strongly depending on temperature, i.e. the higher the temperature, the lower the relaxation time constants and vice versa. Therefore, the relaxation spectrum covers a very wide range on time scale and it is practically impossible to determine the entire function by means of a single stress relaxation measurement. Usually, a set of measurements at several temperatures have to be executed, to create a master curve, based on time - temperature superposition principle (TTS). That means, high effort is required to obtain full information of the stress relaxation behaviour.

With temperature scanning stress relaxation (TSSR) measurements, an alternative strategy has been introduced recently (Vennemann et al., 2001; Vennemann, 2003; Barbe et al., 2005). In contrast to traditional isothermal tests, during TSSR measurements the temperature is not kept constant, but is increasing linearly with a constant heating rate β, e.g. β = 2 K/min. As a result, the non-isothermal relaxation modulus $E_{non\text{-}iso}$ as a function of temperature is obtained. Analogue to isothermal stress relaxation measurements, the spectrum H(T) can be calculated in first approximation by Eq. (3).

$$H(T) = -\Delta T \cdot \left(\frac{dE_{non-iso}}{dT}\right)_{\beta=\frac{\Delta T}{t}=const} \tag{3}$$

$$\text{with } T - T_0 = \Delta T = \beta \cdot t$$

In this equation T_0 stands for the initial temperature at which the test is started and β is the heating rate of the temperature scan. Although this function is not defined on time scale, the relaxation mechanisms the polymer sample undergoes during the test can be identified, clearly, because the relaxation time constant τ decreases monotonously with increasing temperature T. Due to its very strong temperature dependence, the relaxation time constant τ drops down to small values rapidly, within small temperature range. Thus, the entire spectrum is observable on temperature scale within a relative short period of time during a temperature scan of a TSSR test.

Beside stress relaxation, two other phenomena, i.e. thermal expansion and rubber elasticity of the sample, have to be taken into consideration, if a sample, mounted between two sample holders having constant distance, is heated up linearly. In the following sections, those effects will be described in detail.

2.2 Thermal expansion of the sample

Due to thermal expansion of the material the initial length l_0 of the sample is increasing, if a temperature scan is performed starting at initial temperature T_0 up to higher temperatures T. In consequence, the thermal expansion of the sample contributes to a decrease of stress, if a stretched sample is mounted between sample holders with constant distance. The thermally induced variation of strain $\varepsilon = (l - l_0)/l_0$ during TSSR tests can be easily calculated by

$$\varepsilon(T) = \frac{l}{L_0(1+\alpha \cdot \Delta T)} - 1 \quad (4)$$

where L_0 is the initial distance of the sample holders at temperature T_0 and α is the coefficient of linear thermal expansion of the sample. For rubber typical values for α are in the range of 1 to $3 \cdot 10^{-4}\,K^{-1}$ (Gent, 2001). In Figure 2 the relative strain $\varepsilon/\varepsilon_0$ is plotted against temperature for certain values of initial strain ε_0. It becomes obvious, increasing temperature results in decreasing strain. However, the influence of temperature on relative strain is more or less negligible, if the strain is sufficiently high. To minimize the influence of thermal expansion, TSSR experiments should be performed at initial strains not below 50 %.

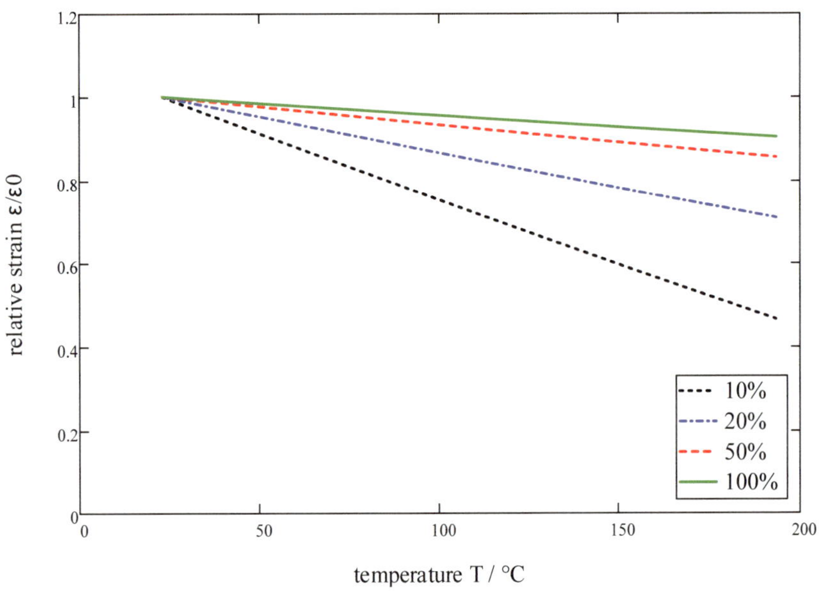

Fig. 2. Influence of thermal expansion of the sample on relative strain. Curves were calculated for certain initial strain values, as indicated, by means of Eq. (4) with $\alpha = 3 \cdot 10^{-4}\,K^{-1}$.

2.3 Rubber elasticity

According to the well known theory of rubber elasticity in case of an ideal rubber network, the mechanical stress σ is proportional to the absolute temperature T and can be expressed by Eq. (5) (Mark, 1981).

$$\sigma = \nu \cdot R \cdot T\left(\lambda - \lambda^{-2}\right) \quad (5)$$

Where v is the crosslink density of the network and R is the universal gas constant. The strain ratio λ is defined as $\lambda = l/l_0$, where l is the length and l_0 the initial length of the sample. According to Eq. (5) the stress should increase with increasing temperature, if the strain ratio λ is kept constant. The slope of the stress - temperature - curve at constant elongation can be obtained from the derivative of stress with respect to temperature which is assigned as temperature coefficient κ in the following.

$$\kappa = \left(\partial\sigma/\partial T\right)_{\lambda} = \nu \cdot R \cdot \left(\lambda - \lambda^{-2}\right) \tag{6}$$

For high elongations, the temperature coefficient κ is positive, however at low strain ratios, i.e. $\lambda < 1.1$, a negative value of κ was found experimentally. The transition from a negative to a positive value of the temperature coefficient is known as thermoelastic inversion (Pellicer, 2001) . The phenomenon of thermoelastic inversion is not predicted by theory but, it was shown early (Anthony et al., 1942) and has been confirmed by own measurements (Vennemann & Heinz, 2008), this apparent contradiction results only from thermal expansion of the sample. Considering the temperature dependence of the strain, as described by Eq. (4), the relation of Eq.(5) can be rewritten as

$$\sigma = \nu \cdot R \cdot T \left(\frac{\lambda_0}{1+\alpha \cdot \Delta T} - \left(\frac{\lambda_0}{1+\alpha \cdot \Delta T} \right)^{-2} \right) \tag{7}$$

where λ_0 is the initial strain ratio at temperature T_0. The influence of thermal expansion on the stress - temperature - curve is shown in Figure 3, where the uncorrected curve as calculated from Eq. (5), is represented in comparison to the corrected curve, as calculated from Eq. (7). Obviously, the initial slope is slightly reduced due to thermal expansion. Furthermore, the corrected function is no longer strictly linear, but exhibits a slight curvature with increasing temperature. From Eq. (7), by derivation with respect to temperature T, the corrected temperature coefficient κ is obtained, which is now also a function of temperature. The initial value of κ at temperature T_0 is given by Eq. (8) (Vennemann et al., 2011).

$$\kappa_0 = \kappa(T_0) = \nu \cdot R \cdot \left[\left(\lambda_0 - \lambda_0^{-2}\right) - T_0 \cdot \alpha \cdot \left(\lambda_0 + 2 \cdot \lambda_0^{-2}\right) \right] \tag{8}$$

2.4 General remarks

It has been shown; macroscopic recognizable increase of stress is recognizable if an elongated piece of rubber is heated up linearly. The macroscopic reaction of the material is caused by the change of entropy on microscopic scale and thus, it becomes possible to easily determine the crosslink density of a rubber sample, which is an important microscopic parameter of the system. But, it is important to notice, that the above equations are only strictly correct for ideal rubber networks. Real systems, such as filled elastomers and thermoplastic elastomers, are more complex, and cannot be fully described by this simple theory. Therefore, further development of theory is required to better understand the behaviour of those materials. Recently, a model has been developed to describe the thermoelastic behaviour of carbon black filled elastomers (Vennemann et al., 2011).

In case of thermoplastic elastomers the situation is even more difficult, because these materials consist of at least two phases and in case of commercial grades additionally of fillers, plasticizer and other additives. Although most elastomers and also thermoplastic elastomers (TPE) exhibit thermoelastic behaviour similar to ideal rubber, calculation of true crosslink density is not possible, but only apparent values because of lack of adequate theory. Nevertheless, the characterization of thermoelastic behaviour by TSSR measurements is very useful, in particular in case of thermoplastic vulcanizates, because properties which are closely related to the structure of the polymer network become recognizable. Furthermore, additional information about the composition, morphology and structure of the sample can be deduced from the entire relaxation spectrum.

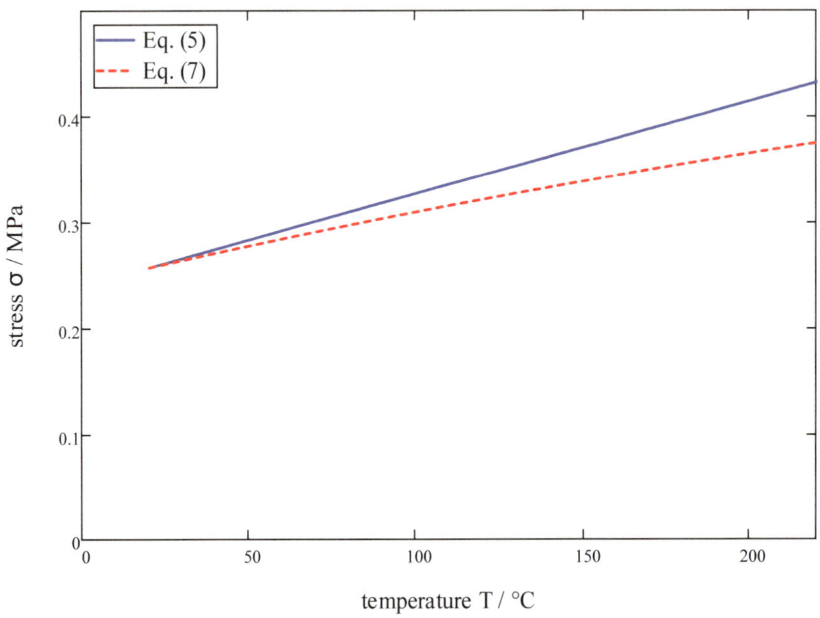

Fig. 3. Theoretical stress - temperature - curves, calculated by use of Eq. (5) and Eq. (7) with $v = 100\ mol/m^3$, $\lambda_0 = 1.5$ and $\alpha = 3 \cdot 10^{-4}\ K^{-1}$.

3. Experimental

3.1 Materials and preparation of the samples

3.1.1 Thermoplastic elastomers based on Styrene Block Copolymers (SBC)

High molecular weight poly(styrene-b-ethylene/butylene-b-styrene) (SEBS) with a polystyrene (PS) content of 33%, a molar mass of the PS-blocks of 29000 g/mol and a total molar mass of M_w = 174000 g/mol were used as the basis for the compounds prepared. In SBC/-polyolefin blends a standard isotactic polypropylene and, alternatively a standard high density polyethylene were used as the polyolefin component of the compounds. In SBC/PPE blends high molecular weight poly(p-phenylene ether) (PPE) with T_g = 215°C and molar mass of M_w = 38900 g/mol was used as the modifier. Additionally, high purity medicinal paraffin oil was used as the extender oil for all compounds and a small amount of stabilizer was added to protect the polymers against degradation during the mixing process.

SBC/polyolefin compounds were produced using a twin-screw extruder (L/D: 32/1, 25 mm diameter; Berstoff GmbH). SBC/PPE compounds were produced by means of a single-screw extruder (Göttfert GmbH, L/D: 15/1). In all cases the ingredients were mixed together prior feeding to the extruder, having a barrel temperature of 260°C. Test plates of 2mm thickness of all compounds were produced in a pneumatic injection moulding press. Further details are described in earlier papers (Vennemann et al., 2004) and (Barbe et al., 2005).

3.1.2 Thermoplastic polyolefin blends (TPO) and dynamic vulcanizates (TPV)

a. Commercial grades based on EPDM/PP

Several commercial grades of thermoplastic vulcanizates based on EPDM/PP covering a wide range of hardness were obtained from Solvay Engineered Polymers (TX/USA) and tested as received. The novel TPV-AP materials were produced via a dynamic vulcanization process using a new curative system and DVA process developed by Solvay Engineered Polymers (Reid et al., 2004). The new cure system results in a material with non-hygroscopic behaviour, white colour, and low odour. Properties of TPV-AP are compared to two other commercially available TPV materials. TPV-HS is a commercial TPV based upon EPDM and PP where the elastomer is crosslinked with a hydrosilation process. TPV-PH is also a commercial TPV based upon EPDM and PP where the elastomer is crosslinked with a phenolic resin curing process. The samples of both TPV-HS and TPV-PH were not produced by Solvay Engineered Polymers, but commercial grades, produced by other suppliers.

b. Model compounds of peroxide cured TPV based on EPDM/PP

Commercial available EPDM rubber and isotactic polypropylene homopolymer (PP) were used as the basis for the dynamic vulcanizates (TPV). The EPDM contains 50 wt % ethylene and 4 wt % ethylidene norbornene (ENB). It has a Mooney viscosity ML(1+4) at 125 °C, of 36. The melt flow rate of the polypropylene, measured at 230 °C and 2.16 kg is 12 g/10 min. The crosslink system consists of di(tert-butylperoxyisopropyl)benzene (abbrev.: DTBPIB) as peroxide and trimethylolpropane trimethacrylate (abbrev.: TRIM) as co-agent. The peroxide and co-agent are supplied commercially on a silica carrier, with active agent content of 40 wt % and 70 wt %, respectively. The TPV samples are designated as TPV1 to TPV6, whereas the total amount of curatives (DTBPIB and TRIM) is increasing from 1 phr to 6 phr in steps of 1 phr. The volume fraction of polypropylene was $\phi_{PP} = 0.23$ in all compounds. An uncured compound of identical EPDM/PP ratio was also produced and tested as reference sample. All samples were produced in a two-step mixing process using a Haake Rheocord 600 laboratory internal mixer (Thermo Electron Corporation, Karlsruhe). Further details of the production process are published elsewhere (Vennemann, 2006).

c. Model compounds of phenolic cured TPV based on EPDM/HDPE

Two different commercially available EPDM rubber and two different grades of high density polyethylene (HDPE) were used as the basis for the dynamic vulcanizates, in this study. Crosslinking of the EPDM in all compounds was performed with a phenolic resin cure system, consisting of stannous chloride ($SnCl_2 \cdot 2\, H_2O$), zinc oxide (ZnO) and SMD 31214. The latter is a commercially available solution of paraffinic mineral oil and 30 wt % of phenolic resin SP 1045. Further details of the composition and preparation of the compounds are published elsewhere (Vennemann, 2009).

3.2 TSSR instrument and test procedure

The temperature scanning stress relaxation tests were performed by use of a commercial available TSSR instrument obtained from Brabender GmbH (Duisburg, Germany). The TSSR instrument (Fig. 4) consists of an electrical heating chamber where the sample, a S2 testing rod, is placed between two clamps. The clamps are connected to a linear drive unit to apply a certain uniaxial extension to the sample. A high quality signal amplifier in combination with a high resolution AD-converter is used to detect and digitize the analogue signals of the high-resolution force transducer and the thermocouple. In order to detect the current temperature the thermocouple is placed near the centre of the sample. All signals are transferred to a personal computer. A special software program is used for treatment and evaluation of the data as well as for the control of the test procedure.

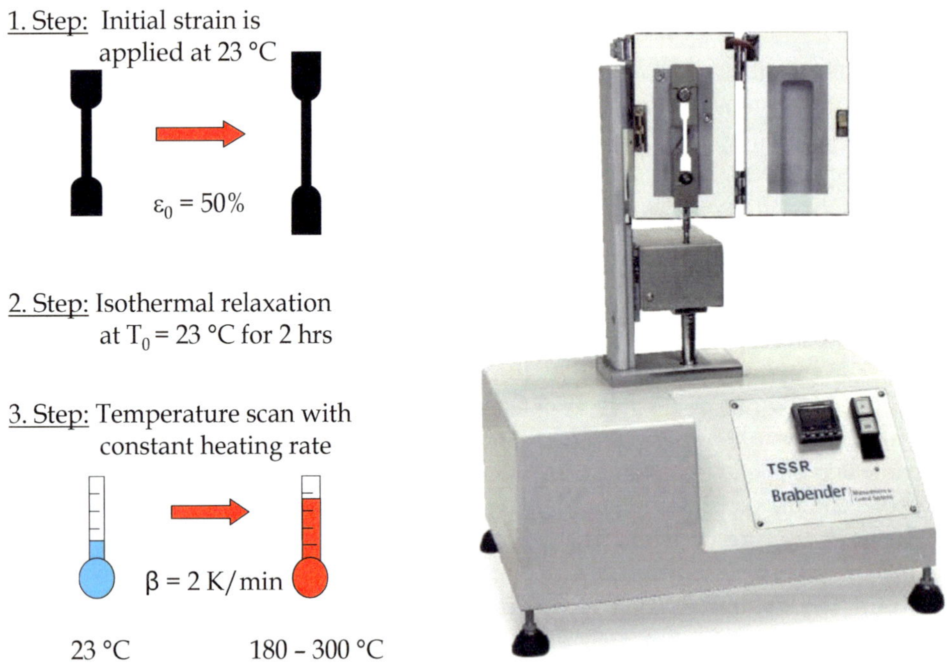

Fig. 4. TSSR instrument and test procedure

The test procedure starts with placing the sample in the electrical heated test chamber, which is controlled at the initial temperature T_0 of 23 °C. After the initial strain of $\varepsilon_0 = 50\%$ is applied, the isothermal relaxation period starts, whereas the temperature remains constant at 23 °C within +/- 0.1 °C. During this time most of the short time relaxation processes occur and the sample reaches a quasi equilibrium state. Then the sample is heated linearly at a constant rate of $\beta = 2$ K/min, until the stress relaxation has been fully completed or rupture of the sample has occurred.

From the obtained force - temperature curve certain characteristic quantities such as T_{10}, T_{50}, T_{90} and the TSSR index RI can be calculated. The temperature T_x stands for the temperature at which the force ratio F/F_0 has decreased about x% with respect to the initial force F_0. The TSSR index RI is a measure of the rubber like behaviour of the material and is calculated from the area below the normalized force - temperature curve, as represented in Figure 5 and given by Eq. (9). Additionally, the temperature coefficient κ and the relaxation spectrum H(T) are calculated from the initial slope and the derivative of the stress–temperature curve, respectively, as described in chapter 2, in more detail.

$$RI = \frac{\int_{T_0}^{T_{90}} F/F_0 \, dT}{T_{90} - T_0} \qquad (9)$$

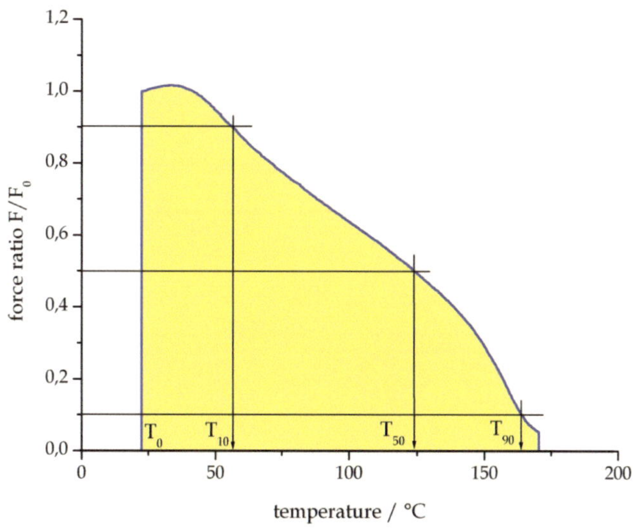

Fig. 5. Normalized force as a function of temperature and determination of characteristic temperatures T_{10}, T_{50} and T_{90}.

4. Results and discussion

4.1 Thermoplastic elastomers based on Styrene Block Copolymers (SBC)

Commercial available TPE-S materials are generally a compound of a styrene block copolymer, commonly poly(styrene-b-ethylene/butylene-b-styrene) (SEBS) or poly(styrene-b-butadiene-b-styrene), and a thermoplastic polymer, mostly polypropylene (PP). Additionally, plasticizer, mineral fillers and other components are used to achieve the demanded properties. In Fig. 6 (left) force - temperature curves and the corresponding relaxation spectra of two different types of SBC - compounds are represented. Up to 110 °C, both materials behave almost identical, but at higher temperatures the force of the SEBS/PE compound drops down to zero close to 120 °C, whereas the force of the SEBS/PP compound decreases more or less slightly until the base line is approached at about 165 °C. In the relaxation spectrum of both materials a significant peak at about 100 °C is observable which corresponds to the glass transition temperature of the styrene hard phase of the SEBS. At higher temperature (120 °C or 160 °C) an additional peak appears which is caused by the melting of the thermoplastic component, i.e. polyethylene or polypropylene, respectively. From these measurements it becomes clearly obvious; the upper service temperature range of SBC compounds is limited by the glass transition of the polystyrene hard phase. An increased upper service temperature limit may result from the existence of a co-continuous phase of a thermoplastic component having a higher melting temperature. In case of polyethylene as the thermoplastic component, an improvement up to 120 °C can be achieved, whereas by use of polypropylene higher temperature, up to a maximum value of 160 °C, is possible.

However, the latter values have to be considered as theoretical maxima. For obvious reasons, the upper limits of the service temperature have to be significantly lower than those maximum values. Because the melting temperature of polypropylene is considerably higher than of polyethylene, PP is favoured as thermoplastic component for commercial TPE-S materials. Normally, it is not possible to detect the glass transition temperature of the PS hard phase of commercial SBC compounds by means of traditional DSC and DMA measurements, because of the complex compound composition and the limited sensitivity of the instruments. In contrast, TSSR measurements are very sensitive with respect to relaxation processes of the hard phase and therefore more suitable, to characterize and improve those materials.

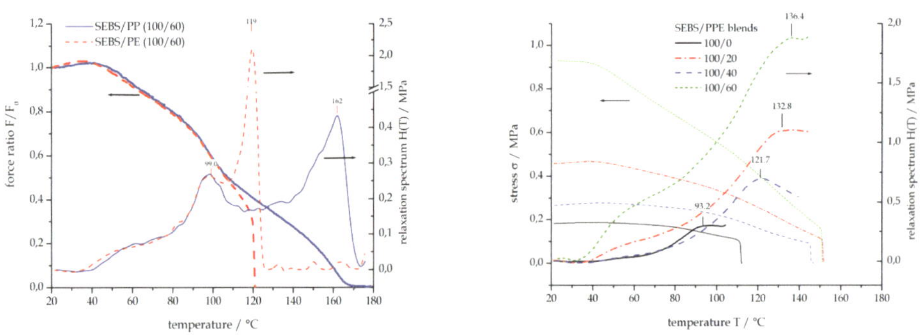

Fig. 6. *Left:* Force - temperature curves and relaxation spectra obtained from SEBS/PE and SEBS/PP blends. *Right:* Stress - temperature - curves and relaxation spectra obtained from SEBS/PPE blends with increasing proportion of PPE.

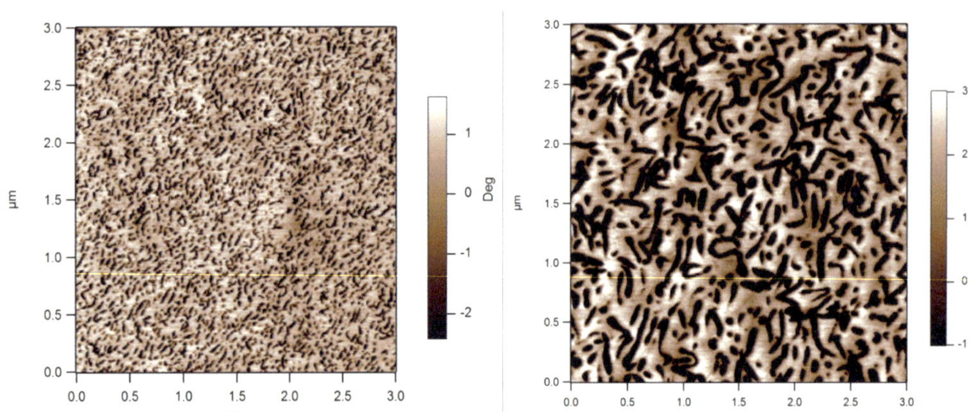

Fig. 7. AFM phase images of pure SEBS (left) and SEBS/PPE blend (right) with blend ratio of 100/20.

An alternative route to improve the heat resistance of TPE-S materials, i.e. TPE based on SBC, exists by blending the SBC with poly(p-phenylene ether) (PPE). PPE and PS are known to be thermodynamically miscible over the entire composition range, i.e., they form a blend with only one glass transition temperature (Tucker et al., 1988). Thus, by blending with PPE the glass transition of the SBC hard phase may be increased up to 150 °C (Barbe et al., 2005).

In Figure 6 (right) a series of stress - temperature - curves and relaxation spectra is also represented, which were obtained from SEBS/PPE blends, containing increasing proportion of PPE. Due to increasing content of PPE, the glass transition of the SEBS hard phase is shifting to higher temperatures and thus makes the hard phase more heat resistant. Furthermore, the level of the stress - temperature - curves is also increasing, caused by the reinforcing effect of the hard domains. As can be seen from the AFM phase images shown in Figure 7, the size of the hard domains (dark) is increasing, after PPE was added to the system. Obviously, the hard domains of the SEBS and also of the SEBS/PPE blends act as physical network junctions and additionally as filler particles. In this case, no co-continuous phase of the thermoplastic component exists, as in case of the SEBS/PP blends. The stress - temperature - curves and the relaxation spectra reveal the failure of the samples occurred slightly above the glass transition temperature of the hard phase. If combining the SEBS/PPE system with an additional thermoplastic blend component, e.g. polypropylene or polyamide 12, the thermal mechanical behaviour of the material can be improved on further (Barbe et al., 2005).

4.2 Thermoplastic polyolefin blends (TPO) and dynamic vulcanizates (TPV)

a. Commercial grades based on EPDM/PP

Most of the commercial available thermoplastic vulcanizates (TPV) are produced from polyolefin blends, in particular EPDM/PP, by the process of dynamic vulcanization. TPV exhibit several advantages over simple thermoplastic polyolefin blends (TPO). Due to selective crosslinking of the EPDM rubber phase, almost all material properties are improved. Especially compression set, creep, stress relaxation and swell behaviour are highly important for automotive applications, e.g. all kinds of sealing systems. As shown in Figure 8 for different types of commercial TPO and TPV, dynamic vulcanization has strong impact on relaxation behaviour (Reid et al., 2004). Whereas the stress - temperature - curve of a simple TPO blend exhibits a strong decrease of stress with increasing temperature, the decrease of stress of a TPV material of comparable hardness is significantly lower. As it becomes also obvious from Figure 8, the differences between TPO and TPV depend on hardness; the lower the hardness, the bigger the differences, and vice versa. Only one significant peak is observable in the relaxation spectra of the TPV, which is assigned to the melting of the PP matrix.

Sample	Shore - A	σ_o MPa	T_{10} °C	T_{50} °C	T_{90} °C	TSSR - Index RI
SEBS/PP (100/60)	69	1.098	63.7	107.5	159.0	0.655
SEBS/PE (100/60)	70	1.106	61.9	106.2	120.5	0.796
SEBS/PPE (100/0)	22	0.183	80.6	111.1	111.8	0.922
SEBS/PPE (100/20)	28	0.269	90.5	128.1	145.7	0.844
SEBS/PPE (100/40)	43	0.460	69.9	125.7	151.3	0.765
SEBS/PPE (100/60)	64	0.929	54.0	112.6	150.7	0.659

Table 1. Results of the TSSR tests as represented in Fig. 6.

In case of low hardness TPO a smaller peak appears at about 40 to 60 °C, which might be related to the α -relaxation process of the PP phase. At higher temperatures, when the PP matrix is melting, no significant peak can be detected in case of TPO, because the stress is almost zero due to relaxation. The results of the TSSR tests represented in Figure 8 are compiled in Table 1. For similar hardness TSSR - Index and T_{50} - values of the TPV are significantly higher than those of TPO materials.

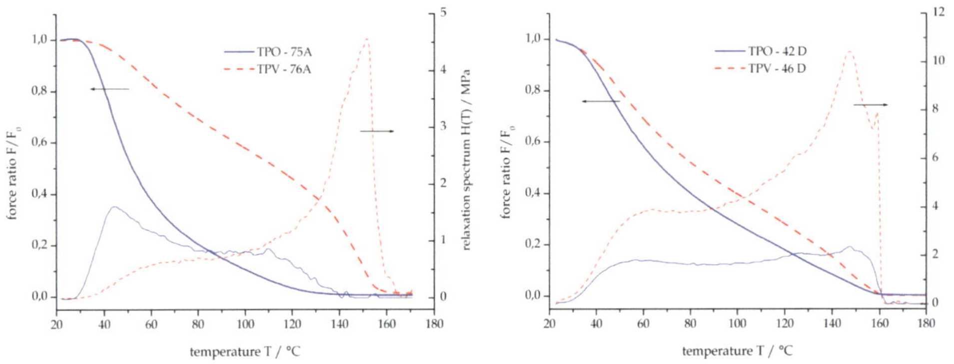

Fig. 8. *Left:* Force - temperature curves and relaxation spectra obtained from commercial TPO and TPV of 75 Shore A hardness. *Right:* Force - temperature curves and relaxation spectra obtained from commercial TPO and TPV of 40 Shore D hardness.

b. Model compounds of peroxide cured TPV based on EPDM/PP

Phenolic resin cure systems are mostly used for the production of commercial TPV materials. Although the phenolic resin cure system normally results in products with better compression set and oil swell than equivalent compositions produced with other cure systems, e.g. hydrosilation, peroxide, or sulfur, peroxide cured TPV materials were introduced into the market, recently (Reid et al., 2004). In EPDM/PP based systems serious problems result from degradation of the PP matrix, which may lead to deterioration of mechanical properties of the TPV. Thus, during the dynamic vulcanization by use of peroxides not only crosslinking of the EPDM rubber phase occurs, but more or less of the free radicals produce degradation of the PP matrix (Loan, 1967; Dickens, 1982), which has to be taken into consideration, also. Therefore, the producers have to find out the right balance between crosslinking of the EPDM phase and degradation of the PP phase; both reactions are initiated by free radicals, created by the curatives.

Sample	Shore -Hardness	σ_0	T_{10}	T_{50}	T_{90}	TSSR - Index
		MPa	°C	°C	°C	
TPV- 76A	76A	0.9	52	114	151	0.66
TPO - 75A	75A	1.2	37	52	101	0.49
TPV - 46D	46D	4.7	41	83	146	0.54
TPO - 42 D	42D	2.0	38	68	136	0.49

Table 2. Results of the TSSR tests as represented in Fig. 8.

Because crosslink density is one of the most important parameters of elastomers and dynamic vulcanizates, the aim of this study was to investigate the crosslink density of peroxide cured TPV based on EPDM/PP. Until now, there are only few methods available for the determination of crosslink density (Eisele, 1979; Grinberg, 1999; Zhao, 2007). Most of these methods require high effort and are not suitable for daily use in product development or quality control. For this reason, a new test method for the determination of the crosslink density of dynamic vulcanizates was introduced. The suitability of this method was examined at the example of polyolefinic model compounds, which were dynamically vulcanized by means of a peroxide cure system (Vennemann, et al., 2006).

The TPV samples were produced in a two-step mixing process. In a first mixing step pre-blends containing EPDM and varying amounts of the crosslink agents were prepared at 100 °C and a rotor speed of 40 rpm. After a mixing time of 3 minutes the pre-blends were removed from the mixer and immediately cooled down to room temperature to avoid scorch of the material. The composition of the different pre-blends is summarized in Table 3. In a second mixing step the TPV compounds were produced, by melt mixing the pre-blends with the PP homopolymer. The composition of the TPV compounds is compiled in Table 4. At a fixed EPDM volume fraction of $\phi_{EPDM} \approx 0.77$ the compound compositions differ only in the content of crosslink agents.

The morphology of the samples was investigated by transmission electron microscopy (TEM). TEM micrographs of the phase morphology of selected TPV samples are shown in Figure 9. Samples for these micrographs were cut parallel to flow from an injection moulded plaque, stained with ruthenium oxide, and cryo-microtomed into thin sections. In Figure 9 (left), the co-continuous phase morphology of the thermoplastic polyolefin blend (TPO) is visible, where the EPDM is shown as dark and the PP is shown as light areas. Phase inversion already occurs at low curative content (1 phr peroxide/1 phr co-agent) as it becomes obvious from the TEM micrograph obtained from TPV2 as presented in Figure 9 (right). Here, the elastomer appears as dark discrete particles with diameters of less than one micron, embedded in the continuous PP phase (lighter colour).

Sample ID	E0	E1	E2	E3	E4	E5	E6
EPDM	100	100	100	100	100	100	100
DTBPIB	-	0.5	1.0	1.5	2.0	2.5	3.0
TRIM	-	0.5	1.0	1.5	2.0	2.5	3.0

Table 3. Composition of the pre-blends in parts per hundred rubber (phr)

Sample ID	TPO	TPV1	TPV2	TPV3	TPV4	TPV5	TPV6
Ex*)	100	100	100	100	100	100	100
PP	30	30	30	30	30	30	30

*) Ex stands for the corresponding pre-blend, e.g. TPV1 contains pre-blend E1 etc.

Table 4. Composition of the TPV compounds in parts per hundred rubber (phr)

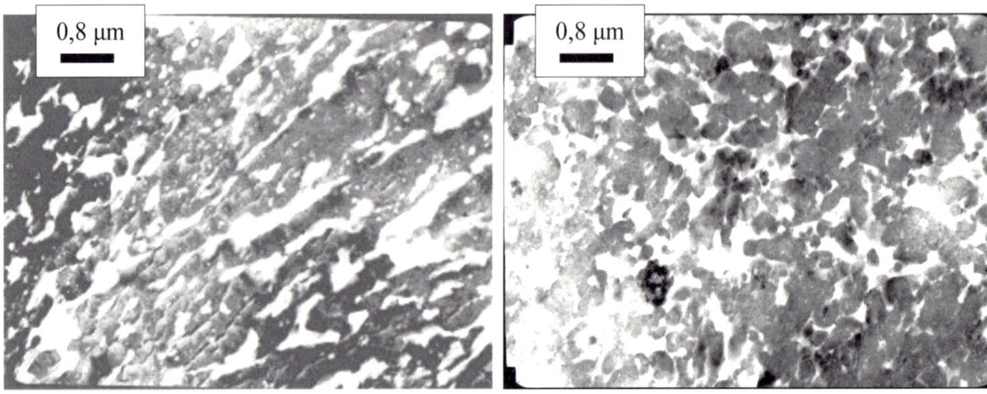

Fig. 9. TEM micrographs obtained from TPO (left) and TPV2 (right).

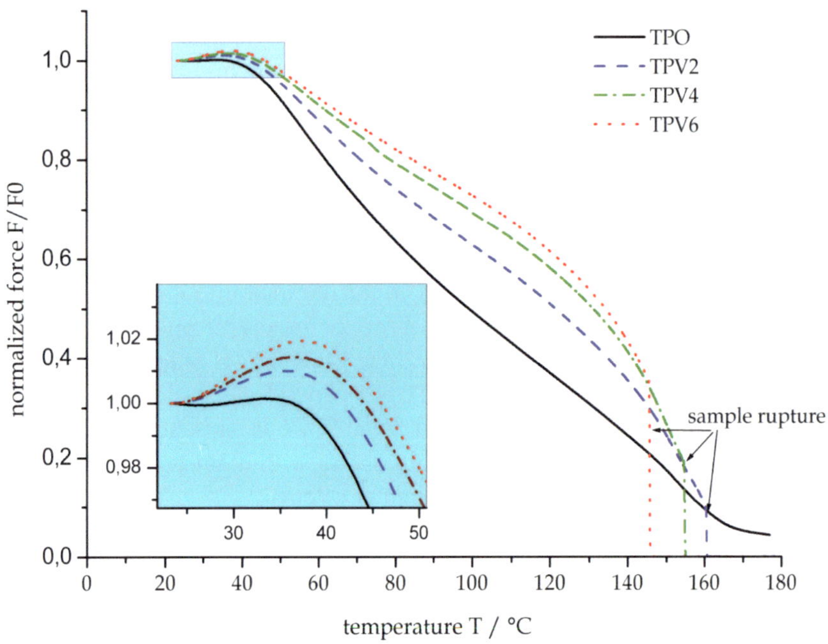

Fig. 10. Selected normalized stress - temperature curves of TPV samples obtained from TSSR measurements. TPO (solid line), TPV2 (dashed line), TPV4 (dashed dotted line) and TPV6 (dotted line).

In Figure 10 selected normalized force – temperature curves obtained from TSSR tests are presented. The influence of dynamic vulcanization is clearly recognizable from the shape of the curves. Whereas the uncrosslinked TPO sample exhibits the strongest stress decrease, this is significantly reduced in the case of the TPV samples and thus, higher values of T_{10} and T_{50} (Table 5) are obtained. This behaviour was expected, because dynamic vulcanization is well known as a process to improve the stress relaxation properties of polyolefin blends. Due to the reduced stress relaxation, the area below the normalized force–temperature curve increases. Consequently, the rubber index RI increases also and is indicating an improvement of "rubber like" - behaviour. In Table 5 are also values of Shore A hardness, compression set, tensile strength and elongation at break summarized. From Figure 10 it can be also observed that rupture of the samples occurs if the peroxide content of the TPV increases. This is accompanied with a decrease of the TSSR T_{90} values. Unlike this, the force-temperature curve of the TPO sample approached zero, without rupture of the sample. The rupture of the samples can be explained with the degradation of the polypropylene matrix by peroxide, which is also a well known phenomenon. Due to the consumption of peroxide by the PP matrix, crosslink density of the dispersed EPDM particles is reduced. Figure 11 contains normalized force - temperature curves obtained from thermoset rubber samples, prepared by static vulcanization of the pre-blends. The behaviour differs strongly from that of the TPV samples, presented in Figure 10, although the crosslink systems are identical.

Sample ID	Shore A	CS 22h / 125°C	tensile strength MPa	elongation @ break	TSSR			
					T_{10} °C	T_{50} °C	T_{90} °C	RI
TPO	61	87.9%	1.9	102%	52.2	99.1	159.4	0.59
TPV1	69	74.2%	3.4	263%	53.7	106.7	159.3	0.62
TPV2	76	64.5%	6.3	328%	57.2	121.3	160.4	0.66
TPV3	74	62.5%	8.1	381%	61.6	131.4	156.9	0.72
TPV4	75	61.2%	8.1	366%	61.7	130.9	154.8	0.73
TPV5	77	60.5%	7.9	335%	63.3	133.3	149.2	0.76
TPV6	77	59.2%	8.1	309%	64.4	134.0	145.7	0.78

Table 5. Mechanical properties and TSSR results of the TPO and TPV samples

As described before, temperature coefficient values κ were determined from the initial part of stress - temperature curves, which were obtained from TSSR measurements at TPV (Figure 10) and thermoset rubber samples (Figure 11). The crosslink densities of all samples were calculated according to Eq. (6) and Eq. (8) and are plotted against the total amount of curatives in Figure 12. It should be noticed, that the compositions of the thermoset rubber samples E1 - E6 are identical with the rubber phase of the corresponding TPV samples. Thus, the crosslink density of TPV1 can be directly compared with sample E1, TPV2 can be compared with E2 and so on. Figure 12 clarifies that with the same amount of curatives a higher crosslink density is achieved in the thermoset rubber than in the rubber phase of the corresponding TPV compound. This result confirms the assumption that the crosslink density in the rubber phase of the TPV is reduced, due to partial consumption of the peroxide by the PP matrix. By comparison of the number of moles of curatives used in the recipe, the crosslink of efficiency of the cure system can be verified, also. The obtained results of crosslink density were also compared with the reciprocal swell ratio of the samples, which is a measure of crosslink density. It was shown that the crosslink density obtained from TSSR measurements, correlates well with the reciprocal swell ratio (Vennemann et al., 2006).

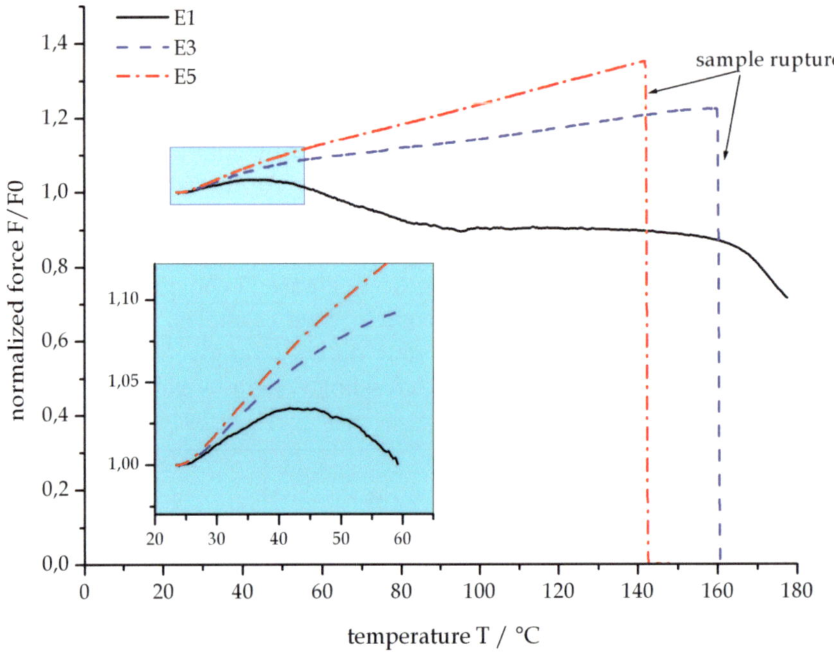

Fig. 11. Selected normalized stress – temperature curves of thermoset rubber samples obtained from TSSR measurements. E1 (solid line), E3 (dashed line) and E5 (dashed dotted line).

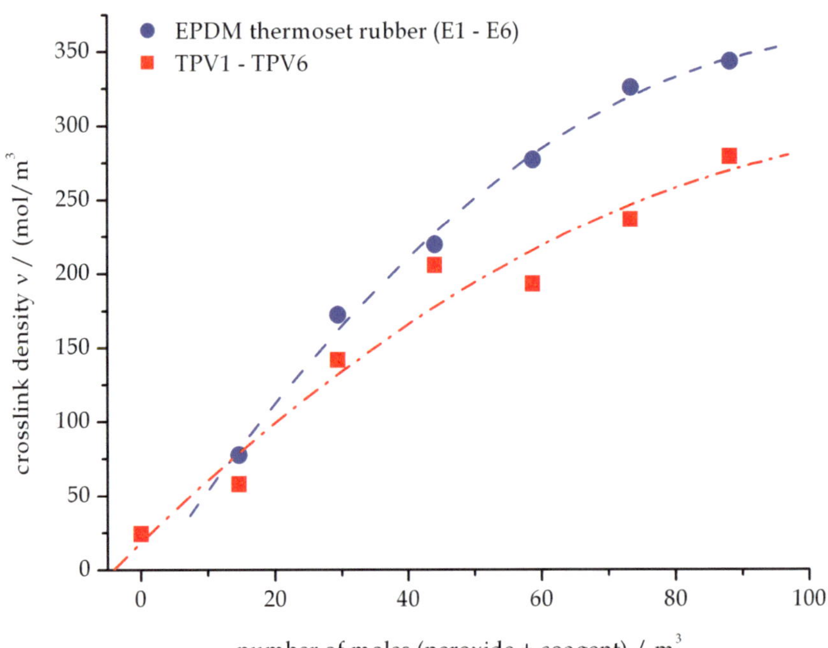

Fig. 12. Crosslink density of thermoset rubber samples E1 - E6 (•) and TPV1 - TPV6 samples (■) as obtained from TSSR – measurements.

c. Advanced TPV based on EPDM/HDPE designed for hart/soft combinations with HDPE

Until now, most of the commercial available thermoplastic vulcanizates are based on the EPDM/PP system. Thus, good adhesion is achieved to components made of polypropylene (PP), due to inherent compatibility of the TPV matrix and commonly used PP materials. In combination with other polymers the bonding strength is more or less poor, particularly in case of more polar engineering thermoplastics. Even with other polyolefins bonding strength to TPV may deteriorate because of incompatibility of both partners. PP and HDPE are generally considered immiscible (Shanksa, et al., 2000). Thus, adhesion at the interface of EPDM/PP based TPV and HDPE or UHMW-PE components is lower than in the case of miscible polymers and may have a negative effect on the functionality of the part.

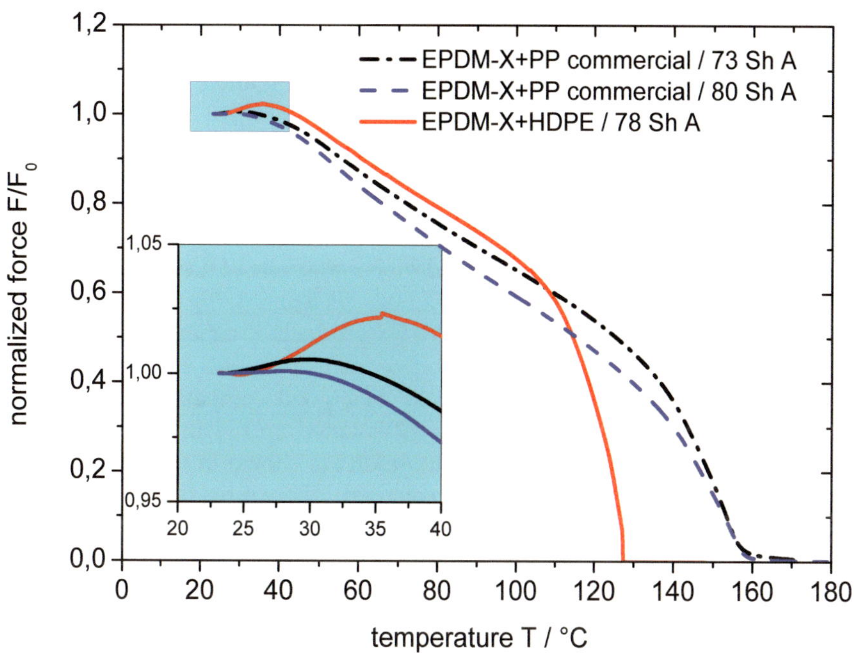

Fig. 13. TSSR force - temperature curves of a model TPV based on EPDM/HDPE (——) and commercial TPVs based on EPDM/PP of 73 (— · –) and 80 (- - -) Shore A.

Only a few papers have been published about dynamic vulcanization of EPDM/HDPE blends (Gosh et al., 1994) (Machado & van Duin, 2005), until now. A strongly increased compound viscosity as a result of dynamic vulcanization has been reported, particularly at higher EPDM contents. Therefore, one aim of this work was to optimize the blend composition of EPDM/HDPE dynamic vulcanizates with respect to mechanical properties and processibility, in order to produce TPVs with improved rubber elasticity and compatibility to rigid HDPE thermoplastics (Vennemann et al., 2009). In Figure 13 TSSR force - temperature curves of a model TPV based on EPDM/HDPE and of commercial TPVs based on EPDM/PP are shown for comparison. From the initial part of the curve it becomes obvious, that the entropy elastic behaviour of the EPDM/HDPE material is more pronounced than of the EPDM/PP based materials.

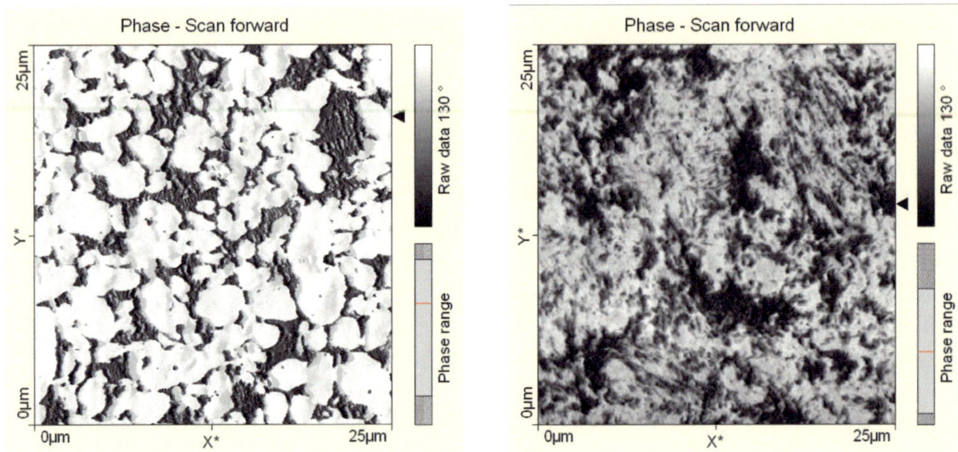

Fig. 14. AFM phase images of TPVs based on EPDM/PP (left) and EPDM/HDPE (right)

This observation corresponds to results from DMA measurements, in particular loss tangent, which is significantly lower at room temperature in case of EPDM/HDPE compared with EPDM/PP (Vennemann et al., 2009). At higher temperatures, above 100 °C, the EPDM/HDPE exhibits stronger stress relaxation due to the melting of the HDPE matrix. But, at moderate temperatures, up to 70 °C, compression set and rubber elasticity of these materials are equal or even better than existing products based on EPDM/PP. Thus, TPVs based on EPDM/HDPE might be an alternative to well known EPDM/PP products if enhanced rubber elastic properties are required. Furthermore, in hard/soft combinations with HDPE these compounds may have advantages over other TPEs due to their good compatibility, which results in excellent adhesion. The phase morphology of TPVs based on EPDM/PP and EPDM/HDPE having the same blend ratio and similar composition are shown in Figure 14. In these AFM phase images the EPDM rubber phase is light coloured, whereas the thermoplastic hard phase appears dark. In the case of the EPDM/PP sample (left), sharp boundaries between the light coloured EPDM domains and the dark appearing PP matrix phase are recognizable. The size of the EPDM domains is less than 5 µm, which is typical for this type of material. The AFM phase image obtained from the EPDM/HDPE sample differs significantly from the EPDM/PP sample. Obviously, micro-phase separation also occurred, but the phase boundaries are not as sharp, as in case of EPDM/PP. Apparently, both polymers are interpenetrating each other at the phase boundaries, because of good compatibility of EPDM and HDPE. This might be also an explanation for higher interaction and higher viscosity of the EPDM/HDPE material, which results in better rubber elasticity but also in poorer processibility.

Recently, a novel powdery EPDM/HDPE material has been developed, which can be produced by means of a special two-step mixing process (Vennemann et al., 2009). Consistency and processibility of the powdery raw material is similar to UHMW-PE powder. That means, the only way of processing is compression moulding. But, due to the almost identical consistency and the very good compatibility of the EPDM/HDPE powder and the UHMW-PE, the production of double-layer or multi-layer plates is possible, by a simple compression moulding process. Thus, hard/soft combinations can be produced easily which combine the extraordinary high toughness and abrasion resistance of UHMW-PE with the rubber like behaviour of a TPV, based on EPDM/HDPE.

4.3 Determination of crosslink density by means of rapid TSSR - tests

Although TSSR measurements are significantly faster and easier to perform than other methods to characterize the crosslink density of thermoset rubber and TPV, further improvement of the method is desired from industry, in particular by reduction of the test duration, without deterioration of the reliability of the results. High potential for the reduction of test duration is included by the isothermal relaxation period. During the isothermal test period, which lasts normally 2 h, the short time relaxation processes occur and the sample reaches a quasi equilibrium state before the non-isothermal test starts. If the isothermal test period is reduced to shorter values, the entropy effect will be partially compensated by stress relaxation and thus the obtained temperature coefficient κ is diminished, systematically. This problem can be solved by considering the effect of stress relaxation on the experimentally observable temperature coefficient κ. Generally, the total value of κ can be divided into two parts, as described in Eq. (10). It should be mentioned, that the entropic part of κ is positive, whereas the contribution of stress relaxation is negative.

$$\kappa = \kappa_{relax} + \kappa_{entropy} \tag{10}$$

Under the condition of sufficiently long duration of the isothermal testing period, the influence of stress relaxation on the initial slope of the non-isothermal stress - temperature curve is negligible. But, if the isothermal test period is shorter, the contribution of κ_{relax} can no longer be neglected. For this reason a theoretical approach was developed, which is suitable to estimate the contribution of the isothermal stress relaxation. In this approach it is considered that under isothermal conditions the decay of stress can be described by the empirical function given in Eq. (11), which fits the experimental values very nicely.

$$\sigma(t) = a + b \cdot t^{-c} \tag{12}$$

Differentiation of Eq. (12) with respect to time t and by considering the linear relationship between t and temperature T leads to Eq. (13), from which the theoretical contribution of stress relaxation κ_{relax} on the experimentally obtained temperature coefficient κ can be estimated.

$$\kappa_{relax} = \frac{-c}{\beta} \cdot b \cdot t_{\max}{}^{-c-1} \tag{13}$$

Where β is the heating rate and t_{max} is the duration of the isothermal test period. The empirical parameters a, b and c of Eq. (12) are easily calculated from least - square fits by non-linear regression.

The entire stress - curve, including the isothermal part, is presented as a function of time in Figure 15, as obtained from a TSSR - test of a standard commercial TPV sample of about 70 Shore A. In contrast to the standard test procedure, this test was performed with reduced duration of the isothermal test period of 60 minutes. As can be seen from the magnified part of the curve in Figure 15 (right), the slope of the stress - curve increases immediately, after the non-isothermal test period has started. It becomes also obvious, that the slope of the non-isothermal stress - curve is partially reduced by the influence of the ongoing stress relaxation.

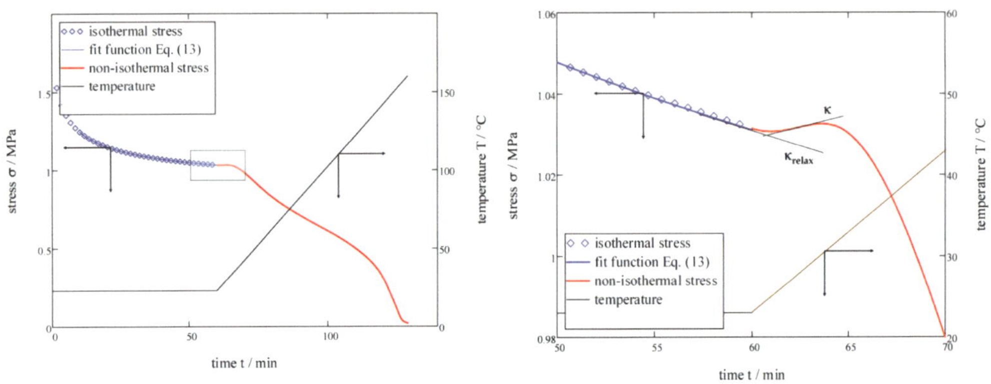

Fig. 15. Entire stress - curve (left) of a TSSR - test and zoomed part (right) of the curve, as indicated by the rectangle.

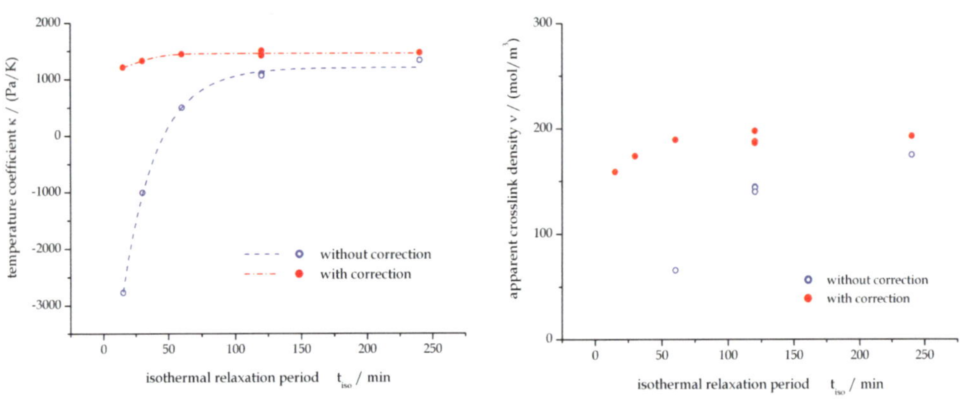

Fig. 16. Temperature coefficient (left) and apparent crosslink density (right), with and without correction, of a commercial TPV.

In order to determine the entropic part $\kappa_{entropy}$ of the experimentally obtained temperature coefficient κ, it is necessary to eliminate the contribution of stress relaxation by means of Eq. (13). Thus, an apparent value of crosslink density of the sample can be calculated from Eq. (8) when κ_0 is replaced by the entropic part $\kappa_{entropy}$ of the temperature coefficient. To demonstrate the influence of the correction, the experimentally obtained values of temperature coefficient are plotted as a function of isothermal relaxation period, with and without correction for comparison, on the left side of Figure 16. On the right hand side of Figure 16 the corresponding values of apparent crosslink density are presented, as calculated from Eq. (8) without and with correction by Eq. (13). Without correction, the values of temperature coefficient and crosslink density vary over a wide range, starting from negative values and approaching an almost constant level at long periods of isothermal relaxation. After correction by Eq. (13), the influence of the duration of the isothermal relaxation period has almost vanished. Therefore, rapid TSSR - tests with strongly reduced isothermal relaxation period give comparable results than standard TSSR - tests, but in shorter time, which might be advantageous, in particular for production control.

5. Conclusion

The aim of this paper is to describe the opportunities of a new test method, especially developed to characterize the stress relaxation behaviour and thermoelastic properties of TPE. In contrast to conventional stress relaxation measurements the new TSSR test method is less time consuming and requires only a minimum on manual effort. Generally, three phenomena have to be considered if a stretched rubber sample is annealed under the conditions of non-isothermal TSSR tests. Stress relaxation or decrease of stress, immediately occurs after the strain has been applied to the sample. Because relaxation time constants strongly decrease with increasing temperature, stress relaxation is accelerated when temperature is scanned during a TSSR test. An opposite effect results from entropy elastic behaviour of the rubber sample. Due to increasing temperature a stretched rubber sample tends to contract and therefore the stress increases if the strain is kept constant. Furthermore, stress relaxation and entropy elastic behaviour will be superimposed by a slight increase of sample length, caused by thermal expansion. It is shown that thermal expansion of the sample is negligible, if the strain is sufficiently high. Basic equations for evaluation have been developed, taking into account the specific conditions of TSSR tests.

The versatility of TSSR measurements has been demonstrated at several examples of commercial TPE and model compounds. The relaxation spectra of commercially available TPE based on SBC exhibit two significant peaks, which can be identified as glass transition temperature of the polystyrene end blocks and the melt temperature of an additional blend component, e.g. polypropylene. Blends of SBC and PPE were investigated to improve the heat resistance of the material. It has been shown that PPE and the polystyrene end blocks of SBC form a mixed phase with elevated glass transition temperature. The corresponding shift of glass transition temperature of the hard phase could be clearly identified from TSSR relaxation spectra. Thus, TSSR measurements are a suitable tool to determine the stress relaxation properties of such complex systems.

Results obtained from investigations of thermoplastic polyolefin blends (TPO) and dynamic vulcanizates (TPV) based on EPDM/PP and EPDM/HDPE, demonstrate the versatile opportunities of TSSR measurements to characterize stress relaxation behaviour and crosslink density. Comparison of commercial TPO and TPV of different hardness clearly show that the relaxation behaviour of the material is significantly improved by crosslinking of the rubber phase. It is also seen, the impact on stress relaxation is more pronounced for materials of lower hardness.

A model system of peroxide cured TPV based on EPDM/PP was investigated to determine the crosslink density of the rubber phase. By varying the amount of curatives the crosslink density of the samples has been altered within certain limits. These samples were subjected to TSSR measurements and from the initial slope of the TSSR stress temperature curves the crosslink densities of the samples were determined, by considering the entropy effect. Additionally, TPV based on EPDM/HDPE which was designed for hard/soft - combinations with HDPE and UHMW-PE. Due to better compatibility of EPDM and HDPE, the phase morphology and also the properties differ from EPDM/PP based TPV. From the results of TSSR measurements the differences with respect to rubber elasticity and heat resistance become clearly obvious.

Although TSSR tests are relatively fast and easy to perform, an accelerated test procedure has been developed for rapid determination of crosslink density of TPV. Based on a theoretical approach a basic equation has been developed to separate the phenomenon of stress relaxation from the initial part of the experimentally observable stress - temperature curve. Thus, reliable values of crosslink density can be obtained even at strongly reduced test duration.

6. References

Alfrey, T. & Doty, P. (1945). J. Appl. Phys., Vol. 16, 700

Anthony, R. L., Caston, R. H. & Guth, E. (1942). Equations of State for Natural and Synthetic Rubberlike Materials. I. *J. Phys. Chem.*, Vol. 46, pp. 826 - 840

Barbe, A., Bökamp, K., Kummerlöwe, C., Sollmann, H., Vennemann, N. & Vinzelberg, S. (2005) Investigation of Modified SEBS-Based Thermoplastic Elastomers by Temperature Scanning Stress Relaxation Measurements. *Polymer Eng. and Science*, Vol. 45, pp. 1498 - 1507

Bittmann, E. (2004) Elastisch und maßgeschneidert. *Kunststoffe,* Vol. 94, No. 12, pp. 109 - 111

Dickens, B. (1982) Thermal-Degradation Study of Isotactic Polypropylene Using Factor-Jump Thermogravimetry. *J. Polym. Sci. Pol. Chem.;* Vol. 20, No. 5, pp. 1169-1183

Eisele, U. (1979) Einflüsse der Molekülstruktur auf Verarbeitungs- und Festigkeitseigenschaften von hauptvalenzmäßig vernetzten Elastomeren. *Progr. Colloid & Polymer Sci.*, Vol. 66, pp. 59 - 72.

Ferry, J. D. (1980). *Viscoelastic Properties of Polymers (Third Edition),* John Wiley & Sons, ISBN 0-471-04894-1, New York

Gent, A. (Ed.) (2001). *Engineering with Rubber: How to Design Rubber Components (Second Edition),* Hanser, ISBN 3-446-21403-8, Munich

Ghosh, P., Chattopadhyay, B. & Sen, A.K. (1994) Thermoplastic elastomers from blends of polyethylene and ethylene-propylene-diene rubber: influence of vulcanization technique on phase morphology and vulcanizate properties. *Polymer,* Vol. 35, pp. 3958-3965

Grinberg, F. , Garbaraczyk, M. & W. Kuhn (1999) Influence of the cross-link density and the filler content on segment dynamics in dry and swollen natural rubber studied by the NMR dipolar-correlation effect. *J. Chem. Phys.*, Vol. 111, No. 24, pp. 11222-11231

Holden, G., Kricheldorf, H. R. & Quirk, R. P. (Eds.) (2004) *Thermoplastic Elastomers (Third Edition)*, Carl Hanser Verlag, ISBN 3-446-22375-4, Munich

Loan, L. D. (1967) Mechanism of Peroxide Vulcanization of Elastomers. *Rubber Chem. Technol.*, Vol. 40, pp. 149 - 177

Machado, A.V. &van Duin, M. (2005) Dynamic vulcanisation of EPDM/PE-based thermoplastic vulcanisates studied along the extruder axis. *Polymer*, Vol. 46, 6575–6586

Mark, J. E. (1981) Rubber elasticity. *J. Chem. Educ.*, Vol. 58, No. 11, pp. 898 - 903

Pellicer, J., Manzanares, J. A., Zúñiga, J., Utrillas, P. & Fernández, J. (2001) Thermodynamics of Rubber Elasticity. *J. Chem. Ed.*, Vol. 78, No. 2, pp. 263 - 267

Reid, Ch. G., Cai, K. G., Tran, H. & Vennemann, N. (2004) Polyolefin TPV for Automotive Interior Applications. *Kautsch. Gummi Kunstst.*, Vol. 57, pp. 227 - 234

Shanks, R.A., Li, J. & Yu, L. (2000) Polypropylene–polyethylene blend morphology controlled by time–temperature–miscibility. *Polymer*, Vol. 41, pp. 2133–2139

Schäfer, E. (2001) Überproportionales Wachstum bei TPE, *Kunststoffe*, Vol. 91, No.1, pp. 38 - 39

Tucker, P. S., Barlow, J. W. & Paul, D.R. (1988) Thermal, mechanical, and morphological analyses of poly(2,6-dimethyl-1,4-phenylene oxide)/styrene-butadiene-styrene copolymer blends. *Macromolecules,* Vol. 21, pp. 1678 - 1685

Tucker, P. S., Barlow, J. W. & Paul, D. R. (1988) Molecular weight effects on phase behavior of blends of poly(phenylene oxide) with styrenic triblock copolymers. *Macromolecules,*Vol. 21, pp. 2794 - 2800.

Vennemann, N., Hündorf, J., Kummerlöwe, C. & Schulz, P. (2001) Phasenmorphologie und Relaxationsverhalten von SEBS/PP-Blends. *Kautsch. Gummi Kunstst.,* Vol. 54, pp. 362 – 367

Vennemann, N. (2003) Praxisgerechte Prüfung von TPE. *Kautsch. Gummi Kunstst.,* Vol. 55 pp. 242 -249

Vennemann, N., et al. (2004). New Test Methods for the Characterization of Thermoplastic Elastomers. *Proceedings of Thermoplastic Elastomers 2004,* ISBN 1 85957 450 5, Brussels (Belgium), September 2004

Vennemann, N., Bökamp. K. & Bröker, D. (2006). Crosslink Density of Peroxide Cured TPV. *Macromol. Symp.,* Vol. 245-246, pp. 641–650

Vennemann, N. & Heinz, M. (2008). Model Analysis and Experimental Investigation of Thermoelastic Behaviour of Filled Elastomers. *Kautsch. Gummi Kunstst.,* Vol. 61, pp. 447 – 454

Vennemann, N. et al. (2009). Advanced TPVs Based on EPDM/HDPE Designed for Hart/Soft Combinations with HDPE and UHMW-PE. *Proceedings of Thermoplastic Elastomers 2009,* ISBN 978-1-84735-397-9, Frankfurt (Germany), November 2009

Vennemann, N., Melcher, E. & Wu, M. (2009). Development of a powdery TPV based on EPDM/HDPE for hard/soft combinations with UHMW-PE. *TPE Magazine,* Vol. 2, pp. 98 - 104

Vennemann, N., Heinz, M. & Wu, M. (2011). Experimental Investigations and Development of a Model for the Description of the Thermoelastic Properties of Carbon Black Filled SBR - Vulcanizates. *Kautsch. Gummi Kunstst.,* Vol. 64, No. 7-8, pp. 40 - 46

Zhao, F., Ping, Z., Zhao, S., Jian, Y. & Kuhn, W. (2007) Characterization of Elastomer Networks by NMR Parameters Part II. *Kautsch. Gummi Kunstst.*, Vol. 60, No. 12, pp. 685 - 688

18

New Thermoplastic Ionic Elastomers Based on MA-g-EPDM with Advanced Characteristics

Anton Airinei[1], Mihaela Homocianu[1],
Daniel Timpu[1] and Daniela Maria Stelescu[2,*]
[1]*"Petru Poni" Institute of Macromolecular Chemistry, Iasi,*
[2]*National Research and Development Institute for Textile, Leather and Footwear Research, Bucharest, Romania*

1. Introduction

Ionic thermoplastic elastomers or ionomers are copolymers involving a major non-polar constituent (which can be crystallized or not) and a minor ionizable constituent, partly or entirely neutralized with mono- and divalent inorganic ions as a salt at a concentration not exceeding 10 mol % (Andrei & Dobrescu, 1987; Stelescu, 2011).

In order to obtain ionic thermoplastic elastomers, one or more of the techniques listed bellow can be used:

a. Synthesis of elastomers containing ionizable monomers or just preparing the thermoplastic ionic elastomers on the same equipment where polymerization, co-polymerization or polycondensation of the basic monomers in the polymer chain are performed, but adding small amounts (5-7%) of other monomers resulting in ionic or ionizable groups in the polymer chain.
b. Converting chemically some elastomers containing double bonds in their macromolecules.
c. Processing elastomers containing ionizable groups with metal salts, metal alkali or metal oxides which can react with the functional groups in the ionizable groups, thus yielding ionic ranges; fillers, ionic plasticizers, polyolefins, antioxidants etc. can be also added in order to obtain ionic thermoplastic elastomer compounds with a large variety of applications depending on their composition (Zuga, 2005).

The ionic elastomer compounds resemble highly with the traditional rubber compounds but there are some differences. Because the former are thermoplastics, there is no curing stage and, therefore, no sulfur and vulcanization accelerators or peroxides are added. Another significant difference is the use of an ionizing agent - ionic plasticizer in preparing ionic thermoplastic elastomers. It plays the role of promoting the ionic break-up of the ionic interactions at high temperatures to enable the shearing flow of the compound; at room

---

*Corresponding Author

temperature it behaves like a filler. Zinc stearate is the most largely used ionic plasticizer but some others can be also used, like calcium stearate, zinc acetate, stearamide (Nahmias & Marco Serra, 2002; Zuga et. al, 2009; Stelescu et al., 2011).

A formulation of ionic thermoplastic elastomer compound consists generally of a neutralized ionomer, ionic plasticizer, non-ionic plasticizer, filler, antioxidants, other polymers, etc. (Zuga & Cincu, 2006a). A significant increase of vitrifying point by incorporating ions into polymers was prouved (Eisenberg, 1977). This increase depends on the nature of the basic ion, the effect being stronger as the ionic forces are larger. They have also shown that the nature of the basic ion influences not only the value of T_g (glass transition temperature), but also the position and shape of the module - temperature curve. The increase of ion content leads to an increase of the module and the increase of the specific plateau of vulcanized rubber.

An important contribution to establishing the relation between properties and the ion content in ionomers has been brought by research on carboxylated elastomers. Ionomer-rubbers have remarkable properties such as: high module, large elongation at break and a constant plateau in the module - temperature curve. This constant plateau can be explained through the presence of a small concentration of stable interchain bonds, called multiplets. Elongation at break has been attributed to the loosening of ionic bonds by exchange reactions between crosslinking bridges of various chains, thus hindering an excessive strain. Finally, due to the presence of ionic aggregates which may act as "filling materials", of reinforcement and of quasi-crosslinking, the initial high module of materials can be explained.

Rheological studies on ionomers in melt (Szymczyk & Roslaniec, 1999) have highlighted a remarkably large increase of melt viscosity, as a result of introducing ions into polymers. This increase depends on the ion concentration or on the neutralization degree and less on the nature of ions. Also, the non-Newtonian character of ionomers in melt is noticed, probably due to the fact that the movement of chain segments and chains implies dissociation of ion pairs in cluster aggregates. This dissociation certainly has a determining influence on rheological properties in melt. In partially neutralized ionomers, hydrogen bonds between the carboxyl functional groups may also have a significant effect on flow properties.

There are a few studies (Zuga & Cincu, 2006b, 2006c; Datta & Kharagpur, 1997; Paeglis & O'Shea, 1988) on ionomers based on maleated and/or sulphonate EPDM rubber indicating the fact that, by introducing neutralization agents (zinc oxide, sodium hydroxide etc.) when a metal base is obtained, modifications of physico-mechanical properties which take place in system are assumed to be due to the rigid phase resulted from the restriction of chain mobility in the ionic aggregate area and a reduction of crystallinity compared to that existing in the initial elastomer is noticed (Stelescu, 2010).

This review gives an overview about our research on ionic thermoplastic elastomers based on maleated ethylene propylene diene terpolymer (EPDM-g-MA). The investigations were aimed to obtain some new generations of ionic thermoplastic elastomers with high technical and processing characteristics intended to be processed on the injection moulding machines, resulting in high quality products complying with the international market requirements. Two types of maleated ethylene propylene terpolymer elastomers (EPDM-g-MA) with various levels (0.5 and 1.0 %) of maleic anhydride were used. The EPDM-g-MA rubbers were modified by neutralizing them with zinc oxide and stearic acid, and then the ionomer has been converted into a thermoplastic ionic elastomer by adding various percentages of ionic plasticizer to the blend. The behaviour of the thermoplastic ionic elastomers resulted by adding fillers like as precipitated silica and carbon black in various percentages was assessed. The extent of the neutralization agent, ionic and non-ionic plasticizers, filler, polyolefin's which influence the characteristics of the resulting products was determined with the purpose of selecting the best types of ionic thermoplastic elastomer based on EPDM-g-AM. All the laboratory prepared compounds were tested for the physico-mechanical characteristics; based on the results, three thermoplastic ionic elastomers with the best characteristics were selected and the applications of these were set up.

To obtain better characteristics, the creation of nanocomposites from a selected EPDM-g-AM compound and organically modified montmorillonite clay (OMMT) was targeted. Nanocomposites based on maleated ethylene propylene diene terpolymer and organically modified montmorillonite (OMMT) has been prepared by melt intercalation procedure. The materials were characterized by X-ray diffraction (XRD), thermogravimetric analysis (TGA), dynamic scanning calorimetry (DSC) and mechanical tests. XRD data show the increase of the distance between the silicate layers indicating the intercalation of polymer chains in the montmorillonite galleries. The incorporation of OMMT in composites determines the increase of melting peak temperature (Tm) and heat of fusion (ΔHf) suggesting a supplementary nucleation increase due to the OMMT presence. The mechanical properties were analyzed as a function of the OMMT level in the composite and the results reveal remarkable improvement relative to the conventional composite.

Ionic thermoplastic elastomer compounds can be used in many applications such as: to manufacture a wide range of consumer goods (hoses, soles etc), as modifier for other materials, to make adhesives, to obtain impermeable flexible thin membranes such as cover membranes etc. (Swapan et al., 2001; Zhao-Hua et al., 2002; Einsenbach et al., 1998; Piere, 2002; Ronatti, 1990; Nachmias & Sera, 2002).

The new types of thermoplastic elastomers can be processed by specific techniques for thermoplastic materials, thus removing the vulcanization stage involving high power expenditure and release of noxious products, improved characteristics (higher values for elasticity, ageing resistance, abrasive resistance, acid and alkali fastness) of these materials can be assured.

The potential users of the new rubber materials will be economic operators processing rubber and plastics, footwear and car component manufactures etc. They can be used in the manufacture of a large range of products like as hoses, gaskets, rubber shoes, protective equipment etc.

2. Preparation

Ionic thermoplastic elastomers can be obtained by processing of ethylene propylene diene elastomers (EPDM) grafted with maleic anhydride (EPDM-g-MA) having different contents of maleic anhydride. The compositions of thermoplastic elastomers contain besides EPDM-g-MA the following elements: neutralizing agents of the ionic groups (zinc oxide in the presence of stearic acid), ionic plasticizers (zinc stearate), nonionic plasticizers (paraffin oil), fillers (precipitated silica, carbon black, chalk), polyolefins (high density polyethylene

(HDPE), polyethylene grafted with maleic anhydride (PE-g-MA), polypropylene (PP)), clays, antioxidants (Irganox 1010).

EPDM-g-MA elastomers exhibit the peculiar features of EPDM elastomers, but they can react with divalent metal oxides salts leading to the crosslinking by ionic bonds. The structure of EPDM-g-MA is shown in Fig. 1. The most used EPDM-g-MA elastomers contain 0,5% MA having semicrystalline structure and 1% MA with amorphous structure. The ionic plasticizer solvates the ionic domains at higher temperatures improving flow and at the room temperature it has the role of a filler.

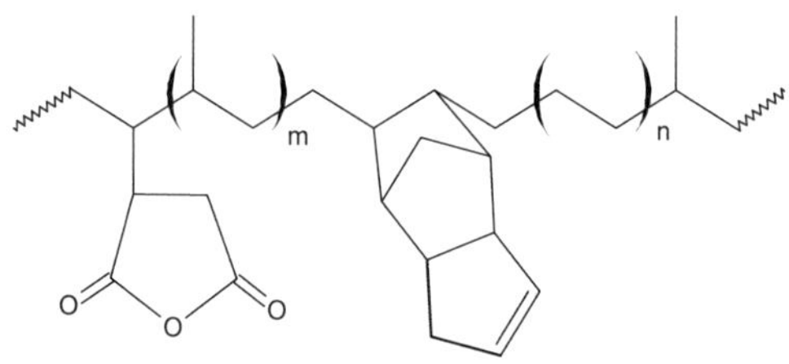

Fig. 1. Chemical structure of EPDM-g-MA elastomers

In the compositions of thermoplastic elastomers, plasticizers assure the decrease of blend stiffness, suitable adherence, improvement of extruding, calendering or elongation, the decrease of freezing temperature of vulcanizates, the enhancing of ozone rezistance and the price reducing. Due to their property to dissolve in rubber, the pasticizers get into the polymer chains diminishing the intermolecular interactions but in the same time new polymer plasticizer interaction appear in system.

The fillers are fine powder materials incorporated in blends in order to improve the composition properties (active fillers) and to reduce price (inactive fillers). The active fillers determine improving of some mechanical properties of elastomer composition namely tensile strength, shearing strength, abrasion resistance, repeated strain resistance.

Carbon black is a filler for many rubber materials having concomitantly the function of reinforcement, filling and coloring material. In rubber blends the particle sizes, structure and pH were affected by the carbon black content. Higher specific surface higher the break and tear resistance of the rubber. The increase specific surface leads to the energy consumption for mixing, rendering difficult the blend processing and the decrease of elasticity of the vulcanized products. The use of carbon black in rubber compositions provides a higher thermal and dimensional stability, increased hardness, enhanced thermal and electrical conductivities, better processing ability.

Polyolefins are utilized in elastomer compositions in order to improve tensile strength, tear strength, chemical reagent resistance.

Thermoplastic compositions were prepared by melt blending technique, on a laboratory electrically heated roller mill equiped with a cooling system. The process flow for the preparation of ionic thermoplastic elastomer compositions is shown in Fig. 2. After the raw materials were tested, the ingredients were weighed according to the processing formulations. The blend constituents were added in the following sequence: roll binding of maleated EPDM (and HDPE) (5-8 min), embedding zinc oxide, stearic acid and antioxidant

(2 min), introducing zinc stearate, paraffin oil and filler (5 min), homogenizing the blend and removing it from roll in 2 mm thick sheets (4 min). The working parameters were: friction 1:1.1 and temperature 150-170°C, for blends without polyolefins.

No.	Ingredient	Quantity (phr)
1.	EPDM-g-MA (Royaltuf 485, Royaltuf 498)	100
2.	Zinc oxide	5-20
3.	Stearic acid	0.5-2
4.	Ionic plasticizer	0-40
5.	Nonionic plasticizer (Paraffin oil)	0-50
6.	Filler (carbon black, precipitated silica, chalk)	0-90
7.	Polyolefins (PE, PP)	0-160
8.	Antioxidants (Irganox 1010)	2.0

Table 1. Mix formulation of thermoplastic composition

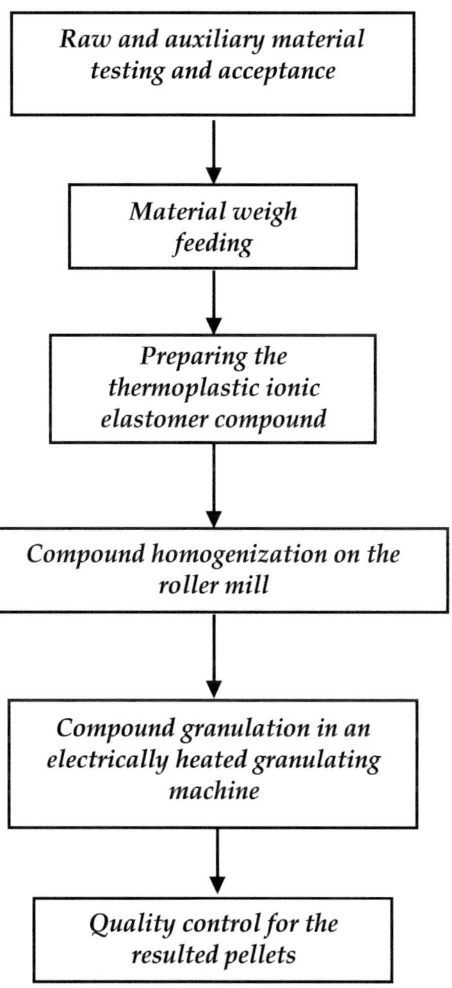

Fig. 2. Laboratory processing stages for ionic thermoplastic elastomers

When the thermoplastic compositions are formulated the resulting compound characteristics depend on the quantity of ionic or nonionic plasticizer, neutralizing degree, the ratio elastomer/filler and filler type, the quantity and type of polyolefins.

A basic formulation for ionic thermoplastic elastomers contains the components from Tab. 1. The quantities from Tab. 1 are expressed in parts per hundred parts of rubber (phr).

The resulting blends were granulated on an extruder-granulator equipped with feeder hopper, screw with three beating zones, granulating device and rotary cutter. Feeding was done using materials under the form of strips with 2-3 mm thickness and 20-30 cm length, obtained as a result of blend homogenizing. The thermal regime of extruder-granulator in obtaining EPDM-g-MA rubber granules is presented in Tab. 2. The maleated EPDM-blend granules were then cooled and homogenized. The temperatures during granulation process are given in Tab. 3. Upon the granulation operation we must be sure that the keeping time of EPDM-g-MA rubber blend in extruder-granulator correspond to that required to achieve ionic crosslinking, determined used an rheometer (20 min).

Types of ionic thermoplastic elastomer granules	Area 1 (°C)	Area 2 (°C)	Area 3 (°C)	Jet (°C)
Blends without polyolefins	145-155	160-170	145-155	150
Blends with polyolefins	150-180	160-190	150-170	160

Table 2. Thermal regime for obtaining EPDM-g-MA rubber granules

No.	Heating zone in granulating machine	Temperature range (°C)
1.	Pre-heating	145-155
2.	Blending	160-170
3.	Cooling	145-155

Table 3. Process variables for granulation of thermoplastic ionic elastomer composition

The pellets needed to determine the physico-mechanical characteristics of resulting ionic thermoplastic elastomer compositions were obtained by injection moulding in a mould with two cavities using an electrically heated injection moulding machine. The test samples were prepared in the following stages: 1) electrical supply, adjusting the process temperature and bringing the machine up to the process temperature; 2) feeding the ionic thermoplastic elastomer pellets in feeder hopper; 3) heating the pellets up to 165-170°C within the injection machine screw; 4) injecting the melt composition in the mould; 5) cooling the plates for 3 min; 6) taking the plates off the mould.

Laboratory tests aim to measurements of tensile strenght, tear strenght, hardness, elasticity, melt flow index, abrasion resistance, flexion resistance, accelerated aging, etc.

3. Influence of the crosslinking degree on EPDM-g-MA ionic elastomer characteristics

In order to study the effect of neutralizing degree on the physico-mechanical properties of EPDM-g-MA rubber, zinc oxide was used as neutralization agent (Stelescu, 2010). Chemical structure of an EPDM-g-MA elastomer which subjected to the neutralization reaction with zinc oxide is presented in Fig. 3, the resulting carboxylic salts can act as ionic crosslinkings (Setua & White, 1991).

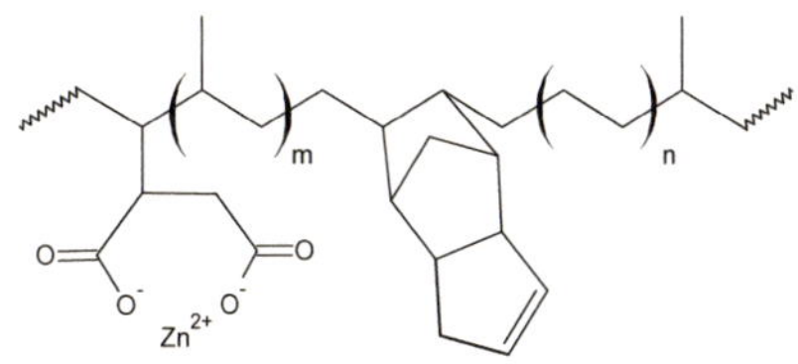

Fig. 3. Schematic representation of ionic elastomers obtained by neutralization with zinc oxide

Analyzing physico-mechanical properties of blends based on EPDM-g-MA (Royaltuf 485 and 498 elastomers) with 0.5 % and 1.0% maleic anhydride, respectively, neutralized with zinc oxide the following are noticed:

Hardness of ionic thermoplastic composition shows an increase from 71 to 74 °ShA when introducing 5 phr of zinc oxide, while for composition with Royaltuf 498, hardness exhibits an increase from 55 to 66 °ShA, then further increasing of zinc oxide quantity does not lead to the improving of this property (Fig. 4). Elasticity has very good values especially in the case of Royaltuf 498 elastomer utilization (Fig. 5), but the elasticity practically does not increase to zinc oxide quantity increasing in blend, values of 60% being obtained for Royaltuf 498 composition.

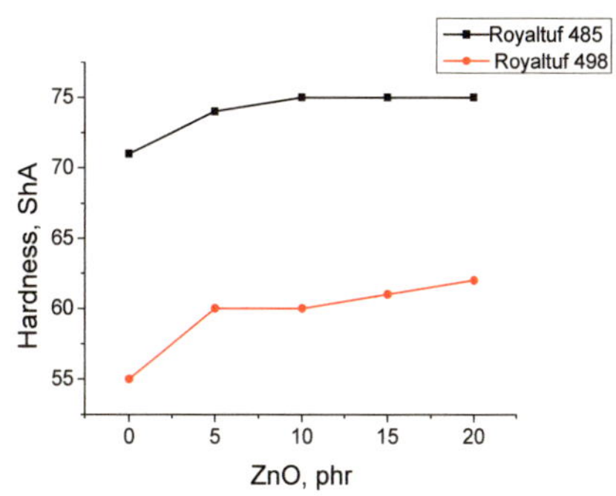

Fig. 4. Dependence of hardness on variation of ZnO level

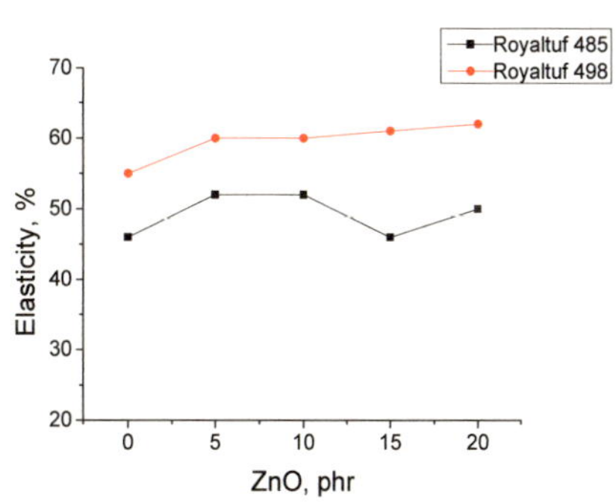

Fig. 5. Elasticity variation depending on ZnO quantity

The tensile strength increase from 10.4 to 18.5 N/mm^2 for Royaltuf 485 blends when the maximum value is observed for 5 phr of zinc oxide, after which it exhibit a slight decrease while the zinc oxide quantity continues to increase (Fig. 6). Module increases together with the increase of neutralization degree in the elastomer compositions (Figs. 6,7). Elongation at break exhibits very high values and a maximum between 5-15 phr of zinc oxide was observed for the two elastomer compositions.

By increasing of neutralization degree this parameter decreases (Fig. 8). Tear strength exhibits a significant and continuous increase for Royaltuf 485 composition when a increase from 38.5 to 53.5 N/mm was observed at the increse of ZnO quantity from 0 to 10 phr. After this ZnO content a slight decrease of tear strength was noticed for Royaltuf 498 composition (Fig. 9).

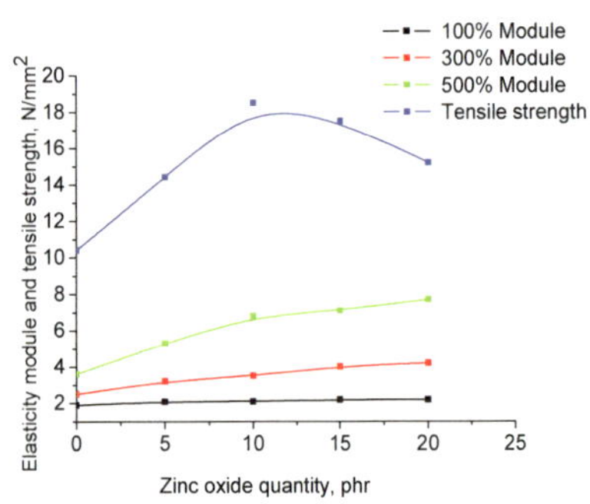

Fig. 6. Tensile strength and module dependencies on ZnO quantity for Royaltuf 485 blends

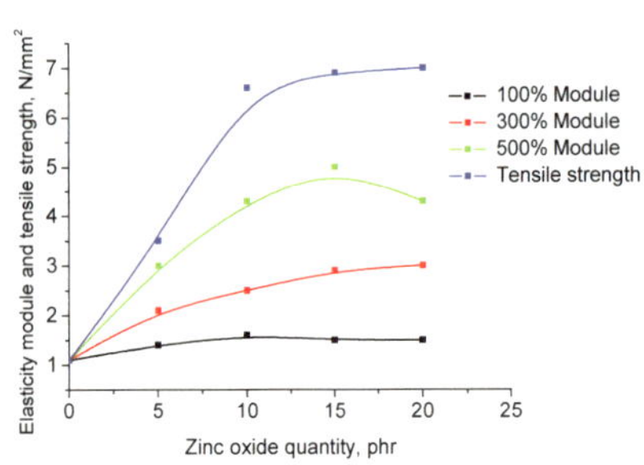

Fig. 7. Tensile strength and module variation with ZnO quantity for Royaltuf 498 blends

Comparing characteristics of blends made with Royaltuf 485 rubber with those made with Royaltuf 498 elastomer, both neutralized with zinc oxide, it is noticed that physico-mechanical properties are better in blends in which Royaltuf 485 rubber was used. This behaviour is determined by the composition and structure of the two types of rubber; thus, blends containing elastomer with semi-crystalline structure (Royaltuf 485) have exhibited higher values of hardness, module, tensile strength, elongation at break and tear strength than blends containing Royaltuf 498 elastomer which has an amorphous structure

The rheological properties of compositions based on EPDM-g-MA elastomer containing different ZnO levels are presented in Tab. 4. It is noticed that zinc oxide replaces sulfur and vulcanization agents in these blends. The optimal time of vulcanization decreases as the quantity of neutralization agent (ZnO) increases, and the minimum and maximum moments increase as a result of the reinforcing effect due to ionic bond formation.

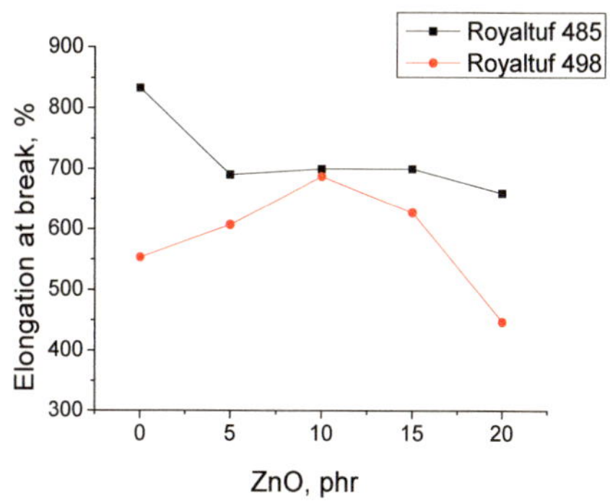

Fig. 8. Elongation at break variation with ZnO quantity

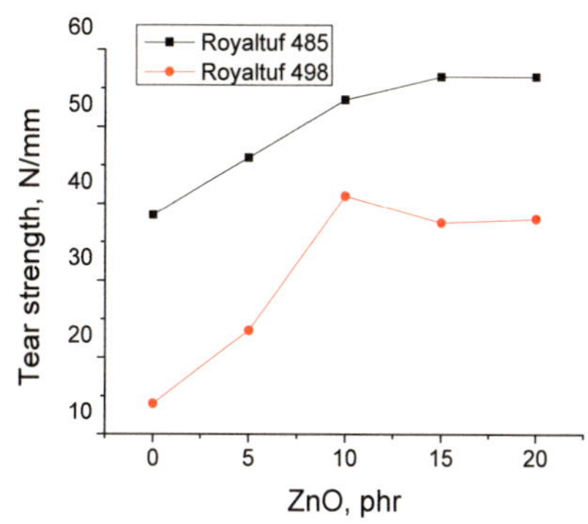

Fig. 9. Tear strength dependence on ZnO level

From the analysis of physico-mechanical properties of blends based on EPDM-g-AM Royaltuf 485 rubber and Royaltuf 498 rubber, respectively, it is noticed that, together with the increase of the neutralization degree, due to the property of groups specific to maleic anhydride existing on the macromolecular chain of reacting with oxides of divalent metals (zinc oxide), ionic bonds form similar to sulphur bridges from vulcanized rubber. This ionic crosslinking has led to an improvement in the value of the module, in tensile and tear strength. It can be inferred that EPDM-g-AM elastomers have reacted chemically with the zinc oxide.

Characteristics / ZnO level	5 phr ZnO	10 phr ZnO	15 phr ZnO	20 phr ZnO
Blends with Royaltuf 485				
Minimum moment, Nm	3.8	5	9	7
Maximum moment, Nm	6	17.5	19	20
Optimal vulcanization time	25′	22′45″	17′30″	16′15″
Blends with Royaltuf 498				
Minimum moment, Nm	4.4	5.6	13.4	11.4
Maximum moment, Nm	9	16.6	16.7	17.5
Optimal vulcanization time	25′	22′30″	16′15″	11′15″

Table 4. Rheological characteristics of EPDM-g-AM blends containing zinc oxide

4. Effect of fillers on the properties of thermoplastic elastomers based on maleated EPDM

Fillers are used to improve some properties of thermoplastic compositions. It is well known that the fillers interact with ionic elastomers leading to the formation of ionic, covalent or hydrogen bonds with them. The appearance of these bonds determines a reinforcing effect of rubber composition and the improving of modulus, tear strength or of tensile strength (Zuga et al., 2004).

The filler nature (carbon black, precipitated silica or chalk) will influence the properties of ionic thermoplastic elastomers and it is very important to select the type of filler and optimum quantity which must be incorporated in a composition designated to be utilized in light industry.

Ingredients/Sample	O^1	$U^{1}1$	$U^{1}2$	$U^{1}3$	$C^{1}1$	$C^{1}2$	$C^{1}3$	$F^{1}1$	$F^{1}2$	$F^{1}3$
EPDM-g-AM Royaltuf 485, g	100	100	100	100	100	100	100	100	100	100
Zinc oxide, g	20	20	20	20	20	20	20	20	20	20
Stearic acid, g	2	2	2	2	2	2	2	2	2	2
Zinc stearate, g	20	20	20	20	20	20	20	20	20	20
Precipitated silica Perkasil, g	-	30	60	90	-	-	-	-	-	-
Chalk, g	-	-	-	-	30	60	90	-	-	-
Carbon black HAF, g	-	-	-	-	-	-	-	30	60	90
Paraffin oil Texpar 22, g	10	10	10	10	10	10	10	10	10	10
Antioxidant Irganox 1010, g	1	1	1	1	1	1	1	1	1	1
Characteristics										
Hardness, °ShA	63	72	81	89	66	67	72	67	77	84
Elasticity, %	34	34	36	34	36	32	30	32	30	26
Tensile strength, N/mm^2	8.8	8.2	5.2	4.7	7	6.9	5.8	9.3	9.6	11
Elongation at break, %	767	580	273	100	567	473	473	633	490	420
Residual elongation, %	83	61	24	9.3	59	49	53	75	60	60
Tear strength, N/mm	36	43.5	46	48	37.5	41	44	53	60	74
Specific weight, g/cm^3	1.07	1.12	1.16	1.23	1.14	1.25	1.35	1.11	1.16	1.22
Abrasion resistance, mm^3	102	135	182	246	165	176	196	149	162	161

Table 5. Formulations and characteristics of rubber blends based on EPDM-g-MA (Royaltuf 485)

Some formulations and properties of rubber blends based on EPDM-g-MA are given in Tab. 5 and 6 in order to reveal the influence of filler nature on the resulting compositions. From these tables it can be seen that the hardness of composition increases together with filler incorporation due to the reinforcing effect of fillers. This improvement of hardness is more pronounced for compositions with active fillers (carbon black, silica) as compared to compositions containing inactive fillers (chalk). For the compositions containing Royaltuf 485 the increase of filler level determines an increase of hardness of 41 % for compositions having precipitated silica and of 33 % for compositions with carbon black, respectively. The same effect was observed for compositions containing Royaltuf 498, where the hardness increase with 34.8% introducing precipitated silica and with 28.8 % for compositions having carbon black as filler. The introducing of chalk determines a lower increase of hardness (about 15 %, Tab. 5).

The elasticity decreases by introducing fillers in composition. For compositions containing Royaltuf 485 the elasticity was diminished from 34 to 26 % in the case of carbon black utilization and lower decrease together with the increase of chalk level (from 34 to 30 %). The elasticity shows a nonuniform variation in compositions with silica or chalk and it decreases till 28% in compositions with carbon black (compositions with Royaltuf 498).

Ingredients / Sample	O^2	U^21	U^22	U^23	C^21	C^22	C^23	F^21	F^22	F^23
EPDM-g-AM Royaltuf 498, g	100	100	100	100	100	100	100	100	100	100
Zinc oxide, g	20	20	20	20	20	20	20	20	20	20
Stearic acid, g	2	2	2	2	2	2	2	2	2	2
Zinc stearate, g	20	20	20	20	20	20	20	20	20	20
Precipitated silica Perkasil, g	-	*30*	*60*	*90*	-	-	-	-	-	-
Chalk, g	-	-	-	-	*30*	*60*	*90*	-	-	-
Carbon black HAF, g	-	-	-	-	-	-	-	*30*	*60*	*90*
Paraffin oil, Texpar oil 22, g	10	10	10	10	10	10	10	10	10	10
Antioxidant Irganox 1010, g	1	1	1	1	1	1	1	1	1	1
Characteristics										
Hardness, °ShA	66	70	83	89	67	70	73	79	80	85
Elasticity, %	32	35	32	30	34	32	32	23	25	23
Tensile strength, N/mm²	7.7	5.5	6.6	6.5	6.6	6.1	6.5	10.6	9.2	8.2
Elongation at break, %	753	553	487	200	660	527	513	460	433	313
Residual elongation , %	77	53	41	12	67	44	49	69	60	51
Tear strength, N/mm	36	34.5	44.5	57	33	49	43.5	59.5	70	57
Specific weight, g/cm³	1.02	1.07	1.16	1.21	1.14	1.18	1.21	1.08	1.15	1.22
Abrasion resistance , mm³	111	162	211	257	147	220	252	150	169	198

Table 6. Formulations and characteristics of rubber blends based on EPDM-g-MA (Royaltuf 498)

Tensile strength also decreases together with the increase of silica and chalk levels for the both compositions. In case of compositions containing carbon black the tensile strength increases with 25% as the filler level increases, while in compositions containing Royaltuf 498 the tensile strength increases with 37.7 % for carbon black level of 30 phr and then it decreases untill 22.6 %. The elongation at break decreases together with the increase of filler level. The most increase was observed in compositions containing precipitated silica. Tear strength increases against filler level for all compositions, the high values were obtained for compositions containing carbon black.

Also, the specific weight and abrasion resistance increase as the filler level becomes higher. However, the values of specific weight for compositions containing more than 60 phr of filler are very high and inadequate for use in shoes fabrication. The values of abrasion resistance confirm that these compounds can be used to manufacture „every day" shoes.

From Tabs. 5 and 6, the best characteristics of the compositions based EPDM-g-MA were obtained using as fillers carbon black and silica, occurring an increase of hardness, tensile strength and tear strength. Taking into account that a higher filler level leads to the decrease of melt composition viscosity and to the increase of specific weight and abrasion resistance, a rubber composition containing 30 phr carbon black or 30 phr precipitated silica could be selected in order to utilize in EPDM-g-MA compounds (Zuga et al., 2004).

5. Influence of plasticizers on the properties of maleated EPDM ionic elastomers

Ionic plasticizers play the role of promoting the ionic break-up of the ionic interactions at high temperatures to enable the shearing flow of the compound; at room temperature they behave like a filler. The most largely used ionic plasticizer is zinc stearate but some others can be also used, like calcium stearate, zinc acetate, stearamide.

Non-ionic plasticizers play the role of solvating the non-ionizing elastomer chains. They are chemically and thermally stable materials which are added to polymers to facilitate their processing, imparting flexibility and softness to the finished products. The major plasticizer functions in the polymer blends are the following: improving the processing and applications of long chain polymers; lowering the polymer processing temperature under their decay temperature; plasticizers decrease the intermolecular polymer forces, like the temperature does; changing the finished product characteristics, enabling the polymers to be used in specific fields requiring conditions which cannot be met by the unplasticized polymers. Plasticizers increase generally the polymer characteristics, such as flexibility, elongation, resistance to low temperatures but also can lower some characteristics like the tensile strength, the dielectric prperties, etc.; enlarging the application field because of their lower costs.

The action of the ionic plasticizer (zinc stearate) and non-ionic plasticizer (paraffin oil) on the ionic thermoplastic elastomer characteristics based on the maleinized ethylene propylene terpolymer rubber was discussed in order to select some optimum compositions for applications.

The prepared blends have shown different physico-mechanical characteristics according to the levels of the added plasticizer. Zinc stearate was used as plasticizer agent in concentrations of 20 and 40 phr, respectively and as nonionic plasticizer, paraffin oil in concentration of 10-50 phr.

Figure 10 shows that adding the plasticizer (paraffin oil) for polymer chains has resulted in a decreased hardness (by 12°ShA) but adding the ionic plasticizer (zinc stearate) has resulted in an increased hardness as the last one plasticizes the ionic groups at a temperature above the its melting point (128°C) and at the room temperature it acts like a filler (increases by 1°ShA up to 11°ShA). Hardness changes as a result of adding zinc stearate are more marked with the blend containing EPDM and 1 % maleic anhydride because of the enlarged ionic ranges.

An increase in elasticity with the increase in the amount of paraffin oil added to the blends was found out (Fig. 11); but the elasticity decreases as the amount of the added ionic plasticizer is increased and this decrease is more marked with blends containing EPDM and 1 % maleic anhydride. The blends containing EPDM and 1 % maleic anhydride show a higher elasticity with the decreasing in crystalline phase to the favour of amorphous phase.

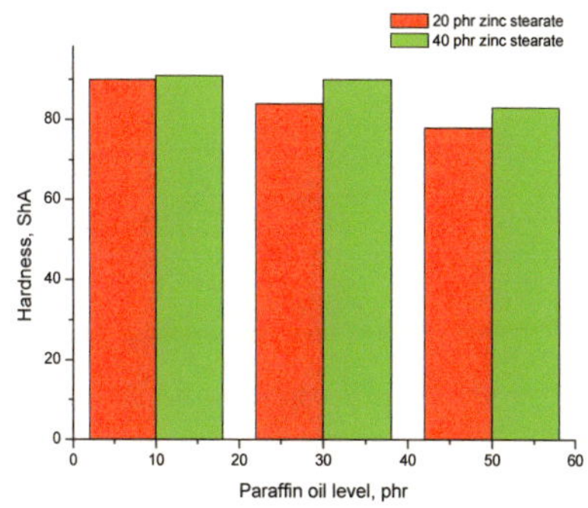

(a)

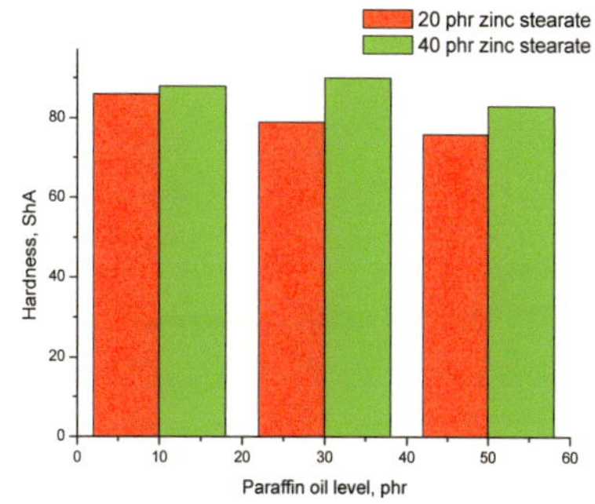

(b)

Fig. 10. Changes in hardness according to the added plasticizer amounts. (a) for blends with Royaltuf 485; (b) for blends with Royaltuf 498

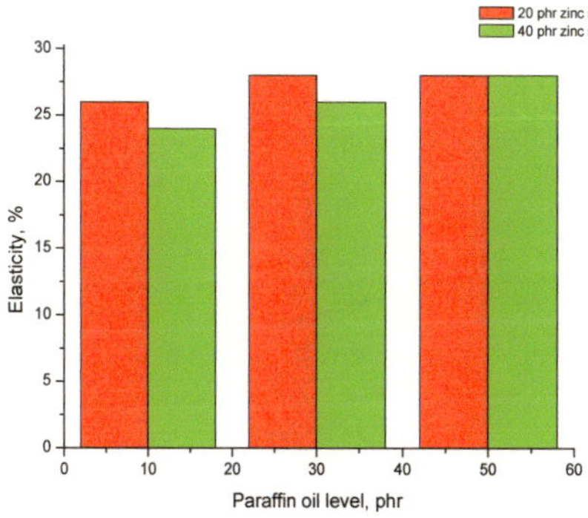

(a)

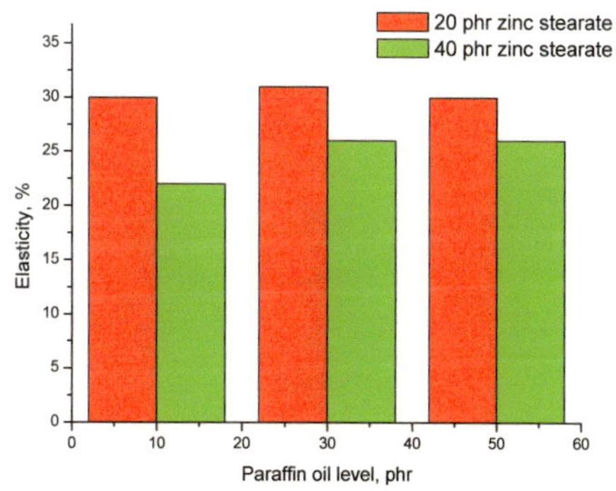

(b)

Fig. 11. Changes in elasticity according to the added plasticizer amounts: (a) for blends with Royaltuf 485 (b) for blends with Royaltuf 498

As can be seen in Figs. 12 and 13, paraffin oil solvates the non-ionic ranges leading to a decrease in elasticity modulus and tensile strength and this decrease is less marked with blends containing higher levels of zinc stearate. Increasing the zinc stearate levels in blends has resulted in an insignificant decrease in the elasticity modulus and tensile strength.

Elongation at break increases as the amount of paraffin oil added to the blends is increased and decreases as the amount of zinc stearate added to the blends is increased (Fig. 14). It decreases with high plasticizer levels.

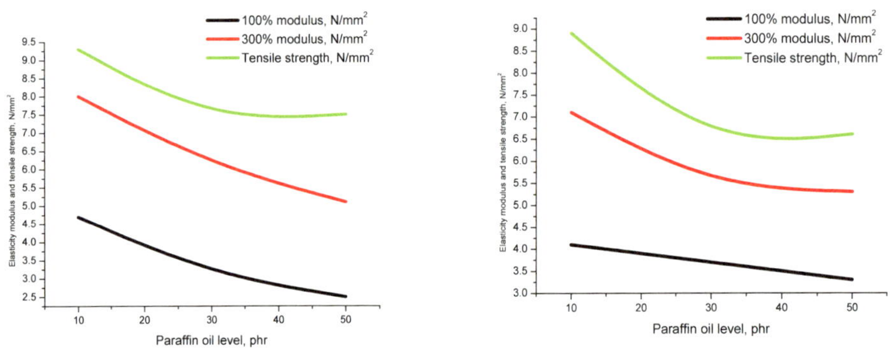

(a) (b)

Fig. 12. Changes in elasticity modulus and tensile strength according to the added amount paraffin oil to the blends of EPDM with 0.5 % maleic anhydride. (a) for the blends with 20 phr zinc stearate; (b) for the blends with 40 phr zinc stearate

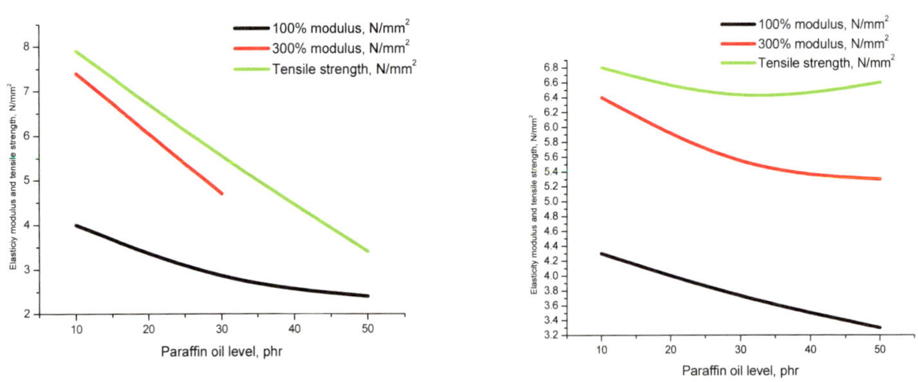

(a) (b)

Fig. 13. Changes in elasticity modulus and tensile strength according to the added amount paraffin oil to the blends of EPDM with 1.0 % maleic anhydride. (a) for the blends with 20 phr zinc stearate; (b) for the blends with 40 phr zinc stearate

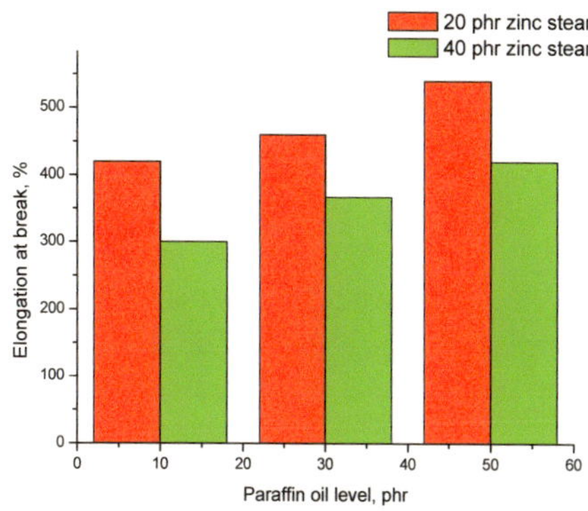

(a)

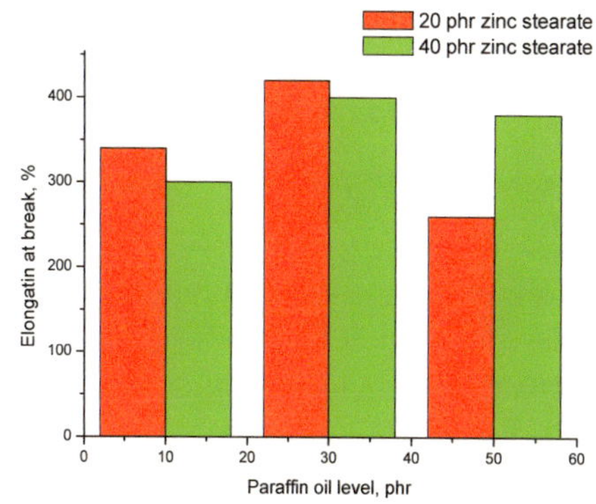

(b)

Fig. 14. Changes in elongation at break according to the added plasticizer amounts. (a) for blends with Royaltuf 485; (b) for blends with Royaltuf 498

Tear strength decreases with the added paraffin oil and also with the increase in zinc stearate level (Fig. 15).

The ionic plasticizer has shown filler characteristics leading to some increase in hardness and decrease in elasticity, elasticity modulus, tensile strength and tear strength. Non-ionic plasticizer has solvated non-ionizable elastomer chains leading to some increase in elasticity and elongation at break and decrease in hardness, tensile strength and tear strength.

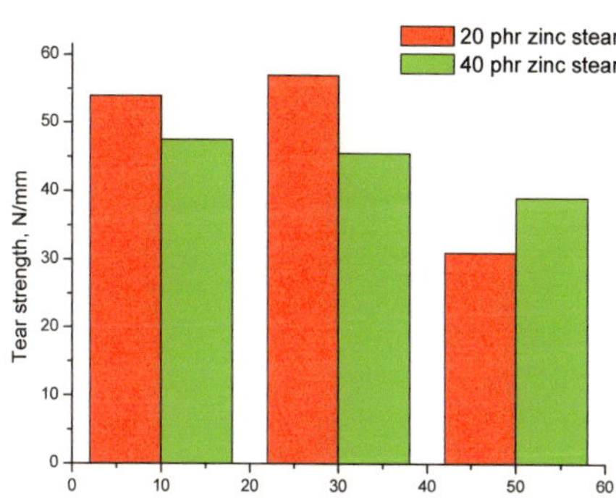

(a)

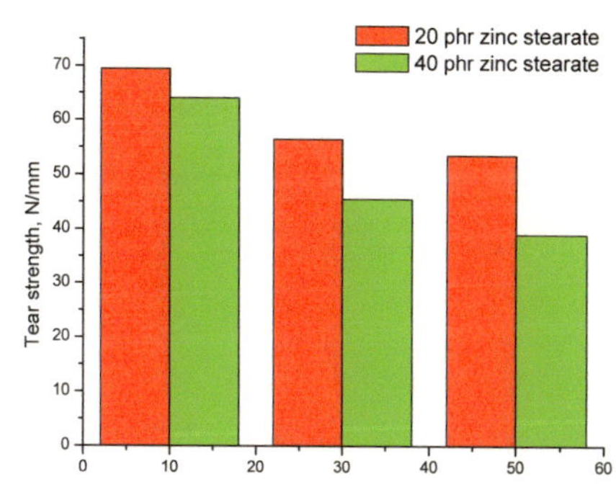

(b)

Fig. 15. Changes in tear strength according to the added plasticizer amounts. (a) for blends with Royaltuf 485; (b) for blends with Royaltuf 498;

6. Influence of polyolefins on the properties of ionic thermoplastic elastomer compositions

Some characteristics of the compositions based on EPDM-g-MA can be improved by adding other polymers in system. The choising of polymer type must be made as a function of its miscibility with the thermoplastic elastomer present already in composition. The selection of polymers to obtain blends with enhanced properties takes into account the following criteria: structure, solubility, crystallinity degree, superficial tension, etrc. (Datta & Lohse, 1996, Utracki, 1998). According to these criteria polymers such as polypropylene, polyethylene (PE), maleated polyethylene (PE-g-MA), HDPE were selected to be used in compositions based on EPDM-g-MA.

Upon analyzing physico-mechanical characteristics of resulting blends, the following are noticed:

- Hardness increases with polyolefin quantity (Fig. 16), and elasticity decreases with the increase of HDPE quantity (Fig. 17). This indicates the fact that with a polyolefin quantity of about 80 phr it has the tendency of becoming a continuous phase and imprints surface properties on the blend.
- 100% modulus (Fig. 18), 300% modulus (Fig. 19) and tear strength (Fig. 20) increase significantly with the increase of polyolefin quantity in the blend, suggesting a good miscibility (link formation) of HDPE with EPDM-g-MA and the properties of the blend are additive.
- Tensile strength (Fig. 21) and elongation at break (Fig. 22) vary unevenly upon the increase of polyolefin amount in blends, but their values are good.
- Residual elongation (Fig. 23) significantly increases as the HDPE quantity introduced in the blends increases, indicating that the obtained blends have both characteristics specific to plastic materials, namely "neck" formation, and properties specific to elastomers (good recovery after applying a force), their share depending on the ratio between the elastomer and polyolefin existing in the blend.

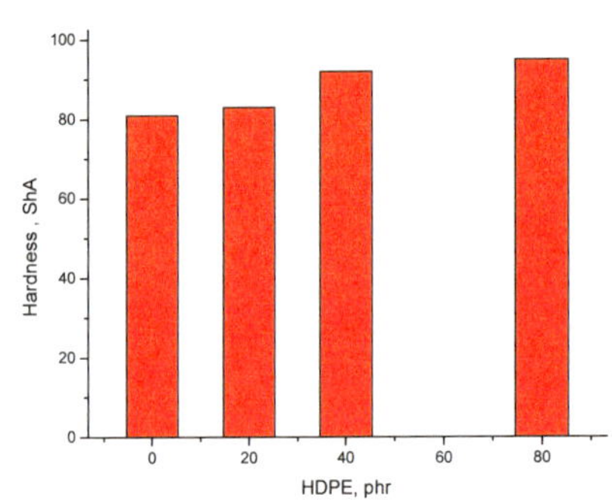

Fig. 16. Hardness versus the HDPE level

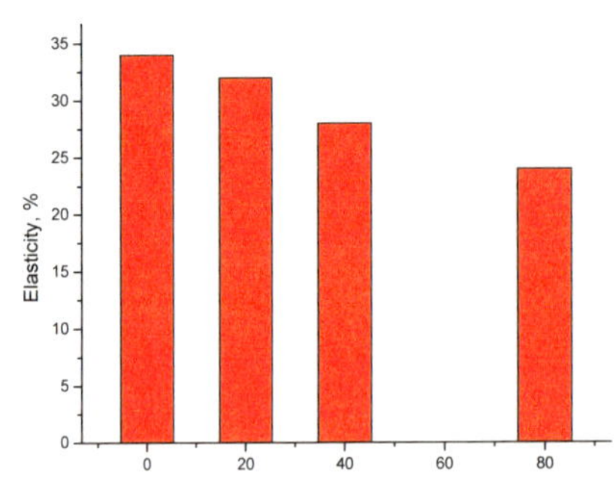

Fig. 17. Elasticity versus the HDPE level

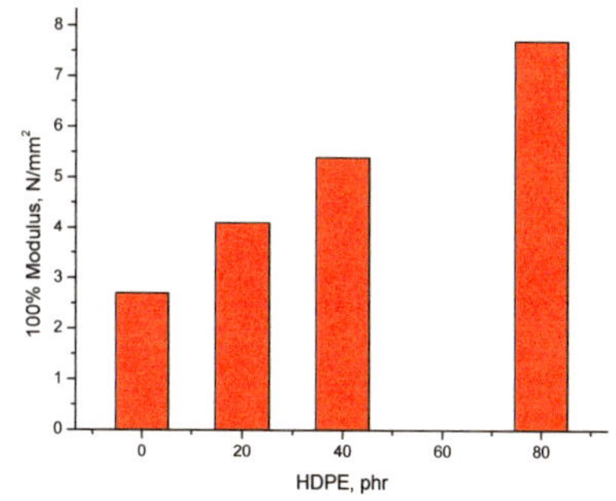

Fig. 18. 100% modulus versus the HDPE level

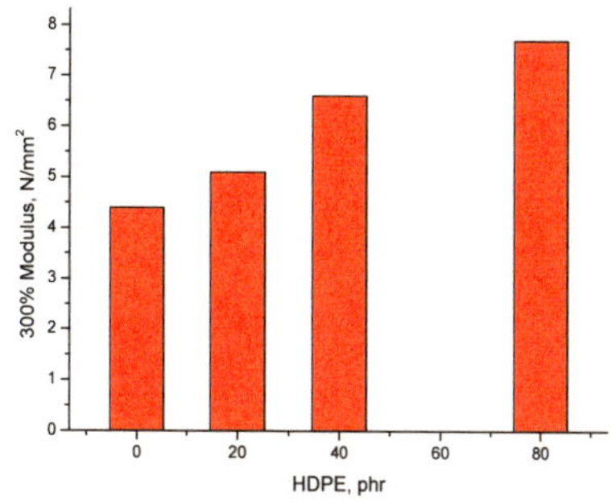

Fig. 19. 300% modulus versus the HDPE level

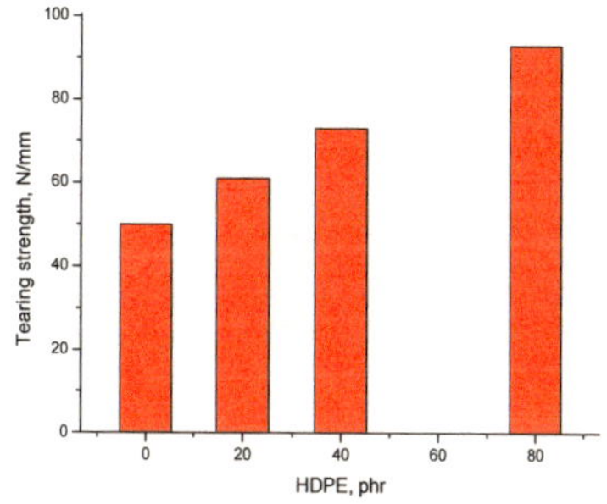

Fig. 20. Dependence of tear strength on HDPE level

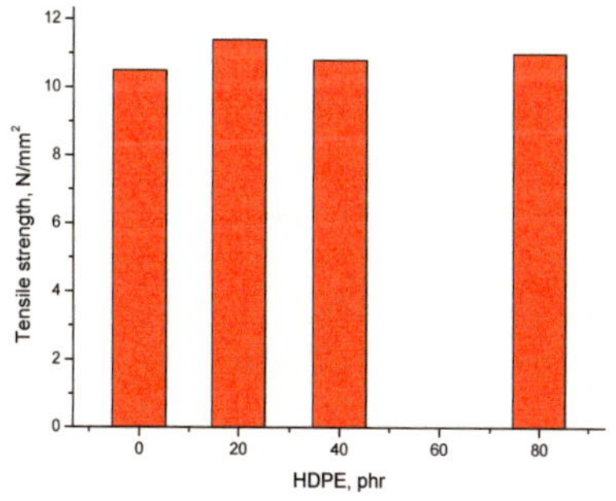

Fig. 21. Dependence of tensile strength on HDPE level

From the results obtained it was noticed that by increasing the quantity of HDPE introduced in EPDM-g-MA blends, hardness, modulus and tear strength increase and elasticity and residual elongation decrease. These effects prove that the properties of polymer blends depend on the characteristics of component polymers and on their molar fractions. The blend containing 80 phr polyethylene was selected, as it had the best values of hardness, modulus, tear strength and good values of tensile strength, elongation at break and elasticity (Zuga & Cincu, 2006c). This blend was used to make ionic thermoplastic elastomer granules in the laboratory extruder-granulator.

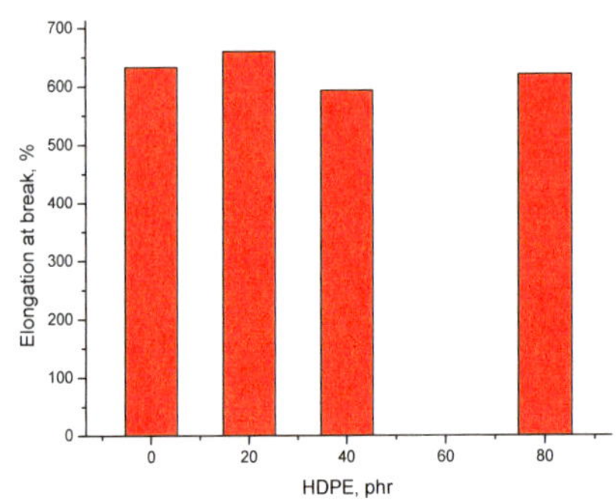

Fig. 22. Elongation at break versus HDPE level

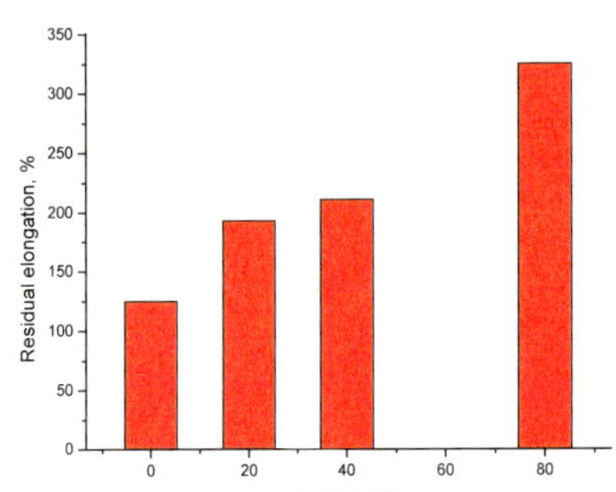

Fig. 23. Residual elongation versus HDPE level

7. Nanocomposites based on maleated ethylene propylene diene monomer and clay

In the last decades, polymer/layered clay nanocomposites (PCN) have attracted considerable attention in both basic research and industry exploitation because they possess a combination of properties that are not available in any of the single components. These systems exhibited improvements in mechanical properties, thermal stability, gas barrier properties, flame retardance, chemical and dimension stability (Stelescu et al, 2010). The clay mineral most used in elastomers/clay nanocomposites is montmorillonite, which have a layered structure consisting of two silicate tetrahedral sheets with an edge shared octahedral sheet of either aluminium or magnesium hydroxide (Ahmadi et al., 2005; Lowe et al., 2011). In order to obtain a nanoscale dispersion of clays in a polymer matrix, the silicates must be pretreated with alkylammonium ions to produce an organoclay (Tjong et al., 2002).

If the organomodified clay particles are dispersed in a polymer matrix three different types of structures can be found in nanocomposites: intercalated, flocculated and exfoliated. In the intercalated structures the polymer chains have penetrated into the layered structure maintaining the well-order multilayered nature. The flocculated structure is similar to the intercalated one, but the intercalated silicate layers sometimes flocculated because of the hydroxylated edge-edge interactions. The exfoliated structures of the silicate nanolayers are randomly dispersed throughout the polymer matrix. The exfoliated structures will determine enhanced properties of nanocomposites due to their higher phase homogeneity as compared to intercalated ones (Kang et al., 2007; Ray et al., 2003). XRD at small angles ($2\theta < 10^{o}$) and transmission electron microscopy (TEM) are effective methods to investigate these structures present in nanocomposites. XRD data reveal some peaks whose position is related to the basal spacing d_{001} and its broadness can give information about the distribution of spacings. In the exfoliated structures the interlayer spacing can be of the order of the gyration radius of the polymer and no peaks in their region are present due to the large d-spacings (10 nm) (Kang et al., 2007; Cole, 2008). Transmission electron microscopy gives the possibility to obtain data on the delaminated layers or intercalated stacks. The intercalation/exfoliation of layered clays is determined by compatibility of components, polymer diffusivity and processing conditions.

Polymer/clay nanocomposites can be obtained by different methods, such as solution intercalation, in-situ intercalation polymerization and polymer melt intercalation. The melt compounding method is more advantageous due to its compatibility with current industrial polymer processing procedures and its environmental benefit determined by the lack of solvents (Kawasumi et al., 1997; Vaia & Giannelis, 1997). The properties of nanocomposites obtained by melt blending technique can be controlled by various parameters: molecular architecture of the alkylammonium cation used in ionic exchange, the presence of additives during silicate modification, processing temperature, the type and the content of compatibilizer and polymer viscosity (Reichert et al., 2000).

Because ethylene propylene diene terpolymer is a widely used material, EPDM/clay composites should be of great application potential. The homogeneous dispersion of silicate clay in polymer matrix occurs with difficulty because EPDM does not include polar groups in the polymer chain. In this case the modification of EPDM with maleic anhydride or the use of a compatibilizing agent can assure a good dispersability of organoclay in EPDM matrix.

The EPDM-gMA compositions were prepared by melt blending method using a Plasti-Corder Brabbender equipment at a temperature of 190 °C for 12 min as mix time. The organoclay (OMMT) was montmorillonite intercalated by octadecyltrimethylamine (Nanomer 128E). The other compounding ingredients such as zinc oxide, stearic acid, zinc strearate and antioxidant (Irganox 1010) were also utilized in formulations. PP-g-MA (1 % MA) was applied as compatibilizing agent (Stelescu et al., 2010). Three compositions containing EPDM-g-MA were prepared according to tab. 7. The compositions contain also compounding ingredients such as zinc oxide (10 g), stearic acid (1 g), zinc stearate (20 g) and Irganox 1010 (2 g). The resultant composites were homogenized on an electrically operated laboratory roller mill at 155-165°C. The test specimens for physico-mechanical measurements were obtained by pressing in an electrical press at 170°C for 5 min and pressure of 150 MPa.

Code	Compound
1	EPDM-g-MA
2	EPDM-g-MA/OMMT
3	EPDM-g-MA/OMMT/PP-g-MA

Table 7. EPDM-g-MA compositions

Sample	2θ (°)	d (nm)
OMMT	3.81	2.32
	5.65	1.56
	19.86	0.45
	26.74	0.33
EPDM-g-MA/OMMT	2.24	3.94
	4.61	1.92
	21.55	0.41
	23.82	0.37
EPDM-g-MA/OMMT/PP-g-MA	2.24	3.94
	4.72	1.87
	17.02	0.52
	21.53	0.41

Table 8. Diffraction pattern characteristics of composites (according to Stelescu et al., 2010)

X-ray diffractometry is a powerful method to study the dispersion of organoclay in polymer matrix. It permitts the precise determination of silicate layer spacing and monitors the intercalation behavior of polymer chains. X-ray diffractograms were collected from a Bruker A8 Advance diffractometer with Ni-filtered Cu-Kα radiation (λ = 0.1541 nm) operating a tube voltage of 40 kV and a tube current of 35 mA. The diffractograms were scanned in 2θ range from 1°-30° at a rate of 1°/min. The interlayer spacing (d001-spacing) was evaluated by the Braag equation: $\lambda = 2d \sin \theta$, where θ is the diffraction angle, λ is the x-ray wavelength and d is the interlayer spacing. The modified organoclay shows two 2θ peaks at 3.81° and 5.65° which correspond to a basal spacing of 2.32 and 1.56 nm, respectively (Fig. 24). For the nanocomposites EPDM-g-MA/OMMT and EPDM-g-MA/OMMT/PP-g-MA a high increase in d-spacing of the layered silicate was observed (Tab. 8) and the corresponding diffraction peak is shifted to lower 2θ angles namely 2.24°. The shift of the sharp diffraction peak of organoclay from 2θ = 3.81° to lower diffraction angles in our composites suggests that EPDM-g-MA has been able to intercalate into the gallery space of layered silicate, expanding the basal spacing of organoclay to 3.94 nm, which can clearly confirm the nanocomposite formation. The basal spacing increase in nanocomposites can be also influenced by the presence of hydrogen bondings between the maleic anhydride groups and the oxygen of silicate. The greater polarity of the EPDM-g-MA facilitates the interdiffusion of the polymer chains into the gallery spaces of the organoclay leading to a better dispersion of the nanoclay in polymer matrix. As given in Fig. 24, the EPDM-g-MA/PP-g-MA nanocomposite exhibits a diffraction peak at the same 2θ with the same basal spacing relating to sample EPDM-g-MA (Tab. 8). It can be observed that the intensity of 2θ diffractive peak at 2.24° is practically the same for nanocomposites EPDM-g-MA/OMMT and EPDM-g-MA/OMMT/PP-g-MA, a slight decrease was found out for the peak at about 4.72° for the composition containing compatibilizer PP-g-MA. The X-ray diffractogram of

EPDM-g-MA sample shows a wide diffraction maximum centered at 2θ = 20.65° due to amorphous phase. For the studied composites a diffraction peak appeared at around 2θ = 6.54°. This peak at small diffraction angle scan be determined by the presence in composites of some diffraction centers formed by the aggregation tendency of ionic species leading to the formation of some microphases enriched in ions within polymer matrix and it can be attributed to interaggregated interface, the distances between ionic aggregates being Braag spaces while other authors assigned this peak to intraaggregated phase (Zuga, M.D. & Cincu, C., 2006b; 2006c).

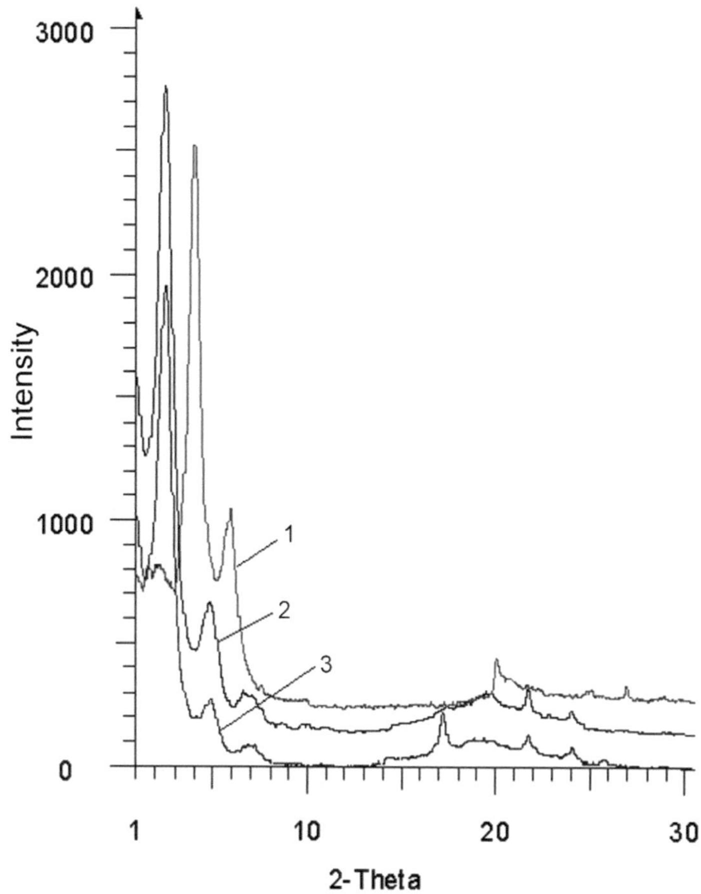

Fig. 24. X-ray diffraction patterns of organoclay (1) and EPDM nanocomposites: 2-EPDM-g-MA/OMMT; 3-EPDM-g-MA/OMMT/PP-g-MA

The evolution of the weight loss of our composites with temperature is depicted in Fig. 25. Thermogravimetric analysis (TGA) was conducted on MOM Derivatograpgh at a heating rate of 10°C/min. The samples were heated in the temperature range from room to 750°C in air.The modified organoclay exhibits two stage of decomposition, the first corresponds to the elimination of absorbed free and interlayer water. In the second stage the degradation of organic material occurs starting at 220°C. The temperature at which the weight loss was 5% can be considered the initial decomposition temperature (T_d).

Tab. 9 summarized the thermal analysis data such as temperature at 5% weight loss, decomposition temperature at maximum weight loss rate (T_{max}), weight loss determined at the end decomposition (R). The three composites exhibit a very sharp weight loss of about 83% between 250 and 490°C, followed by a second short stage with maximum temperature near 510-530°C. As it can be noticed, the thermal stability of composition containing nanoclay and compatibilizer was increased slightly compared to initial sample EPDM-g-MA. Also, the char residue has higher values in the first two compositions (EPDM-g-MA/OMMT and EPDM-g-MA/OMMT/PP-g-MA). The thermal stability of EPDM-g-MA composites did not enhance as much, compared to the simple sample EPDM-g-MA. This small increase in thermal stability can be attributed to the clay nanolayers which can proceed as barriers to reduce the permeability of volatile degradation products from polymer matrix (Hsueh & Chen, 2003; Ahmadi et al., 2005). From Table 8 it can see that the organoclay practically does not decompose during processing or characterization of these materials.

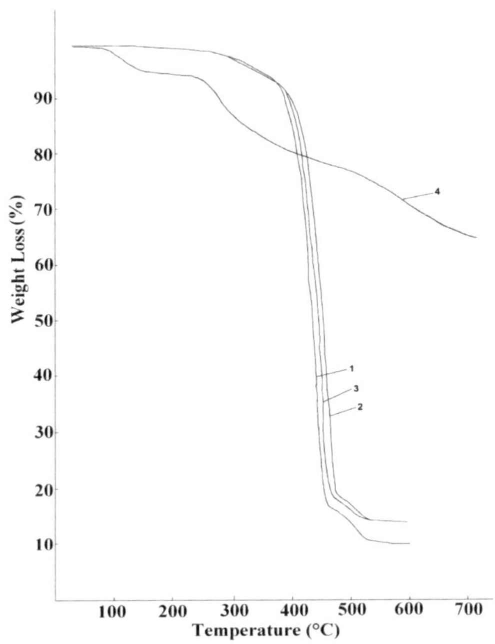

Fig. 25. TGA curves of 1-EPDM-g-MA; 2-EPDM-g-MA/OMMT; 3-EPDM-g-MA/OMMT/PP-g-MA; 4-OMMT (according to Stelescu et al., 2010)

Sample	$T_{5\%}$ (°C)	T_{max} (°C)	R (%)	T_m (°C)	ΔH_f (J/g)	T_c (°C)
EPDM-g-MA	350	475	10.08	119.01	20.57	76.41 103.87
EPDM-g-MA/OMMT	340	460	13.31	120.52	22.07	74.02 95.24
EPDM-g-MA/OMMT/PP-g-MA	335	455	13.91	120.17 162.34	16.45 11.84	74.64 108.97
PP-g -AM				163.23	85.70	118.01

Table 9. Thermal characteristics of composites (according to Stelescu et al., 2010)

DSC determinations were utilized to characterize melting and crystallization behavior of composites based on EPDM-g-MA. Dynamic scanning calorimetry (DSC) measurements were performed on Perkin Elmer Pyris Diamond calorimeter, at a heating rate of 15°C/min. Figures 26 and 27 depict the DSC cooling and reheating melt thermograms of composites under study. Some of representative data from DSC scans are presented in Table 8 such as melting peak temperature (T_m), heat of fusion (ΔH_f), crystallization temperature (T_c). The melting process of EPDM-g-AM (nano)composites shows a broad endotherm peak between 65 and 130°C (Fig. 26).

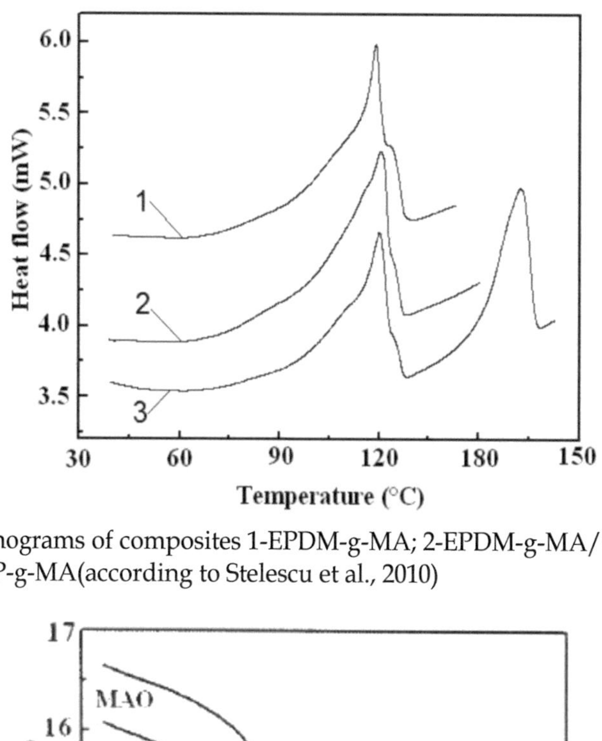

Fig. 26. DSC thermograms of composites 1-EPDM-g-MA; 2-EPDM-g-MA/OMMT; 3-EPDM-g-MA/OMMT/PP-g-MA(according to Stelescu et al., 2010)

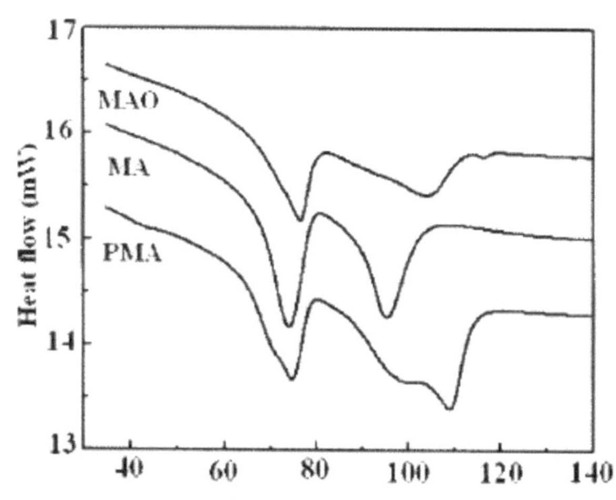

Fig. 27. DSC crystallization properties of composites 1-EPDM-g-MA; 2-EPDM-g-MA/OMMT; 3-EPDM-g-MA/OMMT/PP-g-MA.(according to Stelescu et al., 2010)

The melting temperatures for the first stage were found to be almost the same for all samples. The small decrease of melting temperature with the introduction of PP-g-MA as well as the decrease of exothermic signal intensities can be due to the decrease of maleated polypropylene crystallization in composition in the presence of EPDM-g-MA. Therefore PP-gAM retards the crystallization of PP. Especially as the incorporation of OMMT in composite (sample EPDM-g-MA/OMMT) determines the increase of T_m and of the fusion heat (Table 9) suggesting a supplementary nucleation increase due to the nanoclay presence. The crystallization temperature of PP-g-MA was around 118°C and this signal is shifted to 109°C after addition of OMMT and EPDM-g-MA in composition due to the limitation of chain mobility determined by the formation of hydrogen bonds between maleated EPDM and OMMT.

The mechanical properties of EPDM-g-MA (nano) composites under different OMMT loadings are presented in figures 28-33. Significant improvement in hardness, tensile strength, modulus and tear strength is clearly noticed for EPDM-g-MA/OMMT/PP-g-MA nanocomposites containing compatibilizer with the increasing of OMMT content. Only the elasticity decreases with the increase of OMMT level in composites (fig.30). The modulus and tensile strength increase rapidly with the increase of nanoclay loading (figs. 29 and 31) compared to that of MA sample. The enhancement in modulus and tensile strength shows that a better dispersion of layered silicate in polymer matrix in the presence of PP-g-MA as compatibilizer occurs and that the properties of the resulting polymer blends are additive. The tensile strength and elongation at break reach a maximum at about 2.5 phr of OMMT and then decrease (figs. 31 and 32), taking into account high nanoclay level and the ability of EPDM-g-MA to accept high loadings of clay diminishes. However, tensile strength and elongation at break of the uncompatibilized sample (EPDM-g-MA) show a minimum at 5 phr of OMMT content. This fact can be attributed to the decrease of ductibility while the stiffness become higher by reinforcing effect of OMMT layered silicate.

Tear strength and hardness exhibit a remarkable increase with increasing nanoclay level (figs. 28 and 33).

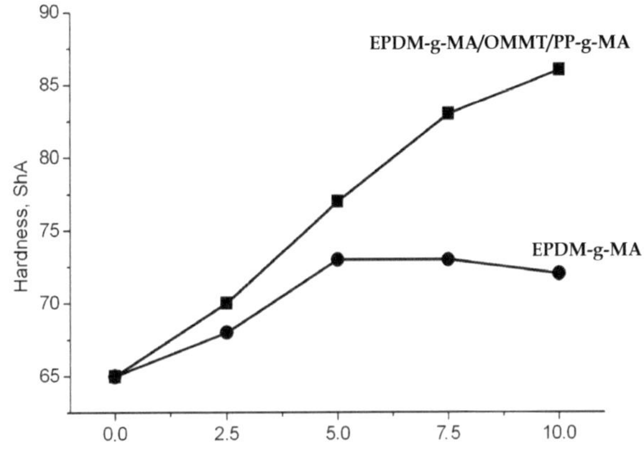

Fig. 28. Dependence of hardness on nanoclay content

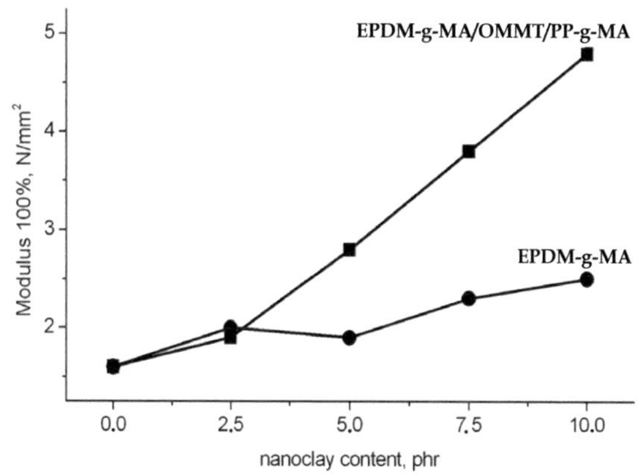

Fig. 29. Dependence of modulus 100% on nanoclay loading

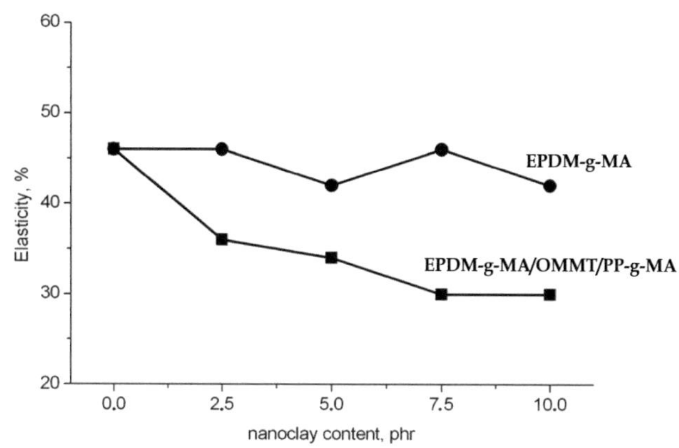

Fig. 30. Dependence of elasticity on nanoclay level

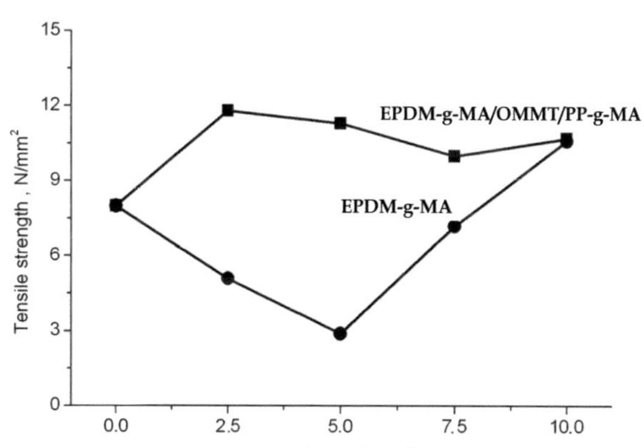

Fig. 31. Tensile strength as a function of nanoclay content

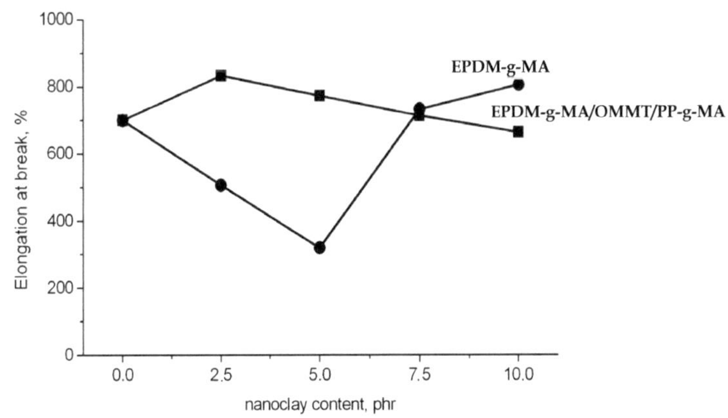

Fig. 32. Effect of nanoclay loading on elongation at break

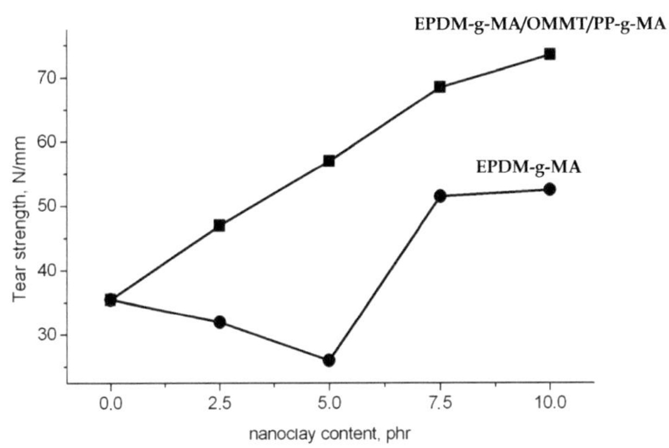

Fig. 33. Tear strength as a function of nanoclay content

This increase is due to the fact that the OMMT clay behaves as a reinforcing agent for the polymer matrix leading to the enhanced hardness.

8. Conclusions

The results confirm that the ionic thermoplastic elastomers based on EPDM-g-MA have properties similar to EPDM-based vulcanized rubber blends, and in addition, they can be easily processed using methods specific for thermoplastic materials, thus removing the vulcanization stage involving high power expenditure and release of noxious compounds determining improved characteristics (higher values for elasticity, ageing resistance, abrasive resistance, acid and alkali fastness) of the resulting materials.

Ionic thermoplastic elastomer granules can be used in various areas, due to specific properties such as resistance to water and diluted or concentrated acid and base solutions, resistance to accelerated aging, abrasion resistance or resistance to repeated bending.

The potential users of the new rubber materials will be economic operators processing rubber and plastics, footwear and car component manufactures etc. They can be used in the manufacture of a large range of products like as hoses, gaskets, rubber shoes, protective equipment etc.

9. References

[1] Ahmadi, S. J., Huang, Y. & Li, W. (2005). Fabrication and physical properties of EPDM-organoclay nanocomposites, Composites Sci. Technol., 65, 1069-1076.

[2] Andrei, C. & Dobrescu, V. (1987). Progrese in chimia si tehnologia poliolefinelor, Ed. Stiintifica si Enciclopedica, Bucuresti, Romania.

[3] Cole, K. C. (2008). Use of infrared spectroscopy to characterize clay exfoliation in polymer nanocomposites, Macromolecules, 41, 834-843.

[4] Datta, S. &. Kharagpur, S. K. (1997). Ionic thermoplastic elastomer based on crystallizable maleated EPDM rubber of high ethylene content, Kautschuk Gummi Runst., 50, 634.

[5] Datta, S. & Lohse, D. J. (1996). Polymer Compatibilizers: Uses and Benefits in Polymer Blends, Hanser, Munnich.

[6] Eisenbach, C. D., Godel, A., Terskan-Reinold, M & Schubert, U. S. (1998). Thermoplastic elastomers through polymer ion complex formation, Kautschuk, Gummi Kunst., 6, 422.

[7] Einsenberg, A. (1977), Ion Containing Polymers, Acad. Press, New York.

[8] Hsuech, H. B. & Chen, C. Y. (2003). Preparation and properties of LDHs/epoxy nanocomposites, Polymer, 44, 5275-5283.

[9] Kang, D., Kim, D., Yoon, S. H., Kim, D., Barry, C. & Mead, J. (2007). Properties and dispersion of EPDM/modified-organoclay nanocomposites, Macromol. Mater. Eng., 28, 329-338.

[10] Kawasumi, M., Hasegawa, N., Kato, M., Usuki, A. & Okada, A. (1997). Preparation and mechanical properties of polypropylene-clay hybrids, Macromolecules, 30, 6333-6338.

[11] Lowe, D. J., Chapman, A. V., Cook, S. & Busfield, J. C. Natural rubber nanocomposites by in situ modification of clay, Macromol. Mater. Eng., 296, 693-702.

[12] Nahmias, N. & Marco Serra, A. (2002). Tires produced from carboxylated rubber compositions vulcanized in the presence of metal oxides or salts, PCT International Application, 62, 594.

[13] Paeglis, A. U. & O'Shea, F. X. (1998). Thermoplastic elastomer compounds from sulfonated EPDM ionomers, Rubber Chem. Technol., 61, 223-238.

[14] Piere, R. (2002). US 6344524.

[15] Ray, S. S., Okamoto, K. & Okamoto, M. (2003). Structure-properties relationship in biodegradable poly(butylene succinate)/layered silicate nanocomposites, Macromolecules, 41, 2355-2367.

[16] Reichert, P., Nitz, H., Klinke, S., Brandsch, R., Thomann, R. & Mulhaupt, R. (2000). Polypropylene/organoclay nanocomposite formation: Influence of compatibilizer functionality and organoclay modification, Macromol. Mater. Eng., 275, 8-17.

[17] Ronatti, L. (1990). EP 0353720.

[18] Setua, D. K. & White, J. L. (1991). Flow visualization of the influence of compatibilizing agents on the mixing of elastomer blends and the effect on phase morphology, Polym. Eng. Sci., 31, 1742-1754.

[19] Stelescu, M. D. (2010), The influence of the crosslinking degree on EPDM-g-MA elastomer characteristics, Proc. 3rd Int. Conf. Adv. Mater. Systems, ICAMS 2010, 137-142, Bucuresti, Romania, Sept. 16-18, 2010.

[20] Stelescu, D. M., Airinei, A., Homocianu, M., Timpu, D. & Grigoras, C. (2010). Characterization of some EPDM-g-MA/OMMT nanocomposites, Mater. Plast., 47, 411-415.

[21] Stelescu, M. D. (2011). Elastomeri termoplastici performanți pe bază de cauciuc terpolimer etilena-propilena (EPDM) care se pot utiliza în industria de încălțăminte, Ed. Performantica, Iasi, Romania.

[22] Stelescu, M. D., Berechet, M. D. & Gurau, D. (2011). The influence of high density polyethylene added to ionic thermoplastic elastomer compounds based on maleinized ethylene-propylene terpolymer rubber, Leather and Footwear J.,11(2), 91-108.

[23] Swapan, K. G., Bhattacharya, A. K., De, P. P., Khastgir, D., De, S. K. &Peiffer, D. G. (2001). Effect of degree of neutralization on the properties of ionic thermoplastic elastomer based on sulfonated and maleated styrene-ethylene/butylene block copolymer, Rubber Chem. Technol., 74, 883-898.

[24] Szymczyk, A. & Roslaniec, Z. (1999). Sulfonated poly(ether-block-ester) ionomers with anions in the polyester hard segments, Polym. Adv. Technol., 10, 579-587.

[25] Tjong, S. C., Meng, Y. Z. & Xu, Y. (2002). Preparation and properties of polyamide 6/polypropylene-vermiculite nanocomposite/polyamide 6 alloys, J. Appl. Polym. Sci., 86, 2330-2337.

[26] Utracki, L. A. (1998). Commercial Polymer Blends, Chapman & Hall, London

[27] Vaia, R. A. & Giannelis, E. P. (1997). Lattice model of polymer melt intercalation in organicaly-modified layered silicates, Macromolecules, 30, 7990-7999.

[28] Zhao-Hua, Z., Jian-Wen, J. & Zhao-Mei, L. (2002). FTIR analysis of the coordination interaction between sulfonated butyl rubber and styrene-4-vinylpyridine copolymer, Gongneng Gaofenzi Xuebao, 15, 32.

[29] Zuga, D., Alexandrescu, L., Berechet, D., Zaharia, C. & Rusen, E. (2004). Determinarea influentei negrului de fum HAF si a cretei asupra caracteristicilor elastomerilor termoplastici ionici pe baza de cauciuc etilen propilen terpolimer maleinizat, Rev. Pielarie Incaltaminte, 4(2), 28-33.

[30] Zuga, D. M. (2005). Behaviour of thermoplastic ionic elastomers based on the maleated ethylene-propylene terpolymer in successive reclaiming, Proc. Romanian Int. Conf. Chem. Chem. Eng., RICCCE XIV, vol. 4(S9), pp. 16-23, Bucuresti, Romania, Sept. 22-24, 2005.

[31] Zuga, D. & Cincu, C. (2006a). The plasticizer action on the physical-mechanical characteristics of thermoplastic elastomers based on maleated ethylene-propylene terpolymer (EPDMm), Sci. Bull. Univ. Politehnica Bucharrest, Series B, 6, 27-34.

[32] Zuga, M. D. & Cincu, C. (2006b). Study regarding new thermoplastic ionic elastomers obtained through chemical changes of the existent products, Mater. Plast., 43, 90.

[33] Zuga, M. D. & Cincu, C. (2006c). Influence of polyolefins on the characteristics of thermoplastic elastomers based on maleated ethylene-propylene terpolymer, Mater. Plast., 43, 194-198.

[34] Zuga, M. D & Alexandrescu, L. (2009). Composition of ionic thermoplastic elastomers comprising maleinized ethylene-propylene terpolymer rubber, charging agents and plasticizer for making soles, RO Patent 122286.

19

Electroactive Thermoplastic Dielectric Elastomers as a New Generation Polymer Actuators

Chong Min Koo
Korea Institute of Science and Technology
& University of Science and Technology
South Korea

1. Introduction

Actuators or transducers, represent devices that directly convert electrical energy to mechanical energy and thus generate a force and motion. The fast growing industries of highly integrated electronics, medicals, and robotics, eagerly demand new types of transducers with flexibility, high energy efficiency and compactness, because conventional actuators including pneumatic actuators, motors, and hydraulic cylinders, have many restrictions such as heavy weight, rigidity, restrictive shape, complex transmission, and limited size. Electroactive dielectric elastomers have garnered much more attention as promising alternative candidates for next generation compact actuators or transducers than other electroactive materials such as electroactive ceramics, shape memory alloys, and even other electroactive polymers like conductive polymers and ionic polymer metal composites, owing to their attractive properties such as large electromechanical strain, fast response, high power to mass ratio, softness, facile proccessibility, and affordability (Pelrine et al. 2000a, 2000b; Shankar et al., 2007a, 2007b). For example, a comparison of the properties of electroactive dielectric elastomers and other widely used transducer materials lists in Table 1. Piezoelectric materials have quite fast and high energy efficient response, but produce a limited strain (Furukawa & Seo, 1990). Shape memory alloys (Lagoudas, 2008), conducting polymers (Bay et al., 2004) and ionic polymer metal composites (Nemat-Nasser & Wu, 2003) are capable of producing relatively large strain, but they suffer pretty slow response and poor coupling efficiency. In contrast, the electroactive dielectric elastomers have much superior actuation properties than others.

A dielectric elastomer film is compressed electrostrictively in the longitudinal direction, and spreads in the transverse planar direction, as an electric field is applied across dielectric film thickness throughout the electrodes, illustrated in Figure 1. A dielectric elastomer film is coated with compliant conductive electrodes such as carbon grease and silver grease. When the external electric field is removed, the film is recovered to original shape owing to its shape memory property. That is to say, the electric actuation of the dielectric elastomer is a reversible process.

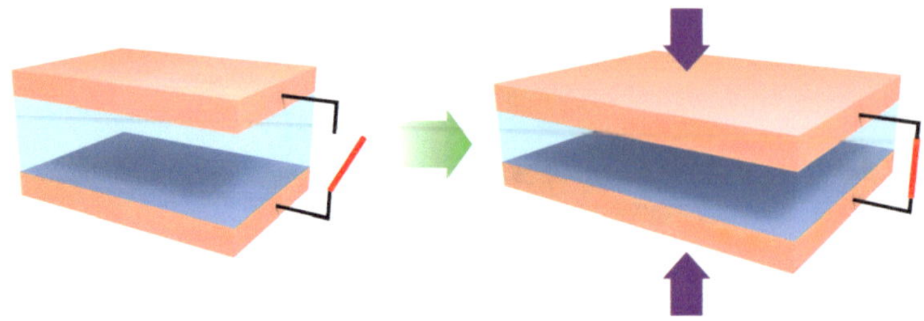

Fig. 1. Illustration of electric actuation of the dielectric elastomer actuators.

Unlike piezoelectricity, which has a linear relationship with applied electric field strength, this electric actuation of dielectric elastomers demonstrates that the total thickness strain, s_z has a quadratic relationship with the applied electric field E, as delineated by the following eq. (1) (Pelrine et al. 2000a, 2000b).

$$s_z = R_{33}E^2 , \quad (1)$$

where a thickness strain, s_z represents a thickness change ($t-t_0$) divided by initial thickness (t_0), and R_{33} represents the sensitivity of the strain response of a material to the applied electric field.

Type	Maximum strain (%)	Elastic energy density (Jcm^{-3})	Coupling efficiency (k^2)	Specific density ($g\ cm^{-3}$)	Response speed	References
Conventional dielectric elastomer						
Acrylic with prestrain	380	3.4	85	1	Fast	[a][b]
Silicone with prestrain	63	0.75	63	1	Fast	[a]
Thermoplastic dielectric elastomer with prestrain	250	0.15	85	1	Fast	[d]
Piezoelectric ceramic (PZT)	0.2	0.1	52	7.7	Fast	[a]
Piezoelectric polymer (PVDF)	0.1	0.0024	7	1.8	Fast	[a]
Shape memory alloy (TiNi)	>5	>100	5	6.5	Slow	[a]
Conducting polymer (PANI)	10	23	<1	1	Slow	[a][c]

[a] Referred from Pelrine et al., 2000b.
[b] Referred from Kornbluh et al., 2000.
[c] Referred from Kornbluh et al., 2002.
[d] Referred from Shankar et al., 2007.

Table 1. Comparison of actuator materials.

Thermoplastic dielectric elastomers, in this book, represent microphase-separated block copolymers such as a poly (styrene-*b*-ethylbutylene-*b*-styrene) (SEBS) or a poly (styrene-*b*-ethylbutylene-*b*-styrene)-*graft*-maleic anhydride (MA) triblock copolymers, described in Figure 2. For the first time, Shankar et al. (2007a, 2007b, 2007c) demonstrated that the SEBS thermoplastic dielectric elastomers with microphase-separated nanostructures had not only the very high electric actuation performance even at low electric field, but also versatile tunability of their properties, even though the SEBS exhibited pretty low dielectric feature. The thermoplastic dielectric elastomers differ from conventional homopolymer dielectric elastomers such as acrylics, silicones, fluoropolymers and natural or synthetic rubbers in several aspects such as nanostructure morphology, shape memory property and electric actuation mechanism(Drobny, 2007; Hamley, 2003; Kim, et al. 2011).

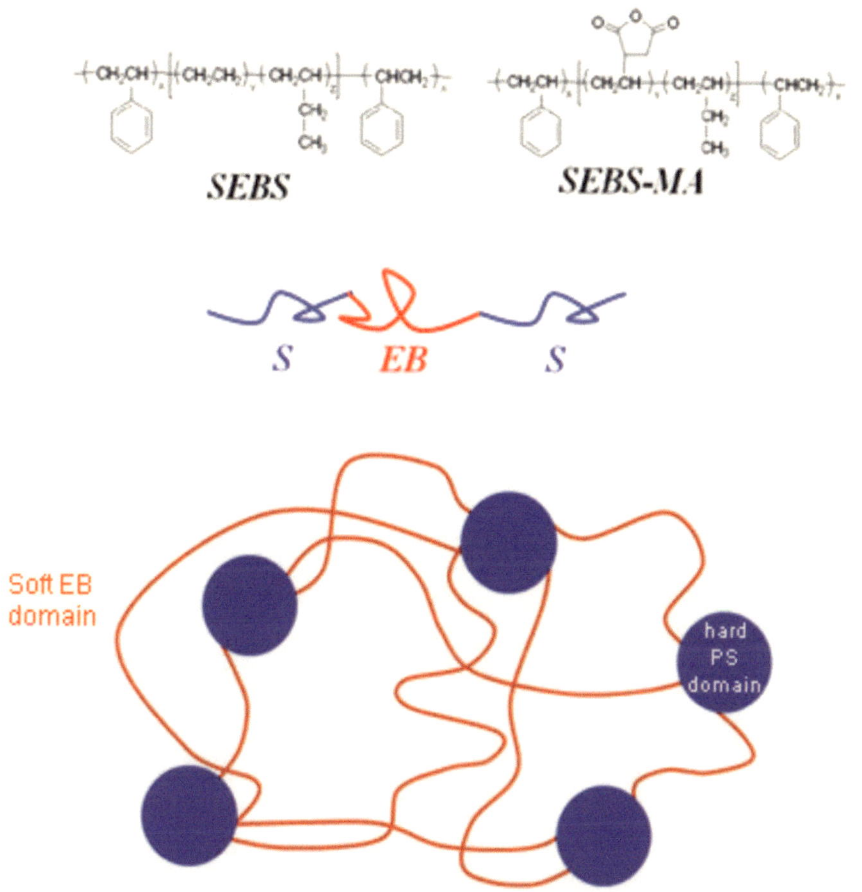

Fig. 2. Molecular and chain structures, and segregation morphology of the thermoplastic dielectric elastomers

Several well-organized reviews and books have already been published and summarized properties, mechanisms and applications of electroactive polymers including the conventional dielectric elastomers (Bar-Cohen, 2004; Brochu & Pei, 2010; Carpi et al., 2008; Carpi & Smela, 2009; Kim & Tadokoro, 2007; O'Halloran et al., 2008; Pons, 2005; Shankar et al., 2007; Shahinpoor et al., 2007). However, electroactive thermoplastic dielectric elastomer, sometimes referred to as electroactive nanostructured polymers (ENP), is a quite new terminology in the field of electroactive polymers (Chmidt et al., 2008; Jang et al., 2011; Kim et al 2010, 2011; Shankar et al., 2007a, 2007b, 2007c, 2008; Vargantwar, 2011). Thus, this book will focus on demonstrating unique electric actuation properties of the thermoplastic dielectric elastomers.

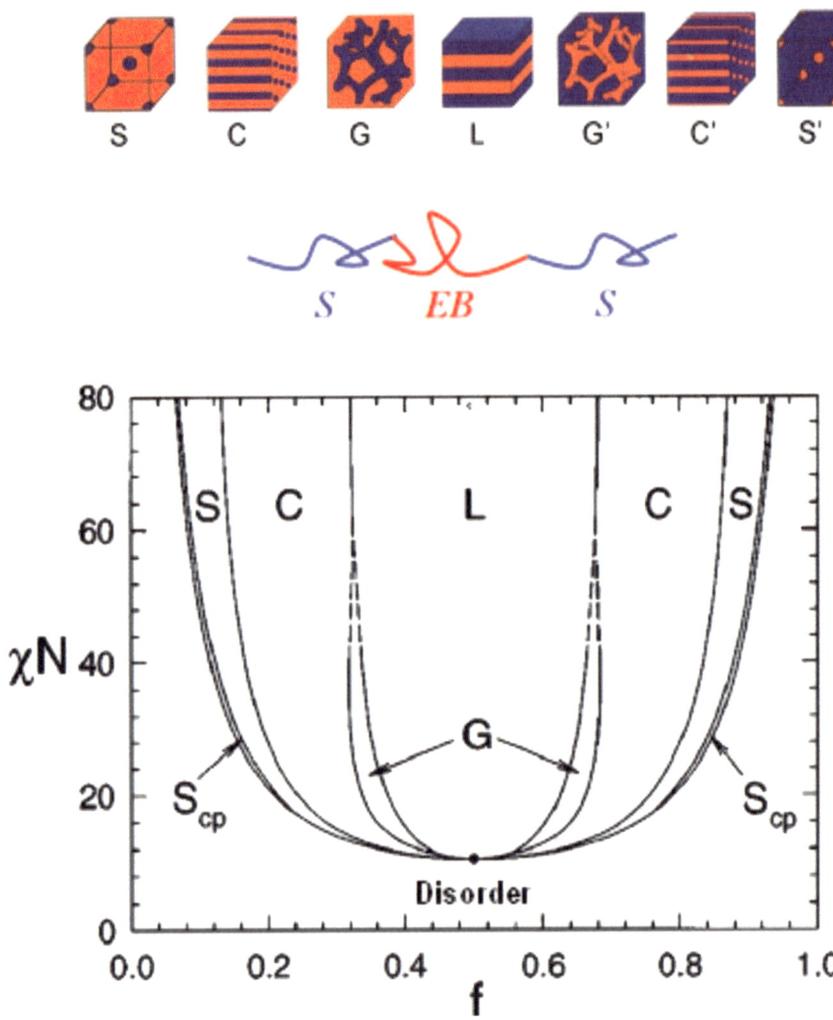

Fig. 3. Phase diagram of two component triblock copolymer. S, C, G and L represent sphere, cylinder, gyroid, lamellae nanostructures, and χ and N represent Flory-Huggins interaction parameter and molecular weight, respectively (Matsen & Bates 1996).

2. Nanostructures of thermoplastic dielectric elastomers

Thermoplastic dielectric elastomers like a poly (styrene-*b*-ethylbutylene-*b*-styrene) (SEBS) triblock copolymer, are composed of hard end blocks and soft middle block in the molecular architecture (Drobny, 2007; Hamley, 2003). Unlike conventional homopolymer dielectric elastomers with one rubbery phase structure, the thermoplastic dielectric elastomers exhibit a microphase-separated multi-phase structures, because strong incompatibility between middle and end blocks forces to be segregated and to form regular microphase-separated agglomerates due to the covalent bonding between blocks. The enthalpy of demixing is proportional to the Flory-Huggins interaction parameter χ, which is inversely proportional to the temperature according to eq. (2).

$$\chi = \frac{A}{T} + B \tag{2}$$

where A and B are system dependant constants, and T is absolute temperature. In theory, at much higher χN than the critical value of 10.4, a block copolymer forms a strongly segregated microphase-separation, while it exhibits a miscible one phase without any phase-separation at lower χN than the critical value. The nanostructure depends not only on the fraction of each component, but also universal parameter χN, where χ and N represent Flory-Huggins interaction parameter and degree of polymerization, respectively. They can exhibit a variety of segregation nanostructures such as sphere (S), body-centered cubic (BCC), hexagonal cylinder (C), gyroid (G), lamellae (L), if they have a narrow molecular weight distribution, as shown in Figure 3 (Hamley, 2003; Leibler, 1980; Matsen & Bates, 1996).

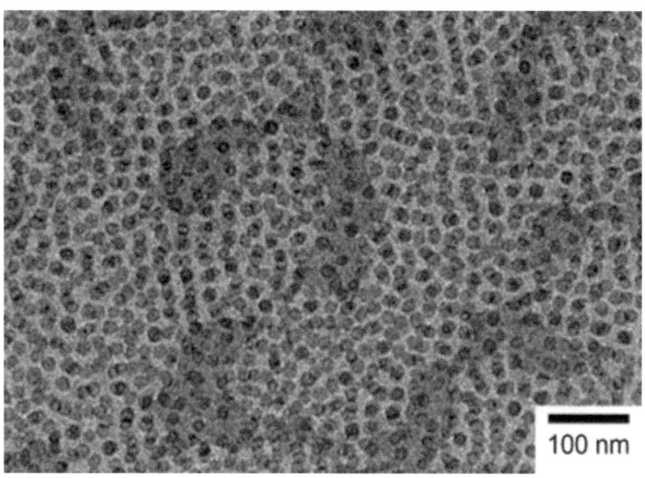

Fig. 4. TEM images of a SEBS thermoplastic dielectric elastomer gel. SEBS was 110kg/mol and the gel was swollen by 80wt% paraffinic mineral oil.

In addition, size of the nanostructure can be simply controlled by the molecular weight. For example, inter-lamellar domain spacing d is expressed be eq. (3),

$$d = a\chi^{1/6}N^{2/3} \tag{3}$$

where a is a constant (Leibler, 1980; Matsen & Bates, 1996). As a degree of polymerization increases, an inter-lamellae domain spacing increases. A number density of the nanostructure is robustly high (higher than 10^{22} ea/m^3 for a sphere (S) nanostructure), because the nanostructure size is on the several tens nanometer scale that depends on the molecular weight of block copolymer. Figure 4 shows a representative transmission electron micrograph of SEBS thermoplastic dielectric elastomer. It has sphere nanostructures on the several tens nanometer scales. Dark sphere is composed of hard polystyrene end blocks in SEBS, and bright matrix of soft polyethybutylene middle block plasticized by mineral oil.

3. Shape memory property

Shape memory property represents an ability to remember its shape or to be recovered to original shape. In the field of dielectric elastomer actuators, the shape memory property relates to the elasticity under compressive deformation. It is one of the most essential requirements of dielectric elastomers. Although applied external electric field forces a dielectric elastomer to deform in a planar extension way, the recovery to original shape should be achieved by the material's own shape memory property.

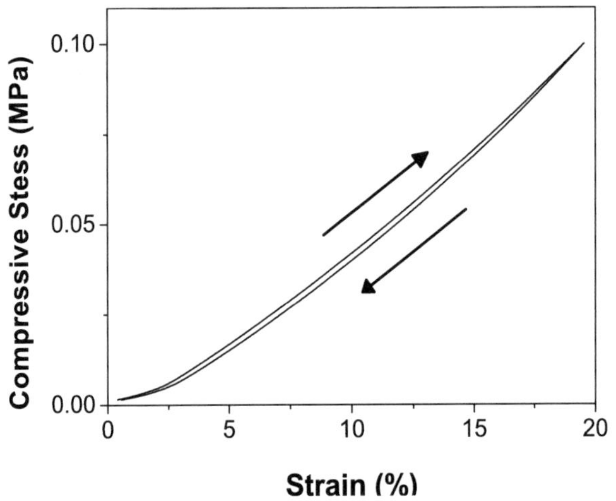

Fig. 5. Cyclic load-unload compression test of a SEBS thermoplastic dielectric elastomer (Kim et al., 2010).

Figure 5 shows shape memory property of the SEBS thermopolastic dielectric elastomer (Kim et al., 2010). With external compressive stress loaded, the thermopolastic dielectric elastomer deforms. After the stress is removed, the sample completely recovers without permanent residual strain observed. As a result, the thermoplastic dielectric elastomer have very good shape memory property.

The shape memory property of thermopolastic dielectric elastomer results from physical cross-linking, which totally differs from that of conventional homopolymer dielectric elastomer that stems from chemical cross-linking as shown in Figure 2 (Daniel et al., 2000; Drobny, 2007; Hamley, 2003; Koo, et al., 2006). The thermoplastic dielectric elastomers form microphase-separated agglomerates in a regular fashion. Both ends of soft middle block are physically pinned by hard end-block segregation domains, because a glass transition temperature T_g of a hard block domain is much higher than room temperature, while a T_g of soft middle block is much lower than room temperature. The physical pinning does not only prohibit the soft middle block chains from experiencing chain slippage when the external stress is applied, but also makes the deformed molecular chains be recovered to original state when the external stress is removed. In addition, the physical crosslinking makes materials processible in much easy and versatile way. The materials can be easily processed and changed into various shapes and domain orientations repeatedly at temperature above T_g of hard block, unlike the chemically crosslinked ones which cannot change the shape or orientation after once chemically crosslinked.

4. Electric actuation mechanism of thermoplastic dielectric elastomers

4.1 Maxwell stress

In general, dielectric elastomers electrically actuate through the two actuation mechanisms of Maxwell stress and true electrostrictive effect, as illustrated in Figure 5. Maxwell stress is caused by the Coulomb interaction between oppositely charged electrodes. When electric field is applied onto the dielectric elastomers, the each electrode is oppositely charged. The degree of charging on the electrodes depends on the dielectric properties of the dielectrics. Maxwell stress is expressed as eq. (4), illustrated in Figure 5a (Pelrine et al, 2000).

$$s_M = R_M E^2 = -\frac{\varepsilon_0 K}{2Y} E^2 \tag{4}$$

where R_M is the Maxwell stress contribution for R_{33}, and Y is a compressive modulus. K and ε_0 are a dielectric constant and a vacuum dielectric permittivity, respectively. As a result, when an electric field is applied across dielectric film, the dielectric film is compressed in the thickness direction, and spreads in the transverse direction. The longitudinal thickness strain s_z has a quadratic relationship with the applied electric field E.

4.2 True electrostriction effect

Meanwhile, the true electrostriction effect that originates from direct coupling between the polarization and mechanical response generally constitutes a contribution to the electrostrictive actuation of a dielectric elastomer as shown in Figure 5b. For a linear dielectric, the strain is induced by a change in the polarization level in the material, as expressed in eq. (5) (Su et al, 1997a, 1997b),

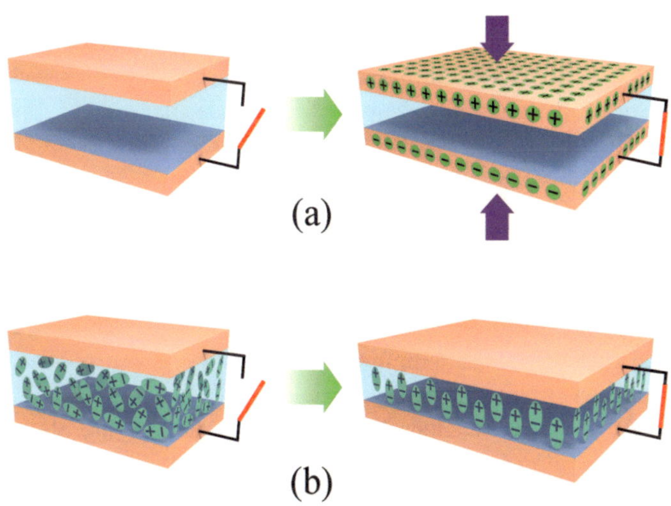

Fig. 6. Schematic illustrations of two electric actuation mechanisms of a dielectric elastomer. (a) Maxwell stress that originates from the Coulomb interaction between oppositely charged compliant electrodes. It is known as a dominant contributor for the electric actuation of a dielectric elastomer. (b) A true electrostrictive effect that originates from direct coupling between the polarization and electromechanical strain response.

$$s_{ES} = R_{ES}E^2 = QP^2 = Q\varepsilon_0^{\ 2}(K-1)^2 E^2 \quad (5)$$

where R_{ES} is a true electrostriction contribution for R_{33}, and P and Q are a polarization and a electrostrictive coefficient of the material, respectively. Here Q is a material related parameter that determines a true electrostrictive strain to a given polarization.

To date, it was a well-known common sense that electric actuation of a dielectric elastomer is usually dominated by Maxwell stress. Thus, many endeavors have mainly focused on enhancing a dielectric constant of dielectric elastomer itself or dispersing high dielectric fillers in order to improve the Maxwell stress effect and thus to enhance the electric actuation (Huang & Zhang, 2004, 2005; Xu et al, 2002; Zhang et al., 2002). Unfortunately, the true electrostriction effect, in particular, Q parameter is rarely taken into careful consideration in the electric actuation of dielectric elastomers.

4.3 Electric actuation of thermoplastic dielectric elastomers

The unique actuation performance of several SEBS thermoplastic dielectric elastomers with same styrene content and different molecular weights was reported by Shankar et al. (2007). Sample films were prepared from their mineral oil swollen gel and their prestrained films were fixed on the circular rigid frame for the actuation test as illustrated in Figure 7. Both side of film surfaces were coated with compliant electrodes such as carbon grease. Actuation strain was evaluated via monitoring the area change of active area in which electric field is applied cross the film thickness direction.

A areal strain represents an active area change relative to an initial active area. The electric areal strains were evaluated from a sequence of digital images acquired from the active area of the SEBS gel film upon exposure to electric field varying in strength. Active area of the sample film increased as the electric field increased. That is, the areal strains increased with the applied electric field. The maximum areal actuation strain increased with decreasing SEBS concentration in the gel and increasing SEBS molecular weight. Break-down electric field decreased sharply with decreasing SEBS concentration, while it increased with increasing size and number density of S(styrene) domain.

In Table 2 and Figure 8, actuation behaviors of SEBS gels were compared with those of other dielectric polymers (Shankar et al., 2007a, 2007b, 2007c, 2008; Vargantwar, 2011). Although the SEBS gel materials possess low-to-moderate energy densities, they exhibit exceptional electroactive behavior owing to a easy composition tunability. The lowest modulus SEBS gel exhibited a 14-55% increase in maximum areal actuation strain relative to the VHB4910 acrylic at markedly lower breakdown electric fields (22-32 V/μm vs 161 V/μm) and 94-103% increase relative to the HS3 silicone elastomer. In particular, the SEBS gels can exhibit substantially higher displacements than other dielectrics at very reduced electric fields. This is a very invaluable feature, because conventional dielectric elastomers generally require much higher actuation voltage in comparison with other electroactive materials such as ionic polymer-metal composites, and conducting polymers. That is, one of the greatest advantages of the SEBS thermoplastics is that not only physical properties of the thermoplastic elastomer gels such as compressive modulus and strength, but also electromechanical actuation properties can be easily tuned via simple changing a oil content in a gel. However, the underlying actuation mechanism of thermoplastic dielectric elastomers were still behind a veil.

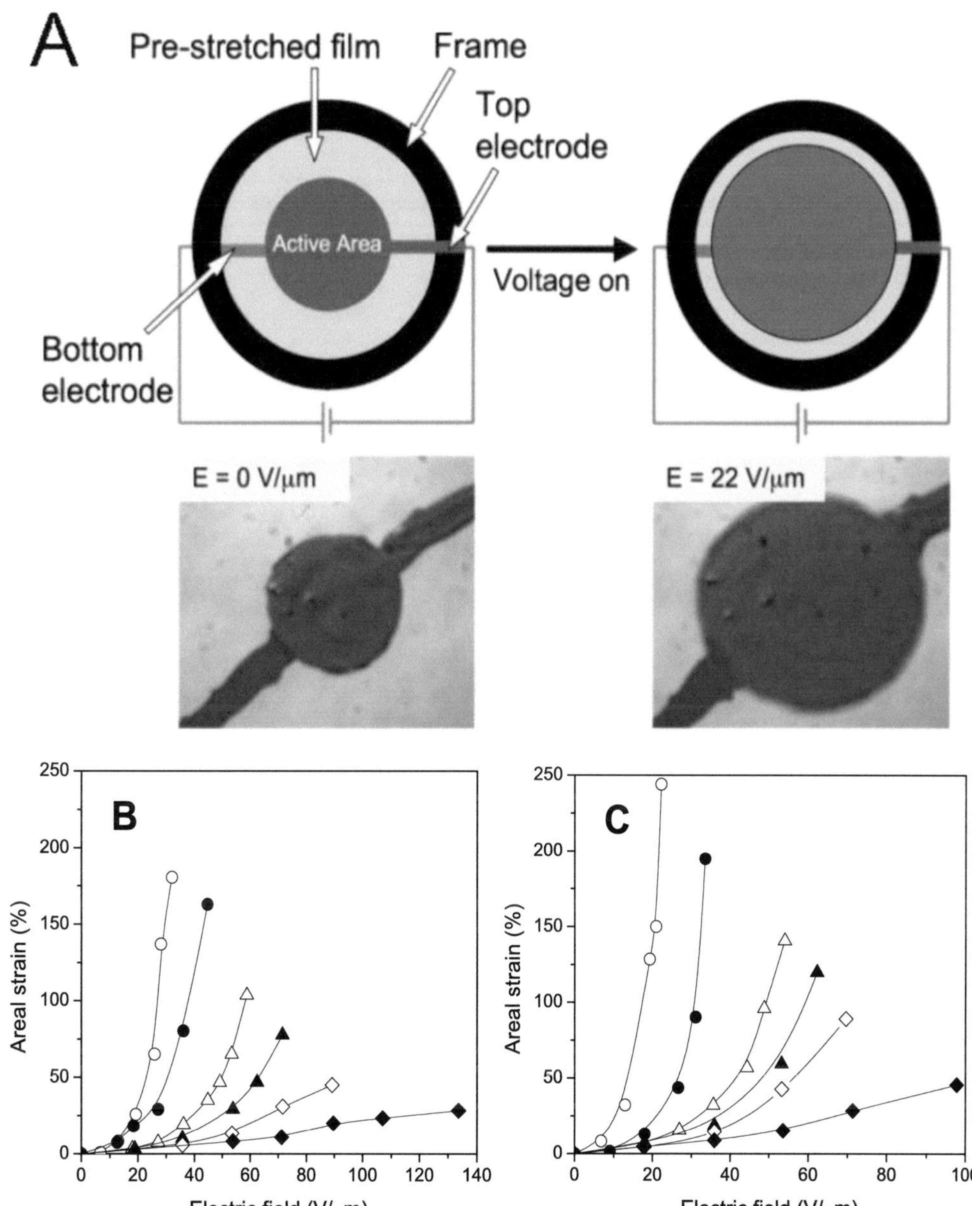

Fig. 7. (a) Circular experimental setup for the strain test of the SEBS gel films and the optical images of the films in the presence of applied electric field. (b) and (c) actuation strains of the SEBS161(Mw=161kg/mol) and SEBS217 (Mw=217kg/mol) triblock copolymer gels at various solid concentrations (in wt%) : 5 (O), 10 (●), 15 (△), 20 (▲), 25(◇), 30 (◆).(Shankar et al., 2007a)

Type	Polymer fraction (%)	Prestrain (x%, y%)	Tensile modulus (kPa)	Maximum areal strain (%)	Breakdown electric field (v μm-1)	Energy density (k Jm-3)	Ref.
Acrylic VHB4910	100	300,300	1000-3000	158	161	3400	[a],[b]
Silicone HS3	100	68,68	130	93	57	98	[a]
ENP							
SEBS161	5-30	300,300	7-163	180-30	32-133	141-151	[c][d]
SEBS217	5-30	300,300	2-133	245-47	22-98	119-139	[c][d]

[a] Referred from Pelrine et al., 2000a.
[b] Referred from Madden et al., 2004.
[c] Referred from Kornbluh et al., 1999.
[d] Referred from Shankar et al., 2007.

Table 2. Actuation behavior of SEBS gels varying in composition and other dielectric elastomers (Shankar et al., 2007a, 2007c).

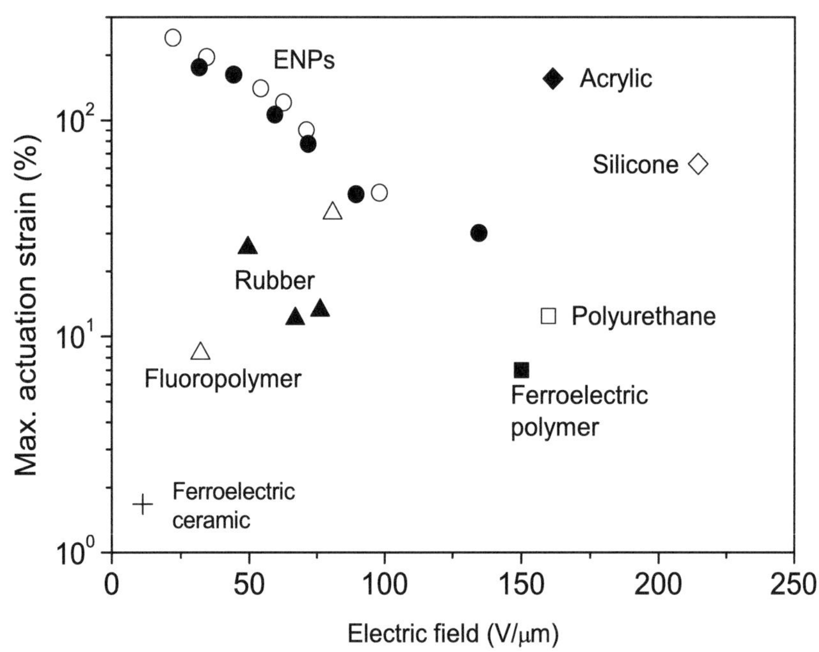

Fig. 8. Maximum actuation strain of several classes of dielectric elastomers as a function of electric field. ENPs indicate the thermoplastic dielectric elastomers such as SEBS (Shankar et al., 2007a).

Kim et al (2011) provided an important clue for the better understanding of the underlying actuation mechanism of thermoplastic dielectric elastomers. They verified via non-contact strain measurement that thermoplastic dielectric elastomers considerably actuate without Maxwell stress contribution. Non-contact represents that a sample film does not have any direct contact with the electrodes as illustrated in Figure 9a. The thermoplastic dielectric elastomer film was sandwiched with the transparent glass on which FTO electrodes were patterned. Insulate polyimide film spacers were inserted between the sample film and the electrodes in order to avoid the contact between the sample film and the electrodes. Maxwell stress results from the accumulated electron charges on the sample surface, which is injected from the electrodes. The non-contact setup prevents charge injection through the electrodes. As a result, non-contact experiment excludes Maxwell stress contribution in the electric actuation of the nanostructured thermoplastic elastomer actuators. A poly (styrene-*b*-ethylbutylene-*b*-styrene) (SEBS) and a poly (styrene-*b*-ethylbutylene-*b*-styrene)-*graft*-maleic anhydride (MA) triblock copolymer gels were used as thermoplastic dielectric elastomers. Figure 9b showed the electromechanical strain responses of the SEBS20 and MA20 with conventional method (contact experiment) and with non-contact experiment at the constant applied electric field strength of 5 V/μm. The SEBS20 and MA20 films without a direct contact with electrodes still experienced the electric actuation under the electric field. The electric actuation strains of the SEBS20 and MA20 gels measured with non-contact mode was comparable with those with contact mode.

Kim et al.(2010, 2011) carefully analyzed the degree of contribution of each actuation mechanism mode to total actuation via a in-situ synchrotron small angle X-ray scattering (SAXS) measurement. They introduced a synchrotron SAXS method for the accurate actuation strain measurement of thermoplastic dielectric elastomers. Accurate strain measurement is no doubt essential for better understanding of its underlying actuation mechanism of the dielectric elastomer. Usually, the thickness strain of the dielectric elastomer film has been evaluated by measuring lateral areal actuation strain under the assumption of an isochoric condition. The areal strain used to be measured by captured real-time video imagery. In some research studies, the thickness strain was directly observed by using a laser displacement sensor which can detect the distance between the sample and the sensor. Both methods are very useful because they provide a non-destructive method of measurement. However, unfortunately, an out-of-plane flexure motion or a bending motion is inevitably observed in the electromechanical actuation of most unstrained dielectric elastomers due to non-uniformity in sample thickness or electrode thickness as well as large expansion in the lateral strain. Fortunately, the well defined nanostructures of the thermoplastic dielectric elastomers make it possible not only to measure the true electromechanical strain from the nanostructure dimension change monitored, but also to monitor the directional dependence of the dimension change in nanoscale order, via an in-situ synchrotron SAXS. On the basis of actuation strains measured via synchrotron SAXS measurement, a comparison between the measured R_{33} and the contribution from the Maxwell stress effect, R_M of SEBS and MA thermoplastic elastomers at 15 V/μm, was presented in Figure 10. The sensitivity of the strain, R_{33}, and the contribution from the Maxwell stress effect, R_M, could be calculated by eq. (1) and eq. (5), respectively, as the field induced strain is proportional to the square of the applied electric field. The contribution from the electrostrictive effect, R_{ES}, can be simply obtained by R_{33}-R_M. They concluded that the true electrostriction contribution, R_{ES}, dominates the electric actuation of the thermoplastic dielectric elastomer.

Table 3 lists electrostrictive coefficients, Q, of the SEBS and MA thermoplastic dielectric elastomers and other dielectric materials, obtained using eq. (5). Both the SEBS and MA had ultra-high Q values, compared with those of conventional ferroelectric materials such as inorganic PZT and organic PVDF, and even compared with a polyurethane dielectric elastomer, a fluoroelastomer, and a polyurethane-based molecular composite with highly improved dielectric permittivity (Kim et al. 2011).

Table 3. Comparison of thermoplastic dielectric elastomers with materials commonly employed in electric actuation (Kim et al., 2011).

Materials	Q (m^4/C^2)	K	Y (MPa)	R_{33} (m^2/N^2)
PZT[a]	0.096	2,000	6.3×10^4	- [f]
PVDF[b]	-2.0	9	3.3×10^3	- [f]
Polyurethane[a]	-150~-450	4~8	20~200	-2.0×10^{-18}
Polyurethane/CuPc/PANI [c]	-5~ -300	100~800	80-140	-2.5×10^{-16}
Fluoroelastomer [d]	-5.2×10^3	12.1	2.5	-7.8×10^{-17}
SEBS [e]	-7.4×10^5~-8.4×10^4	2.32~2.36	0.37~1.23	-1.3×10^{-16}~-3.1×10^{-17}
MA [e]	-1.1×10^6~-1.3×10^5	2.55~2.62	0.36~1.42	-2.4×10^{-16}~-3.4×10^{-17}

Table 3. Comparison of the SEBS and MA thermoplastic dielectric elastomers with materials commonly employed in electromechanical actuation. [a] Referred from Zhang et al., 1997, [b] Referred from Furukawa & Seo, 1990, [c] Referred from Huang & Zhang, 2005, [d] Referred from Pelrine et al, 2000, [e] Referred from Kim et al., 2011, [f] PZT and PVDF follow piezoelectricity that a strain response is linearly proportional to electric field.

It is very interesting that the true electrostriction effect dominates the actuation of the thermoplastic dielectric elastomer. This unique actuation behavior is considerably different from that of a conventional dielectric elastomer, where the Maxwell stress contribution is dominant in a conventional dielectric elastomer (Su et al, 1997a, 1997b). This strong true electrostriction effect might be attributed to the dielectric mismatched periodic nanostructures. The interfaces between nanodomains result in the development of an inhomogeneous electric field across the film thickness (Guiffard et al., 2006; Kim et al., 2011; Li & Rao, 2002; Shankar et al., 1997; Su et al, 1997a, 1997b; Zhang, et al, 1997). The nonuniform field distribution can enhance the strain response because coupling between the strain and electric field in a dielectric elastomer is described by a quadratic relationship, delineated by eq. (4) and eq. (5) (Pelrine et al., 2000). The thermoplastic dielectric elastomers are typical multiphase systems with mismatched dielectric constants, and the sphere domain of polystyrene has a relatively higher permittivity than the matrix domain. The thermoplastic dielectric elastomer includes a ultrahigh density of dielectric mismatched nanodomains, higher than 10^{22} ea/m^3. In case of the dispersion phase has a higher dielectric constant than the matrix phase, the induced dipole moment is parallel to the electric field. The resulting polarized micelles tend to attract each other due to electrostatic interactions, similar to particles in an dielectric medium and an electrorheological fluid (Boker et al., 2002; Boissy et al, 1995; Giacomelli et al, 2008; Morkved, et al., 1996). The additional attraction among adjacent polarized spheres might build a unique characteristic of thermoplastic dielectric elastomer actuation.

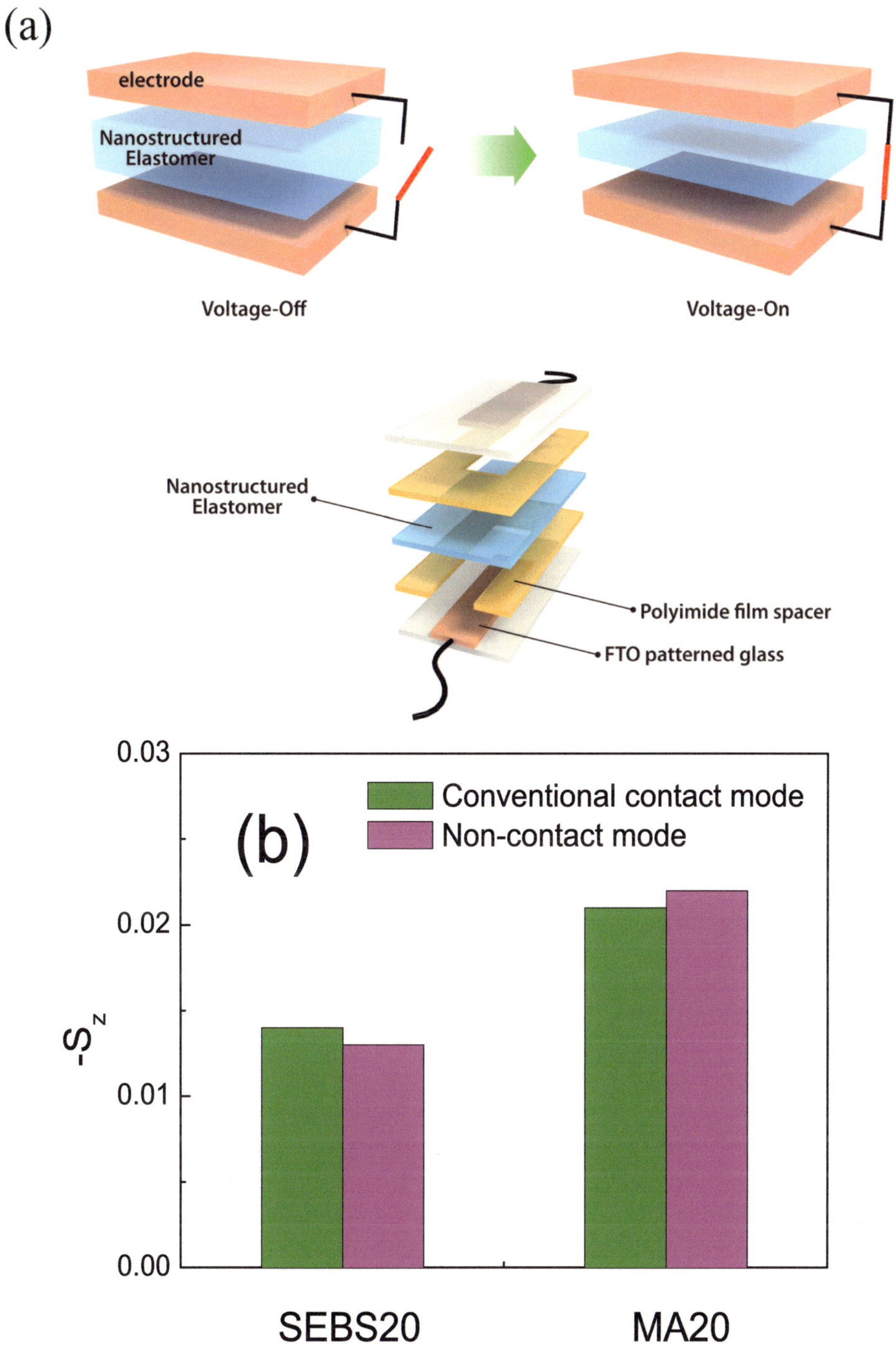

Fig. 9. (a) The illustration of a non-contact strain measurement setup and (b) The electromechanical strain responses of the SEBS20 and MA20 gels at the constant applied electric field strength of 5 V/μm. (Kim et al., 2011)

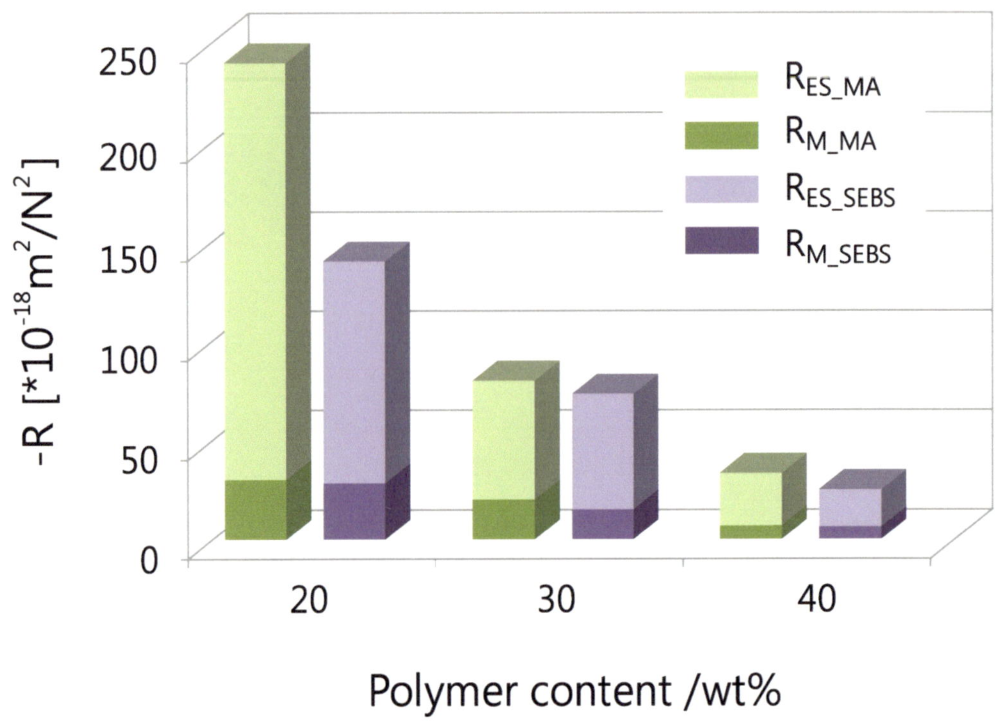

Fig. 10. Comparison between Maxwell stress contribution, R_M and true electrostriction contribution, R_{ES}, of SEBS and MA thermoplastic dielectric elastomer to the strain sensitivity, R_{33} at 15 V/µm. R_{33} represents the sum of R_M and R_{ES} (Kim et al., 2011).

5. Conclusion

In this book, we demonstrated that thermoplastic dielectric elastomers are distinguished from conventional homopolymer dielectric elastomers such as acrylics and silicones in many aspects such as nanostructure morphology, shape memory property and electric actuation mechanism. Unlike conventional homopolymer dielectric elastomers with one phase structure, these thermoplastic dielectric elastomers exhibit a well defined microphase-separated nanostructure owing to the molecular architecture. That kind of nanostructure results in the physical-crosslink-induced shape memory effect which is quite different from chemical-crosslink-induced shape memory effect of conventional dielectric elastomers. Most of all, thermoplastic dielectric elastomers have much larger true electrostrictive coefficients than conventional homopolymer dielectric elastomers, thus show the much fabulous electric actuation properties even at low electric field. Such unique behavior basically stems from the presence of a high density of dielectric mismatched nanostructures. As a result, we consider thermoplastic dielectric elastomers as fascinating actuation materials, although many challenges still remains for the real applications. We hope that, in future, this thermoplastic dielectric elastomer approach contributes to opening an era of polymer transducers.

6. Acknowledgment

This work was financially supported by a grant from the Fundamental R&D Program for Core Technology of Materials funded by the Ministry of Knowledge Economy, Republic of Korea, and partially by a grant from the Nano Hybrids Center of Korea Institute of Science and Technology (KIST).

7. References

Bar-Cohen, Y. (2004). *Electroactive Polymer (EAP) Actuators as Artificial Muscles: Reality, Potential, and Challenges*, SPIE press, Bellingham, ISSN 9780819452979

Bay, L., West, K., Sommer-Larsen, P., Skaarup, S. & Benslimane, M. Z. (2003). A Conductng Polymer Artificial Muscle with 12% Linear Strain, *Adv. Mater.*, Vol. 15, pp. 310- 313. ISSN 0935-9648

Boissy, C., Atten, P. & Foulc, J. N. (1995). On a Negative Electrorheological Effect, *J. Electrostatics*, Vol. 35, pp. 13-20, ISSN 0304-3886

Boker, A., Elbs, H., Hansel, H., Knoll, A., Ludwigs, S., Zettl, H., Urban, V., Abetz, V., Muller, A. H. E. & Krauch, G. (2002). Microscopic Mechanisms of Eletric-Field-Induced Alignment of Block Copolymer Microdomains, *Phys. Rev. Lett.*, Vol. 89, pp. 135502, ISSN 0031-9007

Brochu, P. & Pei, Q. (2010). Advances in Dielectric Elastomers for Actuators and Artificial Muscles, *Macromol. Rapid Commun.*, Vol. 31, pp. 10-36, ISSN 1022-1336

Carpi, F., De Rossi, D., Kornbluh, R., Pelrine, R. & Sommer-Larsen, P. (2008) *Dielectric Elastomers as Electromechanical Transducers,* Elsevier Ltd., Oxford, UK, ISSN 978-0-08-047488-5

Carpi, F. & Smela, E. (2009). *Biomedical Applications of Electroactive Polymer Actuators,* John Wiley & Sons, Chichester, West Sussex, UK, ISSN 9780470773055

Chmidt, K., Schoberth, H. G., Ruppel, M., Zettl, H., Hansel, H., Weiss, T. M., Urban, V., Krausch, G. & Boker, A. (2008). Reversible Tuning of a Block-Copolymer Nanostructure Via Electric Fields, *Nature Materials,* Vol. 7, pp. 142-145, ISSN 1476-1122

Daniel, C., Hamley, I. W. & Mortensen, K. (2000). Effect of Planar Extension of the Strurcture and Mechanical Properties of Polystyrene-poly(ethylene-co-butylene)- polystyrene Triblock Copolymers, *Polymer,* Vol. 41, pp. 9239-9247, ISSN 0032- 3861

Drobny, J. G. (2007). *Handbook of Thermoplastic Elastomers,* William Andrew, New York, NY, USA, ISSSN 978-0-8155-1549-4

Giacomelli, F. C., Riegel, I. C., Petzhold, C. L. & da Silveria, N. P. (2008). Block copolymer solutions under external electric field: Dynamic Behavior Monitored by Light Scattering, *Macromolecules,* Vol. 41, pp. 2677-2682, ISSN 0024-9297

Guiffard, B., Seveyrat, L., Sebald, G. & Guyomar, D. (2006). Enhanced Electric Field-Induced Strain in Non-Percolative Carbon Nanopowder/Polyurethane Composites, *J. Phys. D: Appl. Phys.,* Vol. 39, pp. 3053-3057, ISSN 0021-8979

Furukawa, T. & Seo, N. (1990). Electrostriction as the Origin of Piezoelectricity in Ferroeletric Polymers, *Jpn. J. Appl. Phys.,* Vol. 29, pp. 675-680, ISSN 0021-4922

Huang, C. & Zhang, Q. M. (2004). Enhanced Dielectric and Electromechanical Responses in High Dielectric Constant All-Polymer Percolative Composites, *Adv. Funct. Mater.,*

Vol. 14, pp. 501-506, ISSN 1616-301X

Huang, C. & Zhang, Q. M. (2005). Fully Functionalized High-Dielectric-Constant Nanophase Polymers with High Electromechanical Response, *Adv. Mater.*, Vol. 17, pp. 1153-1158, ISSN. 0935-9648

Jang, Y., Kato, T., Ueki, T. & Hirai, T. (2011). Performance of PMMA-PnBA-PMMA Dielectric Film Actuator with Controllable Phase Morphology, *Sensors and Actuators A: Physical*, Vol. 168, pp. 300-306, ISSN 0924-4247

Kim, B., Park, Y. D., Kim, J., Hong, S. M. & Koo, C. M. (2010). Measuring True Electromechanical Strain of Electroactive Thermoplastic Elastomer Gels Using Synchrotron SAXS, *J. Polym. Sci. Part. B : Polym. Phys.*, Vol. 48, pp. 2392-2398, ISSN 0887-6266

Kim, B., Park, Y. D., Min, K. H., Lee, J. H., Hwang, S. S., Hong, S. M., Kim, B. H., Kim, S. O. & Koo, C. M. (2011). Electric Actuation of Nanostructured Thermoplastic Elastomer Gels with Ultralarge Electrostriction Coefficients, *Adv. Mater.*, Vol. 21, pp. 3242-3249, ISSN. 0935-9648

Kim, K. J. & Tadokoro, S. (2007). *Electroactive Polymers for Robotic Applications: Artificial Muscles and Sensors*, Springer, London, UK, ISSN 184628371X

Koo, C. M., Hillmyer, M. A. & Bates, F. S. (2006). Structure and Properties of Semicrystalline-Rubbery Multiblock Copolymers, *Macromolecules*, Vol. 39, pp. 667-677, ISSN 0024-9297

Kornbluh, R. Pelrine, R., Joseph, J., Heydt, R., Pei, Q. & Chiba, S. (1999). High-Field Electrostriction of Elastomeric Polymer Dielectrics for Actuation, *Proc. SPIE-int. Soc. Opt. Eng.*, Vol. 3669, pp. 149-161, ISSN

Kornbluh, R., Pelrine, R., Pei, Q., Heydt, R., Standord, S., Oh, S. & Eckerle, J. (2002). Electrroelastomers: Applications of Dielectric Elastomer Transducers for Actuation, Generation, and Smart Structures, *Proc. SPIE*, Vol. 4698, pp. 254-270, ISSN

Kornbluh, R., Pelrine, R., Pei, Q., Oh, S. & Joseph, J. (2000). Ultrahigh Strain Response of Field-Actuated Elastomeric Polymers, *Proc. SPIE*, Vol. 3987, pp. 51-64, ISSN

Lagoudas, D. C. (2008). *Shape Memory Alloys: Modeling and Engineering Applications*, Springer, ISSN 0387476849

Leibler, L. (1980).Theory of Microphase Separation in Block Copolymer, *Macromolecules*, Vol. 13, pp. 1602-1617 ISSN 0024-9297

Li, J. Y. & Rao, N. (2002). Dramatically Enhanced Effective Electrostriction in Ferroelectric Polymeric Composites, *Appl. Phys. Lett.*, Vol. 81, pp. 1860-1862, ISSN 0003-6951

Madden, J. D. W., Vandesteeg, N. A., Anquetil, P. A., Madden, P. G. A., Takshi, A., Pytel, R. Z. , Lafontaine, S. R., Wieringa, P. A. & Hunter, I. W (2004). Artificial Muscle Technology: Physical Principles and Naval Prospects, *IEEE L. Oceanic Eng.*, Vol. 29, pp. 706-728, ISSN 0364-9059

Matsen, M. W. & Bates, F. S. (1996). Unifying Weak- and Strong-Segregation Block Copolymer Theories, *Macromolecules*, Vol. 29, pp. 1091-1098, ISSN 0024-9297

Hamley, I. W. (2003). *The Physics of Block Copolymers*, Oxford University Press, Oxford, UK, ISSN 0198502184

Morkved, T. L., Lu, M., Urbas, A. M., Ehrichs, E. E., Jaeger, H. M., Mansky, P. & Russell, T. P. (1996). Local Control of Microdomain Orientation in Diblock Copolymer Thin Films with Electric Fields, *Science*, Vol. 273, pp. 931-933, ISSN 0036-8675

Nemat-Nasser, S. & Wu, Y. X. (2003). Comparative Experimental Study of Ionic Polymer-Metal Composites with Different Backbone Ionomers and in Various Cation Forms, *J. Appl. Phys.*, Vol. 93, pp. 5255-5267, ISSN 0021-8979

O'Halloran, A., O'Malley, F. & McHugh, P. (2008). A Review on Dielectric Elastomer Actuators, Technology, Applications, and Challenges, *J. Appl. Phys.*, Vol. 104, pp. 071101, ISSN 0021-8979

Pelrine, R., Kornbluh, R., Joseph, J., Heydt, R., Pei, Q. & Chiba, S. (2000). High-Field Deformation of Elastomeric Dielectrics for Actuators, *Mater. Sci. Eng. C*, Vol. 11, pp. 89-100, ISSN 0928-4931

Pelrine, R., Kornbluh, R., Pei, Q. & Joseph, J. (2000). High-Speed Electrically Actuated Elastomers with Strain Greater than 100%, *Science*, Vol. 287, pp. 836-839, ISSN 0036-8675

Pons, J. L. (2005). *Emerging Actuator Technologies: A Micromechatronic Approach*, Wiley, NJ, ISSN 978-0-470-09197-5

Shahinpoor, M, Kim, K. J. & Mojarrad, M. (2007). *Artificial Muscles: Applications of Advanced Polymeric Nanocomposites*, CRC press Taylor & Francis Group, FL, ISSN 1584887133

Shankar, R., Ghosh, T. K. & Spontak, R. J. (2007). Electroactive Nanostructured Polymers as Tunable Actuators, *Adv. Mater.*, Vol. 19, pp. 2218-2223, ISSN 0935-9648

Shankar, R., Ghosh, T. K. & Spontak, R. J. (2007). Electromechanical Response of Nanostructured Polymer Systems with no Mechanical Pre-strain, *Macromol. Rapid Commun.*, Vol. 28, pp. 1142-1147, ISSN 1022-1336

Shankar, R., Ghosh, T. K. & Spontak, R. J. (2007). Dielectric Elastomers as Next-generation Polymeric Actuators, *Soft Matter*, Vol. 3, pp. 1116-1129, ISSN 1744-6848

Shankar, R., Krishnan, A. K., Ghosh, T. K. & Spontak, R. J. (2008). Triblock Copolymer Organogels as High Performance Dielectric Elastomers, *Macromolecules*, Vol. 41, pp. 6100-6109, ISSN 0024-9297

Su, J., Ting, R. Y. & Zhang, Q. M. (1997). Space-Charge-Enhanced Electromechanical Responses of Polyurethane Elastomers, *Appl. Phys. Lett.*, Vol. 71, pp. 386-388, ISSN 0003-6951

Su, J., Zhang, Q. M., Kim, C. H., Ting, R. Y. & Capps, R. (1997). Effects of Transitional Phenomena on the Electric Field Induced Strain-Electrostrictive Response of a Segmented Polyurethane Elastomer, *J. Appl. Poly. Sci.*, Vol. 65, pp. 1363-1370, ISSN 0021-8995

Vargantwar, P. H., Shankar, R., Krishnan, A. S., Ghosh, T. K. & Spontak, R. J. (2011). Exceptional Versatility of Solvated Block Copolymer/Ionomer Networks as Electroactive Polymers, *Soft Matter*, Vol. 7, pp. 1651-1655, ISSN 1744-6848

Xu, H., Li, H., Zhang, Q. M., Kavarnos, G. J., Ting, R. Y., Abdel-Sadek, G. & Belfield, K. D. (2002). High Electromechanical a Poly(vinylidene fluoride-trifluoroethylene-chlorofluoroethylene) Terpolymer, *Adv. Mater.*, Vol. 14, pp. 1574-1577, ISSN 0935-9648

Zhang, Q. M., Li, H., Poh, M., Xia, F., Cheng, Z. Y., Xu, H. & Huang, C. (2002). An All-Organic Composite Actuator Material with a High Dielectric Constant, *Nature*, Vol. 419, pp. 284-287, ISSN 0028-0836

Zhang, Q. M., Su, J., Kim, C. H., Ting, R. Y. & Capps, R. (1997). An Experimental Investigation of Electromechanical Responses in a Polyurethane Elastomer, *J. Appl. Phys.*, Vol. 81, pp. 2770-2776, ISSN 0021-8979